T0320745

BLACK HOLES FROM STARS TO GALAXIES
ACROSS THE RANGE OF MASSES

IAU SYMPOSIUM No 238

Cover illustration.

One of the most exciting targets in the sky – the central region of the Galaxy surrounding the massive black hole of Sagittarius A*. The NACO/VLT adaptive optics reveals stars and dust emission seen at 3.8 μm wavelength in the 0.5 × 0.5 square parsec field. Compelling evidence for a massive black hole is provided by the observation of variable emission in the X and NIR spectral domains.

The flares exhibit a surprising fine structure of polarized light on the time-scale of only ten to twenty minutes. The two lines centered on the position of Sagittarius A* cover the range over which the polarization angle has varied between the observations taken in 2004–2005.

Credit: A. Eckart and collaborators. For a detailed description and references, see the article in the Galactic center session of this Volume.

INTERNATIONAL ASTRONOMICAL UNION

UNION ASTRONOMIQUE INTERNATIONALE

BLACK HOLES
FROM STARS TO GALAXIES

ACROSS THE RANGE OF MASSES

PROCEEDINGS OF THE 238th SYMPOSIUM

OF THE

INTERNATIONAL ASTRONOMICAL UNION

HELD IN

PRAGUE, CZECH REPUBLIC

AUGUST 21–25, 2006

Edited by

VLADIMÍR KARAS

Astronomical Institute, Academy of Sciences, Prague, Czech Republic

and

GIORGIO MATT

Dipartimento di Fisica, Università degli Studi Roma Tre, Rome, Italy

CAMBRIDGE
UNIVERSITY PRESS

CAMBRIDGE
UNIVERSITY PRESS

University Printing House, Cambridge CB2 8BS, United Kingdom

One Liberty Plaza, 20th Floor, New York, NY 10006, USA

477 Williamstown Road, Port Melbourne, VIC 3207, Australia

314-321, 3rd Floor, Plot 3, Splendor Forum, Jasola District Centre, New Delhi - 110025, India

79 Anson Road, #06-04/06, Singapore 079906

Cambridge University Press is part of the University of Cambridge.

It furthers the University's mission by disseminating knowledge in the pursuit of education, learning and research at the highest international levels of excellence.

www.cambridge.org
Information on this title: www.cambridge.org/9780521863476

First published 2007

A catalogue record for this publication is available from the British Library

Library of Congress Cataloging in Publication data

ISBN 978-0-521-86347-6 Hardback
ISSN 1743-9213

Table of Contents

Preface xiii

Scientific organizing committee................................. xv

Conference poster... xvi

Conference participants... xvii

Part I. Reviews and contributed talks

Stellar-mass black holes... 1
    Chairs: P. A. Charles & R. Narayan

Observational evidence for stellar-mass black holes 3
    J. Casares

Energy spectra of X-ray quasi-periodic oscillations in accreting black hole binaries   13
    P. T. Życki, M. A. Sobolewska & A. Niedźwiecki

Microquasars: disk–jet coupling in stellar-mass black holes 19
    I. F. Mirabel

Suzaku observation of the black hole transient 4U1630–472: discovery of absorption
    lines 23
    A. Kubota, T. Dotani, J. Cottam, T. Kotani, C. Done, Y. Ueda, A. C.
    Fabian, T. Yasuda, H. Takahashi, Y. Fukazawa, K. Yamaoka,
    K. Makishima, S. Yamada, T. Kohmura & L. Angelini

Hyperaccretion 31
    A. R. King

Sawtooth-like oscillations of black hole accretion disks...................... 37
    R. Matsumoto & M. Machida

Chemical abundances of secondary stars in low mass X-ray binaries 43
    J. I. González Hernández, R. Rebolo & G. Israelian

Formation and evolution of massive black holes 49
    Chairs: K. A. Pounds & F. D. Macchetto

Massive black holes: formation and evolution 51
    M. J. Rees & M. Volonteri

X-ray emission properties of BLAGN in the XMM-2dF Wide Angle Survey 59
    S. Mateos, M. G. Watson, J. A. Tedds, F. J. Carrera, M. Page & A. Corral

Cosmological evolution of the AGN kinetic luminosity function 65
    A. Merloni & S. Heinz

Formation and early evolution of massive black holes...................... 73
    P. Madau

Quasar evolution: black hole mass and accretion rate determination 83
    D. Dultzin-Hacyan, P. Marziani, C. A. Negrete & J. W. Sulentic

Local active black hole mass functions.............................. 87
    J. E. Greene & L. C. Ho

Active galactic nuclei 91
    Chairs: A. Celotti & G. Matt

Mapping the circumnuclear dust in nearby AGN with MIDI................ 93
    K. R. W. Tristram, K. Meisenheimer, W. Jaffe & W. D. Cotton

Constraints on a strong X-ray flare in the Seyfert galaxy MCG–6-30-15 99
    R. W. Goosmann, B. Czerny, V. Karas, M. Dovčiak, G. Ponti &
    M. Mouchet

The soft X-ray spectrum of PG1211+143 105
    K. A. Pounds, K. L. Page & J. N. Reeves

Uncertainties of the masses of black holes and Eddington ratios in AGN....... 111
    S. Collin

Warped discs and the Unified Scheme 117
    A. Lawrence

An accretion disk laboratory in the Seyfert 1.9 galaxy NGC 2992 123
    T. Yaqoob, K. D. Murphy & Y. Terashima

Physical processes near black holes 127
    Chair: V. Karas

Strong gravity effects: X-ray spectra, variability and polarimetry 129
    A. C. Fabian

Black holes and magnetic fields........................... 139
    J. Bičák, V. Karas & T. Ledvinka

Constraining jet physics in weakly accreting black holes 145
    S. Markoff

Black hole accretion: theoretical limits and observational implications......... 153
    D. Heinzeller, W. J. Duschl, S. Mineshige & K. Ohsuga

Search for the event horizon by means of optical observations with high temporal
    resolution.. 159
    G. Beskin, V. Debur, S. Karpov, V. Plokhotnichenko & A. Biryukov

Dynamics of radiatively inefficient flows accreting onto radiatively efficient black
    hole objects..................................... 165
    D. Proga

The Galactic Center 171
    Chairs: Z. Stuchlík & T. Yaqoob

The Galactic Center..................................... 173
    R. Genzel & V. Karas

Variable and polarized emission from Sgr A* 181
    A. Eckart, R. Schödel, L. Meyer, C. Straubmeier, M. Dovčiak, V. Karas,
    M. R. Morris & F. K. Baganoff

The structure of the nuclear stellar cluster of the Milky Way 187
    R. Schödel & A. Eckart

Contents

Accretion of stellar winds onto Sgr A*................................... 191
  J. Cuadra & S. Nayakshin

Short-term variability of Sgr A*...................................... 195
  M. R. Morris, S. D. Hornstein, A. M. Ghez, J. R. Lu, K. Matthews &
  F. K. Baganoff

Kozai resonance model for Sagittarius A* stellar orbits..................... 201
  L. Šubr, V. Karas & J. Haas

Ultraluminous X-ray sources 207
  Chairs: M. A. Abramowicz, B. Czerny

Observational evidence for intermediate-mass black holes: ultra-luminous X-ray
  sources... 209
  K. Makishima

SS433 and the nature of ultra-luminous X-ray sources...................... 219
  P. A. Charles, A. D. Barnes, J. Casares, J. S. Clark, R. Cornelisse,
  C. Knigge & D. Steeghs

The supercritical accretion disk in SS 433 and ultraluminous X-ray sources..... 225
  S. N. Fabrika, P. K. Abolmasov & S. Karpov

The optical counterpart of the ultraluminous X-ray source NGC6946 ULX-1 ... 229
  P. K. Abolmasov, S. N. Fabrika & O. N. Sholukhova

Recipes for ULX formation: necessary ingredients and garnishments 235
  R. Soria

Explosion of very massive stars and the origin of intermediate mass black holes. 241
  S. Tsuruta, T. Ohkubo, H. Umeda, K. Maeda, K. Nomoto, T. Suzuki &
  M. J. Rees

Ultra-luminous X-ray sources: X-ray binaries in a high/hard state? 247
  Z. Kuncic, R. Soria, C. K. Hung, M. C. Freeland & G. V. Bicknell

On the nature of ultra-luminous X-ray sources from optical/IR measurements .. 251
  M. Cropper, C. Copperwheat, R. Soria & K. Wu

Variability of ultraluminous X-ray sources in the Cartwheel Ring............. 255
  A. Wolter, G. Trinchieri & M. Colpi

Supermassive black holes and their galaxies 259
  Chairs: G. Hasinger & S. Collin

The inner workings of early-type galaxies: cores, nuclei and supermassive black
  holes.. 261
  L. Ferrarese & P. Côté

Imaging compact supermassive binary black holes with VLBI................ 269
  G. B. Taylor, C. Rodriguez, R. T. Zavala, A. B. Peck, L. K. Pollack &
  R. W. Romani

Radiatively inefficient accretion disks in low-luminosity AGN................ 273
  F. D. Macchetto & M. Chiaberge

The central 80×200 pc of M83: how many black holes and how massive are they? 277
  H. Dottori, R. Díaz, I. Rodrigues, M. P. Agüero & D. Mast

Folowing the gas flows from nuclear spirals to the accretion disk . . . .　283
　　T. Storchi-Bergmann

A survey of AGN and supermassive black holes in the COSMOS Survey. . . . . . .　287
　　C. D. Impey, J. R. Trump, P. J. McCarthy, M. Elvis, J. P. Huchra,
　　N. Z. Scoville, S. J. Lilly, M. Brusa, G. Hasinger, E. Schinnerer, P. Capak
　　& J. Gabor

Cosmic evolution of black holes and galaxies to $z = 0.4$. . . . .　291
　　J.-H. Woo, T. Treu, M. A. Malkan & R. D. Blanford

Black holes across the mass spectrum. . . . .　295
　　Chair: I. F. Mirabel

The thermal–viscous disk instability model in the AGN context. . . . .　297
　　J.-M. Hameury, J.-P. Lasota & M. Viallet

Radiation hydrodynamic simulations of super-Eddington accretion flows. . . . . . .　301
　　K. Ohsuga

Synchrotron outbursts in Galactic and extragalactic jets, any difference? . . . . . .　305
　　M. Türler & E. J. Lindfors

Black Holes: from Stars to Galaxies. . . . .　309
　　I. F. Mirabel

Part II. Poster papers

X-ray variability of accreting black hole systems: propagating-fluctuation scenario　319
　　P. Arévalo, P. Uttley & I. McHardy

Monte Carlo simulations of dusty gas discs around supermassive black holes . . .　321
　　M. Baes & S. Piasecki

Dipole–vortex structure of the obscuring tori in active galactic nuclei . . . .　323
　　E. Yu. Bannikova & V. M. Kontorovich

Black hole masses in type II AGNs from the Sloan Digital Sky Survey . . . .　325
　　W. Bian, Q. Gu & Y. Zhao

Variations of geometrical and physical characteristics of innermost regions of active
　　galactic nuclei on time-scale of years . . . .　327
　　N. G. Bochkarev & A. I. Shapovalova

Accretion in the broad line region of active galactic nuclei . . . .　329
　　E. Bon, L. Č. Popović & D. Ilić

Triaxial orbit-based model of NGC 4365 . . . .　331
　　R. C. E. van den Bosch, G. van de Ven, M. Cappellari & P. Tim de Zeeuw

Modulation of high-frequency quasi-periodic oscillations by relativistic effects . .　333
　　M. Bursa

XMM-Newton RGS spectra of four Seyfert 1 galaxies . . . .　335
　　M. V. Cardaci, M. Santos-Lleó & A. I. Díaz

Analytical binary black hole model of Sgr A* and its implications . . . .　337
　　T. K. Chatterjee

Evolution of black-hole intermediate-mass X-ray binaries . . . .　339
　　W.-C. Chen & X.-D. Li

Relationship between X-shaped radio sources and double-double radio galaxies .   341
    X. Chen & F. K. Liu

General relativistic results for a galactic disc in a multidimensional space-time. .   343
    C. H. Coimbra-Araújo & P. S. Letelier

Formation history of supermassive black holes in a viable cosmological window .   345
    C. H. Coimbra-Araújo & A. C. S. Friaça

The central cluster and X-ray emission from Sgr A*   347
    R. F. Coker & J. M. Pittard

Upper limits on the mass of supermassive black holes from HST/STIS archival
    data   349
    E. M. Corsini, A. Beifiori, E. Dalla Bontà, A. Pizzella, L. Coccato, M. Sarzi
    & F. Bertola

The nuclear properties of early-type galaxies in the Virgo and Fornax clusters . .   351
    P. Côté

Accreting corona model of the X-ray variability in soft state GBH and AGN . . .   353
    B. Czerny & A. Janiuk

Supermassive black holes in Brightest Cluster Galaxies   355
    E. Dalla Bontà, L. Ferrarese, J. Miralda-Escudé, L. Coccato, E. M. Corsini
    & A. Pizzella

The VSOP Survey: final aggregate results   357
    R. Dodson, S. Horiuchi, W. Scott, E. Fomalont, Z. Paragi, S. Frey, K. Wiik,
    H. Hirabayashi, P. Edwards, Y. Murata, G. Moellenbrock, L. Gurvits,
    Z. Shen & J. Lovell

Polarization from an orbiting spot   359
    M. Dovčiak, V. Karas & G. Matt

Optical, ultraviolet and X-ray analysis of the black hole candidate BG Geminorum   361
    P. Elebert, P. J. Callanan, L. Russell & S. E. Shaw

Dynamical evolution of rotating globular clusters with embedded black holes . . .   363
    J. Fiestas & R. Spurzem

A near-infrared view of the 3CR: properties of hosts and nuclei   365
    D. Floyd, M. Chiaberge, E. S. Perlman, B. Sparks, F. D. Macchetto,
    J. Madrid, S. Baum, C. O'Dea, D. Axon, A. Quillen, A. Capetti, G. Miley
    & S. Tinarelli

Generation of shocked hot regions in black hole magnetosphere   367
    K. Fukumura, M. Takahashi & S. Tsuruta

The gravitational redshift in the broad line region of the active galactic nucleus
    Mrk 110   369
    N. Gavrilović, L. Č. Popović & W. Kollatschny

Activity type of galaxies in HyperLeda   371
    N. Gavrilović, A. Mickaelian, C. Petit, L. Č. Popović & P. Prugniel

Evidence for the AGN nature of LINERs   373
    O. González-Martín, J. Masegosa, I. Márquez & E. Jiménez-Bailón

AGN polarization modeling with Stokes   375
    R. W. Goosmann, C. M. Gaskell & M. Shoji

The entropy increase during the black hole formation   377
    T. Hara, K. Sakai, S. Kunitomo & D. Kajiura

Uncertainty principle for the entropy of black holes, de Sitter and Rindler spaces   379
    T. Hara, K. Sakai, S. Kunitomo & D. Kajiura

Intensity and polarization light-curves from radiatively-driven clouds   381
    J. Horák & V. Karas

Physical properties of emitting plasma near massive black holes: the Broad Line
    Region   383
    D. Ilić, G. La Mura, L. Č. Popović, A. I. Shapovalova, S. Ciroi,
    V. H. Chavushyan, P. Rafanelli, A. N. Burenkov & A. Marcado

Fe K line profile in PG quasars: the averaged shape and Eddington ratio dependence   385
    H. Inoue, Y. Terashima & L. C. Ho

Radiation in models with cosmological constant   387
    H. Kadlecová & J. Podolský

The results of X-ray binary Cyg X-1 investigations based on the optical photometry
    and high-resolution spectroscopy   389
    E. A. Karitskaya

Observational manifestations of accretion onto isolated black holes of different
    masses   391
    S. Karpov & G. Beskin

AGN types as different states of massive black holes   393
    B. V. Komberg

Relativistic jets and non-thermal radiation from collapse of stars to black holes .   395
    V. Kryvdyk & A. Agapitov

Long-term photometry of blazars at Abastumani Observatory   397
    O. M. Kurtanidze, M. G. Nikolashvili, G. N. Kimeridze, L. A. Sigua,
    B. Z. Kapanadze & R. Z. Ivanidze

Optical variability of X-ray selected blazars   399
    O. M. Kurtanidze, M. G. Nikolashvili, G. N. Kimeridze, L. A. Sigua &
    B. Z. Kapanadze

Rotation curves, dark matter and general relativity   401
    P. S. Letelier

The capture of main sequence stars and giant stars by a massive black hole   403
    Y. Lu, Y. F. Huang, Z. Zheng & S. N. Zhang

Low-frequency, one-armed oscillations in black hole accretion flows obtained from
    direct 3D MHD simulations.   405
    M. Machida & R. Matsumoto

The orbiting spot model gives constraints on the parameters of the supermassive
    black hole in the Galactic Center   407
    L. Meyer, A. Eckart, R. Schödel, M. Dovčiak, V. Karas & W. J. Duschl

Masses of radiation pressure supported stars in extreme relativistic realm.   409
    A. Mitra

Radiation spectra from MHD simulations of low angular momentum flows 411
   M. Moscibrodzka, D. Proga & B. Czerny

The K-correction for LMXBs: an application to X1822-371 and GX339-4 413
   T. Muñoz-Darias, J. Casares & I. G. Martínez-Pais

Proper motions of thin filaments at the Galactic Center 415
   K. Mužić, A. Eckart, R. Schödel, L. Meyer & A. Zensus

Two-dimensional MHD simulations of accretion disk evaporation 417
   K. E. Nakamura

Long-term and intra-day variability of BL Lacertae since the last great outburst 419
   M. G. Nikolashvili & O. M. Kurtanidze

Steady models of optically thin, magnetically supported two-temperature accretion
   disks around a black hole 421
   H. Oda, K. E. Nakamura, M. Machida & R. Matsumoto

QPOs expected in rotating accretion flows around a supermassive black hole . . . 423
   T. Okuda, V. Teresi & D. Molteni

Power spectra from spotted accretion discs 425
   T. Pecháček, M. Dovčiak & V. Karas

Brownian motion of black holes in stellar systems with non-Maxwellian distribution
   of the field stars 427
   I. T. Pedron & C. H. Coimbra-Araújo

XMM-Newton study of the spectral variability in NLS1 galaxies 429
   G. Ponti, M. Cappi, B. Czerny, R. W. Goosmann & V. Karas

Can gravitational microlensing be used to probe geometry of a massive black-hole? 431
   L. Č. Popović & P. Jovanović

Exact solutions for discs around stationary black holes 433
   N. Požár, O. Semerák, J. Šácha, M. Žáček & T. Zellerin

IR contamination in quiescence X-Ray Novae 435
   M. T. Reynolds, P. J. Callanan & C. Kelleher

First e-VLBI observations of GRS 1915+105 437
   A. Rushton, R. E. Spencer, M. Strong, R. M. Campbell, S. Casey,
   R. P. Fender, M. A. Garrett, J. C. A. Miller-Jones, G. G. Pooley,
   C. Reynolds, A. Szomoru, V. Tudose & Z. Paragi

Detection of IMBHs from microlensing in globular clusters 439
   M. Safonova & S. Rahvar

Episodic ejection from super-massive black holes 441
   L. Saripalli, R. Subrahmanyan & R. W. Hunstead

Detailed structure of the X-ray jet in 4C 19.44 (PKS1354+195) 443
   D. A. Schwartz, D. E. Harris, H. Landt, A. Siemiginowska, E. S. Perlman,
   C. C. Cheung, J. M. Gelbord, D. M. Worrall, M. Birkinshaw, S. G. Jorstad,
   A. P. Marscher & L. Stawarz

Color variations of gravitationally lensed quasars as a tool for accretion disk size
   estimation 445
   V. N. Shalyapin

Outburst of the unique X-ray transient CI Cam and its impact on the system .. 447
V. Šimon, C. Bartolini, A. Guarnieri, A. Piccioni & D. Hanžl

Marginally stable thick discs orbiting the Kerr–de Sitter black holes: the mass estimates 449
P. Slaný & Z. Stuchlík

An $f(R)$ gravitation for galactic environments 451
Y. Sobouti

Chilled disks in ultraluminous X-ray sources............................. 453
R. Soria, Z. Kuncic & A. C. Gonçalves

A ULX and a giant cloud collision in M 99 455
R. Soria & D. S. Wong

Near-IR integral field spectroscopy of the NLR of ESO428-G14: the role of the radio jet.. 457
T. Storchi-Bergmann, R. A. Riffel, F. K. B. Barbosa & C. Winge

Humpy LNRF-velocity profiles in accretion discs orbiting rapidly rotating Kerr black holes: a possible relation to epicyclic oscillations.................. 459
Z. Stuchlík, P. Slaný & G. Török

The X-ray–radio association in RATAN and RXTE monitoring of microquasar GRS 1915+105 461
S. A. Trushkin, T. Kotani, N. Kawai, M. Namiki & S. N. Fabrika

Flaring activity of microquasars from multi-frequency daily monitoring program with RATAN-600 radio telescope 463
S. A. Trushkin, N. A. Nizhelskij, N. N. Bursov & E. K. Majorova

Peculiar outbursts of a black hole X-ray transient, V4641 Sgr............... 465
M. Uemura, T. Kato, D. Nogami, A. Imada & R. Ishioka

Self-gravitating warped disks around nuclear black holes.................. 467
A. Ulubay-Siddiki, O. Gerhard & M. Arnaboldi

The modelling of the line-locking effect in broad absorption line QSOs 469
E. Vilkoviskij & S. Yefimov

Observational properties of relativistic black hole winds 471
K. Watarai

The relation between the properties of the NLR in narrow-line Seyfert 1 galaxies and the accretion rate 473
D. Xu & S. Komossa

Measuring the supermassive black hole parameters with space missions........ 475
A. F. Zakharov

Doppler boosting and de-boosting effects in relativistic jets of AGNs and GRBs 477
J. Zhou & Y. Su

Author Index 479

Preface

Astronomy is a very ancient science, divinely entrusted to mankind from the time of Adam Protoplastus. It is by far the most prestigious, insofar as what is celestial and sublime certainly surpasses what is terrestrial and inferior. This divine astronomy, I say, drawing its origin from the very senses of our eyes, observing the wandering movements of the stars to the outer reaches of the world, has vexed both the geniuses and the abilities of the most eminent men for the long span of time since the beginning of things. The majesty of the highest and thrice-greatest God is surely so great that the wisdom of his works can not be exhausted by any of his creatures.

But since the gaze of the eye alone was unable to capture all the mysteries of the celestial theater – which are wonderful beyond measure – and all the intricate-appearing differences, for which subtlety and accuracy was needed, various craftsmen in every age have contrived means and instruments by which their vision might be aided in perceiving the obscure motions of the stars. Hence, there are those columns, which Joseph, the writer about Jewish matters, reports the descendants of Adam had built in Syria and had their discoveries inscribed on them, so that those who came after them would remember...†

The main motivation to organize this Symposium was to bring together observers and theoreticians working in black hole astrophysics – from the stellar-mass black holes to the supermassive ones residing in centres of galaxies, including the Milky Way. The symposium was divided in fourteen sessions with review lectures, contributed talks and posters. The programme is closely reflected in this book.

Evolutionary tracks leading to the formation of black holes must be very different for so vastly diverse types of objects. Stellar-mass black holes are born after core-collapse events possibly related to gamma-ray bursts, whereas the origin of the supermassive black holes is an open question. However, we know that the birth and growth of supermassive black holes must be linked with the history of the host galaxy, as demonstrated by the correlation of the black hole mass with the bulge luminosity and velocity dispersion.

The best evidence of the existence of supermassive black holes is provided by the center of our own Galaxy. The extremely low level of its current activity along with possible evidence of an intermittent moderate activity in the recent past make the Galactic center an ideal test case to study galactic nuclei in quiescence. In particular, the low luminosity radiatively inefficient accretion modes can be examined in detail. The Galactic centre is such an important system that we dedicated a full session to it.

Black holes provide a unique opportunity to probe gravity in the strong-field regime. Spectral, temporal and polarimetric properties of the emerging radiation are all influenced by the prresence of a black hole. Effects related to stellar motion were examined during the conference, also for other galactic nuclei – including the impact of the interstellar environment on the form of stellar cusps. Indeed, it would be highly desirable to understand better the mutual interaction between the stellar population and the diluted gaseous component around massive black holes.

† Preface to the Noblest Emperor Rudolph II, which Tycho Brahe wrote for his *Astronomiae instauratae mechanica*, published in Wansdeck shortly before he joined the Prague Court. Historical prints are preserved in the libraries throughout the world, including the Library of the Astronomical Institute in Ondřejov, which kindly provided us with illustrations adorning this book. Tycho Brahe served as the Imperial Mathematician and invited Johannes Kepler to work as his assistant. The excerpt is from the revised and commented translation by Alena Hadravová, Petr Hadrava and Jole Shackelford (Prague 1996).

Our Symposium was held as part of the 26th General Assembly of the International Astronomical Union in Prague, Czech Republic, and we are grateful to everyone who helped us to make this meeting a fruitful and enjoyable one. First of all, we thank the members of the Scientific Organizing Committee for their support and useful ideas and suggestions since the early stages of the programme preparation.

Needless to say, we very much appreciate the efficient support of the local organizers working under chairmanship of Dr. Cyril Ron.

We greatly benefited from the enthusiastic and efficient assistance of students and our colleagues at the Astronomical Institute, who helped us with the conference logistics and organization. Special thanks go to Michal Bursa for drawing the conference poster and processing the video recordings of the lectures; Michal Dovčiak for taking pictures during sessions; Ivana Stoklasová for collecting the lecture presentations from the speakers; René Goosmann, Hedvika Kadlecová, Jaroslava Schovancová, Jiří Svoboda and Markéta Tukinská for their help with the proceedings files; and Ondřej Karas, Filip Munz, Tomáš Pecháček and the entire student staff of the Forum Hall where our Symposium took place. We thank the authors of Atlas Coeli Novus for providing us with the electronic version of the star-field and Andrew Hamilton for his drawing of the Kruskal diagram we used in the conference poster.

Last but not least, we thank all participants for coming and creating the lively and active atmosphere of the meeting. We have captured the main points of the discussions at the end of lectures (our apologies if some parts had to be abridged). Some of the more detailed and vigorous comments have not made it to the printed version; however, video recordings of the sessions are available – if not for eternity then surely till the lifetime of the current technology allows. Those who could not come in Prague or missed some sessions will surely appreciate that all speakers kindly allowed us to post their lectures on the conference website, where they can be consulted together with final versions of the papers published in these Proceedings.

The large interest of the general public in this meeting came as no surprise, given the long-lasting history of physics, astrophysics and astronomy research and popularization in this country. On the break of 16th and 17th century, Tycho Brahe and Johannes Kepler wrote several key chapters of astronomy in Bohemia. In a modern reminiscence, Albert Einstein spent a year of his first professorship in Prague and his wordline intersected here with some of the most distinguished intellectuals of the first decades of 20th century – Philipp Frank, Gerhard Kowalewski, Ernst Mach, Franz Kafka, Christian von Ehrenfels, Max Brod, Hugo Bergmann... And in August 2006, our brief meeting has brought together theorists and observers working in this rapidly developing field of black hole astrophysics. It is not difficult to smell the touch of centuries meeting modern times in this city.

Vladimír Karas and Giorgio Matt
On behalf of the Scientific Organizing Committee
Prague and Rome, 31 December 2006

SCIENTIFIC ORGANIZING COMMITTEE

Roger D. Blandford (USA)
Annalisa Celotti (Italy)
Philip A. Charles (South Africa)
Bożena Czerny (Poland)
Andrew C. Fabian (UK)
Reinhard Genzel (Germany)
Günther Hasinger (Germany)
Vladimír Karas (co-chair, Czech Republic)
Kazuo Makishima (Japan)
Giorgio Matt (co-chair, Italy)
I. Félix Mirabel (Chile)
Kenneth A. Pounds (UK)
Martin J. Rees (UK)
Jean H. Swank (USA)
Tahir Yaqoob (USA)
Shuang Nan Zhang (China)

This symposium was sponsored and hosted by the International Astronomical Union and the Academy of Sciences of the Czech Republic.

The meeting was endorsed by the Division XI (Space and High Energy Astrophysics) of the International Astronomical Union, and it was supported by several Commissions: Galaxies (Commission 28), Radial Velocities (Commission 30), Radio Astronomy (Commission 40), and Close Binary Stars (Commission 42).

The Symposium took place as part of the 26th General Assembly of the International Astronomical Union. The Local Organizing Committee was operated under the auspices of the Academy of Sciences of the Czech Republic.

We acknowledge additional funding from the Czech Science Foundation (project No 205/07/0052) and the ESA Plan for European Cooperating States (project No 98040) which supported the preparation of these Proceedings and the associated video archive where the lectures are recorded and made available on-line.

These Proceedings are published both electronically and in the book form. Colour figures are reproduced in colour in the electronic version.

Black Holes
From Stars to Galaxies
across the range of masses

IAU Symposium 238
Prague, 21-25 August 2006

Conference participants †

Pavel K. **Abolmasov**, Special Astrophys. Obs., Nizhnij Arkhyz, Russian Fed.　　pasha@sao.ru
Marek A. **Abramowicz**, Gothenburg Univ., Sweden　　marek@fy.chalmers.se
Patricia **Arevalo**, Southampton Univ., Southampton, United Kingdom　　patricia@astro.soton .ac.uk
Magda **Arnaboldi**, European Southern Observatory, Germany　　marnabol@eso.org
Maarten. **Baes**, Univ. Gent, Belgium　　maarten.baes@ugent.be
Elena Yu. **Bannikova**, Inst. of Astronomy, National Univ., Kharkov, Ukraine　　bannikova@astron.kharkov.ua
Aaron J. **Barth**, Univ. of California, Irvine, USA　　barth@uci.edu
Corrado **Bartolini**, Univ. Bologna, Italy　　corrado.bartolini@unibo.it
Mitchell C. **Begelman**, Univ. of Colorado, USA　　mitch@jila.colorado.edu
Gregory **Beskin**, Special Astrophys. Obs., Russian Fed.　　beskin@sao.ru
Weihao H. **Bian**, Dept. of Physics, Nanjing Normal Univ., China　　whbian@njnu.edu.cn
Jiří **Bičák**, Charles Univ., Prague, Czech Republic　　bicak@mbox.troja.mff.cuni.cz
Geoffrey V. **Bicknell**, Australian National Univ., Australia　　geoff@mso.anu.edu.au
Nikolai G. **Bochkarev**, Sternberg Astron. Inst., Russian Fed.　　boch@sai.msu.ru
Edi **Bon**, Astronomical Observatory, Belgrad, Republic of Serbia　　ebon@aob.bg.ac.yu
Remco van den **Bosch**, Leiden Observatory, Netherlands　　bosch@strw.leidenuniv.nl
Michal **Bursa**, Astronomical Institute ASCR, Prague, Czech Republic　　bursa@astro.cas.cz
Mónica V. **Cardaci**, Universidad Autónoma de Madrid, Spain　　monica.cardaci@uam.es
Jorge **Casares**, Instituto de Astrofisica de Canarias, Spain　　jcv@iac.es
Annalisa **Celotti**, SISSA, Trieste, Italy　　celotti@sissa.it
Philip A. **Charles**, SAAO, South Africa　　pac@saao.ac.za
George **Chartas**, Penn State Univ., USA　　chartas@astro.psu.edu
Tapan K. **Chatterjee**, Univ. of the Americas (UDLA), Mexico　　chtapan@yahoo.com
Xian **Chen**, Peking Univ., China　　chenxian_83@pku.edu.cn
Carlos H. **Coimbra-Araujo**, Campinas State Univ., Brazil　　carlosc@ifi.unicamp.br
Robert F. **Coker**, Los Alamos National Laboratory, USA　　robc@lanl.gov
Suzy **Collin-Zahn**, Observatoire de Paris-Meudon, France　　suzy.collin@obspm.fr
Enrico Maria **Corsini**, Universita' di Padova, Italy　　corsini@pd.astro.it
Patrick **Cote**, Herzberg Institute of Astrophysics, Canada　　Patrick.Cote@nrc-cnrc.gc.ca
Mark **Cropper**, MSSL/UCL, United Kingdom　　msc@mssl.ucl.ac.uk
Jorge **Cuadra**, Max Planck Inst. f. Astrophysik, Germany　　jcuadra@mpa-garching.mpg.de
Bożena **Czerny**, N. Copernicus Astron. Center, Warsaw, Poland　　bcz@camk.edu.pl
Xu **Dawei**, National Astronomical Observatories of China, China　　dwxu@bao.ac.cn
Horacio **Dottori**, Univ. Federal do Rio Grande do Sul, Brazil　　dottori@if.ufrgs.br
Michal **Dovčiak**, Astronomical Institute ASCR, Prague, Czech Republic　　dovciak@astro.cas.cz
Deborah **Dultzin-Hacyan**, Univ. Nacional Autonoma de Mexico, Mexico　　deborah@astroscu.unam.mx
Renato **Dupke**, Univ. of Michigan, Ann Arbor, USA　　rdupke@umich.edu
Andreas **Eckart**, Univ. of Cologne, Germany　　eckart@ph1 .uni-koeln.de
Patrick **Elebert**, Univ. College Cork, Ireland　　p.elebert@ucc.ie
Andrew C. **Fabian**, Univ. of Cambridge, United Kingdom　　acf@ast.cam.ac.uk
Sergei N. **Fabrika**, Special Astrophys. Obs., Nizhnij Arkhyz, Russian Fed.　　fabrika@sao.ru
Xiaohui H. **Fan**, Univ. of Arizona, USA　　fan@as.arizona.edu
Laura F. **Ferrarese**, Herzberg Institute of Astrophysics, Canada　　laura.ferrarese@nrc-cnrc.gc.ca
Jose F. **Fiestas**, Astronomisches Rechen-Institut, Germany　　fiestas@ari.uni-heidelberg.de
David J. E. **Floyd**, Space Telescope Science Inst., USA　　floyd@stsci.edu
Sara **Fogelström**, Gothenburg Univ., Sweden　　sara.fogelstrom@rudorna.se
Sándor **Frey**, FÖMI, Satellite Geodetic Obs., Budapest, Hungary　　frey@sgo.fomi.hu
Natasa **Gavrilovic**, Astronomical Obs., Belgrade, Republic of Serbia　　ngavrilovic@aob.bg.ac.yu
Reinhard **Genzel**, Max Planck Inst. f. Extraterrestrial Physics, Germany　　genzel@mpe.mpg.de
Ortwin **Gerhard**, Max Planck Inst. f. Extraterrestrial Physics, Germany　　gerhard@mpe.mpg.de
Omaira **Gonzalez-Martin**, Instituto de Astrofisica de Andalucia, Spain　　omaira@iaa.es
René W. **Goosmann**, Astronomical Institute ASCR, Prague, Czech Republic　　rene.goosmann@obspm.fr
Jenny E. **Greene**, Department of Astronomy, Princeton University, USA　　jgreene@cfa.harvard.edu
Leonid I. **Gurvits**, Joint Inst. for VLBI in Europe, Netherlands　　lgurvits@jive.nl
Alexander V. **Halevin**, Odessa National Univ., Ukraine　　halevin@astronomy.org.ua
Jean-Marie **Hameury**, Strasbourg Observatory, France　　hameury@astro.u-strasbg.fr
Tetsuya **Hara**, Kyoto Sangyo Univ., Japan　　hara@cc.kyoto-su.ac.jp
Dan A. **Harris**, Smithsonian Astrophys. Observatory, USA　　harris@cfa.harvard.edu
Günther **Hasinger**, Max Planck Inst. f. Extraterrestrial Physics, Germany　　ghasinger@mpe.mpg.de
Dominikus **Heinzeller**, Inst. for Theoretical Astrophys., Germany　　dh@ita-uni-heidelberg.de
Jonay Isai Gonzales **Hernandez**, Observatoire de Paris, France　　Jonay.Gonzalez-Hernandez@obspm.fr
Luis C. **Ho**, The Obs. of the Carnegie Inst. of Washington, Pasadena, USA　　lho@ociw.edu
Jiří **Horák**, Astronomical Institute ASCR, Prague, Czech Republic　　horak@astro.cas.cz
Richard W. **Hunstead**, Univ. of Sydney, Australia　　rwh@physics.usyd.edu.au
Dragana **Ilic**, Faculty of Mathematics, Univ. of Belgrade, Rep. of Serbia　　dilic@matf.bg.ac.yu
Hirohiko **Inoue**, ISAS/JAXA, Japan　　hirohiko@astro.isas.jaxa.jp
Makoto **Inoue**, National Astron. Obs. of Japan, Japan　　inoue@nro.nao.ac.jp
Takeshi I. **Itoh**, Univ. of Tokyo, Japan　　titoh@amalthea.phys.s.u-tokyo.ac.jp
Natalia **Ivanova**, CITA, Canada　　nata@cita.utoronto.ca
Paul C. **Joss**, MIT, USA　　joss@space.mit.edu
Bruno **Jungwiert**, Astronomical Institute ASCR, Prague, Czech Republic　　bruno@ig.cas.cz
Stephen **Justham**, Oxford Univ., United Kingdom　　sjustham@astro.ox.ac.uk
Hedvika **Kadlecová**, Charles Univ., Prague, Czech Republic　　hedvika.kadlecova@centrum.cz
Seiji **Kameno**, Kagoshima Univ., Japan　　kameno@sci.kagoshima-u.ac.jp
Vladimír **Karas**, Astronomical Institute ASCR, Prague, Czech Republic　　vladimir.karas@cuni.cz
Eugenia A. **Karitskaya**, Institute of Astronomy RAS, Russian Fed.　　karitsk@sai.msu.ru
Sergey **Karpov**, Special Astrophys. Obs., Nizhnij Arkhyz, Russian Fed.　　beskin@sao.ru
Andrew R. **King**, Univ. of Leicester, United Kingdom　　ark@astro.le.ac.uk

† The 26th General Assembly of the International Astronomical Union was attended by 2412 astronomers, out of which 651 people indicated their interest in our Symposium, "Black Holes: from Stars to Galaxies". Many of the General Assembly participants attended lectures and other scientific activities connected with this Symposium, which took place during the second week of the General Assembly. Here we list active participants of the meeting: those who presented review lectures, oral contributions and posters.

Anton M. **Koekemoer**, Space Telescope Science Inst., USA — koekemoer@stsci.edu
Boris V. **Komberg**, Astro Sp. Center, Lebedev Phys. Inst., Russian Fed. — bkomberg@asc.rssi.ru
Victor M. **Kontorovich**, Inst. of Radio Astronomy of NAS, Ukraine — vkont@ira.kharkov.ua
Volodymyr **Kryvdyk**, National Taras Shevchenko Univ., Kyiv, Ukraine — kryvdyk@univ.kiev.ua
Aya **Kubota**, Institute of Physical and Chemical Research (RIKEN), Japan — aya@crab.riken.jp
Zdenka **Kuncic**, Univ. of Sydney, Australia — z.kuncic@physics.usyd.edu.au
Hideyo **Kunieda**, Nagoya Univ., Japan — kunieda@u.phys.nagoya-u.ac.jp
Pawel **Lachowicz**, N. Copernicus Astron. Center, Warsaw, Poland — paulo@camk.edu.pl
Andrew **Lawrence**, Univ. of Edinburgh, United Kingdom — al@roe.ac.uk
Patricio S. **Letelier**, State Univ. of Campinas, Brazil — letelier@ime.unicamp.br
Lina **Levin**, Gothenburg Univ., Sweden — linalevin@hotmail.com
Xiangdong **Li**, Nanjing Univ., China — lixd@nju.edu.cn
Fukun **Liu**, Peking Univ., China — fkliu@bac.pku.edu.cn
Ye **Lu**, The National Astron. Obs. of China, China — ly@bao.ac.cn
Ferdinando D. **Macchetto**, Space Telescope Science Inst. & ESA, USA — macchetto@stsci.edu
Mami **Machida**, NAOJ, Japan — mami@th.nao.ac.jp
Witold **Maciejewski**, Univ. of Oxford, United Kingdom — witold@astro.ox.ac.uk
Piero **Madau**, Univ. of California, USA — pmadau@ucolick.org
Kazuo M. **Makishima**, Univ. of Tokyo, Japan — maxima@phys.s.u-tokyo.ac.jp
Sera B. **Markoff**, Univ. of Amsterdam, Netherlands — sera@science.uva.nl
Cristopher **Martin**, Oberlin College, USA — Chris.Martin@oberlin.edu
Silvia **Mateos**, Univ. of Leicester, United Kingdom — sm279@star.le.ac.uk
Ryoji **Matsumoto**, Chiba Univ., Japan — matumoto@astro.s.chiba-u.ac.jp
Giorgio **Matt**, Università Roma Tre, Rome, Italy — matt@fis.uniroma3.it
Claire E. **Max**, Univ. of California at Santa Cruz, USA — max@ucolick.org
Andrea **Merloni**, Max Planck Inst. f. Astrophysik, Germany — am@mpa-garching.mpg.de
Leonhard **Meyer**, Univ. of Cologne, Germany — leo@ph1.uni-koeln.de
Areg **Mickaelian**, Byurakan Astrophysical Observatory, Armenia — aregmick@apaven.am
Felix **Mirabel**, ESO, Santiago, Chile — fmirabel@eso.org
Abhas **Mitra**, Bhabha Atomic Research Centre, India — abhasmitra@rediffmail.com
Mark R. **Morris**, Univ. of California, Los Angeles, Los Angeles, CA, USA — morris@astro.ucla.edu
Monika Agnieszka **Moscibrodzka**, N. Copernicus Astron. Center, Warsaw, Poland — mmosc@camk.edu.pl
Martine **Mouchet**, Univ. Paris 7, Paris, France — martine.mouchet@obspm.fr
Richard **Mushotzky**, NASA/GSFC, USA — richard@xray-5.gsfc.nasa.gov
Koraljka **Muzic**, Univ. of Cologne, Germany — muzic@ph1.uni-koeln.de
Kenji **Nakamura**, Matsue College of Technology, Japan — nakamrkn@matsue-ct.jp
Ramesh **Narayan**, Harvard-Smithsonian Center for Astrophysics, USA — rnarayan@cfa.harvard.edu
Daisaku **Nogami**, Kyoto Univ., Japan — nogami@kwasan.kyoto-u.ac.jp
Achille A. **Nucita**, Univ. of Lecce, Italy — nucita@le.infn.it
Hiroshi **Oda**, Chiba Univ., Japan — oda@astro.s.chiba-u.ac.jp
Seungkyung **Oh**, Kyung Hee Univ., Yongin, Korea — skoh@ap1.khu.ac.kr
Ken **Ohsuga**, Department of Physics, Rikkyo Univ., Tokyo, Japan — k_ohsuga@rikkyo.ac.jp
Toru **Okuda**, Hakodate Campus, Hokkaido Univ. of Education, Japan — okuda@cc.hokkyodai.ac.jp
Victor **Orlov**, Sobolev Astron. Inst., St. Petersburg State Univ., Russian Fed. — vor@astro.spbu.ru
Stéphane **Paltani**, INTEGRAL Science Data Centre, Switzerland — Stephane.Paltani@obs.unige.ch
Tomáš **Pecháček**, Astronomical Institute ASCR, Prague, Czech Republic — pechacek_t@seznam.cz
Chien Y. **Peng**, Space Telescope Science Inst., USA — cyp@sts ci.edu
Gabriele **Ponti**, Univ. Bologna, Italy — ponti@iasfbo.inaf.it
David **Pooley**, Univ. of California at Berkeley, USA — dave@astron.berkeley.edu
Luka C. **Popovic**, Astronomical Observatory, Belgrade, Republic of Serbia — lpopovic@aob.bg.ac.yu
Kenneth A. **Pounds**, Univ. of Leicester, United Kingdom — kap@le.ac.uk
Daniel **Proga**, Univ. of Nevada, Las Vegas, USA — dproga@physics.unlv.edu
Sohrab **Rahvar**, Sharif Univ. of Technology, Islamic Rep. of Iran — rahvar@sina.sharif.edu
Travis A. **Rector**, Univ. of Alaska, Anchorage, USA — rector@uaa.alaska.edu
Martin J. **Rees**, Cambridge Univ., United Kingdom — mjr@ast.cam.ac.uk
Mark T. **Reynolds**, Univ. College Cork, Ireland — m.reynolds@ucc.ie
Barry **Rothberg**, Space Telescope Science Inst., USA — rothberg@stsci.edu
Anthony P. **Rushton**, Jodrell Bank Observatory, United Kingdom — arushton@jb.man.ac.uk
Bassem M. **Sabra**, Notre Dame Univ.-Louaize, Lebanon — bsabra@ndu.edu.lb
Margarita **Safonova**, Indian Institute of Astrophysics, India — rita@iiap.res.in
Lakshmi **Saripalli**, Raman Research Institute, India — lsaripal@rri.res.in
Oldřich **Semerák**, Charles Univ., Prague, Czech Republic — semerak@mbox.troja.mff.cuni.cz
Vyacheslav N. **Shalyapin**, Inst. for Radiophysics and Electronics, Ukraine — vshal@ire.kharkov.ua
Rainer **Schödel**, Universitaet zu Koeln, Germany — rainer@ph1.uni-koeln.de
Dan A. **Schwartz**, Smithsonian Astrophysical Observatory, USA — das@cfa.harvard.edu
Ayse Ulubay **Siddiki**, Max Planck Inst. f. Extraterrestrial Physics, Germany — siddiki@exgal.mpe.mpg.de
Vojtěch **Šimon**, Astronomical Institute ASCR, Ondřejov, Czech Republic — simon@asu.cas.cz
Petr **Slaný**, Silesian Univ., Opava, Czech Republic — sla10uf@axpsu.fpf.slu.cz
Rikard **Slapak**, Gothenburg Univ., Sweden — orendien@hotmail.com
Roberto **Soria**, Harvard-Smithsonian Center for Astrophysics, USA — rsoria@cfa.harvard.edu
Rainer **Spurzem**, Univ. of Heidelberg, Germany — spurzem@ari.uni-heidelberg.de
Ivana **Stoklasová**, Astronomical Institute ASCR, Prague, Czech Republic — ivana@sirrah.troja.mff.cuni.cz
Thaisa **Storchi-Bergmann**, Instituto de Fisica – UFRGS, Brazil — thaisa@ufrgs.br
Zdeněk **Stuchlík**, Silesian Univ., Opava, Czech Republic — zdenek.stuchlik@fpf.slu.cz
Ladislav **Šubr**, Argelander Inst. f. Astronomie, University of Bonn, Germany — lsubr@astro.uni.bonn.de
Ravi **Subrahmanyan**, Raman Research Institute, India — rsubrahm@rri.res.in
Greg B. **Taylor**, Univ. of New Mexico, Albuquerque, USA — gbtaylor@unm.edu
Jonathan **Tedds**, Univ. of Leicester, United Kingdom — jat@star.le.ac.uk
Gabriel **Török**, Silesian Univ., Opava, Czech Republic — terek@volny.cz
Ginevra **Trinchieri**, INAF – Osservatorio di Brera, Italy — ginevra.trinchieri@brera.inaf.it
Konrad R. W. **Tristram**, MPI für Astronomie Germany — tristram@mpia.de
Sergei A. **Trushkin**, Special Astrophys. Obs., Russian Fed. — satr@sao.ru
Sachiko **Tsuruta**, Montana State Univ., USA — uphst@gemini.msu.montana.edu
Marc **Türler**, INTEGRAL Science Data Centre, Switzerland — marc.turler@obs.unige.ch
Yoshihiro **Ueda**, Kyoto Univ., Japan — ueda@kusastro.kyoto-u.ac.jp
Makoto **Uemura**, Hiroshima Univ., Japan — uemuram@hiroshima-u.ac.jp
Emmanuil **Vilkoviskij**, Fesenkov Astrophys. Inst., Almaty, Kazakhstan — vilk@aphi.kz
Kenya **Watarai**, Osaka Kyoiku Univ., Japan — watarai@cc.osaka-kyoiku.ac.jp
Anna **Wolter**, INAF – Osservatorio di Brera, Italy — anna@brera.mi.astro.it
Jong-Hak **Woo**, Univ. of California at Santa Barbara, USA — woo@physics.ucsb.edu
Dawei **Xu**, National Astronomical Observatories of China, China — dwxu@bao.ac.cn
Tahir **Yaqoob**, Johns Hopkins Univ./NASA GSFC, USA — yaqoob@pha.jhu.edu
Alexander F. **Zakharov**, Inst. of Theor. and Exp. Physics, Russian Fed. — zakharov@itep.ru
Tim De **Zeeuw**, Leiden Observatory, Netherlands — dezeeuw@strw.leidenuniv.nl
Andreas **Zezas**, Smithsonian Astrophysical Observatory, USA — azezas@cfa.harvard.edu
Shuang Nan **Zhang**, Tsinghua Univ., China — zhangsn@mail.tsinghua.edu.cn
Jianfeng **Zhou**, Tsinghua Univ., China — zhoujf@tsinghua.edu.cn
Piotr T. **Życki**, N. Copernicus Astron. Center, Warsaw, Poland — ptz@camk.edu.pl

Stellar Mass Black Holes I

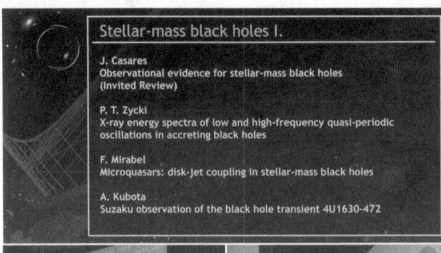

Stellar-mass black holes I.

J. Casares
Observational evidence for stellar-mass black holes
(Invited Review)

P. T. Zycki
X-ray energy spectra of low and high-frequency quasi-periodic
oscillations in accreting black holes

F. Mirabel
Microquasars: disk-jet coupling in stellar-mass black holes

A. Kubota
Suzaku observation of the black hole transient 4U1630-472

Queen Anna's Summer Pavilion in
the Royal Garden of Prague Castle.
Tycho Brahe and Johannes Kepler
carried out observations from here.

Black Holes from Stars to Galaxies – Across the Range of Masses
Proceedings IAU Symposium No. 238, 2006　　　　　　　© 2007 International Astronomical Union
V. Karas & G. Matt, eds.　　　　　　　　　　　　　　　doi:10.1017/S1743921307004590

Observational evidence for stellar-mass black holes

Jorge Casares

Instituto de Astrofísica de Canarias, 38200 – La Laguna, Tenerife, Spain
email: jcv@iac.es

Abstract. Radial velocity studies of X-ray binaries provide the most solid evidence for the existence of stellar-mass black holes. We currently have 20 confirmed cases, with dynamical masses in excess of 3 $M_\odot$. Accurate masses have been obtained for a subset of systems which gives us a hint at the mass spectrum of the black hole population. This review summarizes the history of black hole discoveries and presents the latest results in the field.

Keywords. Accretion, accretion disks – black hole physics – X-rays: binaries

1. Introduction

Theoretical black holes (BHs) have been considered for almost a century, although their first observational evidence was not found until quite recently, during the past 3 decades. Since then BHs have become essential ingredients in the construction of modern astrophysics on all scales, from binaries to galaxies and AGNs. It is stellar-mass BHs that offer us the best opportunity to study these objects in detail. Their implications are wide ranging, from late evolution of massive stars, supernovae models, production of high-energy radiation, relativistic outflows, chemical enrichment in the Galaxy, etc. But still the most compelling evidence for the existence of stellar-mass BHs relies on dynamical arguments, which will be the focus of this review. Section 2 summarizes the first observations of historical BH "candidates", whereas Sect. 3 reviews the most recent discoveries. In Sect. 4 we deal with BH demography and their mass distribution. Finally, we present our conclusions in Section 5.

2. Early discoveries: the first BH "candidates"

In the late 1960s X-ray detectors onboard satellites revolutionized astronomy with the discovery of an unexpected population of luminous X-ray sources in the Galaxy. The energetics (with $L_x \sim L_{\text{Edd}}$) together with their short timescale variability (down to milliseconds) lend support to an interacting binary model where X-rays are supplied by accretion onto a collapsed object (Shklovskii 1967). It is now well established that there are two main populations of X-ray binaries: the high mass X-ray binaries (HMXBs), containing O-B supergiant donor stars, and the low mass X-ray binaries (LMXBs), with typically short orbital periods and K-M donors. The optical flux in LMXBs is triggered by reprocessing of the X-rays into the accretion disc whereas in HMXBs it is dominated by the hot supergiant star (see reviews in Charles & Coe 2006 and McClintock & Remillard 2006).

Therefore, it was not surprising that one of the first optical counterparts to be identified was the 9th magnitude supergiant star HD 226868, associated with the HMXB Cyg X-1. But, remarkably, it showed radial velocity variations which made it a prime candidate for a stellar-mass BH (Webster & Murdin 1972, Bolton 1972). The supergiant star was shown to move with a velocity amplitude of ~ 64 km s^{-1} (later refined to 75 km s^{-1}) in a 5.6 day orbit due to the gravitational influence of an unseen companion (see Fig. 1). The orbital period P_{orb} and the radial velocity amplitude K, combine in the mass function equation $f(M_x) = K^3 P_{orb}/2\pi G = M_x^3 \sin^3 i/(M_x + M_c)^2$ which relates the mass of the compact object M_x with that of the companion star M_c and the inclination angle i. The mass function $f(M_x)$ is a lower limit to M_x and, for the case of Cyg X-1 is $0.25 M_\odot$. The key factor here is M_c which, for a HMXB, is a large number and has a wide range of uncertainty. If the optical star were a normal O9.7Iab its mass would be $\sim 33 M_\odot$ which, for an edge-on orbit ($i = 90°$), would imply a compact object of $\sim 7 M_\odot$. However, the optical star is likely to be under-massive for its spectral type as a result of mass transfer and binary evolution, as has been shown to be the case in several neutron star binaries (e.g. Rappaport & Joss 1983). In fact, it could be under-massive by as much as a factor of 3 given the uncertainty in distance, $\log g$ and T_{eff}. A plausible lower limit of $10 M_\odot$, combined with an upper limit to the inclination of $60°$, based on the absence of X-ray eclipses, leads to a compact object of $> 4 M_\odot$ (Bolton 1975).

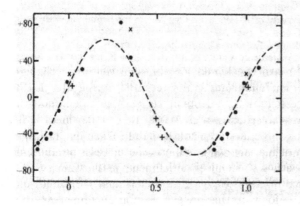

Figure 1. Radial velocity curve of HD 226868, the O9.7Iab companion star in the HMXB Cyg X-1, folded on the 5.6 day orbital period. After Webster & Murdin (1972).

The importance of this result rests on the fact that there is a maximum mass for neutron stars (NS) to be stable against gravitational collapse (Oppenheimer & Volkoff 1939). This maximum depends on the equation of state (EoS), which is uncertain in the high density regime because of the poorly constrained many-body interactions. However, Rhoades & Ruffini (1974) showed that an upper limit of $\sim 3.2 M_\odot$ can be derived assuming that causality holds beyond densities where the EoS starts to be uncertain which, at the time, was 1.7 times the density of nuclear matter $\rho_{nm} = 2.7 \times 10^{14}$ g cm^{-3}. More recently, a new limit of $2.9 M_\odot$ was obtained using modern EoS which are accurate up to $2 \times \rho_{nm}$ (Kalogera & Baym 1996). This can be further boosted by up to $\sim 25\%$ if the NS rotates close to break-up (Friedman & Ipser 1987). Therefore, the compact object in Cyg X-1, with $M_x \geqslant 4 M_\odot$, is a very strong BH candidate.

In 1975 the satellite Ariel V detected A0620-00, a new X-ray source which displayed an increase in X-ray flux from non-detection to a record of ~ 50 Crab. It belongs to the class of X-ray transients (XRTs, also called X-ray Novae), a subclass of LMXBs which undergo dramatic episodes of enhanced mass-transfer or "outbursts" triggered by viscous-thermal instabilities in the disc (e.g. King 1999). During outburst, the companion remains undetected because it is totally overwhelmed by the intense optical light from the X-ray heated disc. However, the X-rays switch off after a few months of activity, the reprocessed flux drops several magnitudes into quiescence and the companion star becomes the dominant source of optical light. This offers a very special opportunity to perform radial velocity studies of the cool companion and unveil the nature of the compact star. The first detection of the companion in A0620-00 revealed a mid-K star moving in a 7.8 hr period with velocity amplitude of 457 km s^{-1}. The implied mass function was 3.2 $\pm0.2M_\odot$, the largest ever measured (McClintock & Remillard 1986). An absolute 3σ lower limit to the mass of the compact star of $3.2M_\odot$ was established by assuming a very conservative low-mass companion of $0.25M_\odot$ and $i < 85°$, based on the lack of X-ray eclipses. This exceeds the maximum mass allowed for a stable NS and hence it also became a very compelling case for a BH.

3. From "candidates" to confirmed BHs

In the 1980s there was a hot debate about the real existence of BHs. On one hand there were 3 strong candidates, the HMXBs Cyg X-1 and LMC X-3 and the transient LMXB A0620-00, all with lower limits to M_x very close to the maximum mass for NS stability. On the other, alternative scenarios were proposed to avoid the need for BHs such as multiple star systems (Bahcall *et al.* 1974) or non-standard models invoking exotic EoS for condensed matter. An example of the latter are Q-stars, where neutrons and protons are confined by the strong force rather than gravity and can be stable for larger masses (Bahcall *et al.* 1990). In this context, it was proposed that the *"holy grail in the search for black holes is a system with a mass function that is plainly $5M_\odot$ or greater"* (McClintock 1986).

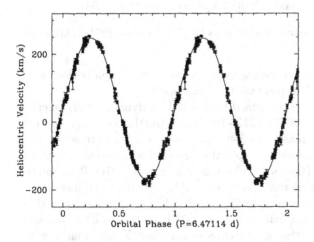

Figure 2. Radial velocity curve of the K0 companion in the transient LMXB V404 Cyg during quiescence. This graph contains velocity points obtained between 1991 and 2005.

In 1989, the X-ray satellite Ginga discovered a new XRT in outburst named GS 2023+338 (=V404 Cyg). Its X-ray properties drew considerable attention because of the exhibition of a possible luminosity saturation at $L_x \simeq 10^{39}$ erg s^{-1} and dramatic variability (Życki, Done & Smith 1999). Spectroscopic analysis during quiescence revealed a K0 star moving with a velocity amplitude of 211 km s^{-1} in a 6.5 day orbit (see Fig. 2). The mass function implied by these numbers is $6.3 \pm 0.3M_\odot$, and hence the compact object must be more massive than $6M_\odot$, independent of any assumption on M_c and i (Casares *et al.* 1992). This remarkable result established V404 Cyg as the "holy grail" BH for almost a decade. Since then, many other BHs have been unveiled through dynamical studies of XRTs in quiescence, seven others with mass functions also in excess

Table 1. Confirmed black holes and mass determinations

System	P_{orb} [days]	$f(M)$ [$M_\odot$]	Donor Spect. Type	Classification	M_x [†] [$M_\odot$]
GRS 1915+105[a]	33.5	9.5 ± 3.0	K/M III	LMXB/Transient	14 ± 4
V404 Cyg	6.471	6.09 ± 0.04	K0 IV	,,	12 ± 2
Cyg X-1	5.600	0.244 ± 0.005	09.7 Iab	HMXB/Persistent	10 ± 3
LMC X-1	4.229	0.14 ± 0.05	07 III	,,	> 4
XTE J1819-254	2.816	3.13 ± 0.13	B9 III	IMXB/Transient	7.1 ± 0.3
GRO J1655-40	2.620	2.73 ± 0.09	F3/5 IV	,,	6.3 ± 0.3
BW Cir[b]	2.545	5.74 ± 0.29	G5 IV	LMXB/Transient	> 7.8
GX 339-4	1.754	5.8 ± 0.5	–	,,	
LMC X-3	1.704	2.3 ± 0.3	B3 V	HMXB/Persistent	7.6 ± 1.3
XTE J1550-564	1.542	6.86 ± 0.71	G8/K8 IV	LMXB/Transient	9.6 ± 1.2
4U 1543-475	1.125	0.25 ± 0.01	A2 V	IMXB/Transient	9.4 ± 1.0
H1705-250	0.520	4.86 ± 0.13	K3/7 V	LMXB/Transient	6 ± 2
GS 1124-684	0.433	3.01 ± 0.15	K3/5 V	,,	7.0 ± 0.6
XTE J1859+226[c]	0.382	7.4 ± 1.1	–	,,	
GS2000+250	0.345	5.01 ± 0.12	K3/7 V	,,	7.5 ± 0.3
A0620-003	0.325	2.72 ± 0.06	K4 V	,,	11 ± 2
XTE J1650-500	0.321	2.73 ± 0.56	K4 V	,,	
GRS 1009-45	0.283	3.17 ± 0.12	K7/M0 V	,,	5.2 ± 0.6
GRO J0422+32	0.212	1.19 ± 0.02	M2 V	,,	4 ± 1
XTE J1118+480	0.171	6.3 ± 0.2	K5/M0 V	,,	6.8 ± 0.4

[†] Masses compiled by Orosz (2003) and Charles & Coe (2006).
[a] New photometric period of 30.8 ± 0.2 days recently reported by Neil, Bailyn & Cobb (2006). The implied mass function, assuming constant velocity amplitude, would be 8.7 $M_\odot$.
[b] Updated after Casares *et al.* (2007).
[c] Period is uncertain, with another possibility at 0.319 days (see Zurita *et al.* 2002). This would drop the mass function to 6.18 $M_\odot$.

of $5M_\odot$. This has been possible thanks to the improvement in spectrograph performance on a new generation of 10-m class telescopes over the last decade.

Table 1 presents the current list of 20 confirmed BHs based on dynamical arguments, ordered by orbital period. The case of GRS 1915+105 is noteworthy, not only because of its long orbital period and large mass function; also because IR spectroscopy was needed to overcome the > 25 magnitudes of optical extinction and reveal the radial velocity curve of the companion star (Greiner, Cuby & McCaughrean 2001). It should be noted that, although some systems have mass functions $< 3M_\odot$, solid constraints on the inclination and/or M_c can be set which result in $M_x > 3M_\odot$. The great majority of BHs are transients and only 3 show persistent behaviour, the HMXBs Cyg X-1, LMC X-1 and LMC X-3. From the list we also note that a new class of transient X-ray binaries with A-F companions is starting to emerge, the so-called Intermediate Mass X-ray Binaries (IMXBs). It has been proposed that LMXBs may descend from IMXBs through a phase of thermal mass-transfer (Pfahl, Rappaport & Podsiadlowski 2003), although it is still unclear whether IMXBs and LMXBs represent an evolutionary sequence.

3.1. *A novel technique: fluorescence emission from the irradiated donor*

In addition to the dynamical BHs, there are ~ 20 other BH candidates based on their X-ray temporal and spectral behaviour (McClintock & Remillard 2006). Unfortunately, they have never been seen in quiescence, or they simply become too faint for an optical detection of the companion star. However, a new strategy was devised to allow the extraction of dynamical information in these systems during their X-ray active states. It utilises narrow high-excitation emission lines powered by irradiation on the companion star, in particular the strong CIII and fluorescence NIII lines from the Bowen blend at $\lambda\lambda4630\text{-}40$ (see Fig. 3). This technique was first applied to the NS LMXB Sco X-1 and the Doppler shift of the CIII/NIII lines enabled the motion of the donor star to be traced for the first time (Steeghs & Casares 2002). This was also attempted during the 2002 outburst of the BH candidate GX 339-4, using high-resolution spectroscopy to resolve the sharp NIII/CIII lines. The right panel in Figure 3 shows the radial velocity curve of the NIII lines folded on the 1.76 day orbital period. The orbital solution yields a velocity semi-amplitude of 317 km s^{-1} which defines a strict lower limit to the velocity amplitude of the companion star because these lines arise from the irradiated hemisphere and not the center of mass of the donor. Therefore, a solid lower bound to the mass function is $5.8M_\odot$ which provides compelling evidence for a BH in GX 339-4 (Hynes *et al.* 2003). And most important, this technique opens an avenue to extract dynamical information from new XRTs in outburst and X-ray persistent LMXBs, which hopefully will help increase the number of BH discoveries.

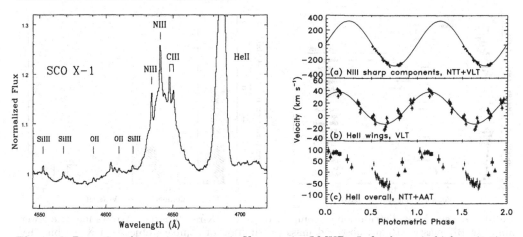

Figure 3. Detecting the companion star in X-ray active LMXBs. Left: the main high excitation emission lines due to by irradiation of the donor star in Sco X-1. Adapted from Steeghs & Casares (2002). Right: Radial velocities of the NIII lines (top), HeII $\lambda4686$ (bottom) and the wings of HeII $\lambda4686$ (middle) in GX 339-4 folded on the 1.76 day orbital period. After Hynes *et al.* (2003).

3.2. *Further BH signatures*

Aside from dynamical arguments, there are other observations that lend support to the absence of a solid surface and, hence, the BH nature of these accreting compact objects, namely:

• Lack of pulses and Type I X-ray bursts. This has been quantified by Remillard *et al.* (2006) using observations of dynamical BHs over 9 years of RXTE data. The probability that the non-detection of bursts were consistent with a solid surface is found to be $\sim 2 \times 10^{-7}$.

• The classic colour-colour diagram of X-ray binaries at high accretion rates shows a clear separation in the evolution of NS and BH binaries (Done & Gierliński 1999). This has been ascribed to the presence of a boundary layer in NS which gives rise to an additional thermal component in the spectrum and drags NS outside the BH region.

• For a given orbital period, quiescent BH binaries are ~ 100 times dimmer than quiescent NS binaries (Menou *et al.* 1999a). This difference is interpreted as thermal radiation from the NS surface which, for BHs, is advected through the event horizon.

4. BH demography

Despite the handful of confirmed BHs one can try to account for selection effects and estimate the size of the underlying galactic population. To start with, dynamical studies of XRTs indicate that about 75% contain BHs, i.e. $M_x > 3M_\odot$. Also, the extrapolation of the number of BH XRTs detected since 1975, with outburst duty cycles of $\sim 10-100$ years, suggests that there is a dormant population of $\sim 10^3$ BH binaries. (Romani 1998 and references therein). This is likely to be an underestimate if one accounts for systems with longer recurrence times and a likely population of faint persistent BH LMXBs (Menou *et al.* 1999b). Even so, incidentally, these numbers are in reasonable agreement with recent population synthesis calculations of BH binaries (Yungelson *et al.* 2006). On the other hand, stellar evolution models predict a population of about 10^9 stellar-mass BHs in the Galaxy (Brown & Bethe 1994). Therefore, our observed sample of BH XRTs is just the tip of the iceberg of a large hidden population of which nothing is known. Is it still possible to extract some meaningful information from such a limited sample?

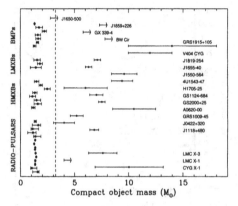 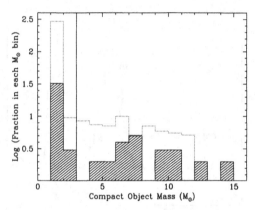

Figure 4. Left: Mass distribution of compact objects in X-ray binaries. Arrows indicate lower limits to BH masses. Right: observed mass distribution of compact objects in X-ray binaries (shaded histogram), compared to the theoretical distribution computed in Fryer & Kalogera (2001) for the "Case C + Winds" scenario (dotted line). Mass-loss rates by Woosley, Langer & Weaver (1995) were used in the computations. The model distribution has been re-scaled for clarity.

The main property of the BH population is the mass distribution. However, in order to get accurate BH masses, in addition to the mass function one needs to determine the binary inclination and the mass of the companion star or a related quantity such as the mass ratio. This information can be extracted from two experiments: (i) resolving the rotational broadening $V_{rot} \sin i$ of the companion's absorption lines, which is correlated with the binary mass ratio and (ii) fitting synthetic models to ellipsoidal lightcurves from which the inclination angle can be determined. This is the classic method to derive masses which has been reviewed in several papers e.g. Casares (2001).

Following this prescription, reliable BH masses have been determined in 15 binaries. These are listed in the last column of Table 1 and displayed in Figure 4, relative to NS masses compiled by Stairs (2004) and Lattimer & Prakash (2004). The figure nicely shows a mass segregation between the 2 populations of collapsed objects, with NS clustering around $1.4M_\odot$ and BH masses scattering between 4 and $14M_\odot$. Many more BH masses and errorbars $\leqslant 10\%$ are needed before fundamental questions can be addressed such as (i) do BHs masses cluster at a particular value? (ii) what are the edges of the distribution, i.e the minimum and maximum masses for BHs to be formed in X-ray binaries? (iii) is there a continuum distribution of masses between NS and BHs? All these questions are intimately related to models of SN and close binary evolution.

Figure 4 also shows the histogram of the current distribution of compact remnant masses, relative to a model distribution computed by Fryer & Kalogera (2001) which includes mass-loss through winds and binarity effects. The model shows a mass cut at $12M_\odot$ but this seems to be challenged by the compact objects in V404 Cyg and GRS 1915+105, with $12 \pm 2M_\odot$ and $14 \pm 4M_\odot$ respectively. However, the conflict may be spurious since recent estimates of mass-loss rates in Wolf-Rayet stars suggests that previous determinations were biased too low (Nugis & Lamers 2000). Also, the model predicts a continuum distribution between NS and BH whereas observations seem to show a paucity of objects at $\sim 3\text{--}4M_\odot$. This may be caused by selection effects since low-mass BHs are expected to be persistent. Recently, compact object masses in the range $2\text{--}4M_\odot$ have been reported in 4U1700-37 (Clark *et al.* 2002), V395 Car (Shahbaz *et al.* 2004) and LS 5039 (Casares *et al.* 2005). Although they are assumed to contain NS, none of them has shown X-ray bursts nor pulses. Therefore, they may well be members of the missing low-mass BH population. We are clearly dominated by low number statistics and more observations are required.

5. Conclusions

The best observational evidence for the existence of stellar-mass BHs is provided by dynamical studies of X-ray binaries The first solid *candidates* were the classic X-ray binaries Cyg X-1 and, in particular, A0620-00. BHs became *confirmed* with the discovery of mass in excess of $5\text{--}6M_\odot$, the first case being V404 Cyg. Their global X-ray properties (such as the lack of pulsations/bursts, weak quiescent Lx, etc.) also support the presence of an event horizon in these objects.

XRTs are the best hunting ground for new stellar-mass BHs with 17 cases currently known and estimated masses between $4\text{--}14M_\odot$. These are only the tip of the iceberg of an estimated dormant population of $\sim 10^3$ BH binaries and $\sim 10^9$ stellar-mass BHs in the Galaxy. Clearly many more discoveries and better statistics are essential to derive useful constraints on BH formation models.

New strategies, aimed at unveiling more quiescent BH transients, and novel techniques, such as the detection of the irradiated donor using high-excitation reprocessed lines, need to be exploited.

Acknowledgements

I acknowledge useful comments from Phil Charles. I am also grateful for support from the Spanish MCYT grant AYA2002-0036.

References

Bahcall J. N., Dyson F. J., Katz J. I. & Pacyńsky B. 1974, ApJ, 189, L17
Bahcall S., Bryan W. & Selipsky S. B. 1990, ApJ, 362, 251
Bolton C. T. 1972, Nature Phys. Sci., 235, 271
Bolton C. T. 1975, ApJ, 200, 269
Brown G. E. & Bethe H. A. 1994, ApJ, 423, 659
Casares J. 2001, in: F. C. Lazaro & M. J. Arevalo (eds.), Binary Stars: Selected Topics on Observations and Physical Processes, LNP 563, p. 277
Casares J., Charles P. A. & Naylor T. 1992, Nature, 355, 614

Casares J. *et al.*, 2007, in preparation

Casares J., Ribó M., Ribas I., Paredes J. M., Martí J. & Herrero A. 2005, MNRAS, 364, 899

Charles P. A. & Coe M. J. 2006, in: W. H. G. Lewin & M. van der Klis (eds.), Compact Stellar X-ray Sources, Cambridge Astrophysics Series No. 39 (Cambridge: Cambridge University Press), p. 215

Clark J. S., Goodwin S. P., Crowther P. A., Kaper L. Fairbairn M., Langer N. & Brocksopp C. 2002, A&A, 392, 909

Done C. & Gierliński M. 1999, MNRAS, 342, 1041

Friedman J. L. & Ipser J.R. 1987, ApJ, 314, 594

Fryer C. L. & Kalogera V. 2001, ApJ, 554, 548.

Greiner J., Cuby J. G. & McCaughrean M. J. 2001, Nature, 414, 522

Hynes R. I., Steeghs D., Casares J., Charles P. A. & O'Brien K. 2003, ApJ, 583, L95

Jonker P. G., Steeghs D., Nelemans G. & van der Klis M. 2005, MNRAS, 356, 621

Kalogera V. & Baym G. 1996, ApJ, 470, L61

King A. R. 1999, Phys. Rev., 311, 337

Lattimer J. M. & Prakash M. 2004, Science, 304, 536

McClintock J. E. 1986, in: K. O. Mason, M. G. Watson & N. E White (eds.), The Physics of Accretion onto Compact Objects, (Heidelberg: Springer), vol. 266, p. 211

McClintock J. E. & Remillard R. A. 1986, ApJ, 308, 110

McClintock J. E. & Remillard R. A. 2006, in: W. H. G. Lewin & M. van der Klis (eds.), Compact Stellar X-ray Sources (Cambridge University Press: Cambridge), p. 157

Menou K., Esin A. A., Narayan R., Garcia M. R., Lasota J.-P. & McClintock J. E. 1999a, ApJ, 520, 276

Menou K., Narayan R., & Lasota J.-P. 1999b, ApJ, 513, 811

Neil E. T., Bailyn C. D. & Cobb B. E. 2006, ApJ, in press (astro-ph/0610480)

Nugis T. & Lamers H. J. G. L. M. 2000, A&A, 360, 227

Oppenheimer J. R. & Volkoff G. M 1939, Phys. Rev., 55, 374

Orosz, J. A. 2003, in: K. A. van der Hucht, A. Herrero & C. Esteban (eds.) A Massive Star Odyssey, from Main Sequence to Supernova, Proc. IAU Symp. No. 212 (San Francisco: Astronomical Society of the Pacific) p. 365

Pfahl E., Rappaport S. & Podsiadlowski P. 2003, ApJ, 597, 1036

Rappaport S. A. & Joss P. C.. 1983, in: W. H. G. Lewin & E. P. J.. van den Heuve (eds.) Accretion-driven X-ray Sources, (Cambridge: Cambridge University Press), p. 33

Remillard R. A., Dacheng L., Cooper R. L. & Narayan R. 2006, ApJ, 646, 407

Rhoades C. E. & Ruffini R. 1974, Phys. Rev. Lett., 32, 324

Romani R. W. 1998, A&A, 333, 583

Shahbaz T. *et al.* 2004, ApJ, 616, L123

Shklovskii I. S. 1967, Astron. Zhur., 44, 930

Stairs I. H. 2004, Science, 304, 547

Steeghs D. & Casares J. 2002, ApJ, 568, 273

Woosley S. E., Langer N. & Weaver T. A. 1995, ApJ, 448, 315

Webster B. L. & Murdin P. 1972, Nature, 235, 37

Yungelson L. R. *et al.* 2006, A&A, 454, 559

Zurita C. *et al.* 2002, MNRAS, 334, 999

Życki P. T., Done C. & Smith D. A. 1999, MNRAS, 309, 561

RAMESH NARAYAN: When you make use of the ellipsoidal model to determine the binary inclination, how do you avoid contamination of the stellar light by the disc light?

JORGE CASARES: The example I showed is a "textbook" lightcurve of J1655–40 where the hot F6 III companion star totally dominates the optical light. However, for the great majority of black hole binaries (which contain much cooler companions in shorter orbital periods) the contamination by the accretion disc light certainly starts to show up in the form of flaring activity and superhumps. It is commonly assumed that the active contribution of the accretion disc light decreases at longer wavelengths and, hence, IR lightcurves are used to estimate the orbital inclination.

TSEVI MAZEH: For HMXB, is it a necessary assumption that the star fills its Roche lobe? If not, the ellipsoidal effect is difficult to model.

JORGE CASARES: Roche-lobe filling is not a general assumption of the models because HMXBs are normally wind-fed systems. Normally, the ellipsoidal fits are applied with an extra free parameter which is the star filling-factor.

XIAOPEI PAN: Physical parameters of black holes have huge uncertainties. The Space Interferometry Mission (SIM) can provide high-accuracy determination of inclination, mass, distance, etc. The precisions of black-hole masses can be directly determined to 5%. The inclination precision can reach 2%, and the distance of black holes can be measured to better that 2% by the SIM project. We hope in more collaborations on black hole issues.

JORGE CASARES: The error budget in black-hole mass determination is clearly dominated by uncertainties in the inclination angle and indeed the improvement that the SIM project may provide will be extremely useful.

ANDREW KING: Comment: I would like to reinforce your point about the black holes we detect being the top of an iceberg. Long-period LMXBs have wide separations, big discs, and are all transient. However, the interval between outbursts is so long that we will never detect them at all. We can detect the endpoints of these LMXBs when they contain neutron stars as very wide binaries with millisecond pulsars, but not when they contain black holes.

AD
AVGVSTISSIMVM IMPERATOREM
RVDOLPHVM SECVNDVM

TYCHONIS BRAHE
PRÆFATIO.

ASTRONOMIA ſcientia antiquiſsima, Divinitùs inde ab Adamo Proto-
plaſto humano generi conceſſa, longéque præſtantiſsima, in quantum nimi-
rùm Cœleſtia & ſublimia hæc terrena & inferiora ſuperant: Hæc inquàm Di-
vina Aſtronomia ab ipſis ſenſibus oculorum, multivagas ſiderum viciſsitudi-
nes animadvertentibus, quantum ad exteriora originem trahens, multo tem-
pore indè à rerum conditu, excellentiſsimorum hominum fatigavit & genios & ingenia.
Tanta nimirum eſt Dei Optimi & ter maximi majeſtas, ut à nullis creaturis ipſius operum
ſapientia exhauriri queat. Cum verò ſolus ocularis intuitus omnia illa ſuprà modum admi-
randi theatri Cœleſtis myſteria, intricatasq́; varietates apparentes, eâ, quâ opus erat ſubtilitate
& accuratione capere nequiret, excogitârunt omnib; ſeculis varij Artifices media & organa,
quibus viſus in percipiendis Siderum abſtruſis motibus juvaretur. Hinc ſunt ille columnæ,
quas Ioſephus Iudaicarum rerum ſcriptor refert, Adæ Nepotes in Syria extruxiſſe, ijsq; ſua
inventa memoriæ cauſa ad Poſteros inſcripſiſſe. Huc pertinent Ægyptiorum & aliarum
gentium altiſsimæ & ſumptuoſiſsimæ Pyramides; multæq́; aliæ machinæ, à Regibus anti-
quiſsimis in hunc uſum conſtructæ, quales in India, Syriâ, Arabia, Chaldæâ, Æthiopiâ, Ægy-
pto, præſertim iſthic in porticu Alexandrino, tum quoq; alibi in circumjacentibus Regio-
nibus, ubi homines Siderali ſcientiæ ſedulò intenti degebant, olim ſpectabantur. Id enim o-
mnium primum eſt, ut in Aſtronomicis obſervationes plurimæ & diutinæ idoneis & erro-
ri non obnoxijs Organis Cælitus capiantur: quæ poſteà per Geometriam excogitatis con-
venientibus Hypotheſibus in quantitates continuas & motum circularem ac uniformem
(quem Cœleſtia Naturaliter, & abſq; intermiſsione appetunt atq; exercent) digeruntur;
per Arithmeticam verò in diſcretas: ut ad quævis tempora conſtent Cœleſtium corporum
circuitus & loca. Inter omnes verò qui hâc in re ſtrenuè laborarunt, ad nos ſaltem per-
venerunt ea, quæ à Timochare, Hipparcho, Ptolemæo, Albategnio, Rege Alphonſo, &
ſuperiori ævo Copernico conſignata ſunt: quamvis duorum antecedentium in his traditi-
ones ſaltem ex æquali Ptolemæi relatione conſtent. Quibus verò Organis hi in dimetiendis Siderum Phænomenis potiſsimum uſi ſint, ex ſcriptis eorum utcunq; liquet. Inter quæ hæc tria præcipua invenio: Regulas Parallaticas, Armillas Zodiacales atq; Torque-
tum, quod Arabibus (uti & Aſtrolabia plana) potius in uſu erat: cætera minoris ſunt mo-
menti. Multa tamen forte alia fuere, quæ literis non prodita, ad nos haud pervene-
runt; quæ irritata tamq́; crebrâ Mundanæ ſcenæ confuſione & mutatione, tot bellis &
devaſtationibus ſubinde irrepentibus facilè (quod deplorandum eſt) interire poterant. Re-
centiores Quadrantem, Radium & Annulum Aſtronomicum addidere, tum quoq; quæ-
dam minoris adhuc æſtimationis. Ex quo autem Siderum motus accuratè noſtro æ-
vo conſiderati nequaquam ita ſe exhibeant, uti fert calculus ex illorum ſive veterum, ſive
recentium Artificum Obſervationibus derivatus, non immeritò ſuſpicionem movet, me-

):(2   dia &

Black Holes from Stars to Galaxies – Across the Range of Masses
Proceedings IAU Symposium No. 238, 2006
V. Karas & G. Matt, eds.
© 2007 International Astronomical Union
doi:10.1017/S1743921307004607

Energy spectra of X-ray quasi-periodic oscillations in accreting black hole binaries

P. T. Życki,[1] M. A. Sobolewska[2] and A. Niedźwiecki[3]

[2]Nicolaus Copernicus Astronomical Center, Bartycka 18, 00-716 Warsaw, Poland

[2]Department of Physics, University of Durham, South Road, Durham DH1 3LE, UK

[3]Łódź University, Department of Physics, Pomorska 149/153, 90-236 Łódź, Poland

email: ptz@camk.edu.pl

Abstract. We investigate the energy dependencies of X-ray quasi-periodic oscillations in black hole X-ray binaries. We analyze RXTE data on both the low- and high-frequency QPO. We construct the low-f QPO energy spectra, and demonstrate that they do not contain the thermal disk component, even though the latter is present in the time averaged spectra. The disk thus does not seem to participate in the oscillations. Moreover the QPO spectra are harder than the time averaged spectra when the latter are soft, which can be modeled as a result of modulations occurring in the hot plasma. The QPO spectra are softer than the time averaged spectra when the latter are hard. The absence of the disk component in the QPO spectra is true also for the high-frequency (hecto-Hz) QPO observed in black hole binaries. We compute the QPO spectra expected from the model of disk resonances.

Keywords. Accretion, accretion disks – relativity – X-rays: binaries

---

1. Introduction

Quasi-periodic components of X-ray variability are often seen in power density spectra (PDS) from accreting compact objects, signifying the existence of "clocks" operating on a specific time scale, among the broad-band variability. The nature of these periodic processes is unknown, but they are usually associated with various modes of oscillations of standard accretion disks present in those systems.

Energy dependencies of the QPO can be an important clue to the origin of these features, similarly to analyzes of time averaged energy spectra, which provide information about physical processes of generation of X-rays. It was noted quite early on into the study of QPO that their r.m.s. variability seems to increase with energy (review in van der Klis 2006). Here we report on more detailed investigations of QPO energy dependencies in black hole X-ray binaries. Firstly, we construct observed low-f (1–10 Hz) QPO energy spectra in the 2–20 keV range using RXTE data. Secondly, we construct models of spectral variability of inverse-Compton emission, producing a given type of QPO spectra. We also analyze data on high-f (hecto-Hz) QPO and construct models of energy spectra of these QPO in the model of disk resonances (Abramowicz & Kluźniak 2001).

2. Low-frequency QPO

2.1. *Observations*

We analyzed observations of four sources: GRS 1915+105, XTE 1550-564, 4U1630-47 and XTE 1859+226 performed by RXTE in observing modes providing good energy and timing resolution. Details of data reduction and analysis procedures are given in Sobolewska

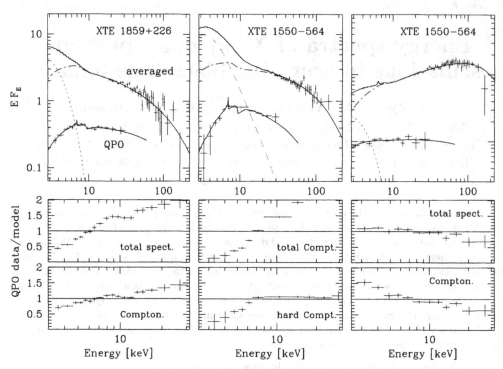

Figure 1. Comparison of time averaged spectra and the QPO spectra (upper panels). Lower panels show ratios of the QPO spectra to various components from the time averaged spectra.

& Życki (2006). Firstly, we construct time-averaged energy spectra in the 3–200 keV energy range using PCA and HEXTE data. These are fit with a model containing two or three continuum components (disk black body, comptonization) and the reprocessed component (Fe Kα line and the Compton reflected continuum). Then we construct the QPO energy spectra: we compute PDS in each energy channel, then fit it with a model consisting of a broad band continuum (broken power law or a number of Lorentzians) and one or two narrow Lorentzian peaks representing the QPO, possibly with a harmonic. The QPO Lorentzians are then integrated over frequency. This, expressed as variance (rather than the usual r.m.s./mean) is the QPO energy spectrum. The meaning of such a function is that it would correspond to energy spectrum of a variable component, if it was a separate spectral component which would be responsible to the QPO.

The results are presented in Fig 1. We show both the time averaged spectra and the QPO spectra. We also show the ratios of QPO spectra to the time averaged models. The main result is the lack of the soft disk component in the QPO spectra. Presence of this is excluded in all analyzed spectra at high confidence level. This can be seen for example in the spectra of XTE 1859+226, where the QPO spectra are much closer to the Comptonized component only than to total time averaged spectra. Secondly, we analyzed the slope of the QPO spectra relative to the time averaged spectra. In soft spectral state the QPO spectra are clearly harder than the time averaged spectra (left), while the opposite is true for the hard state (right). During the two observations of 4U1630-47 the time averaged spectra are intermediate in slope between values typical for the soft and hard states ($\Gamma \approx 2$). We find that the QPO spectra are similar to the time averaged spectra in these two observations.

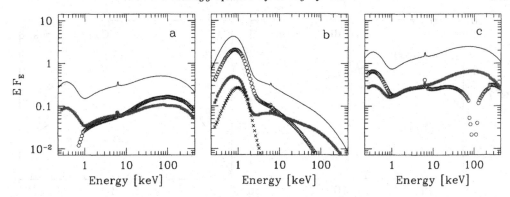

Figure 2. Model predictions for the QPO energy spectra (red circles) in various scenarios for modulations: a) heating; b) cooling; c) covering factor of cold matter. Green squares are the broad band noise spectra, blue stars: QPO harmonic, solid line shows time averaged spectra.

2.2. Models

In an attempt to understand the results presented in the previous Section, we performed simulations of spectral variability of inverse-Compton component. Such spectra are described by three main parameters: the plasma heating rate, l_h, the flux of cooling soft photons, l_s, and the temperature of the input soft photons, T_0. Irrespectively on the physical mechanism of QPO production, it is one of the three parameters that has to be modulated if the observed QPO corresponds to modulation of intrinsic luminosity (other possibility is a periodic modulation of geometry affecting, e.g., relativistic effect influencing the observed luminosity; see below).

The slope of the time averaged Comptonized spectrum depends (almost) only on the ratio l_h/l_s, however the mode of spectral variability depends on which parameter is modulated. Details of this investigation are given in Życki & Sobolewska (2005). To summarize it: modulation of the heating rate produces variability spectrum with largest amplitude of variability at high energies, which means that the variability spectrum is *harder* than the time averaged spectrum. Modulation of the cooling rate or the soft photons temperature produces variability spectrum *softer* than the time averaged spectrum. We considered also modulation of the amplitude of the reprocessed component, which was assumed to relate the cooling and heating rate (see, e.g., Poutanen & Fabian 1999). In this case we also find that the QPO are softer than the time averaged spectra, but, additionally, a very strong Fe Kα line is seen in the QPO spectrum.

Summarizing, properties of low-f QPO spectra in the soft state strongly suggest that these QPO are generated as a result of modulations in the hot plasma rather than in the standard cold accretion disk. This is contrary to most theoretical ideas where oscillations of the cold disk are thought to be the primary driver. However, the physical mechanism does seem to require both accreting plasma phases to work, since these QPO are associated with changes of the geometry of the system. It is also interesting to note, that long-time scale variability of Cyg X-1 in the soft state also seems to be driven by instabilities in the hot Comptonizing plasma (Zdziarski *et al.* 2002).

3. High-frequency QPO

Data on high-frequency QPO in black hole X-ray binaries are of much poorer quality than data for the low-f QPO. It is nevertheless clear that they share at least one characteristics with their low-f counterparts, namely the soft disk component is absent

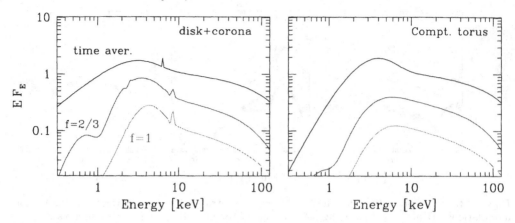

Figure 3. QPO spectra from the model of epicyclic oscillations, with relativistic effects modulating the X-ray flux, for $a = 0.998$ and inclination $\mu_{obs} = 0.3$. Left: geometry of accretion disk with an active corona produces strong disk component in the QPO spectra. Right: hot oscillating torus produces QPO spectrum resembling a comptonization component.

in the QPO spectra. More detailed energy dependencies (i.e., the spectral slope) are not possible to determine with current data.

The fact that pairs of QPO with 3:2 frequency ratio are sometimes observed stimulated development of models invoking non-linear resonances between various modes of oscillations of accretion disks. In particular, Abramowicz & Kluźnika (2001) postulate that the relevant modes are the vertical and radial epicyclic motions. Importantly, they suggest also that the X-ray flux modulation occurs due to changes of geometry and, consequently, changes in the strength of the relativistic effects (Doppler and gravitational shifts; light bending), with the primary emission being constant. This is thus the only current model making definite predictions for the more detailed properties of the QPO, for example, their energy spectra. We used the code of Życki & Niedźwiecki (2005) for photon transfer in Kerr metric and modified it to implement the source motion corresponding to disk epicyclic oscillations (details are in Życki et al., in preparation).

We have performed computations of the relativistic effects expected from such disks at radii corresponding to 3:2 resonance between the oscillation frequencies, and we have computed energy spectra of QPO expected in this model. They are presented in Fig. 3 in two variants of geometry: (1) accretion disk extending to the last stable orbit with an active X-ray corona, and (2) a truncated accretion disk with inner hot Comptonizing flow.

In the first scenario the QPO spectrum contains a strong disk component, since the disk emission is modified by the relativistic effects. In the second scenario the disk is, by assumption, not present at the QPO radius and, as a consequence, only the Comptonized component is modulated. The Fe Kα line is also assumed to be produced at the QPO radius (in the accretion disk with a corona geometry) and it does appear in the QPO spectrum, at positions corresponding to maximum redshift expected from the oscillatory motion.

Acknowledgements

This work was partially supported by grant 2P03D01225 from Polish Ministry of Science (MNIiI).

References

Abramowicz M. A. & Kluźniak W. 2001, A&A, 374, L19

Poutanen J. & Fabian A. C. 1999, MNRAS, 306, L31

Sobolewska M. A. & Życki P. T. 2006, MNRAS, 370, 405

van der Klis M. 2006, in: W. H. G. Lewin & M. van der Klis (eds.), Compact Stellar X-ray Sources (Cambridge University Press: Cambridge), p. 39

Zdziarski A. A., Poutanen J., Paciesas W. S., Wen L. 2002, ApJ, 578, 357

Życki P. T. & Niedźwiecki A. 2005, MNRAS, 359, 308

Życki P. T. & Sobolewska M. A. 2005, MNRAS, 364, 891

SUZY COLLIN: According to your model, should QPOs be visible in AGN?

PIOTR ŻYCKI: Formulation of the model I described is too general to resolve the question whether QPOs should appear and be visible in AGN. I understand there are mainly observational issues – AGN data are only marginally good to search for QPOs. There are also claims that QPO detectability depends on the inclination angle but again this is beyond the scope of the present model.

MITCHELL BEGELMAN: How much information can you extract about properties of the Comptonizing cloud from the energy dependency of the QPO signal?

PIOTR ŻYCKI: Information about optical depth or temperature are better constrained from time averages spectra, because statistics is much better. We deal mostly with the question of what actually oscillates – whether it is hot plasma or whether it is a cold disc that oscillates. This kind of issues we address. If the data are very good, we could in principle distinguish whether it is optical depth or temperature changes that are causing the oscillations.

It seems that low-frequency oscillations are produced by processes in a corona rather than a cold accretion disc. A mechanism that produces oscillations in a hot corona is implied by our model.

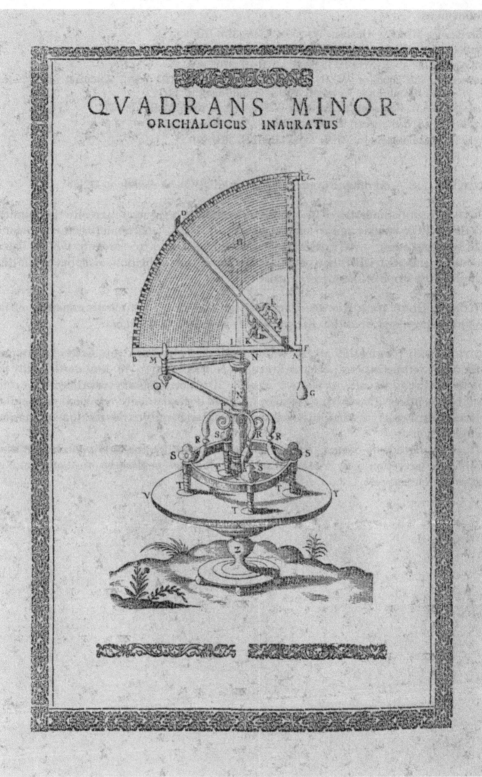

Black Holes from Stars to Galaxies – Across the Range of Masses
Proceedings IAU Symposium No. 238, 2006
V. Karas & G. Matt
© 2007 International Astronomical Union
doi:10.1017/S1743921307004619

Microquasars: disk–jet coupling in stellar-mass black holes

I. Félix Mirabel †

European Southern Observatory, Alonso de Cordova 3107, Santiago, Chile
email: fmirabel@eso.org

Abstract. Microquasars provide new insights into: 1) the physics of relativistic jets from black holes, 2) the connection between accretion and ejection, and 3) the physical mechanisms in the formation of stellar-mass black holes. Furthermore, the studies of microquasars in our Galaxy can provide in the future new insights on: 1) a large fraction of the ultraluminous X-ray sources in nearby galaxies, 2) gamma-ray bursts (GRBs) of long duration in distant galaxies, and 3) the physics in the jets of blazars. If jets in GRBs, microquasars and Active Galactic Nuclei (AGN) are due to a unique universal magnetohydrodynamic mechanism, synergy of the research on these three different classes of cosmic objects will lead to further progress in black hole physics and astrophysics.

Keywords. Black hole physics – X-rays – binaries: jets – stars: general

1. Introduction

The physics in all systems that contain black holes is essentially the same, and it is governed by the same scaling laws. The main differences derive from the fact that the scales of length and time of the phenomena are proportional to the mass of the black hole. If the lengths, masses, accretion rates, and luminosities are expressed in units such as the gravitational radius ($R_g = GM/c^2$), the solar mass, and the Eddington luminosity, then the same physical laws apply to stellar-mass and supermassive black holes (Sams *et al.* 1998; Rees 2003). For a black hole of mass M the density and mean temperature in the accretion flow scale with M^{-1} and $M^{-1/4}$, respectively. For a given critical accretion rate, the bolometric luminosity and the length of the relativistic jets are proportional to the mass of the black hole. The maximum magnetic field at a given radius in a radiation dominated accretion disk scales with $M^{-1/2}$, which implies that in the vicinity of stellar-mass black holes the magnetic fields may be 10^4 times stronger than in the vicinity of supermassive black holes (Sams, Eckart & Sunyaev 1998). In this context, it was proposed (Mirabel *et al.* 1992; Mirabel & Rodríguez 1998) that supermassive black holes in quasars and stellar-mass black holes in X-ray binaries should exhibit analogous phenomena. Based on this physical analogy, the word "microquasar" (Mirabel *et al.* 1992) was chosen to designate compact X-ray binaries that are sources of relativistic jets (see Figure 1).

2. Superluminal motions in microquasars and quasars

A galactic superluminal ejection was observed for the first time in the black hole X-ray binary GRS 1915+105, at the time of a sudden drop at 20–100 keV. Since then, relativistic jets with comparable bulk Lorentz factors $\Gamma = 1/(1 - \beta^2)^{1/2}$ as in quasars have been observed in several other X-ray binaries (Mirabel & Rodríguez 1999; Fender

† On leave from CEA-Saclay, France.

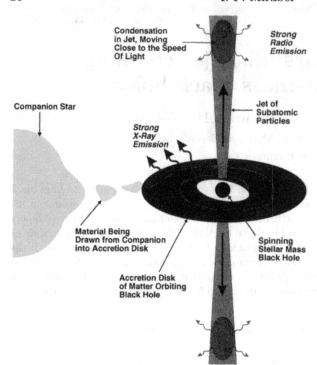

Condensation
in Jet, Moving
Close to the Speed
Of Light

Strong
Radio
Emission

Companion Star

Jet of
Subatomic
Particles

Strong
X-Ray
Emission

Material Being
Drawn from Companion
into Accretion Disk

Spinning
Stellar Mass
Black Hole

Accretion Disk
of Matter Orbiting
Black Hole

Figure 1. This diagram illustrates current ideas of what microquasars might be: compact objects – black holes and neutron stars – that are accreting mass from a donor star in X-ray binary systems and ejecting plasma at relativistic speeds. The diagram not to scale.

2002; Paredes 2005). At present, it is believed that all X-ray accreting black hole binaries are jet sources.

Galactic microquasar jets usually move in the plane of the sky $\sim 10^3$ times faster than quasar jets and can be followed more easily than the latter (see Figure 2). Because of their proximity, in microquasars two-sided jets can be observed, which together with the distance provides the necessary data to solve the system of equations, gaining insight on the actual speed of the ejecta. On the other hand, in AGN located at $\leqslant 100$ Mpc, the jets can be imaged with resolutions of a few times the gravitational radius of the supermassive black hole, as was done for M 87 (Biretta, Junor & Livio 2002). This is presently not possible in microquasars, since such a precision in terms of the gravitational radius of a stellar-mass black hole would require resolutions of a few hundreds kilometers. Then, in terms of the gravitational radius in AGN we may learn better how the jets are collimated close to the central engine. In summary, some aspects of the relativistic jet phenomena associated to accreting black holes are better observed in AGN, whereas others can be better studied in microquasars. Therefore, to gain insight into the physics of relativistic jets in the universe, synergy between the knowledge of galactic and extragalactic black hole is needed.

3. Accretion–jet connection in microquasars and quasars

Microquasars have allowed to gain insight into the connection between accretion disk instabilities and the formation of jets. In ~ 1 hour of simultaneous multi-wavelength observations of GRS 1915+105 during the frequently observed 30-40 min X-ray oscillations in this source, the connection between sudden drops of the X-ray flux from the accretion disk and the onset of jets were observed on several occasions (Mirabel *et al.* 1998; Eikenberry *et al.* 1998); see Figure 3.

Quasar 3C279   Microquasar GRS1915+105

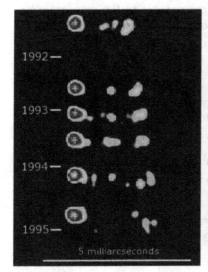 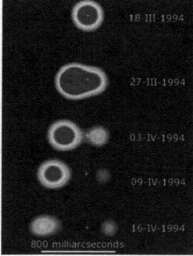

Figure 2. Apparent superluminal motions observed in the microquasar GRS 1915+105 at 8.6 GHz and in the quasar 3C 279 at 22 GHz.

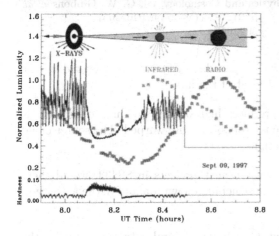

Figure 3. Direct evidence for the disk–jet connection in the microquasar GRS 1915+105 (Mirabel *et al.* 1998). When the hot inner accretion disk disappeared, its X-ray brightness abruptly diminished. The ensuing X-ray recovery documented the inner disk's replenishment, while the rising infrared and radio emission showed plasma being ejected in a jet-forming episode. The sequence of events shows that material indeed was transfered from the disk to the jets. Similar transitions have been observed in the quasar 3C 120 (Marscher *et al.* 2002), but on time-scales of years, rather than minutes.

From these observations we have learned the following:

a) the jets appear after the drop of the X-ray flux;

b) the jets are produced during the replenishment of the inner accretion disk;

c) the jet injection is not instantaneous. It can last up to ∼ 10 min;

d) the time delay between the jet flares at wavelengths of 2μm, 2cm, 3.6cm, 6cm, and 21cm are consistent with the model of adiabatically expanding clouds that had been proposed to account for relativistic jets in AGN (van der Laan 1966).

e) synchrotron emission is observed up to infrared wavelengths and probably up to X-rays. This would imply the presence of electrons with TeV energies in jets;

f) VLBA images during this type of X-ray oscillations (Dhawan, Mirabel & Rodríguez 2000) showed that the ejecta consist of compact collimated jets with lengths of ∼ 100 AU;

g) there is a time delay of ∼ 5 min between the large drop of the X-ray flux from the accretion disk and the onset of the jets. These ∼ 5 minutes of silence suggest that the compact object in GRS 1915+105 has a space-time border, rather than a material border, namely, a horizon as expected in relativistic black holes. However, the absence of evidence of a material surface in these observations could have alternative explanations.

After the observation of this accretion disk–jet connection in a microquasar, an analogous connection was observed in the quasar 3C 120 (Marscher *et al.* 2002), but on scales of years rather than minutes. This time-scale ratio is comparable to the mass ratio between the supermassive black hole in 3C 120 and the stellar black hole in GRS 1915+105, as expected in the context of the black hole analogy.

References

Biretta J. A., Junor W., Livio M. 2002, NewAR, 46, 239

Dhawan V., Mirabel I.F., Rodríguez L.F. 2000, ApJ, 543, 373

Eikenberry S. S., Matthews K., Morgan E. H., Remillard R. A., Nelson R. W. 1998, ApJ, 494, L61

Fender R. 2002, Lecture Notes in Physics, 589, 101

Marscher A. P., Jorstad S. G., Gómez J. L. *et al.* 2002, Nature, 417, 625

Mirabel I. F., Dhawan V., Chaty S. *et al.* 1998, A&A, 330, L9

Mirabel I. F. & Rodríguez L. F. 1998, Nature, 392, 673

Mirabel I. F. & Rodríguez L. F. 1999, ARAA, 37, 409

Mirabel I. F., Rodríguez L. F., Cordier B. *et al.* 1992, Nature, 358, 215

Paredes J. M. 2005, in Radio Astronomy from Karl Jansky to Microjansky, eds. L. I. Gurvits, S. Frey & S. Rawlings (EAS Publ. Ser.: Budapest), vol. 15, pp. 187–206 (astro-ph/0402671)

van der Laan H. 1966, Nature, 211, 1131

Rees M. J. 2003, in Future of Theoretical Physics and Cosmology, ed. G. W. Gibbons *et al.* (Cambridge University Press: Cambridge), pp. 217–235 (astro-ph/0401365)

Sams B. J., Eckart A., Sunyaev R. 1998, Nature, 392, 673

PHIL CHARLES: Could the apparent "dark jet" paradox of SS433 be answered by the high inclination which causes obscuration of most of the flux?

FELIX MIRABEL: Yes, the accretion disc produces high opacity to the X-rays and, as shown in the talk by Andrew King, the disc is probably highly warped. The source could be very bright if seen from a different direction.

VIRGINIA TRIMBLE: Suppose there have never been those jets in SS433, would there still be a supernova remnant there, or is the host nebula blown mainly by the activity of central source?

FELIX MIRABEL: This is an open question. Clearly, the lateral extentions seen in the nebula W 50 that hosts SS433, have been blown away by the jets.

GREGORY BESKIN: Is it possible to create jets without a black hole in the centre of the disc, for example by a neutron star?

FELIX MIRABEL: Yes indeed, there are jets in binaries with confirmed neutron stars. Clear cases are Scorpius X-1 and Circinus X-1.

GLORIA DUBNER: Comment on the previous question by V. Trimble: There is a way to probe if the whole bubble was created by a compact object. I believe that the whole nebula had been created by a supernova remnant and its shape was then distorted by the action of SS433's jets. The way to address this issue is by searching for spectral changes in the radio emission – the jets have a different spectrum than the rest of the bubble.

Black Holes from Stars to Galaxies – Across the Range of Masses
Proceedings IAU Symposium No. 238, 2006 © 2007 International Astronomical Union
V. Karas & G. Matt, eds. doi:10.1017/S1743921307004620

Suzaku observation of the black hole transient 4U1630–472: discovery of absorption lines

Aya Kubota,[1] Tadayasu Dotani,[2] Jean Cottam,[3] Taro Kotani,[4]
Chris Done,[2,5] Yoshihiro Ueda,[6] Andy C. Fabian,[7] Tomonori Yasuda,[8]
Hiromitsu Takahashi,[8] Yasushi Fukazawa,[8] Kazutaka Yamaoka,[9]
Kazuo Makishima,[1,10] Shinya Yamada,[10] Takayoshi Kohmura,[11]
Lorella Angelini[3] and the Suzaku team

[1]Inst. of Physical and Chemical Research (RIKEN), 2-1 Hirosawa, Wako, Saitama 351-0198

[2]Institute of Space and Astronautical Science, Japan Aerospace Exploration Agency, 3-1-1
Yoshinodai, Sagamihara, Kanagawa 229-8510

[3]Exploration of the Universe Div., NASA GSFC, Greenbelt, MD 20771, USA

[4]Department of Physics, Tokyo Tech, 2-12-1 O-okayama, Meguro, Tokyo 152-8551, Japan

[5]Department of Physics, University of Durham, South Road, Durham, DH1 3LE, UK

[6]Department of Astronomy, Kyoto University, Sakyo-ku, Kyoto 606-8502, Japan

[7]Institute of Astronomy, Madingley Road, Cambridge CB3 0HA, UK

[8]Department of Physics, Hiroshima Univ., Higashi-Hiroshima, Hiroshima 739-8526

[9]Department of Physics, Aoyama Gakuin University, Sagamihara, Kanagawa 229-8558, Japan

[10]Department of Physics, Univ. of Tokyo, 7-3-1 Hongo, Bunkyo-ku, Tokyo 113-0033, Japan

[11]Physics Department, Kogakuin University 2665-1, Nakano-cho, Hachioji, Tokyo, 192-0015

email: aya@crab.riken.jp

Abstract. We present the results of six Suzaku observations of the recurrent black hole transient 4U 1630 − 472 during its decline from its most recent outburst in 2006. All observations show the typical high/soft state spectral shape in the 2–50 keV band, roughly described by an optically thick disk spectrum in the soft energy band plus a weak power-law tail.

The disk temperature decreases from 1.4 keV to 1.2 keV as the flux decreases by a factor 2, consistent with a constant radius as expected for disk-dominated spectra. All the observations reveal significant absorption lines from highly ionized (H-like and He-like) iron $K\alpha$ at 7.0 keV and 6.7 keV.

The energies of these absorption lines suggest a blue shift with an outflow velocity of ~ 1000 km s^{-1}. The H–like iron $K\alpha$ equivalent width remains approximately constant at ~ 30 eV over all the observations, while that of the He–like $K\alpha$ line increases from 7 eV to 20 eV. Thus the ionization state of the material decreases, as expected from the decline in flux.

The data constrain the velocity dispersion of the absorber to 200–2000 km s^{-1}, and the size of the plasma as $\sim 10^{10}$ cm assuming a source distance of 10 kpc.

Keywords. Black hole physics – absorption lines

1. Introduction

In recent years, a growing number of X-ray binaries have been found to exhibit absorption lines from highly ionized elements (e.g., Boirin *et al.* 2004 and references therein). These systems range from microquasars including GRO J1655−40 (e.g., Ueda *et al.* 1998; Yamaoka *et al.* 2001; Miller *et al.* 2006) to low-mass X-ray binaries including GX 13+1 (Ueda *et al.* 2004; Sidoli *et al.* 2002). These systems are viewed at high inclination angles. The absorption features are therefore thought to originate in material that is associated with and extends above the outer accretion disk. The recurrent black hole transient, 4U 1630−472, was observed 6 times with Suzaku (Mitsuda *et al.* 2006) from 2006 February 8 through March 23 during its most recent outburst as part of a program to study discrete spectral structures as a function of the changing accretion conditions (Kubota *et al.* 2006). We monitored the source during its decay from outburst, detecting significant absorption lines from highly-ionized iron throughout the observations.

2. Continuum shape

Figure 1 shows the 2–50 keV HXD/PIN (Takahashi *et al.* 2006; Kokubun *et al.* 2006) and XIS (Koyama *et al.* 2006) spectra of 4U 1630 − 472 divided by those of Crab spectra, which has an approximately power-law spectrum with $\Gamma = 2.1$. This plot reveals several noticeable features in the broad band source spectra. The source spectra are characterized by a dominant soft thermal component in the XIS band, while the harder X-ray PIN data clearly show a weak power-law tail. The dominant soft component is generally observed from black hole binaries in the high/soft state (Tanaka & Lewin 1995), and is believed to be emission from the optically thick standard accretion disk (Shakura & Sunyaev 1973). In fact, this component is successfully reproduced with the multi-color disk model (DISKBB; Mitsuda *et al.* 1984; Makishima *et al.* 1986) modified by large interstellar absorption of $N_H \sim 8 \times 10^{22}$ cm^{-2}. For the two months observations, the disk inner temperature, kT_{in}, changed from 1.39 keV to 1.18 keV by keeping an apparent inner radius constant at 25 km for an assumed distance of 10 kpc and inclination angle of 70°. Corresponding disk bolometric luminosity, L_{disk}, decreased from 2.8×10^{38} erg s^{-1} to 1.6×10^{38} erg s^{-1}.

3. Iron absorption lines

The Crab ratios also reveal the presence of complex absorption structures in the iron K band. Figure 2b shows an enlargement of the 6–9 keV XIS spectra. The most obvious features are two narrow dips at 7.0 keV and 6.8 keV, and weaker absorption structures are found at 7.8 keV and 8.2 keV. We thus fit the 2–9 keV data with the absorbed DISKBB model and 4 negative gaussians with fixed line width at $\sigma = 10$ eV. The time histories of the individual absorption-line equivalent widths and their energies are plotted in the left panel of figure 2. The time averaged center energies of the two strong absorption lines are estimated to be 6.987 ± 0.005 keV and 6.714 ± 0.009 keV, which are consistent with H-like and He-like iron Kα suggesting a blue shift velocity of 900 ± 200 km s^{-1} and 700 ± 400 km s^{-1}, respectively. The energies of two weaker absorption lines are almost consistent with H-like and He-like iron Kβ, though they can be contaminated by nickel Kα. While the equivalent width of the H-like line was almost constant throughout the observations, that of the He-like line increased by a factor of two between the first and second observations, and then perhaps increased slightly more through the subsequent observations. These imply that the ionization state of the absorbing gas decreased over time, as expected from the declining source luminosity.

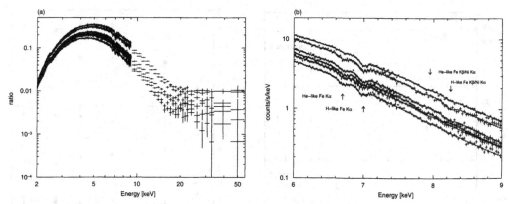

Figure 1. (a) The ratio of 4U 1630 − 472 spectra to the Crab spectrum. Data are obtained with the XIS023 and the HXD PIN. (b) The XIS 6–9 keV spectra concentrating the absorption line structure. The data and the best fit models are shown.

We next fit the absorption lines with Voigt profile (e.g., Ueda *et al.* 2004) to estimate the absorbing ion column density, N_{ion}, under assumed velocity dispersion, b. Although we cannot measure b from line profiles, its upper and lower bounds may be obtained. The upper limit is set at $b \sim 2000$ km s^{-1} by the upper limits on the line width, $\sigma \sim 30$–40 eV (§ 3.2). The lower limit is estimated to be ~ 200 km s^{-1} by comparing the curve-of-growth between Kα and Kβ. Here, the ion column density estimated from the equivalent width of Kα with low velocity dispersion of < 200 km s^{-1} requires too high equivalent width for the Kβ of the same ion. The top three panels of figure 2(right) show the time histories of the ion column, the ratio of the He-like to H-like ions, and estimated blue shift, z, assuming $b = 500$ km s^{-1} which is based on the (marginally) resolved line width in the Chandra HETGS data of GX 13+1 Ueda *et al.* (2004). The column density of the H-like iron, $N_{Fe\ XXVI}$, is almost constant at $\sim 1 \times 10^{18}$ cm^{-2} while that of the He-like iron, $N_{Fe\ XXV}$, increases significantly from the first to the second observation. The weighted average blue shift is about $z = (3.3 \pm 0.7) \times 10^{-3}$, corresponding to an outflow velocity of 1000 ± 200 km s^{-1}.

4. Physical parameters of the absorber

Through the analyses, we found that iron is always predominantly H–like, requiring that the ionization parameter, $\xi = L/nr^2$ (where n and r are the number density and the distance of the absorbing material from the illuminating source), be very high. In this section, we use the XSTAR photoionization code (version 2.1kn5; Kallman & Bautista 2001) to calculate the ionization balance of the line-producing material, to convert the ion column densities into physical parameters including total column density, number density, and location, for each observation. We calculate the relative ion populations under illumination by a DISKBB spectrum, and use this to predict a theoretical ratio of $N_{Fe\ XXV}$ to $N_{Fe\ XXVI}$. The observed ratio from the data then enables us to estimate the ξ-parameter, and so convert the observed ion column densities of H-like and He-like irons into a total column density, N_{tot}.

The inferred time histories of ξ and N_{tot} are also shown in the right panel of figure 2 assuming $b = 500$ km s^{-1}, together with upper and lower bounds corresponding to $b = 200$ and 2000 km s^{-1}. Assuming that b stays constant at 500 km s^{-1} then there is a marginal decrease in N_{tot} as the luminosity decreases, from $\sim 1 \times 10^{23}$ cm^{-2} in the first observation to $\sim 7 \times 10^{22}$ cm^{-2} in the final dataset, as well as a marginal decrease in ξ

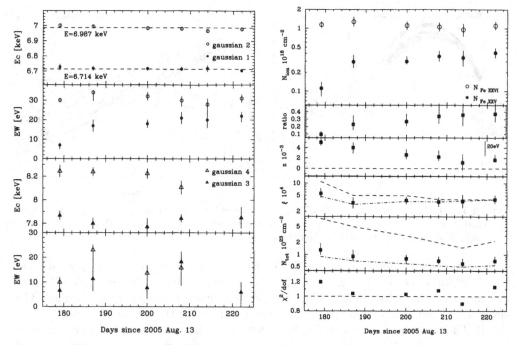

Figure 2. Time histories of absorption line parameters. Left: Based on the negative gaussian fit in the range of 6–10 keV. The first two panels show line center energies and equivalent width of the two strongest lines. The bottom two panels show those of weaker absorption features. Right: Based on the Voigt profile for $b = 500$ km s^{-1}. Based on the ratio, ξ-parameter and $N_{\rm tot}$ are shown in the next two panels. The best estimation of ξ and $N_{\rm tot}$ for $b = 200$ km s^{-1} (dashed line) and 2000 km s^{-1} (dash-dot line) are also shown in the same panels.

from $\sim 6 \times 10^4$ to $\sim 4 \times 10^4$. We can roughly estimate n and r of the absorbing plasma, by using the definitions of $N_{\rm tot} = \bar{n}\Delta R$ with a mean number density $\bar{n}$ and a characteristic thickness of the absorber ΔR.

Using $\xi = L/\bar{n}R^2$, the average distance of the absorber, R, is thus estimated as $R \sim 3 \times 10^{10} \cdot (\Delta R/R) \cdot D_{10}{}^2$ and $\sim 4 \times 10^{10} \cdot (\Delta R/R) \cdot D_{10}{}^2$ cm, for the first and the last observations, respectively. The densities are also estimated as $\bar{n} \sim 5 \times 10^{12} \cdot (\Delta R/R)^{-1} \cdot D_{10}{}^{-2}$ and $\sim 2 \times 10^{12} \cdot (\Delta R/R)^{-1} \cdot D_{10}{}^{-2}$ atoms cm^{-3}. Assuming $\Delta R/R \sim 1$ by following the previous research, the values of R and $\bar{n}$ are estimated to be on the order of $R \sim 10^{10} \cdot D_{10}{}^2$ cm and $\bar{n} \sim 10^{12} \cdot D_{10}{}^{-2}$ atoms cm^{-3}.

These parameters are almost consistent with those obtained in other binary systems (e.g., Ueda *et al.* 1998; Kotani *et al.* 2000). This source joins the growing number of galactic binaries with such absorption, showing that winds are probably a generic feature of bright accretion disks.

References

Boirin L., Parmar A. N., Barret D., Paltani S., Grindlay J. E. 2004, A&A, 418, 1061
Kallman T. R., Angelini L., Boroson B., & Cottam J. 2003, ApJ, 583, 861
Kallman T. & Bautista M. 2001, ApJS, 133, 221
Kubota A. *et al.* 2006 PASJ, in press
Kuulkers E., van der Klis M., Parmar A. N. 1997, ApJ, 474, L47
Kokubun M. *et al.* 2006, PASJ, in press
Kotani T. *et al.* 2000, ApJ, 539, 413

Koyama K. *et al.* 2006, PASJ, in press
Makishima K. *et al.* 1986, ApJ, 308, 635
Miller J. M. *et al.* 2006, Nature, 441, 953
Mitsuda K. *et al.* 1984, PASJ, 36, 741
Mitsuda K. *et al.* 2006, PASJ, in press
Shakura N. I. & Sunyaev R. A. 1973, A&A, 24, 337
Sidoli L., Parmar A. N., Oosterbroek T., Lumb D. 2002, A&A, 385, 940
Takahashi T. *et al.* 2006, PASJ, in press
Tanaka Y., Lewin W. H. G. 1995, in X-ray Binaries, eds. W. H. G. Lewin, J. van Paradijs, and
    W. P. J. van den Heuvel (Cambridge University Press, Cambridge), p. 126
Ueda Y. *et al.* 1998, ApJ, 492, 782
Ueda Y., Murakami H., Yamaoka K., Dotani T., Ebisawa K. 2004, ApJ, 609, 325
Yamaoka K. *et al.* 2001, PASJ, 53, 179

THOMAS MACCARONE: If the inclination angle of the system is such that most of the wind comes towards the observer, how would that change the estimate of the fraction of the accreted mass lost in the wind?

AYA KUBOTA: Outflow rate was estimated as $\rho v_{\text{wind}} 4\pi r^2 c_f$ (c_f is covering fraction of the wind). And the estimation assumes $c_f = 0.1$, which corresponds to angles of 10–15°. So the estimated loss rate of $0.3 M_{\text{acc}}$ can be a secure lower limit for constant $\dot{\rho}$ being independent of i for the covering fraction. But if the absorbing matter is mostly from a very small angle $\Delta i \ll 10°$, it can make the outflow rate smaller. This consideration is model dependent. But roughly a factor three decrease in $\dot{M}_{\text{flow}}$ can be expected based on MHD.

Stellar Mass Black Holes II

Stellar-mass black holes II.

A. R. King
Matter accretion and ejection in black-hole systems
(Invited Review)

P. C. Joss
Formation of rapidly rotating black holes in massive binary stellar systems

R. Matsumoto
Sawtooth-like oscillations of black hole accretion disks

J. I. Gonzalez Hernandez
On the origin of the black hole in the X-ray binary XTE J1118+480

St Vitus Cathedral, the largest temple in Prague and the place of coronations of Czech kings (detail of the mosaic on the south facade).

St Nicholas Church with spectacular decorations, frescoes and an organ, which W. A. Mozart played during his stay in Prague.

Black Holes from Stars to Galaxies – Across the Range of Masses
Proceedings IAU Symposium No. 238, 2006
V. Karas & G. Matt, eds.
© 2007 International Astronomical Union
doi:10.1017/S1743921307004644

Hyperaccretion

Andrew R. King

Theoretical Astrophysics Group, University of Leicester, Leicester LE1 7RH, UK
email: ark@astro.le.ac.uk

Abstract. Accretion at rates exceeding the Eddington limit is common in close binaries. I summarize the arguments leading to the conclusion that such stellar–mass systems appear as ultraluminous X-ray sources when viewed close the the inner accretion disc axis, and like SS433 when viewed from other angles.

I show that AGN are unlikely to achieve electron–scattering Eddington ratios as high as ULXs, so there are few ULX analogues among quasars. However hyperaccretion of *dusty* matter is common among AGN. The resulting outflow naturally has a toroidal geometry, and may well be the origin of the dusty torus invoked in unified AGN schemes.

Keywords. Black hole physics – accretion, accretion discs – X-rays: binaries – galaxies: active

---

1. Introduction

Accretion is the most efficient way of using normal matter to release radiant energy. The very first papers on the subject (e.g. Salpeter 1964) already recognised that radiation pressure imposes a natural limit on the accretion luminosity, usually expressed as the Eddington limit $L_{\rm Edd} \simeq 4\pi GMc/\kappa$, where κ is the opacity and M is the accretor mass. However there is no reason why the source of the accreting matter should know about this limit. This raises the obvious question of what happens in hyperaccreting systems, i.e. ones where the accretor is fed matter at rates $\dot{M}$ far above the value $\dot{M}_{\rm Edd} = L_{\rm Edd}/\epsilon c^2$ which would produce $L_{\rm Edd}$, where $\epsilon \sim 0.1$ is the efficiency. We note that for electron scattering opacity $\kappa \simeq 0.34$ we have $\dot{M}_{\rm Edd} \simeq 10^{-7} {\rm M}_\odot\ {\rm yr}^{-1}(M/10{\rm M}_\odot)$.

2. Hyperaccreting binaries

This question is particularly acute for accreting binary systems, since here there are several generic situations in which mass transfer naturally proceeds at values $\gg \dot{M}_{\rm Edd}$. An extremely common case occurs when a massive early–type star fills its Roche lobe and transfers mass to a less massive accretor. All high–mass supergiant X-ray binaries will pass through this stage for example. It is easy to show (e.g. King & Ritter 1999) that the donor then transfers its mass M_2 mass to the accretor on its thermal timescale $t_{\rm KH}$, i.e. $\dot{M} = -\dot{M}_2 \sim M_2/t_{\rm KH}$, leading to rates $\dot{M} \sim 10^{-5} - 10^{-3}{\rm M}_\odot\ {\rm yr}^{-1}$ or more, i.e. $\dot{M} \gtrsim 10^2 - 10^4 \dot{M}_{\rm Edd}$.

Such accretion rates mean that the accreting luminosity reaches the Eddington value even before the matter has fallen all the way down the accretor's gravitational potential well, that is, well above its surface. At this point radiation pressure must start to blow some of the accreting matter away, presumably self-limiting to the point that accretion within this point proceeds only at the (decreasing) Eddington rate corresponding to each radius. This is essentially the picture derived by Shakura & Sunyaev (1973), who call the critical point the spherization radius, with $R_{\rm sph} \sim (\dot{M}/\dot{M}_{\rm Edd})R_s$, where R_s is the accretor's Schwarzschild radius. This is essentially the same as the trapping radius defined by some other authors (e.g. Begelman 1979).

Near $R_{\rm sph}$ matter must be driven out in a dense, roughly spherical outflow. King & Begelman (1999) show that in most cases of thermal–timescale mass transfer the dense region near $R_{\rm sph}$ is still well away from interacting directly with the donor star, which might possibly trigger common–envelope evolution. Despite the highly supercritical mass transfer rate, the binary evolves in a fairly normal manner when we take account of the angular momentum lost in the outflow (cf King & Ritter 1999). Within $R_{\rm sph}$ the local accretion rate decreases roughly as $\dot{M}(R) \simeq \dot{M}_{\rm Edd}(R/R_{\rm sph})$, implying a total accretion luminosity

$$L_{\rm acc} \simeq L_{\rm Edd} \left(1 + \ln \frac{-\dot{M}_2}{\dot{M}_{\rm Edd}} \right) \qquad (2.1)$$

which can be as large as $\sim 10 L_{\rm Edd}$ for mass transfer rates $-\dot{M}_2 \sim 10^4 \dot{M}_{\rm Edd}$.

Clearly we would like to know how this thermal–timescale phase appears to observation. In particular one might imagine that it should have a fairly spectacular manifestation, given the high energies and prodigious outflows involved. Begelman, King & Pringle (2006) show that the answer depends on our viewing angle. If we observe the system from an angle close to the axis of the inner accretion disc (which is probably close to the spin axis of a black hole accretor) we see an ultraluminous X-ray source (ULX). These sources appear very bright partly because they are intrinsically somewhat super–Eddington (cf eqn 2.1), and also because this luminosity can only escape along the axis of the dense outflow. Given collimation solid angles $\Omega = 4\pi b$ with $b \sim 0.1$ we see that a $10 M_\odot$ black hole can have apparent luminosity $L \sim L_{\rm acc}/b \sim 10^{41}$ erg s^{-1}.

Interestingly, Begelman et al. (2006) did not set out to investigate ULXs, but reached this conclusion by considering the famous source SS433, which turns out to be what a hyperaccreting binary looks like when viewed 'from the side', i.e. away from the inner disc axis. Here the dense outflow obscures our view of the most energetic part of the accretion flow. Our only evidence for this in SS433 is the relativistic jets, which themselves carry mechanical energy $\gtrsim L_{\rm Edd}$. The clue allowing insight here is the fact that these precessing jets also show a nodding motion at one–half of the orbital period. This is clear evidence of the $m = 2$ tidal torque acting on the outer accretion disc. Begelman et al. (2006) show that the only plausible way of conveying this torque to the inner regions of the disc involved in jet formation is the dense outflow. The jet are presumably launched along the local disc axis, and the outflow imprints the long precession period and the nodding motion on them by deflecting them along the outflow axis, which already has these motions.

3. Hyperaccreting AGN

The last Section shows that the hyperaccretion problem is specified purely by $\dot{M}_{\rm Edd}$ and R_s. We can obviously extend all this work to hyperaccreting AGN. The major difference is that AGN have a much more limited Eddington factor, i.e. $\dot{M}/\dot{M}_{\rm Edd}$ does not attain such large values for AGN as for binaries. We can see this from a simple argument. Whatever process fuels the supermassive black holes in active galactic nuclei (e.g. minor mergers with small satellite galaxies) presumably feeds gas in dispersed form into orbit near the hole before it gets accreted. The maximum likely accretion rate then comes from assembling the largest gas mass which can orbit, and assuming that some triggering event causes it to fall freely on to the black hole.

We can assume that the host galaxy is roughly represented by an isothermal sphere of 'gas' (= all normal matter) and dark matter. Then the 'gas' mass inside radius R is

$$M_g(R) = \frac{2\sigma^2 R f_g}{G} \tag{3.1}$$

where σ is the velocity dispersion and f_g the gas fraction.

The highest plausible accretion rate is given by getting all this mass to fall inwards on the dynamical timescale R/σ, i.e.

$$\dot{M}_{\max} \simeq \frac{2\sigma^3}{G} f_g. \tag{3.2}$$

Note that this is independent of R. Now the black hole mass is related to σ by

$$M = \frac{f_g \kappa}{\pi G^2} \sigma^4 \tag{3.3}$$

(King 2003, 2005) which implies an Eddington luminosity for gas–rich matter

$$L_{\mathrm{Edd}} = \frac{4c f_g}{G} \sigma^4 \tag{3.4}$$

and thus an Eddington accretion rate

$$\dot{M}_{\mathrm{Edd}} = \frac{4 f_g}{\epsilon G c} \sigma^4. \tag{3.5}$$

where ϵ is the accretion efficiency. Hence we get the Eddington ratio

$$r_E = \frac{\dot{M}_{\max}}{\dot{M}_{\mathrm{Edd}}} = \frac{\epsilon c}{2\sigma} \tag{3.6}$$

For typical values ($\epsilon = 0.1, \sigma = 150 M_8^{1/4}$ km s^{-1}) we have

$$r_E \sim 100 M_8^{-1/4} \tag{3.7}$$

where $M_8 = M/10^8 M_\odot$.

As it stands this argument concerns 'grand design' inflows occupying a major part of the galaxy. One might try to circumvent it with flows targeted within the black hole's sphere of influence $R_i = 2GM/\sigma^2$, when the problem is more like Bondi accretion. However it is easy to show that this gives essentially the same limit. This $\sim \sigma^3/G$ limit probably always appears if the mass reservoir used for the accretion was in any kind of equilibrium with the host galaxy or the black hole before accreting. The most likely way around it is to bring in a mass reservoir which is is self–gravitating, like a star. If this can be disrupted very close to the hole one can presumably get higher $\dot{M}$.

Evidently in practice one rarely gets accretion at a rate $> \sigma^3/G$, and thus AGN Eddington ratios are generally modest (cf eqn 3.7). Since $r_E < 100$ the bolometric luminosity exceeds the formal L_{Edd} by $< \ln r_E \sim 4.6$. One would expect to see outflow at velocities $< c/r_E^{1/2} \sim 0.1c$, which is indeed about the maximum seen, e.g. in PG1211+143 (Pounds et al. 2003). It seems plausible that the collimation of the radiation by the outflow is weaker at such low r_E. Hence quasars probably only rarely appear very super–Eddington, i.e. there are few extreme ULX analogues among them.

4. The dusty torus

So far we have assumed (in eqn 3.4) that the accreting matter is gas, with electron scattering opacity κ. However at sufficient distance from the active nucleus there must be dusty matter, with opacity $\kappa_d > \kappa$. Then the local Eddington accretion rate is smaller by $k = \kappa/\kappa_d$, which can be $\sim 10^{-2}$, so that the effective r_E increases by $1/k$. This would give a dusty outflow with spherization radius

$$R_d \sim 500 \, \frac{R_s}{k} \sim 5 \times 10^4 R_s = 0.5 M_8 \text{ pc.} \tag{4.1}$$

The radiation temperature of the AGN is

$$T_{\text{rad}} \leqslant 3.5 \times 10^5 \, (l M_8)^{1/4} \left(\frac{R_s}{R}\right)^{1/2} \text{ K,} \tag{4.2}$$

where l is the (electron–scattering) Eddington ratio. If dust sublimates at 1200 K, the sublimation radius is at

$$R_{\text{sub}} = 9 \times 10^4 \, (l M_8)^{1/2} \, R_s = 0.9 l^{1/2} M_8^{3/2} \text{ pc.} \tag{4.3}$$

Hence for $l^{1/3} M_8 < 1$ or so the dust survives.

Equations (4, 5) of Begelman *et al.* (2006) can now be used to show that the dust optical depth τ_d across the outflow increases by $1/k^{3/2}$ to values $\leqslant 2 \times 10^4$. It is clear that $\tau_d > 1$ even for quite modest amounts of matter near the AGN. We thus have a dusty outflow from typical radius R_d, which creates a natural toroidal geometry for the obscuring matter near the AGN. This appears to offer a natural explanation for the dusty torus invoked in unified AGN schemes (Antonucci 1993; Urry & Padovani 1995). Note that there is no reason to assume that the dusty torus must supply the matter needed to power the AGN. Indeed the very long viscous times at such large radii suggest that the latter probably results from an independent accretion flow on much smaller scales. Hence this geometry is likely to be similar in most AGN, regardless of their central accretion rates.

Acknowledgments

I acknowledge a Royal Society Wolfson Research Merit Award. I thank Jim Pringle for illuminating discussions.

References

Antonucci R. 1993, ARA&A, 31, 473
Begelman M. C. 1979, MNRAS, 187, 237
Begelman M. C., King A. R., Pringle J. E. 2006, MNRAS, 370, 399
King A. R. 2003, ApJ, 596, L27
King A. R. 2005, ApJ, 635, L121
King A. R. & Begelman M. C. 1999, ApJ, 519, L169
King A. R. & Ritter H. 1999, MNRAS, 309, 253
King A. R., Taam R. E., Begelman M. C. 2000, ApJ, 530, L25
Pounds K. A., Reeves J. N., King A. R., Page K. L., O'Brien P. T., Turner M. J. L. 2003, MNRAS, 345, 705
Salpeter E. E. 1964, ApJ, 140, 796
Shakura N. & Sunyaev R. 1973, A&A, 24, 337
Urry C. & Padovani P. 1995, PASP, 107, 116

MIKE DOPITA: Comment: There is a lot of evidence to support radiation-pressure dominated outflows in dusty gas around AGN. For example NGC 1068 shows outflowing "comets" with velocity ~ 2000 km/s. However, in AGN there may be two flows – a dusty one like you suggest, and an inner thermal flow like in SS443. Many of these (speculative) ideas were published in Dopita (1997, PASA, 14, 230) which emphasized super-Eddington outflows in the context of AGN.

FELIX MIRABEL: Is there a connection between hyper-accretion and the presence of hadrons in the jets of SS433?

ANDREW KING: I think it is possible that in the collision with the outflows the jets entrain a lot of material and are slowed down. So hyper-accretion tends to load the jets with baryons.

FUKUN LIU: You suggest in your model that the outer part of the disc is affected by irradiation from the inner region. Do you have any observational evidence for it? Because it might seem more reasonable that the outer part of the disc is coplanar with the orbital plane and the inner part is warped due to Bardeen–Peterson effect.

ANDREW KING: No observational evidence; it is possible that the Bardeen–Petterson effect leads to the warp of the inner region of accretion disc. But the inner warped disc would become coplanar or flat on the several precessing timescales.

QVADRANS ALIVS ORICHAL.
CICVS ETIAM AZIMVTHALIS.

EXPLI-

Black Holes from Stars to Galaxies – Across the Range of Masses
Proceedings IAU Symposium No. 238, 2006   © 2007 International Astronomical Union
V. Karas & G. Matt, eds.   doi:10.1017/S1743921307004656

Sawtooth-like oscillations of black hole accretion disks

Ryoji Matsumoto[1] and Mami Machida[2]

[1]Department of Physics, Faculty of Science, Chiba University 1-33, Yayoi-cho, Inage-ku, Chiba, 263-8522 Japan

[2]Division of Theoretical Astronomy, National Astronomical Observatory of Japan, 2-21-1, Osawa, Mitaka, Tokyo 181-8588, Japan

email: matumoto@astro.s.chiba-u.ac.jp, mami@th.nao.ac.jp

Abstract. We studied the origin of quasi-periodic oscillations (QPOs) of X-rays in black hole candidates by three-dimensional global resistive magnetohydrodynamic simulations of accretion disks. Initial state is a rotating disk threaded by weak toroidal magnetic fields. General relativistic effects are simulated by using the pseudo-Newtonian potential. When the temperature of the outer disk decreases, the accreting matter accumulates into an inner torus.

We found that the inner torus is deformed into a crescent shape and that it shows sawtooth-like oscillations of magnetic energy with frequency 3-5Hz when the mass of the black hole is $10M_\odot$. The magnetic energy inside the torus is amplified until magnetic reconnection suddenly releases the accumulated magnetic energy. A new cycle of the oscillation starts when magnetic energy is amplified again. We found that high frequency QPOs with frequency around 100Hz in stellar mass black holes are excited when sawtooth-like oscillation appears in the inner torus.

Keywords. accretion, accretion disks – MHD – QPO

1. Introduction

Quasi-periodic oscillations (QPOs) are sometimes observed in black hole candidates when the source is in low/hard state or during the transition from low/hard state to high/soft state. The frequency of QPOs in low/hard state (1-10Hz in stellar mass black holes) increases as the source luminosity increases. Figure 1 schematically shows the evolution of accretion disks in low/hard state and during the hard-to-soft state transition. Solid curves show thermal equilibrium curves of accretion disks obtained by Abramowicz *et al.* (1995). When the accretion rate exceeds $0.1\dot{M}_E$, where $\dot{M}_E$ is the Eddington accretion rate, optically thin solution disappears.

Machida *et al.* (2006) showed by three-dimensional global resistive magnetohydrodynamic (MHD) simulations that the outer region of the disk where radiative cooling exceeds the heating shrinks in the vertical direction. The disk remains optically thin because magnetic pressure supports the disk. High frequency QPOs (HFQPOs) with frequency ~ 100Hz appear in such X-ray hard, luminous disks. HFQPOs accompany low frequency QPOs (LFQPOs). In this paper we report the results of global three-dimensional resistive MHD simulations of black hole accretion disks during the hard-to-soft state transition.

2. Numerical methods

We numerically solved resistive MHD equations in cylindrical coordinates (ϖ, φ, z) by using a modified Lax-Wendroff method with artificial viscosity. We assumed anomalous resistivity $\eta = \eta_0[\max(v_d/v_c - 1, 0)]^2$ where $v_d = j/\rho$ is the electron-ion drift speed

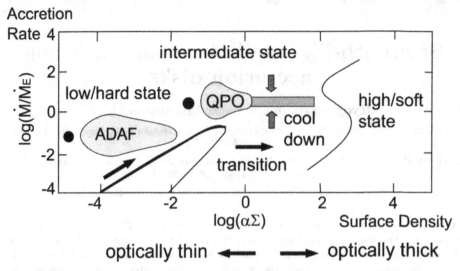

Figure 1. A schematic picture showing the hard-to-soft state transition. Solid curves denote thermal equilibrium curves of accretion disks in $\Sigma - \dot{M}$ plane, where Σ is surface density and $\dot{M}$ is the accretion rate. When mass accretion rate exceeds the critical value for the onset of the cooling instability, outer regions of the disk shrinks in the vertical direction due to cooling.

and v_c is the threshold above which the anomalous resistivity sets in. Here j and ρ are current density and matter density, respectively. Radiative cooling is neglected. General relativistic effects are simulated by using the pseudo-Newtonian potential $\phi = -GM/(r - r_s)$ where M is the mass of the black hole, $r = (\varpi^2 + z^2)^{1/2}$, and r_s is the Schwarzschild radius.

The initial state is a single temperature, polytropic disk with polytropic index $n = 3/2$ and the angular momentum $L \propto \varpi^{0.46}$ threaded by weak toroidal magnetic fields. The initial strength of magnetic field is parametrized by the ratio of gas pressure to magnetic pressure $\beta_0 = P_{\mathrm{gas}}/P_{\mathrm{mag}}$ at the initial pressure maximum of the disk at $(\varpi, z) = (\varpi_0, 0)$. We take $\varpi_0 = 35r_s$ and $\beta_0 = 100$. The initial temperature of the disk at the pressure maximum is $T_0 = 10^9 K$. This temperature is an order of magnitude smaller than that of our previous simulations (Machida & Matsumoto 2003). The initial disk is embedded in a low density, spherical, hot isothermal halo.

The magnetic Reynolds number $R_m = cr_s/\eta_0$ and the threshold for the anomalous resistivity are taken to be $R_m = 2000$ and $v_c = 0.9c$, respectively. In the following, the unit of length is r_s, and the unit of time is $t_0 = r_s/c = 10^{-4} M_{10}$s where $M_{10} = M/(10M_\odot)$.

The number of grid points is 250 in radial direction, 32 in azimuthal direction and 384 in vertical direction. The grid size is $\Delta\varpi = \Delta z = 0.1$ when $0 < \varpi, z < 10$, and otherwise increases with ϖ and z. We imposed absorbing boundary condition at $r = 2r_s$. Other boundaries are free boundary where waves can be transmitted.

3. Numerical results

Figure 2 shows the isosurface of density in the inner region ($\varpi < 25$) at $t = 58000t_0$. As the disk infalls due to the angular momentum transport by MRI (magnetorotational instability) driven turbulent magnetic field, an inner torus is formed around $\varpi = 8r_s$. We found that the inner torus is deformed into a crescent shape.

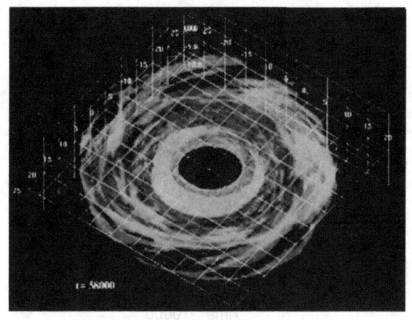

Figure 2. Isosurface of density at $t = 58\,000 t_0$.

Figure 3 shows the time evolution of the magnetic energy (solid curves) and the Joule heating rate (dashed curves) integrated in the inner torus. They show a sawtooth-like oscillation with period $\sim 2500 rs/c = 0.25 M_{10}$s. The Joule heating rate increases when the magnetic energy decreases. It indicates that magnetic energy is dissipated. Sawtooth like oscillations are excited in nonlinear systems when an instability and dissipation co-exists. When the dissipation is large, the system approaches to a quasi-steady state. On the other hand, when the dissipation is small, the energy accumulated by the growth of the instability is suddenly released by catastrophic events such as magnetic reconnection. When the energy release brings the system back to the original state, the accumulation and release of the energy creates a sawtooth-like oscillation. Sawtooth oscillations are observed in Tokamak fusion reactors, in which plasma is disrupted by magnetic recon-nection.

In the inner torus formed in our simulation, magnetic energy is amplified when one-armed ($m = 1$ where m is the azimuthal mode number), non-axisymmetric density distribution appears. Since the dense blob in a differentially rotating disk stretches the magnetic field lines in azimuthal direction and deform them into a bisymmetric spiral shape, current sheets are formed inside the spiral channel (Machida & Matsumoto 2003). When the current density exceeds the critical density for the onset of the anomalous resistivity, magnetic reconnection taking place in the current sheet releases the accumu-lated magnetic energy. Subsequently, the $m = 1$ mode and the magnetic energy grows again.

Figure 4 shows the power density spectrum (PDS) of the time variation of the mass accretion rate measured at $\varpi = 2.5 r_s$ when $M = 10 M_{\odot}$. A broad peak around 5Hz corresponds to the sawtooth-like oscillation. We found that when the large-amplitude sawtooth oscillation appears, two high frequency QPOs with frequency ratio 2:3 appear around 100Hz. The large amplitude sawtooth-like oscillation excites the high frequency oscillations of the inner torus.

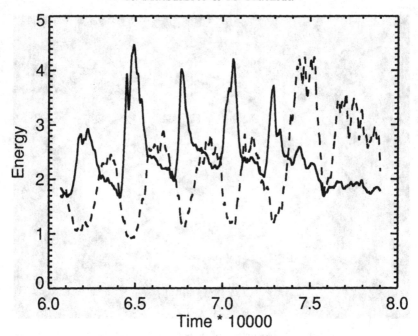

Figure 3. Time evolution of the magnetic energy (solid curve) and the Joule heating rate (dashed curve).

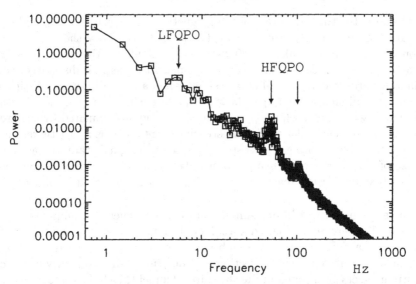

Figure 4. Power density spectrum of the time variation of the mass accretion rate at $\varpi = 2.5r_s$ when the mass of the black hole is $10M_\odot$.

4. Summary and discussion

We showed that when the temperature of the outer disk in black hole accretion flows decreases due to cooling, sawtooth-like oscillation is excited in the inner torus formed inside $10r_s$. The frequency of the sawtooth oscillation is 3–5Hz and comparable to that of LFQPOs observed in black hole candidates. The period of the sawtooth oscillation is determined by the growth time of magnetic fields, typically 10 rotation period of the inner torus. During the growth of magnetic fields, the inner torus is deformed into a crescent shape. When the magnetic energy becomes comparable to the thermal energy, magnetic energy accumulated in the inner torus is released by magnetic reconnection. The sawtooth oscillation excites high frequency oscillations of the inner torus with frequency around 100Hz. Our numerical results are consistent with that HFQPOs are observed during the hard-to-soft state transitions in black hole candidates and that they accompany LFQPOs.

Acknowledgements

Numerical computations were performed by using VPP5000 at National Astronomical Observatory, Japan. This work is supported in part by the Grants-in-Aid for Scientific Research of the Ministry of Education, Culture, Sports, Science and Technology (RM: 17030003) and JSPS Fellowships for young scientists (MM: 17-1907, 18-1907).

References

Abramowicz, M. A., Chen, X., Kato, S., Lasota, J.-P., Regev, O. 1995, ApJ, 438, L37
Machida, M., Matsumoto, R. 2003, ApJ, 585, 429
Machida, M., Nakamura, K. E., Matsumoto, R. 2006, PASJ, 58, 193

ELISABETE DE GOUVEIA DAL PINO: Can the sawtooth instability feed a dynamo action in the disc, so that a poloidal magnetic field component can be amplified from the large scale toroidal component and then raise from the disc into the corona?

RYOJI MATSUMOTO: Magnetorotational instability drives the dynamo. During the growing state of the sawtooth oscillation magnetic fields are amplified up to $\beta \sim 1$, where β is the ratio of gas pressure to magnetic pressure. Since our simulation includes vertical gravity, magnetic fields buoyantly escape from the disc and generate poloidal magnetic fields. Although we solve the full magnetohydrodynamic equations instead of reduced dynamo equations, the amplification and maintenance of magnetic fields in discs are similar to those in classical dynamo theory.

FUKUN LIU: I would like to ask two questions. 1. What's the dependence of QPOs on the radial extent of torus? 2. Can you find multiple QPOs with harmonic relationship e.g. 2 : 3, firstly discovered by Marek Abramowicz and collaborators, and in AGNs by our recent work (see poster S238–129)?

RYOJI MATSUMOTO: 1. We did the simulations with different radial extent of torus of $20R_g$ and $10R_g$. The frequency is independent of the radial size of torus. 2. Yes, we identified two QPOs with 2 : 3 frequency ratio in power-spectral density.

ALEXANDER ZAKHAROV: Did you try to fit the jet formation with your approach?

RYOJI MATSUMOTO: Our simulations produce funnel-wall outflows and magnetically driven disc winds. The speed of the disc wind is typically $\sim 0.05c$.

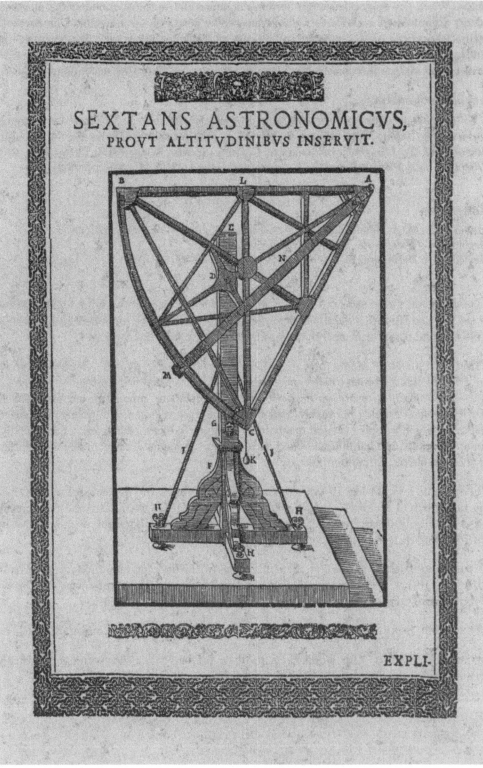

Black Holes from Stars to Galaxies – Across the Range of Masses
Proceedings IAU Symposium No. 238, 2006
V. Karas & G. Matt, eds.
© 2007 International Astronomical Union
doi:10.1017/S1743921307004668

Chemical abundances of secondary stars in low mass X-ray binaries

Jonay I. González Hernández,[1,2,3] Rafael Rebolo[1] and Garik Israelian[1]

[1] Instituto de Astrofísica de Canarias, E-38205 La Laguna, Tenerife, Spain
[2] CIFIST Marie Curie Excellence Team
[3] Observatoire de Paris-Meudon, GEPI, 5 place Jules Janssen, 92195 Meudon Cedex, France
email: jonay@iac.es, rrl@iac.es, gil@iac.es

Abstract. Low mass X-ray binaries (LMXBs) offer us an unique opportunity to study the formation processes of compact objects. Secondary stars orbiting around either a black hole or a neutron star could have captured a significant amount of the ejected matter in the supernova explosions that most likely originated the compact objects. The detailed chemical analysis of these companions can provide valuable information on the parameters involved in the supernova explosion such us the mass cut, the amount of fall-back matter, possible mixing processes, and the energy and the symmetry of the explosion. In addition, this analysis can help us to find out the birth place of the binary system. We have measured element abundances of secondary stars in the LMXBs A0620–00, Cen X-4, XTE J1118+480 and Nova Sco 94. We find solar or above solar metallicity for all these systems, what appears to be independent on their locations with respect to the Galactic plane. A comparison of the observed abundances with yields from different supernova explosion together with the kinematic properties of these systems suggest a supernova origin for the compact objects in all of them except for A0620–00, for which a direct collapse cannot be discarded.

Keywords. black hole physics — stars: abundances — stars: evolution — stars: individual (XTE J1118+480, A0620-00, Nova Sco 94, Cen X-4) — stars: neutron — supernovae: general — X-rays: binaries

1. Introduction

A low mass X-ray binary (LMXB) consists of a secondary star orbiting a compact object, either a black hole or a neutron star. An accretion disk forms when the secondary star transfers matter onto the compact object. These accretion processes that take place in the vicinity of the compact object are responsible of X-ray emission of these systems.

A particularly interesting subclass of X-ray binaries are the soft X-ray transients (SXTs), which show recurrent outbursts followed by periods of quiescence lasting several decades, when the stellar radiation dominates over the disk emission in optical and infrared wavelengths. Spectroscopic analysis of SXTs performed at quiescence allows us to study the dynamical and chemical properties of companion stars.

At the beginning of the 90s, this project was initiated after the discovery of a black hole in the X-ray binary V404 Cygni (Casares *et al.* 1992), from the analysis of the radial velocity curve of the secondary star. The inspection of the spectrum of this star also provided the measurement of an unexpectedly high Li abundance (Martín *et al.* 1992), which later on was also found in some other systems (Martín *et al.* 1994a).

Israelian *et al.* (1999) published the chemical analysis of the secondary star in the SXT Nova Scorpii 1994. They intended to search for any possible signature of a supernova (SN) explosion that originated the black hole in this system and found indeed α-elements significantly enhanced in the secondary star. Afterwards, the element abundances were compared with a variety of SN models which brought in new insights on the parameters involved in the explosion (Podsiadlowski *et al.* 2002).

2. Observations and chemical analysis

Spectroscopic observations of the selected sample of SXTs have been carried out using the 8.2 m Very Large Telescope (VLT) with the UV–Visual Echelle Spectrograph (UVES) at a resolving power $\lambda/\delta\lambda \sim 43,000$, except for XTE J1118+480, for which we used the 10-m Keck II telescope equipped with the Echellette Spectrograph and Imager (ESI; Sheinis *et al.* 2002) at medium-resolution ($\lambda/\delta\lambda \sim 6,000$). Individual spectra were corrected from radial velocities and combined to improve the signal-to-noise ratio.

Table 1. Stellar and veiling parameters in SXTs

Parameter	A0620–00	Cen X-4	XTE J1118+480	Nova Sco 94
T_{eff} (K)	4900 ± 150	4500 ± 100	4700 ± 100	6100 ± 200
$\log(g/\mathrm{cm\ s^2})$	4.2 ± 0.3	3.9 ± 0.3	4.6 ± 0.3	3.7 ± 0.2
[Fe/H]	0.25 ± 0.10	0.4 ± 0.15	0.2 ± 0.2	-0.1 ± 0.1
f_{4500}	0.25 ± 0.05	1.85 ± 0.10	0.85 ± 0.20	0.15 ± 0.05
$m_0/10^{-4}$	-1.4 ± 0.2	-7.1 ± 0.3	-2 ± 1	-1.2 ± 0.3

We obtained the stellar parameters of secondary stars by comparing synthetic spectra with the average spectrum of secondary stars, taking into account the effect of the veiling from the accretion disk on the stellar features (see González Hernández *et al.* 2004, 2005b, 2006 for more details). We selected several spectral features of Fe I and using a grid of LTE models of atmospheres provided by Kurucz (1993) and the LTE code MOOG from Sneden (1973), we generated a grid of synthetic spectra for these features in terms of five free parameters, three to characterize the star atmospheric model (effective temperature, T_{eff}, surface gravity, $\log g$, and metalicity, [Fe/H]) and two further parameters to take into account the effect of the accretion disk emission on the stellar spectrum. This was assumed to be a linear function of wavelength and thus characterized by two parameters: veiling at 4500 Å, $f_{4500} = F_{\mathrm{disk}}^{4500}/F_{\mathrm{cont,star}}^{4500}$, and the slope, m_0.

We compared, using a $\chi2$-minimization procedure, this grid with 1000 realizations of the observed spectrum of these SXTs. Using a bootstrap Monte-Carlo method, we defined the confidence regions for the five free parameters whose most likely values are given in

Table 2. Chemical abundances in SXTs

[X/H]	A0620–00	Cen X-4	XTE J1118+480	Nova Sco 94
Fe	0.14 ± 0.20	0.23 ± 0.10	0.18 ± 0.17	-0.11 ± 0.09
Li*	2.41 ± 0.21	2.98 ± 0.29	< 1.8	< 2.1
Al	0.40 ± 0.12	0.30 ± 0.17	0.60 ± 0.20	0.05 ± 0.18
Ca	0.10 ± 0.20	0.21 ± 0.17	0.15 ± 0.23	-0.02 ± 0.14
Mg	0.40 ± 0.16	0.35 ± 0.17	0.35 ± 0.25	0.69 ± 0.09
Ni	0.27 ± 0.10	0.35 ± 0.17	0.30 ± 0.21	0 ± 0.21
Ti	0.37 ± 0.23	0.40 ± 0.17	–	0.27 ± 0.22
Si	–	–	–	0.58 ± 0.08
O	–	–	–	0.91 ± 0.08
S	–	–	–	0.66 ± 0.11
Na	–	–	–	0.31 ± 0.25

* Li abundance is expressed as: $\log \epsilon(\mathrm{Li})_{\mathrm{NLTE}} = \log[N(\mathrm{Li})/N(\mathrm{H})]_{\mathrm{NLTE}} + 12$

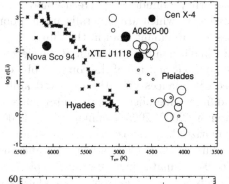

Figure 1. Li abundances of secondary stars in SXTs (filled circles) and rotating Pleiades dwarfs whose age is $\sim 1.2 \times 10^8$ yr (open circles, García López *et al.* 1994), and Hyades stars whose age is $\sim 7 \times 10^8$ yr (asterisks, Thorburn *et al.* 1993), versus effective temperature. The sizes of the circles are related to $v \sin i$.

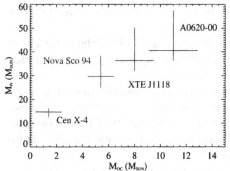

Figure 2. Relationship between the masses of the progenitor star and the compact object in the SXTs of the sample. The size of the crosses indicates the length of the error bars.

Table 1. Average surface gravities were found to be lower than typical values in main sequence stars which indicates that these stars are expanded, filling their Roche lobes.

Using the derived five parameters we analyzed several spectral features of Fe, Li, Mg, Al, Ca, Ti and Ni, except for XTE J1118+480 for which we could not measure the Ti abundance due to the relatively poorer quality of the observed spectrum in the region where the Ti lines are located. In Nova Sco 94 we also measured the element abundances of O, S, Si and Na, using not only the VLT/UVES but also the Keck/HIRES spectrum published by Israelian *et al.* (1999) (see Table 2). These preliminary results of the analysis of the new VLT/UVES data of Nova Sco 94 will be published in González Hernández *et al.* (2007, in preparation).

3. Discussion and conclusions

3.1. *Li abundances in SXTs*

The Li abundances of the secondary stars in the SXTs (see Table 2) of the sample appear to be unexpectedly high in comparison with typical values found in stars with the same spectral type, except for the secondary star of spectral type F in Nova Sco 94. This indicates either that these systems are relatively young ($\lesssim 8 \times 10^7$ years old, see Fig. 1) and the rotational velocity might have partially inhibited the Li depletion (Maccarone *et al.* 2005), or that there exists a mechanism able to enrich the atmospheres of these stars (see e.g. Martín *et al.* 1994b).

3.2. *Nucleosynthesis in the progenitors of compact objects*

The abundances of *heavy* elements in the secondary stars of SXTs are typically solar or higher than solar abundances (see Table 2), independent of their current locations with respect to the Galactic plane. This might not be expected since two of these systems, namely Cen X-4 (González Hernández *et al.* 2005a) and XTE J1118+480 (Mirabel *et al.*

2001) are moving in halo regions, and therefore, they abundance pattern could have resembled that of Halo stars and globular cluster stars which are generally metal poor (with heavy elements between 10 and 1000 times less abundant than in the Sun).

Secondary stars in these system could have captured a significant amount of the ejected matter in the supernova explosions that gave rise to the formation of the compact objects in these SXTs. We have compared the observed abundances with the expected abundances from yields of a variety of SN models of different metalicities, progenitor masses, geometries and explosion energies (Umeda & Nomoto 2002, 2005 Tominaga *et al.* 2006, in preparation Maeda *et al.* 2002). Here we summarize the conclusions of these studies:

(i) The strong overabundance of oxygen, sulphur and other alpha-elements and the kinematics support that the black hole in Nova Sco 94 originated in a supernova explosion. However, the relatively low Al abundance is not reproduced by current SN models (see González Hernández *et al.* 2007, in preparation)

(ii) The chemical abundances of A0620-00 could be explained only if the outer layers of the progenitor He core were ejected although a direct collapse cannot be discarded (see González Hernández *et al.* 2004).

(iii) The observed abundances combined with the peculiar kinematics of Cen X-4 and the halo black hole binary XTE J1118+480 support an origin in a supernova explosion. The massive progenitors were most likely born in the Galactic disc. In XTE J1118+480, an asymmetric explosion appears necessary to explain the kinematics and the abundance pattern (see González Hernández *et al.* 2005b, 2006).

(iv) Finally, this analysis could provide a preliminary relationship between the masses of the compact objects and their massive progenitors (see Fig. 2). We shall remark that the abundance analysis together with the comparison with SN models allowed us to estimate the most likely masses of the He core progenitors. However, the free parameter of the fraction of the amount of captured matter by the secondary star produces a degeneracy and it is only possible to establish a lower limit of the masses of the He cores. The upper limit is estimated from the maximum ejected mass which still keep both components of the system gravitationally bound with each other after the SN explosion.

Acknowledgements

We would like to thank Jorge Casares for his collaboration in this project. We are also grateful to Keiichi Maeda, Hideyuki Umeda, Ken'ichi Nomoto and Nozomu Tominaga for their contribution with the Supernova explosion models and for helpful discussions. This work has been partially financed by the Spanish Ministry project AYA2005-05149.

References

Casares J., Charles P. A., Naylor T. 1992, Nature, 355, 614

García López R. J., Rebolo R., Martín E. L. 1994, A&A, 282, 518

González Hernández J. I., Rebolo R., Israelian G., Casares J., Maeder A., Meynet G. 2004, ApJ, 609, 988

González Hernández J. I., Rebolo R., Peñarrubia J., Casares J., Israelian G. 2005a, A&A, 435, 1185

González Hernández J. I., Rebolo, R., Israelian G., Casares J., Maeda K., Bonifacio P., Molaro P. 2005b, ApJ, 630, 495

González Hernández J. I., Rebolo R., Israelian G., Harlaftis E. T., Filippenko A. V., Chornock R. 2006, ApJL, 644, L49

Kurucz R. L. 1993, ATLAS9 Stellar Atmospheres Programs and 2 km s^{-1} Grid, CD-ROM (Smithsonian Astrophysical Observatory: Cambridge)

Israelian G., Rebolo R., Basri G., Casares J. & Martín E. L. 1999, Nature, 401, 142

Maccarone T. J., Jonker P. G., Sills A. I. 2005, A&A, 435, 671

Maeda K., Nakamura T., Nomoto K., Mazzali P. A., Patat F., Hachisu I. 2002, ApJ, 565, 405

Martín E. L., Rebolo R., Casares J., Charles P. A. 1992, Nature, 358, 129

Martín E. L., Rebolo R., Casares J., Charles P. A. 1994, ApJ, 435, 791

Martín E. L., Spruit H. C., van Paradijs J. 1994b, A&A, 291, L43

Mirabel I. F., Dawan V., Mignani R. P., Rodrigues I. & Guglielmetti F. 2001, Nature, 413, 139

Sheinis A. I. *et al.* 2002, PASP, 114, 851

Sneden C. 1973, PhD Dissertation (University of Texas: Austin)

Thorburn J. A., Hobbs L. M., Deliyannis C. P. & Pinsonneault M. H. 1993, ApJ, 415, 150

Umeda H. & Nomoto K. 2002, ApJ, 565, 385

Umeda H. & Nomoto K. 2005, ApJ, 619, 427

ANDREW KING: Did you consider the idea that the companion was more massive in the past, nuclear-evolved, and then captured by angular momentum losses to a short period? An analysis like this was published by Haswell *et al.* (2003).

JONAY GONZALEZ HERNANDEZ: The important point is if the secondary star was less massive in the past than its massive companion. In that case, considering a secondary star of two solar masses or one solar mass does not change the implications of the abundances measured. Among the parameters we considered in the supernova explosion model, the capture efficiency can be adjusted to reproduce the observed abundances. In this sense, one should compare the result of the pollution by supernova explosion with self-pollution of the secondary if it is an evolved star. To answer this question it is necessary to measure abundances of C, N and O in the secondary star.

FELIX MIRABEL: Comment: Besides the chemistry, the kinematics shows that black holes can be formed in energetic supernova explosions that lead directly to the star death.

JONAY GONZALEZ HERNANDEZ: Kinematics provides additional information on the process of formation of compact objects. However, kinematics alone might not be enough to establish the origin of a compact object. Both the kinematics and the chemical abundances of the secondary star allow us to make stringent constraints on the formation of compact objects.

Formation and Evolution of Massive Black Holes I

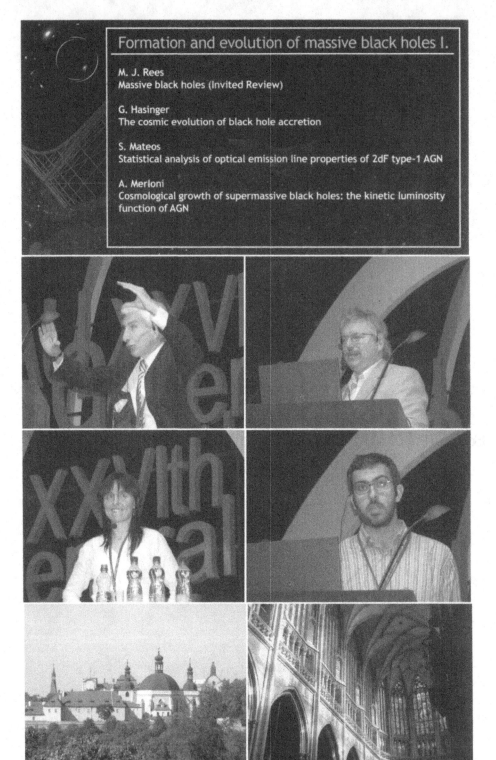

Formation and evolution of massive black holes I.

M. J. Rees
Massive black holes (Invited Review)

G. Hasinger
The cosmic evolution of black hole accretion

S. Mateos
Statistical analysis of optical emission line properties of 2dF type-1 AGN

A. Merloni
Cosmological growth of supermassive black holes: the kinetic luminosity function of AGN

Church of Our Lady's Assumption and Charles the Great, with Classical and Byzantine elements and a star-shaped interior (view from the conference site).

Gothic interior of St Vitus Cathedral, erected in the place of Romanesque rotunda within Prague Castle and containing the tombs of Bohemian kings.

Black Holes from Stars to Galaxies – Across the Range of Masses
Proceedings IAU Symposium No. 238, 2006   © 2007 International Astronomical Union
V. Karas & G. Matt, eds.   doi:10.1017/S1743921307004681

Massive black holes: formation and evolution

Martin J. Rees[1] and Marta Volonteri[2]

[1] Institute of Astronomy, University of Cambridge, Madingley Road, Cambridge CB3 0HA, UK

[2] Department of Astronomy, University of Michigan, 500 Church Street,
Ann Arbor, MI 48109, USA

mjr@ast.cam.ac.uk, email: martav@umich.edu

Abstract. Supermassive black holes are nowadays believed to reside in most local galaxies. Observations have revealed us vast information on the population of local and distant black holes, but the detailed physical properties of these dark massive objects are still to be proven. Accretion of gas and black hole mergers play a fundamental role in determining the two parameters defining a black hole: mass and spin. We briefly review here the basic properties of the population of supermassive black holes, focusing on the still mysterious formation of the first massive black holes, and their evolution from early times to now.

Keywords. Black hole physics – gravitation – radiation mechanisms: general

---

1. Introduction

Black holes (BHs) are the theme of this conference, spanning the full range of masses that we encounter in the astrophysical context: from tiny holes not much more massive than our sun, to monsters weighing by themselves almost as much as a dwarf galaxy. Notwithstanding the several orders of magnitude difference between the smallest and the largest BH known, we believe that all of them can be described by only three parameters: mass, spin and charge (which is not astrophysically relevant, anyway). These in principle simple systems, however, are implicated in the power output from active galactic nuclei (AGNs), and in the production of relativistic jets that energise strong radio sources. Simple inner engines give rise to very messy and complicated energetic phenomena.

By the early 1970s the astrophysically-relevant theoretical properties of black holes – the Kerr metric, the 'no hair' theorems, and so forth – were well established. By now, the evidence points insistently towards the existence of dark objects, with deep potential wells and 'horizons' through which matter can pass into invisibility.

We expect the same mathematical and physical properties to describe both the black holes of a few solar masses and the supermassive black holes (SMBHs) in the mass range of million to billion solar masses, but while the formation path of stellar mass BHs as remnants of supernovae is by now commonly accepted, the formation and evolution of the supermassive variety is far less understood.

It is worth mentioning that in between the observed populations of stellar mass BHs (up to a few tens solar masses) and supermassive BHs (the smallest, detected in the dwarf Seyfert 1 galaxy POX 52, is $\sim 10^5 \, M_\odot$; Barth *et al.* 2004), there might exist an intermediate league, in the range of hundreds or thousands solar masses, bridging the gap.

In the following we will focus on the formation and evolution of supermassive black holes, but we remark once again that stellar-mass and supermassive holes differ for their histories, but not for their physical properties.

2. Probing the properties of Massive Dark Objects: black holes? spinning?

Observers cannot yet definitively confirm the form of the metric in the strong-gravity region, in order to prove that BHs are indeed described by the Kerr metric. The flow patterns close to the hole offer, in principle, a probe of the metric. Until X-ray interferometry is developed, actually 'imaging' the inner discs is beyond the capabilities of current instruments. However, the motion of material in a relativistic potential is characterized by large frequency shifts: substantial gravitational redshifts would be expected, as well as large doppler shifts. These large frequency-shifts can be revealed spectroscopically. The dominant spectral feature in the X-ray spectrum (2–10 keV) is the Fe Kα line at 6.4 keV, which is typically observed with broad asymmetric profile indicative of a relativistic disc (Fabian *et al.* 1989; Laor 1991). The iron line can in principle constrain also the value of black hole spins. Let us define the dimensionless black hole spin, $\hat{a} \equiv J_h/J_{max} = c J_h/G M_{BH}^2$, where J_h is the angular momentum of the black hole. The value of $\hat{a}$ affects the location of the inner radius of the accretion disc (corresponding to the innermost stable circular orbit in the standard picture), which in turn has a large impact on the shape of the line profile, because when the hole is rapidly rotating, the emission is concentrated closer in, and the line displays larger shifts. There is some evidence that this must be the case in some local Seyfert galaxies (Miniutti, Fabian & Miller 2004; Streblyanska *et al.* 2005). The assumption that the inner disc radius corresponds to the innermost stable circular orbit is not a trivial one, however, especially for thick discs (e.g. Krolik 1999, but see Afshordi & Paczynski 2003).

The spin of a hole affects the efficiency of 'classical' accretion processes themselves; the value of $\hat{a}$ in a Kerr BH also determines how much energy is in principle extractable from the hole itself. Assuming that relativistic jets are powered by rotating black holes through the Blandford–Znajek mechanism, the orientation of the spin axis may be important in relation to jet production. Spin-up is a natural consequence of prolonged disc-mode accretion: any hole that has (for instance) doubled its mass by capturing material with constant angular momentum axis would end up with spinning rapidly, close to the maximum allowed value (Thorne 1974). A hole that is the outcome of a merger between two of comparable mass would also, generically, have a substantial spin. On the other hand, a BH which had gained its mass from capturing many low-mass objects (holes, or even stars) in randomly-oriented orbits, would keep a small spin (Figure 1; see Moderski, Sikora & Lasota 1998; Volonteri *et al.* 2005; Volonteri, Sikora & Lasota 2007).

Although observations of the iron line with the Chandra and XMM X-ray satellites are extending the studies of the innermost regions of BHs, aiming at probing black hole properties, interpretation of these studies is impeded by the inherent 'messiness' of gas dynamics. A far cleaner probe would be a compact mass in a precessing and decaying orbit around a massive hole. The detection of gravitational waves from a stellar mass BH, or even a white dwarf, or neutron star, falling into a massive black hole (Extreme Mass Ratio Inspiral, or EMRI) can provide a unique tool to constrain the geometry of spacetime around BHs, and as a consequence, BH spins. Detecting and monitoring such a system may have to await the launch of the planned *Laser Interferometer Space Antenna (LISA)*. Indeed the spin is a measurable parameter, with a very high accuracy, in the gravitational waves *LISA* signal (Barack & Cutler 2004; Berti *et al.* 2005, 2006; Lang & Hughes 2006; Vecchio 2004). Gravitational waves from an EMRI can be used to map the spacetime of the central massive dark object. The resulting 'map' can tell us if the standard picture for the central massive object, a Kerr BH described by general relativity, holds.

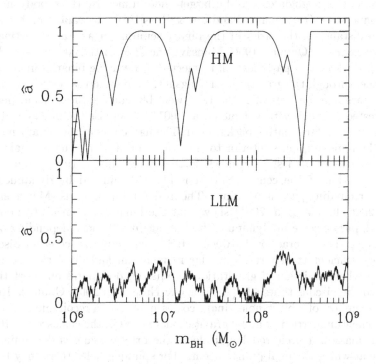

Figure 1. Evolution of a BH spin during a series of accretion episodes lasting for a total of a Hubble time. Initial mass $m_{BH} = 10^6 M_\odot$, initial spin $\hat{a} = 10^{-3}$. Lower panel: the accreted mass at every accretion episode is constrained to be less than 0.01 of the BH mass (LLM). Upper panel: the accreted mass is randomly extracted in the range 0.01–10 times the BH mass (HM). Adapted from Volonteri, Sikora & Lasota (2007).

3. Hierarchies of galaxies and black holes

The demography of massive dark objects, which we will continue to refer to as black holes in the following, has been clarified in the last ten years by studies of the central regions of relatively nearby galaxies (mainly with quiescent nuclei). Strikingly, the centres of all observed galactic bulges host SMBHs (Richstone 2004). The mass of the quiescent SMBHs detected in the local Universe scales with the bulge luminosity – or stellar velocity dispersion – of their host galaxy (Ferrarese & Merritt 2000; Gebhardt *et al.* 2000; Tremaine *et al.* 2002). In the currently favoured cold dark matter cosmogonies (Spergel *et al.* 2006), present-day galaxies have been assembled via a series of mergers, from small-mass building blocks which form at early cosmic times. In this paradigm galaxies experience multiple mergers during their lifetime. If most galaxies host BHs in their centre, and a local galaxy has been made up by multiple mergers, then a black hole binary is a natural evolutionary stage. After each merger event, the central black holes already present in each galaxy would be dragged to the centre of the newly formed galaxy via dynamical friction, and then if/when they get close (≈ 0.01–0.001 pc) the black hole binary would coalesce via emission of gravitational radiation.

The efficiency of dynamical friction decays when the BHs get close and form a binary, when the binary separation is around 0.1–1 pc (for $M_{BH} \simeq 10^5$–$10^8 M_\odot$). Emission of gravitational waves becomes efficient at binary separations about two orders of magnitude smaller. In gas-poor systems, the subsequent evolution of the binary, while gravitational radiation emission is negligible, may be largely determined by three-body interactions with background stars (Begelman, Blandford & Rees 1980), by capturing the stars that pass within a distance of the order of the binary semi-major axis and ejecting them at much higher velocities (Quinlan 1996; Milosavljevic & Merritt 2000; Sesana, Haardt & Madau 2006). Dark matter particles will be ejected by decaying binaries in the same way as the stars, i.e. through the gravitational slingshot. In minihalos a numerous population of low-mass stars may be present if the IMF were bimodal, with a second peak at 1–2 $M_\odot$, as suggested by Nakamura & Umemura (2001). Otherwise the binary will be losing orbital energy to the dark matter background. The hardening of the binary modifies the density profile, removing mass interior to the binary orbit, depleting the galaxy core of stars and dark matter, and slowing down further decay. In gas rich systems, however, the orbital evolution of the central SMBH is likely dominated by dynamical friction against the surrounding gaseous medium. The available simulations (Mayer et al. 2006; Dotti et al. 2006; Escala et al. 2004) show that the binary can shrink to about parsec or slightly sub-parsec scale by dynamical friction against the gas, depending on the gas thermodynamics. The interaction between a BH binary and an accretion disc can also lead to a very efficient transport of angular momentum, and drive the secondary BH to the regime where emission of gravitational radiation dominates on short timescales, comparable to the viscous timescale (Armitage & Natarajan 2005, Gould & Rix 2000).

When the members of a black hole binary coalesce, there is a recoil due to the non-zero net linear momentum carried away by gravitational waves in the coalescence. If the holes have unequal masses, a preferred longitude in the orbital plane is determined by the orbital phase at which the final plunge occurs. For spinning holes there may be a rocket effect perpendicular to the orbital plane, since the spins break the mirror symmetry with respect to the orbital plane. This recoil could be so violent that the merged hole breaks loose from shallow potential wells, especially in the small mass pregalactic building blocks.

A single big galaxy can be traced back to the stage when it was in hundreds of smaller components with individual internal velocity dispersions as low as 20 km s^{-1}. Did black holes form with the same efficiency in small galaxies (with shallow potential wells), or did their formation have to await the buildup of substantial galaxies with deeper potential wells? This issue is important because it determines the expected event rate detected by LISA, and whether there is a population of high-z miniquasars. Of particular interest is whether the merger history can be traced back to 'seed' holes, and be used to distinguish between seed formation scenarios. One first possibility is the direct formation of a BH from a collapsing gas cloud (Haehnelt & Rees 1993; Loeb & Rasio 1994; Eisenstein & Loeb 1995; Bromm & Loeb 2003; Koushiappas, Bullock & Dekel 2004; Begelman, Volonteri & Rees 2006; Lodato & Natarajan 2006). When the angular momentum of the gas is shed efficiently, on timescales shorter than star formation, the gas can collect in the very inner region of the galaxy. The loss of angular momentum can be driven either by (turbulent) viscosity or by global dynamical instabilities, such as the "bars-within-bars" mechanism (Shlosman, Frank & Begelman 1989). The infalling gas can therefore condense to form a central massive object. The mass of the seeds predicted by different models vary, but typically are in the range $M_{\rm BH} \sim 10^4$–10^6 $M_\odot$.

Alternatively, the seeds of SMBHs can be associated with the remnants of the first generation of stars, formed out of zero metallicity gas. The first stars are believed to form at $z \sim 20$–30 in halos which represent high-σ peaks of the primordial density field. The main coolant, in absence of metals, is molecular hydrogen, which is a rather inefficient coolant. The inefficient cooling might lead to a very top-heavy initial stellar mass function, and in particular to the production of an early generation of 'Very Massive

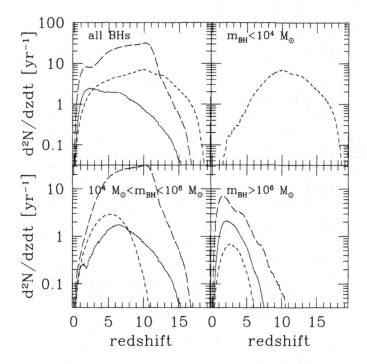

Figure 2. Predicted rate of BH binary coalescences per unit redshift, in different BH mass intervals. Solid curve: BH seeds from Pop-III stars. Long dashed curve: BH seeds from direct collapse (Koushiappas *et al.* 2004). Short dashed curve: BH seeds from direct collapse (Begelman, Volonteri & Rees 2006). Adapted from Sesana *et al.* 2006.

Objects' (VMOs) (Carr, Bond, Arnett 1984). If very massive stars form above 260 $M_\odot$, they would rapidly collapse to massive BHs with little mass loss (Fryer, Woosley, & Heger 2001), i.e., leaving behind seed BHs with masses $M_{BH} \sim 10^2$-$10^3 \, M_\odot$ (Madau & Rees 2001; Volonteri, Haardt & Madau 2003).

LISA in principle is sensible to gravitational waves from binary BHs with masses in the range 10^3-$10^6 \, M_\odot$ basically at any redshift of interest. A large fraction of coalescences will be directly observable by *LISA*, and on the basis of the detection rate, constraints can be put on the BH formation process. Different theoretical models for the formation of BH seeds and dynamical evolution of the binaries predict merger rates that largely vary one from the other (Figure 2). *LISA* will be a unique probe of the formation and merger history of BHs along the *entire* cosmic history of galactic structures.

4. Accretion

Accretion is inevitable during the 'active' phase of a galactic nucleus. Observations tell us that AGN are widespread in both the local and early Universe. All the information that we have gathered on the evolution of SMBHs is indeed due to studies of AGN, as we have to await for *LISA* to be able to "observe" quiescent SMBHs in the distant Universe. A key issue is the relative importance of mergers and accretion in the build-up of the largest holes in giant ellipticals.

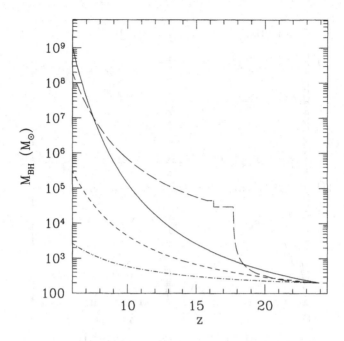

Figure 3. Growth of a BH mass under different assumption for the accretion rate and efficiency. Eddington limited accretion: $\epsilon = 0.1$ (solid line), $\epsilon = 0.2$ (short dashed), $\epsilon = 0.4$ (dot-dashed line). Radiatively inefficient super-critical accretion, as in Volonteri & Rees 2005 (long dashed line).

The accretion of mass at the Eddington rate causes the black hole mass to increase in time as

$$M(t) = M(0) \exp\left(\frac{1-\epsilon}{\epsilon} \frac{t}{t_{\rm Edd}}\right), \tag{4.1}$$

where $t_{\rm Edd} = 0.45\,\mathrm{Gyr}$ and ϵ is the radiative efficiency. The classic argument of Soltan (1982), compares the total mass of black holes today with the total radiative output by known quasars, by integration, over redshift and luminosity, of the luminosity function of quasars (Yu & Tremaine 2002; Elvis, Risaliti & Zamorani 2002; Marconi et al. 2004). The total energy density can be converted into the total mass density accreted by black holes during the active phase, by assuming a mass-to-energy conversion efficiency, ϵ (Aller & Richstone 2002; Merloni, Rudnick & Di Matteo 2004). The similarity of the total mass in SMBHs today and the total mass accreted by BHs implies that the last 2–3 e-folds of the mass is grown via radiatively efficient accretion, rather than accumulated through mergers or radiatively inefficient accretion. However, most of the e-folds (corresponding to a relatively small amount of mass, say the first 10% of mass) could be gained rapidly via, e.g., radiatively inefficient accretion. This argument is particularly important at early times.

The Sloan Digital Sky survey detected luminous quasars at very high redshift, $z > 6$, when the Universe was less than 1 Gyr old. Follow-up observations confirmed that at least some of these quasars are powered by SMBHs with masses $\simeq 10^9\,M_\odot$ (Barth et al. 2003; Willott et al. 2005). Given a seed mass $M(0)$ at $z = 30$ or less, the higher the efficiency, the longer it takes for the BH to grow in mass by (say) 10 e-foldings. If accretion is radiatively efficient, via a geometrically thin disc, the alignment of a SMBH with the angular momentum of the accretion disc tends to efficiently spin holes up (see section 2) , and radiative efficiencies can therefore approach 30–40%. With such a high efficiency, $\epsilon = 0.3$, it can take longer than 2 Gyr for the seeds to grow up to a billion solar masses.

Let us consider the extremely rare high redshift (say $z > 15$) metal–free halos with virial temperatures $T_{vir} > 10^4$K where gas can cool even in the absence of H_2 via neutral hydrogen atomic lines. The baryons can therefore collapse until angular momentum becomes important. Afterward, gas settles into a rotationally supported dense disc at the center of the halo (Mo, Mao & White 1998; Oh & Haiman 2002). This gas can supply fuel for accretion onto a BH within it. Estimating the mass accreted by the BH within the Bondi–Hoyle formalism, the accretion rate is initially largely above the Eddington limit (Volonteri & Rees 2005). When the supply is super-critical the excess radiation can be trapped, as radiation pressure cannot prevent the accretion rate from being super-critical, while the emergent luminosity is still Eddington limited in case of spherical or quasi-spherical configurations (Begelman 1979; Begelman & Meier 1982). In the spherical case, though this issue remains unclear, it still seems possible that when the inflow rate is super-critical, the radiative efficiency drops so that the hole can accept the material without greatly exceeding the Eddington luminosity. The efficiency could be low either because most radiation is trapped and advected inward, or because the flow adjusts so that the material can plunge in from an orbit with small binding energy (Abramowicz & Lasota 1980). The creation of a radiation-driven outflow, which can possibly stop the infall of material, is also a possibility. If radiatively inefficient supercritical accretion requires metal-free conditions in exceedingly rare massive halos, rapid early growth, therefore, can happen only for a tiny fraction of BH seeds. These SMBHs are those powering the $z = 6$ quasars, and later on to be found in the most biased and rarest halos today. The global BH population, instead, evolves at a more quiet and slow pace.

References

Abramowicz M. A. & Lasota J. P. 1980, Acta Astronomica, 30, 35
Afshordi N. & Paczynski B. 2003, ApJ, 594, 71
Aller M. C. & Richstone D. 2002, AJ, 124, 3035
Armitage P. J. & Natarajan P. 2005, Apj, 634, 921
Barack L. & Cutler C. 2004, PhRvD, 70l2002
Barth A. J., Martini P., Nelson C. H. & Ho L. C. 2003, ApJ, 594, L95
Barth A. J., Ho L. C., Rutledge R. E. & Sargent W. L. W. 2004, ApJ, 607, 90
Begelman M. C. 1979, MNRAS, 187, 237
Begelman M. C., Blandford R. D. & Rees M. J. 1980, Nature, 287, 307
Begelman M. C. & Meier D. L. 1982, ApJ, 253, 873
Begelman M. C., Volonteri M. & Rees M. J. 2006, MNRAS, submitted (astro-ph/0602363)
Berti E., Buonanno A. & Will C. M. 2005, Phys. Rev. D71, 084025
Berti E., Cardoso V. & Will C. M. 2006, Phys. Rev. D73, 064030
Bromm V. & Loeb A. 2003, ApJ, 596, 34
Carr B. J., Bond J. R. & Arnett W. D. 1984, ApJ, 277, 445
Dotti M., Colpi M. & Haardt F. 2006, MNRAS, 367, 103
Eisenstein D. J. & Loeb A. 1995, ApJ, 443, 11
Elvis M., Risaliti G. & Zamorani G. 2002, ApJ, 565, L75
Escala A., Larson R. B. & Coppi P. 2004, ApJ, 607, 765
Fabian A. C., Rees M. J., Stella L. & White N. E. 1989, MNRAS, 238, 729
Ferrarese L. & Merritt D. 2000, ApJL, 539, 9
Fryer C. L., Woosley S. E. & Heger A. 2001, ApJ, 550, 372
Gebhardt K. *et al.* 2000, ApJL, 543, 5
Gould A. & Rix H.-W. 2000, ApJ, 532, 29

Haehnelt M. G. & Rees M. J. 1993, MNRAS, 263, 168

Krolik J. H. 1999, ApJ, 515, 73

Koushiappas S. M., Bullock J. S. & Dekel A. 2004, MNRAS, 354, 292

Lang R. N. & Hughes S. A. 2006, gr-qc/0608062

Laor A. 1991, ApJ, 376, 90

Lodato G. & Natarajan P. 2006, astro-ph/0606159

Loeb A. & Rasio F. A. 1994, ApJ, 432, 52

Madau P. & Rees M. J. 2001, ApJ, 551, L27

Marconi A. et al. 2004, MNRAS, 351, 169

Mayer L. et al. 2006, astro-ph/0602029

Merloni A., Rudnick G. & di Matteo T. 2004, MNRAS, 354, 37

Milosavljevic M. & Merritt D. 2001, ApJ, 563, 34

Miniutti G., Fabian A. & Miller J. M. 2004, MNRAS, 351, 466

Moderski R., Sikora M. & Lasota J.-P. 1998, MNRAS, 301, 142

Mo H. J., Mao S. & White S. D. M. 1998, MNRAS, 295, 319

Nakamura F. & Umemura M. 2001, ApJ, 548, 19

Natarajan P. & Pringle J. E. 1998, ApJ, 506, 97

Oh S. P. & Haiman Z. 2002, ApJ, 569, 558

Quinlan G. D. & Hernquist L. 1997, New Astronomy, 2, 533

Richstone D. 2004, Supermassive black holes: Demographics and implications; in Coevolution of Black Holes and Galaxies, Carnegie Obs. Ast. Ser. Vol. 1, ed. L. C. Ho (Cambridge Univ. Press: Cambridge), p. 281

Sesana A., Haardt F. & Madau P. 2006, ApJ, 651, 392

Sesana A., Volonteri M. & Haardt F. 2007, MNRAS submitted

Shlosman I., Frank J. & Begelman M. C. 1989, Nature, 338, 45

Soltan A. 1982, MNRAS, 200, 115

Spergel D. N., Bean R., Dore' O., Nolta M. R., Bennett C. L. et al. 2006, ApJ, in press (astro-ph/0603449)

Streblyanska A. et al. 2005, A&A, 432, 395

Thorne K. S. 1974, ApJ, 191, 507

Tremaine S. et al. 2002, ApJ, 574, 740

Vecchio A. 2004, Phys. Rev. D70, 042001

Volonteri M. et al. 2005, ApJ, 620, 69

Volonteri M., Haardt F. & Madau P. 2003, ApJ, 582, 559

Volonteri M. & Rees M. J. 2005, Apj, 633, 624

Volonteri M., Sikora M. & Lasota J.-P. 2007, MNRAS submitted

Willott C. J. et al. 2005, ApJ, 626, 657

Yu Q. & Tremaine S. 2002, MNRAS, 335, 965

VLADIMIR KOCHAROVSKY: What is the role of dark matter in the process of formation of first massive black holes and in their further growing?

MARTIN REES: The role is indirect. Dark matter halo makes the baryonic matter more dense, but it does not participate in the gravitational collapse at the final step of the massive black hole formation. Nevertheless, gravitational potential is very important for the rocket effect in the binary massive black hole mergers.

MITCHELL BEGELMAN: In the hierarchical merger picture, how important is the three-body Newtonian ejection compared to the ejection due to gravitational rocket effect?

MARTIN REES: It depends crucially on the most uncertain part of the dynamics, which is the radius where the binary gets hung up. So the question is whether dynamical friction can bring the binary close enough to gravitational radiation to take over. I think Piero Madau will be discussing that.

Black Holes from Stars to Galaxies – Across the Range of Masses
Proceedings IAU Symposium No. 238, 2006
V. Karas & G. Matt, eds.
© 2007 International Astronomical Union
doi:10.1017/S1743921307004693

X-ray emission properties of BLAGN in the XMM-2dF Wide Angle Survey

S. Mateos,[1] M. G. Watson,[1] J. A. Tedds,[1] F. J. Carrera,[2] M. Page[3] and A. Corral[2]

[1] Department of Physics and Astronomy, University of Leicester, LE1 7RH, UK
[2] Instituto de Física de Cantabria (CSIC-UC), Avenida de los Castros, 39005 Santander, Spain
[3] Mullard Space Science Laboratory, UCL, Holmbury St. Mary, Surrey, RH5 6NT, UK
email: sm279@star.le.ac.uk

Abstract. We present the preliminary results of the X-ray spectral analysis of one of the largest samples of X-ray selected BLAGN assembled so far from the XMM-2dF Wide Angle Survey. The sample, with 641 spectroscopically identified BLAGN, provides a unique resource to carry out a statistical analysis of the emission properties of these objects over a broad range of X-ray luminosities and redshifts. The X-ray spectra of the majority of the objects were best fitted with a power law with a near constant mean spectral photon index. No obvious trend of this spectral parameter with X-ray luminosity or redshift was found.

We measured the mean photon index of our objects to be ∼1.96±0.05 with an intrinsic dispersion σ=0.22±0.03. X-ray absorption was detected in ∼8% of the sources, with no preferred luminosity or redshift and having typical values of the absorbing column density $\leqslant 10^{22}\,\mathrm{cm}^{-2}$.

Keywords. X-rays: general – X-ray surveys – galaxies: active

1. Introduction

The *Chandra* and *XMM-Newton* deepest pencil beam surveys have resolved more than 80% of the Cosmic X-ray background (XRB) into discrete objects confirming the discrete nature of this cosmic emission (see for example Tozzi *et al.* 2006) and that the XRB is dominated by Active Galactic Nuclei (AGN). Extensive statistical X-ray spectral studies of the sources detected in these surveys are providing valuable information on the nature and cosmic evolution of the population of faint AGN, however many of these results are limited by the lack of large enough samples of objects. In order to understand the accretion history in the Universe, a good knowledge of the X-ray emission properties of the population of AGN and their dependence with the luminosity and redshift of the objects is necessary.

One of the open questions that still needs to be addressed is the existing correlation between X-ray absorption and optical obscuration in AGN. AGN unification models have been successful in explaining the observational differences between different classes of AGN, however X-ray spectral analyses show that ∼10% of BLAGN are absorbed in X-rays (see Mateos *et al.* 2005b), questioning the validity of AGN unification models in these objects. Shallower but wide area surveys are an important complement to pencil beam surveys, as they are more efficient in compiling large samples of sources at intermediate fluxes where the bulk of the XRB emission originates. Hence these surveys are an ideal database to study some of the remaining open questions regarding the AGN population.

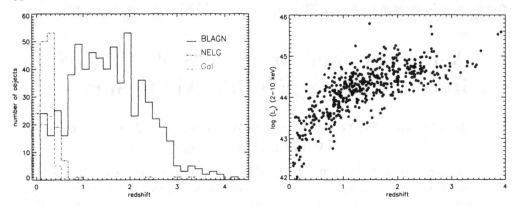

Figure 1. Left panel: redshift distributions of the different samples of extragalactic objects in the XMM-2dF Wide Angle Survey. Right panel: 2–10 keV luminosity vs redshift for the sample of BLAGN.

2. The XMM-2dF Wide Angle Survey

The XMM-2dF Wide Angle Survey contains 68 *XMM-Newton* fields with typical exposure times of a few tens of ksec, and therefore it is a medium-depth wide angle (total sky area of 15 deg^2) survey. More than 3000 *XMM-Newton* serendipitously discovered objects with X-ray fluxes $> 10^{-14}$ erg cm^{-2} s^{-1} and SUPER-COSMOS optical counterparts brighter than V~21 have been observed and reduced. At the time of this analysis 978 sources have been spectroscopically identified. Most of the objects have been identified as BLAGN, 641 (65%) while 157 objects were found to be NELG (16%) and 57 Galaxies (5.8%). The sample contains also 123 Stars (12.6%). The redshift distributions of the different types of extragalactic sources is shown in Figure 1 (left), while the 2–10 keV luminosity versus redshift distribution of sources in shown in Figure 1 (right).

3. Spectral analysis

For all objects with spectroscopic identifications, X-ray spectra from each EPIC-camera and observation were extracted. The spectra were then combined to obtain a co-added MOS and pn spectrum for each source. From the 641 objects in the BLAGN sample, 496 have X-ray spectra of sufficient quality (at least 5 bins in the co-added spectra) to characterise the emission properties of the objects from an analysis of their X-ray spectra. As a starting point of our study, all spectra were fitted with a power law and an absorbed power law, both absorbed by the column density of the Galaxy.

BLAGN mean photon index The 0.2–12 keV X-ray spectra of most BLAGN in our sample were best fitted with a simple power law. The values of the X-ray photon index were found to be near constant and consistent in most cases with the canonical value of ~1.9. However, a clear dispersion on the individual values was evident as can be seen from the distribution of best fit spectral slopes in Figure 2 (left). Using a maximum likelihood analysis (see Maccacaro *et al.* 1988) we measured the mean continuum shape of our BLAGN to be 1.96±0.05 with an intrinsic dispersion σ=0.22±0.03. However, for a large number of objects in our sample the best fit photon index was found to be significantly different from the mean value. We can explain this result as being due to the fact that we are not detecting all X-ray absorption present in the X-ray spectra of our sources, probably due to the low signal of some of the data. This is confirmed by the fact that the mean photon index of our objects becomes harder at the faintest soft (0.5–2 keV) fluxes (see Figure 2), an effect that has been attributed to undetected X-ray absorption

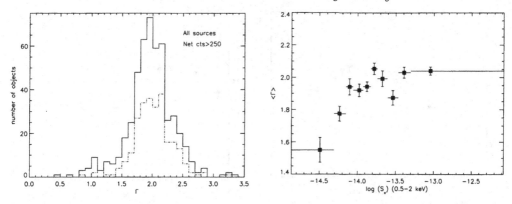

Figure 2. Left panel: distribution of best fit spectral slopes for our sample of BLAGN (solid). The corresponding distribution for the objects with best spectral quality is indicated with a dashed line. Right panel: dependence of the mean continuum shape with the soft (0.5–2 keV) flux of the objects.

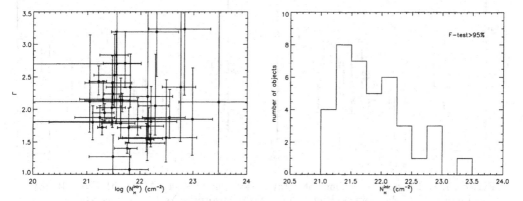

Figure 3. Left panel: correlation of the spectral photon index with X-ray absorption. Right panel: distribution of X-ray absorbing column densities (rest-frame) in our sample of BLAGN.

(Mateos *et al.* 2005a,b). In addition, our spectral models do not contain a soft excess component.

Intrinsic X-ray absorption Intrinsic X-ray absorption was detected in 38 objects ($\sim$8% of the sample) with an F-test significance $\geqslant$95%. No correlation of the power law shape with the detected absorption was found as we see in Figure 3 (left). X-ray absorption in our BLAGN does not seem to occur at any particular redshift or luminosity. The distribution of absorbing column densities (rest-frame) of our sample of absorbed BLAGN is shown in Figure 3 (right). Typical absorbing column densities are $< 10^{22}$ although a significant fraction of type-1 AGN have detected absorbing column densities above this value but with large uncertainties in the values. We did not find the fraction of absorbed BLAGN to depend on the X-ray luminosity of the sources or that X-ray absorption occurs at any preferred redshift.

4. Dependence of emission properties with luminosity and redshift

It has been previously reported that the X-ray photon index of AGN depends on the luminosity of the objects; see Dai *et al.* (2004). On the other hand recent spectral analyses based on sources detected with *Chandra* and *XMM-Newton* claim contradictory results on the cosmic evolution of the photon index of AGN: some suggest a hardening

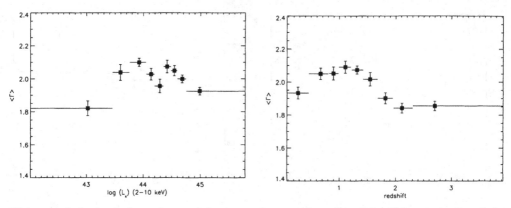

Figure 4. Left panel: dependence of the mean photon index of our BLAGN with the 2–10 keV luminosity of the sources. Right panel: the mean photon index of our BLAGN vs redshift.

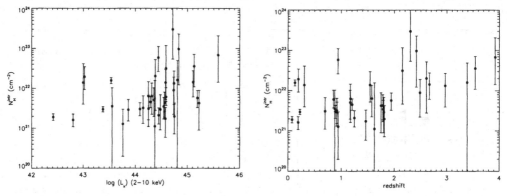

Figure 5. Left panel: dependence of X-ray absorption of our BLAGN with the 2–10 keV luminosity of the sources. Right panel: X-ray absorption of BLAGN vs redshift.

at high redshifts (Bechtold *et al.* 2003) while others suggest a softening (Grupe *et al.* 2006) or no significant evolution (Vignali *et al.* 2005). We do not see any clear trend of the mean photon index of our objects with X-ray luminosity in the range of luminosities $\sim 10^{43}$–10^{45} erg s^{-1} (see Figure 4). Finally we did not found evidence for a correlation of absorption and redshift/luminosity (see Figure 5) in agreement with the findings of Dwelly & Page (2006).

Acknowledgements

This work was supported by funding from the *XMM-Newton* Survey Science Centre.

References

Bechtold J., Siemiginowska A., Shields J., Czerny B. *et al.* 2003, ApJ, 588, 119
Dai X., Chartas G., Eracleous M. & Garmire G. P. 2004, ApJ, 605, 45
Dwelly T. & Page M. J. 2006, MNRAS, 1122
Grupe D., Mathur S., Wilkes B. & Osmer P. 2006, AJ, 131, 55
Maccacaro T., Gioia L. M., Wolter A. *et al.* 1988, ApJ, 326, 680
Mateos S., Barcons X., Carrera F. J., Ceballos M. T. *et al.* 2005, A&A, 433, 855
Mateos S., Barcons X., Carrera F. J., Ceballos M. T. *et al.* 2005, A&A, 444, 79
Tozzi P., Gilli R., Mainieri V., Norman C. *et al.* 2005, A&A, 451, 457
Vignali C., Brandt W. N., Schneider D. P. & Kaspi S. 2005, AJ, 129, 2519

RICHARD MUSHOTZKY: Since you show many of the optically selected broad-line objects have column densities between 10^{21}–10^{22}, these would be normally expected highly reddened. Is this evident or absent in the optical data? That would mean a strong correction to the standard optical luminosity function data.

SILVIA MATEOS: We have some evidence that the absorbed objects are redder, but no firm conclusions can be made at this moment.

BRIAN BOYLE: Comment: Regarding the optical selection in the 2dF survey, the optical colours of low-redshift objects might have might kicked them out of the survey. This might account for the fact that you don't see an increase of absorption in the X-ray broad line AGN towards lower redshift, which seems slightly inconsistent with the picture that Günther Hasinger has proposed.

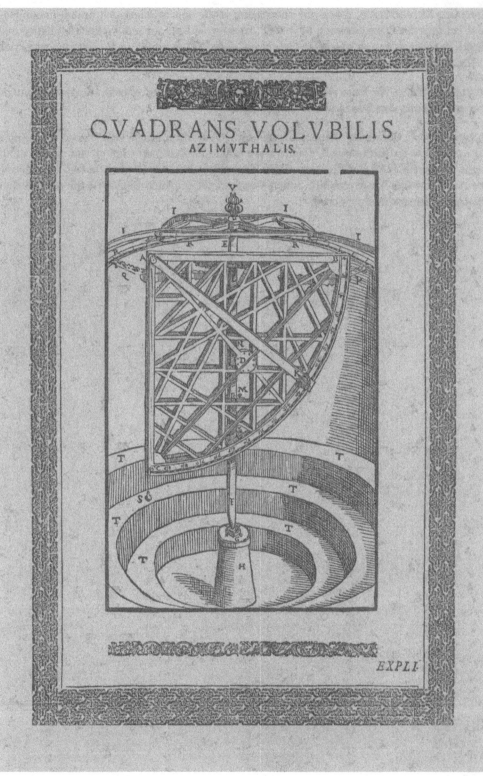

Black Holes from Stars to Galaxies – Across the Range of Masses
Proceedings IAU Symposium No. 238, 2006
V. Karas & G. Matt, eds.
© 2007 International Astronomical Union
doi:10.1017/S174392130700470X

Cosmological evolution of the AGN kinetic luminosity function

Andrea Merloni[1] and Sebastian Heinz[2]

[1] Max Planck Institut für Astrophysik, Garching, Germany

[2] Department of Astronomy, University of Wisconsin, Madison, WI, USA

email: am@mpa-garching.mpg.de, heinzs@astro.wisc.edu

Abstract. We present a first attempt to derive the cosmological evolution of the kinetic luminosity function of AGN based on the joint evolution of the flat spectrum radio and hard X-ray selected AGN luminosity functions. An empirical correlation between jet power and radio core luminosity is found, which is consistent with the theoretical assumption that, below a certain Eddington ratio, SMBH accrete in a radiatively inefficient way, while most of the energy output is in the form of kinetic energy.

We show how the redshift evolution of the kinetic power density from such a low-$\dot{m}$ mode of accretion makes it a good candidate to explain the so-called "radio mode" of AGN feedback as outlined in many galaxy formation schemes.

Keywords. black hole physics – galaxies: active – galaxies: jets

1. Introduction: the radio mode of AGN

Supermassive black holes in the nuclei of many nearby bulges are fundamental constituents of their parent galaxies. The observed correlations between black hole mass and large scale bulge properties (Marconi & Hunt 2003, and references therein) suggest there must have been an epoch in which the central black hole had a direct impact on the galaxy's structure. Recent sustained theoretical efforts have indeed identified the period of most rapid black hole growth as the most likely epoch when such an impact was more dramatically felt. As it is now well established that most of this growth is to be identified with episodes of accretion, when galactic nuclei become "active" (Yu and Tremaine 2002; Marconi *et al.* 2004), AGN feedback is regarded as a necessary ingredient in all models of galaxy formation and evolution.

In particular, numerical simulations of merging galaxies with SMBH growth and radiative feedback (Di Matteo *et al.* 2005) demonstrated the effectiveness of quasars in quenching subsequent episodes of star formation. However, semi-analytic schemes of galaxy formation (Croton *et al.* 2006; Bower *et al.* 2006) as well as indirect evidence of AGN-induced heating in clusters of galaxies both seem to indicate that a different mode of accretion onto SMBH, not directly linked to bright quasar phases (and therefore termed "radio mode") must be effective in the most massive system at late times in order not to over-predict the observed abundances of high mass galaxies as well as to explain the absence of vigorous cooling flows in the center of clusters.

2. A simple scaling for the jet kinetic energy

A mode of AGN feedback dominated by kinetic energy is in fact expected on theoretical grounds (Rees *et al.* 1982) in black holes of low luminosity (in units of the Eddington one $L_{\rm Edd} = 1.3 \times 10^{38} M/M_\odot$ erg s^{-1}). Observational evidence for such "jet dominated" modes of accretion come from either black holes of stellar mass in the so-called low/hard or quiescent states (Fender, Gallo & Jonker 2003; Gallo et al. 2006) and low-luminosity AGN (Merloni, Heinz and Di Matteo 2003 (MHD03); Falcke, Körding & Markoff 2004). In both cases, the observed correlation between the (unresolved) radio luminosity of the jet core and its X-ray luminosity can be interpreted as a by-product of the coupling between a radiatively inefficient accretion flow and a self-absorbed (quasi-conical) jet, provided that the kinetic luminosity of the jet obeys $L_{kin} \propto \dot{M}_{out}c^2$, where $\dot{M}_{out}$ is the accretion rate at the outer boundary of the accretion flow, roughly coincident with the Bondi radius in the case of AGN. The constant of proportionality in the above relationship cannot be directly inferred from the observed radio-X-ray correlations, but has to be determined by direct measurements of jet kinetic energy. On the basis of a very sparse collection of measurements of this sort, Heinz and Grimm (2005) originally proposed the following scaling:

$$L_{kin} = 6 \times 10^{37} (L_R/10^{30})^{0.7} \text{erg/s}, \tag{2.1}$$

where L_R is the radio luminosity of the core at 5 GHz measured in ergs s^{-1}. In fact, it can be shown that such a normalization corresponds to an extreme case in which essentially all accretion power $\eta \dot{M}_{out}c^2$, for $\eta \simeq 0.1$, is released as kinetic energy, on the basis of the observed "fundamental plane" relation of MHD03.

One way to test the above relation is to look for alternative, indirect ways to estimate the kinetic power of the AGN jet. This has been done recently by Allen et al. (2006) and Rafferty et al. (2006) by measuring PdV work done by the jet on the intracluster gas. In Figure 1 we show these observational data points, together with a few estimates of kinetic power from direct modeling of the radio lobe emission (stars) plotted as a function of the radio core luminosity, either measured directly or, preferentially, from the black hole mass and 2–10 keV nuclear luminosity and the MHD03 relation (see Heinz & Grimm 2005 for references). In the same figure we also plot relation (2.1). Indeed, the observational points do not scatter too far away from the predicted relation, even more so if one considers that the kinetic power measurements from X-ray cavities (circles and squares), although averages, are most likely lower limit to the intrinsic jet power, due to the additional presence of weak shocks and low-contrast cavities (Nulsen 2006).

3. The kinetic luminosity evolution

With the aid of eq. (2.1), we are now in the position to attempt a measurement of the kinetic luminosity function of AGN. First, we need to derive the intrinsic 5GHz jet core luminosity function of AGN. This can be done by analytically de-beaming the observed luminosity function of flat spectrum radio quasars and blazars (dominated by relativistically beamed sources), via the formalism described in detail in Urry and Padovani (1991). The analysis is greatly simplified if a narrow distribution of Lorentz factor is assumed for the jets, and, for the sake of simplicity, we assume this holds true here.

One can then use the technique developed in Merloni (2004) to unveil the accretion history of SMBH based on the combined evolution of the hard X-ray and radio cores luminosity function coupled with the fundamental plane relationship. Here, differently from Merloni (2004), we adopt an intrinsic radio core luminosity function evolution, based on the observed FSRQ/Blazar luminosity function of De Zotti et al. (2005), to which we added local constraints on the faint-end slope from Filho, Barthel & Ho (2006). We adopt the scaling (2.1) for all sources accreting below a critical Eddington fraction, that we have fixed at about 3% (see Merloni 2004). Such objects, by construction, will lie on the fundamental plane of black hole activity and represent the majority of radio sources

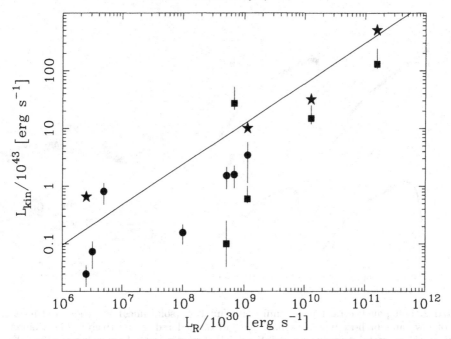

Figure 1. Estimated jet kinetic luminosity as a function of the 5GHz core radio luminosity. Stars represent direct kinetic energy estimates based on radio lobe modeling for NGC 4636, M87, Per A and Cyg A, in order of increasing power. Filled circles are the kinetic energy estimates from Allen *et al.* (2006), filled squares those from Rafferty *et al.* (2006). The solid line is the theoretical scaling $L_{\text{Kin}} = 6 \times 10^{37} (L_R/10^{30})^{0.7}$.

below the FRI/FRII divide. We assume here that only a small fraction of the accreting matter (powering the jet) makes its way onto the central black holes, i.e. we assume that black holes are always radiatively efficient (*with respect to the accreted gas*, rather than to the liberated energy; see the discussion in Gallo *et al.* 2006). For object whose accretion rate is above 3%, we assume that the vast majority (90%) are radio quiet and radiatively efficient (with $\epsilon_{\text{rad}} = \eta = 0.1$), while for the remaining 10% the liberated energy is equally divided into radiative and kinetic. Those powerful radio loud quasar dominate the high end of the radio luminosity function (FR II). It is worth noticing here that the above recipe to include powerful radio sources is highly arbitrary, pending a more complete knowledge of the physical nature of powerful radio loud quasars. Therefore, all our conclusions regarding the evolution of this class of sources, and of their kinetic energy output, should be considered as indicative only.

Given such a recipe for the accretion modes (or, equivalently, for the radiative and kinetic efficiencies as functions of $\dot{m}$), the local SMBH mass function is evolved backwards in time according to the corresponding accretion rate distribution, and the procedure is reiterated until acceptable fits to the observed hard X-ray and radio core luminosity functions are found.

Figure 2 shows, on the left panel, the evolution of the computed kinetic luminosity function of AGN at four different redshifts. By integrating these luminosity functions one can obtain the redshift evolution of the kinetic energy density, ρ_{kin}, provided by SMBH. This is shown in the right panel of fig. 2. Here the uncertainties in ρ_{kin} include not only the jets' Lorentz factor distribution, but also uncertainties on the faint end slope of the radio core luminosity function. The redshift evolution of ρ_{kin} differs substantially from that of the accretion rate (BHAR) density, strongly resembling instead the required

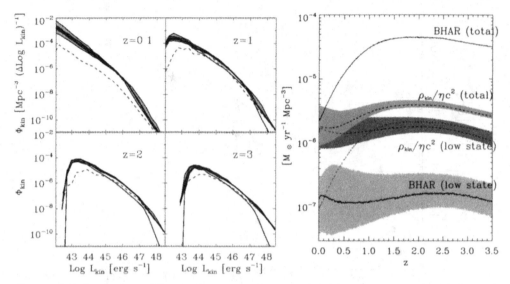

Figure 2. Left panel: total kinetic luminosity function (solid lines) at various redshifts as the sum of low- and high-$\dot{m}$ modes (solid blue and dashed red, respectively). The shaded areas highlight the uncertainties due to variation of the average jets' Lorenz factors ($2 < \Gamma < 10$, black solid line corresponds to $\Gamma = 5$). The low luminosity downturns are due to incompleteness in the high-redshift radio and X-ray luminosity functions adopted here. Right panel: redshift evolution of the kinetic energy density mass equivalent (calculated assuming a energy conversion efficiency of $\eta = 0.1$; black dashed line, orange shaded area), decomposed into the sum of the low- (blue dashed line, cyan shaded area) and high-$\dot{m}$ (red dashed line) modes, and compared with the mass accretion rate onto the black holes (black solid lines).

"radio mode" evolution of galaxy formation models (Croton *et al.* 2006; Bower *et al.* 2006). A more detailed study of these results is underway and will be presented elsewhere (Heinz *et al.*, Merloni *et al.*, in preparation).

4. Conclusions

A separate mode of SMBH growth in which most of the energy is released in kinetic form is postulated both on theoretical grounds and from very general arguments of galaxy formation theory. Here we have discussed how direct observational evidence of such a mode emerging from recent studies of X-ray binary black holes can be generalized to the case of SMBH, when the accretion rate (in Eddington units) is low (less than a few percent).

We have then presented a scaling that provides a simple method to estimate the kinetic energy output of black holes growing at sub-Eddington rates. Using such a scaling, we have derived the kinetic luminosity function of AGN and its redshift evolution. Overall, we show conclusively that the kinetic power output of the low-$\dot{m}$ mode of SMBH growth has a very different redshift dependency from the radiative one, and matches the phenomenological requirements put forward by semi-analytic schemes of galaxy formation: it is in fact more effective at late times (and for the more massive systems).

Our results suggest that the so-called "radio mode" of AGN feedback is simply a jet-dominated *accretion mode*, and that its physical and evolutionary properties are dictated by the physics of accretion in the vicinity of the SMBH.

References

Allen S. W., Dunn R. J. H., Fabian A. C., Taylor G. B. & Reynolds C. S. 2006, MNRAS, 372, 21

Bower R., Benson A. J., Malbon R., Helly J. C., Frenk C. S., Baugh C. M., Cole S. & Lacey C. G. 2006, MNRAS, 370, 645

Croton D. J., Springel V., White S. D. M., De Lucia G., Frenk C. S. & Gao L. *et al.* 2006, MNRAS, 365, 11

De Zotti G., Ricci R., Mesa D., Silva L., Mazzotta P., Toffolatti L. & González-Nuevo J. 2005, A&A, 431, 893

Di Matteo T., Springel L. & Hernquist L. 2005, Nature, 433, 604

Falcke H., Körding E. & Markoff S. 2004, A&A, 414, 895

Fender R. P., Gallo E. & Jonker P. 2003, MNRAS (Letters), 343, L99

Filho M. E., Barthel P. D. & Ho L. C. 2006, A&A, 451, 71

Gallo E., Fender R. P., Miller-Jones J. C. A., Merloni A., Jonker P. G. & Heinz S. *et al.* 2006, MNRAS, 370, 1351

Heinz S., Grimm H.-J. 2005, ApJ, 633, 384

Marconi A., Hunt L. K. 2003, ApJL 589, L21

Marconi A., Risaliti G., Gilli R., Hunt L. K., Maiolino R. & Salvati M. 2004, MNRAS, 351, 169

Merloni A., Heinz S. & Di Matteo T. 2003, MNRAS, 345, 1057 (MHD03)

Nulsen P. E. J. 2006, Proceedings of "Heating and Cooling in Clusters of Galaxies", H. Boehringer, P. Schuecker, G. W. Pratt, A. Finoguenov (eds.)

Rafferty D. A., Mc Namara B. R., Nulsen P. E. J. & Wise M. W. 2006, ApJ, astro-ph/0605323

Rees M. J., Begelman M. C., Blandford R. D. & Phinney E. S. 1982, Nature, 295, 17

Urry M., Padovani P. 1991, ApJ, 371, 60

Yu Q., Tremaine S. 2002, MNRAS, 335, 965

GÜNTHER HASINGER: There is a upper limit to the kinetic feedback given by the efficiency of the black hole accretion. Since radiation is already at the level of 10% (see my talk), kinetic energy can not be much higher then that.

ANDREA MERLONI: Indeed, taking into account the average accretion efficiency of the evolving supermassive black hole population, the results I presented here would suggest that the efficiency with which the rest-mass energy of the black hole is used to power the kinetic feedback does not exceed ~ 1.6–2%.

FELIX MIRABEL: Where in your picture stand the high soft state sources?

ANDREA MERLONI: If one wishes to push the analogy between AGN and X-ray binaries one step forward, one should take into account the complex phenomenology of bright (near-Eddington) stellar-mass accretors. In particular, we know that they can launch powerful relativistic jets episodically, and they can even launch strong winds. This occurs predominantly in the so-called very high state. Such a transient phase is probably related to violent disc instabilities close to the Eddington rate and it might be the origin of radio loud quasars.

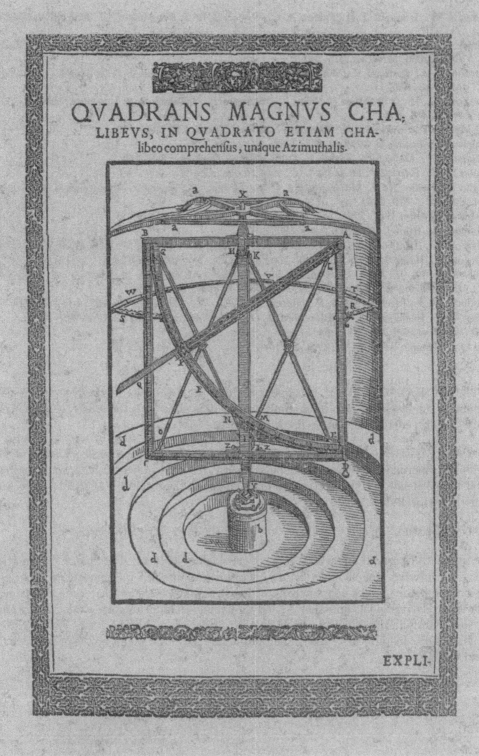

Formation and Evolution of Massive Black Holes II

Formation and evolution of massive black holes II.

P. Madau
Formation and evolution of supermassive black holes, black-hole binary merging (Invited Review)

D. Dultzin-Hacyan
Quasar evolution: black hole mass and accretion rate determination

G. Chartas
X-ray spectral evolution of quasars

J. E. Greene
Black hole growth in the local universe

Church of St Giles with ceiling frescoes and an adjoining monastery (founded 1371, located in Husova Str).

Church of Our Lady Before Týn (inside is the tombstone, dating 1601, of Tycho Brahe, the astronomer of Rudolf II Habsburg).

Black Holes from Stars to Galaxies – Across the Range of Masses
Proceedings IAU Symposium No. 238, 2006   © 2007 International Astronomical Union
V. Karas & G. Matt, eds.   doi:10.1017/S1743921307004723

Formation and early evolution of massive black holes

Piero Madau

Department of Astronomy and Astrophysics, University of California,
Santa Cruz, CA 95064, USA

email: pmadau@ucolick.org

Abstract. The astrophysical processes that led to the formation of the first seed black holes and to their growth into the supermassive variety that powers bright quasars at $z \sim 6$ are poorly understood. In standard ΛCDM hierarchical cosmologies, the earliest massive holes (MBHs) likely formed at redshift $z \gtrsim 15$ at the centers of low-mass ($M \gtrsim 5 \times 10^5 \, M_\odot$) dark matter "minihalos", and produced hard radiation by accretion. FUV/X-ray photons from such "miniquasars" may have permeated the universe more uniformly than EUV radiation, reduced gas clumping, and changed the chemistry of primordial gas. The role of accreting seed black holes in determining the thermal and ionization state of the intergalactic medium depends on the amount of cold and dense gas that forms and gets retained in protogalaxies after the formation of the first stars. The highest resolution N-body simulation to date of Galactic substructure shows that subhalos below the atomic cooling mass were very inefficient at forming stars.

Keywords. Black hole physics – cosmology: theory – galaxies: formation – intergalactic medium – Galaxy: halo

1. Introduction

The strong link observed between the masses of MBHs at the center of most galaxies and the gravitational potential wells that host them (Ferrarese & Merritt 2000; Gebhardt *et al.* 2000) suggests a fundamental mechanism for assembling black holes and forming spheroids in galaxy halos. In popular cold dark matter (CDM) hierarchical cosmologies, small-mass subgalactic systems form first to merge later into larger and larger structures. According to these theories, some time beyond a redshift of 15 or so, the gas within halos with virial temperatures $T_{\rm vir} \gtrsim 10^4$ K – or, equivalently, with masses $M \gtrsim 10^8 \, [(1+z)/10]^{-3/2} \, M_\odot$ – cooled rapidly due to the excitation of hydrogen Lyα and fragmented. Massive stars formed with some initial mass function (IMF) and synthesized heavy elements. Such early stellar systems, aided perhaps by a population of accreting black holes in their nuclei, generated the ultraviolet radiation and mechanical energy that reheated and reionized the cosmos. It is widely believed that collisional excitation of molecular hydrogen may have allowed gas in even smaller systems – virial temperatures of a thousand K, corresponding to masses around $5 \times 10^5 \, [(1+z)/10]^{-3/2} \, M_\odot$ – to cool and form stars at even earlier times (e.g. Tegmark *et al.* 1997).

The first generation of seed MBHs must have formed in subgalactic units far up in the merger hierarchy, well before the bulk of the stars observed today: this is in order to have had sufficient time to build up via gas accretion a mass of several $\times 10^9 \, M_\odot$ by $z = 6.4$, the redshift of the most distant quasars discovered in the *Sloan Digital Sky Survey* (SDSS) (Figure 1; Volonteri & Rees 2005). In hierarchical cosmologies, the ubiquity of MBHs in nearby luminous galaxies can arise even if only a small fraction of halos harbor seed

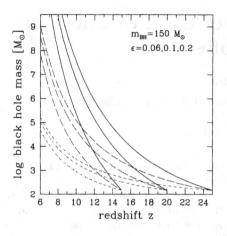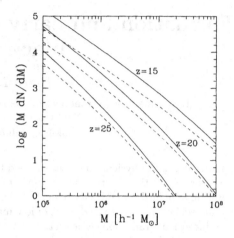

Figure 1. Left: Growth of MBHs from early epochs down to $z = 6$, the redshift of the most distant SDSS quasars. The three sets of curves assume Eddington-limited accretion with radiative efficiency $\epsilon = 0.06$ (solid lines), 0.1 (long-dashed lines), and 0.2 (short-dashed lines). Gas accretion starts at $z = 15, 20, 25$ onto a seed black of mass $m_{BH} = 150$ M$_\odot$. Right: Mass function of progenitor dark matter halos of mass M at $z = 15, 20, 25$ which, by the later time z_0, will have merged into a larger host of mass M_0. Solid curves: $z_0 = 0.8$, $M_0 = 10^{12} h^{-1}$ M$_\odot$ ("Milky Way" halo). Dashed curves: $z_0 = 3.5$, $M_0 = 2 \times 10^{11} h^{-1}$ M$_\odot$ (older "bulge"). If one seed hole formed in each $\sim 10^6$ M$_\odot$ minihalo collapsing at $z \sim 20$ (and triple hole interactions and binary coalescences were neglected), several thousands relic seed black holes and their descendants would be orbiting within the halos of present-day galaxies. (From Madau & Rees 2001.)

holes at very high redshift (Menou, Haiman & Narayanan 2001). The origin and nature of this seed population remain uncertain. Numerical simulations of the fragmentation of primordial molecular clouds in hierarchical cosmologies all show the formation of Jeans unstable clumps with masses exceeding a few hundred solar masses, with the implication that the resulting initial mass function is likely to be biased to very massive "Population III" stars (Abel, Bryan & Norman 2002; Bromm, Coppi & Larson 2002). Barring any fine tuning of the IMF, intermediate-mass black holes – with masses above the 4–18 M$_\odot$ range of known stellar-mass holes – may then be the inevitable end-product of the first episode of pregalactic star formation.

2. Massive black holes as Population III remnants

The first stars in the universe formed out of metal-free gas, in dark matter "minihalos" collapsing from the high-σ peaks of the primordial density field at redshift $z > 15$ or so. Gas condensation in the first baryonic objects was possible through the formation of H$_2$ molecules, which cool via roto-vibrational transitions down to temperatures of a couple hundred Kelvin. At zero metallicity mass loss through radiatively-driven stellar winds or nuclear-powered stellar pulsations is expected to be negligible, and Pop III stars will likely die losing only a small fraction of their mass (except for $100 < m_* < 140$ M$_\odot$). Nonrotating very massive stars in the mass window $140 \lesssim m_* \lesssim 260$ M$_\odot$ will disappear as pair-instability supernovae (Bond, Arnett & Carr 1984), leaving no compact remnants and polluting the universe with the first heavy elements (Oh *et al.* 2001).† Stars with $40 < m_* < 140$ M$_\odot$ and $m_* > 260$ M$_\odot$ are predicted instead to collapse to black holes with masses exceeding half of the initial stellar mass (Heger & Woosley 2002). Since they form in high-σ rare density peaks, relic seed black holes are expected to cluster in the bulges of present-day galaxies as they become incorporated through a series of mergers into larger and larger systems (see Figure 1). The presence of a small cluster of Population III holes in galaxy nuclei may have several interesting consequences associated with tidal captures of ordinary stars (likely followed by disruption), capture by the central supermassive hole, and gravitational wave radiation from such coalescences. Accreting pregalactic seed holes may be detectable as ultra-luminous, off-nuclear X-ray sources (Madau & Rees 2001).

3. The first miniquasars

Physical conditions in the central potential wells of young and gas-rich protogalaxies may have been propitious for black hole gas accretion. The presence of accreting black holes powering Eddington-limited miniquasars at such crucial formative stages in the evolution of the universe may then present a challenge to models of the epoch of first light and of the thermal and ionization early history of the intergalactic medium (IGM), as reheating from the first stars and their remnants likely played a key role in structuring the IGM and in regulating gas cooling and star formation in pregalactic objects. Energetic photons from miniquasars may make the low-density IGM warm and weakly ionized prior to the epoch of reionization breakthrough (Madau *et al.* 2004; Ricotti, Ostriker & Gnedin 2005). X-ray radiation may boost the free-electron fraction and catalyze the formation of H_2 molecules in dense regions, counteracting their destruction by UV Lyman-Werner radiation (Haiman, Abel & Rees 2000; Machacek, Bryan & Abel 2003). Or it may furnish an entropy floor to the entire IGM, preventing gas contraction and therefore impeding rather than enhancing H_2 formation (Oh & Haiman 2003). Photoionization heating may evaporate back into the IGM some of the gas already incorporated into halos with virial temperatures below a few thousand Kelvin (Barkana & Loeb 1999). The detailed consequences of all these effects are poorly understood. In the absence of a UV photodissociating flux and of ionizing X-ray radiation, 3D simulations show that the fraction of cold, dense gas available for star formation or accretion onto seed holes exceeds 20% for halos more massive than 10^6 M$_\odot$ (Figure 2; Machacek *et al.* 2003; Kuhlen & Madau 2005). Since a zero-metallicity progenitor star in the range $40 < m_* < 500$ M$_\odot$ emits about 80,000 photons above 1 ryd per stellar baryon (Schaerer 2002), the ensuing ionization front may overrun the host halo, photoevaporating most of the surrounding gas. Black hole remnants of the first stars that created H II regions are then unlikely to accrete significant mass until new cold material is made available through the hierarchical merging of many gaseous subunits.

High-resolution hydrodynamics simulations of early structure formation in ΛCDM cosmologies are a powerful tool to track in detail the thermal and ionization history of a clumpy IGM and guide studies of early reheating. In Kuhlen & Madau (2005) we used a modified version of ENZO, a grid-based hybrid (hydro+N-body) code developed by Bryan & Norman (see http://cosmos.ucsd.edu/enzo/) to solve the cosmological hydrodynamics equations and simulate the effect of a miniquasar turning on at very high redshift. The

---

† Since primordial metal enrichment was intrinsically a local process, pair-instability SNe may occur in pockets of metal-free gas over a broad range of redshifts. The peak luminosities of typical pair-instability SNe are only slightly greater than those of Type Ia, but they remain bright much longer ($\sim$ 1 year) and have hydrogen lines (Scannapieco *et al.* 2005).

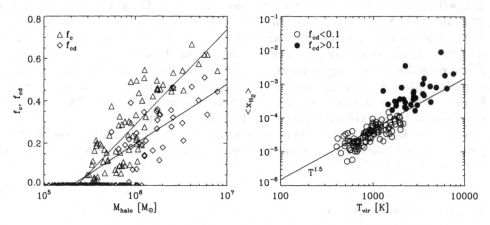

Figure 2. Left: Fraction of cold and cold+dense gas within the virial radius of all halos identified at $z = 17.5$ with $T_{\rm vir} > 400$K, as a function of halo mass. Triangles: f_c, fraction of halo gas with $T < 0.5\,T_{\rm vir}$ and $\delta > 1000$ that cooled via roto-vibrational transitions of H_2. Diamonds: f_{cd}, fraction of gas with $T < 0.5\,T_{\rm vir}$ and $\rho > 10^{19}\,M_\odot\,{\rm Mpc}^{-3}$ that is available for star formation. The straight lines represent mean regression analyses of f_c and f_{cd} with the logarithm of halo mass. Right: Mass-weighted mean H_2 fraction as a function of virial temperature for all halos at $z = 17.5$ with $T_{\rm vir} > 400$ K and $f_{cd} < 0.1$ (empty circles) or $f_{cd} > 0.1$ (filled circles). The straight line marks the scaling of the temperature-dependent asymptotic molecular fraction. (From Kuhlen & Madau 2005.)

simulation follows the non-equilibrium chemistry of the dominant nine species (H, H^+, H^-, e, He, He^+, He^{++}, H_2, and H_2^+) in primordial gas, and includes radiative losses from atomic and molecular line cooling. At $z = 21$, a miniquasar powered by a 150 $M_\odot$ black hole accreting at the Eddington rate is turned on in the protogalactic halo. The miniquasar shines for a Salpeter time (i.e. down to $z = 17.5$) and is a copious source of soft X-ray photons, which permeate the IGM more uniformly than possible with extreme ultraviolet (EUV, $\geqslant 13.6\,{\rm eV}$) radiation. After one Salpeter timescale, the miniquasar has heated up the simulation box to a volume-averaged temperature of 2800 K. The mean electron and H_2 fractions are now 0.03 and 4×10^{-5}: the latter is 20 times larger than the primordial value, and will delay the buildup of a uniform UV photodissociating background. The net effect of the X-rays is to reduce gas clumping in the IGM by as much as a factor of 3. While the suppression of baryonic infall and the photoevaporation of some halo gas lower the gas mass fraction at overdensities δ in the range 20-2000, enhanced molecular cooling increases the amount of dense material at $\delta > 2000$. In many halos within the proximity of our miniquasar the H_2-boosting effect of X-rays is too weak to overcome heating, and the cold and dense gas mass actually decreases. There is little evidence for an entropy floor preventing gas contraction and H_2 formation: instead, molecular cooling can affect the dynamics of baryonic material before it has fallen into the potential well of dark matter halos and virialized. Overall, the radiative feedback from X-rays enhances gas cooling in lower-σ peaks that are far away from the initial site of star formation, thus decreasing the clustering bias of the early pregalactic population, but does not appear to dramatically reverse or promote the collapse of pregalactic clouds as a whole.

Figure 3. Projected dark matter density-squared map of our simulated Milky Way-size halo ("Via Lactea") at the present epoch. The image covers an area of 800 × 600 kpc, and the projection goes through a 600 kpc-deep cuboid containing a total of 110 million particles. The logarithmic color scale covers 20 decades in density-square. (From Diemand, Kuhlen & Madau 2006.)

4. Near-field cosmology and dark satellites

Despite much recent progress in our understanding of the formation of early cosmic structure and the high-redshift universe, many fundamental questions related to the astrochemistry of primordial gas and the formation and evolution of halo+MBH systems remain only partially answered. N-body+hydrodynamical simulations are unable to predict, for example, the efficiency with which the first gravitationally collapsed objects lit up the universe at cosmic dawn, and treat the effects of the energy input from the earliest generations of sources on later ones only in an approximate way. The *Wilkinson Microwave Anisotropy Probe* 3-year data require the universe to be fully reionized by redshift $z_{\rm ion} = 11 \pm 2.5$ (Spergel *et al.* 2006), an indication that significant star-formation activity started at early cosmic times. We infer that Population III massive stars and perhaps miniquasars must have been shining when the universe was less than 350 Myr old, but we remain uncertain about the nature of their host galaxies and the impact they had on their environment and on the formation and structure of more massive systems.

While the above cosmological puzzles can be tackled directly by studying distant objects, it has recently become clear that many of today's "observables" within the Milky Way and nearby galaxies relate to events occurring at very high redshifts, during and soon after the epoch of reionization (see Moore *et al.* 2006 and references therein). In

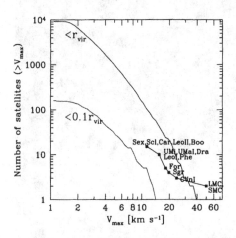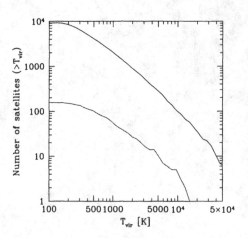

Figure 4. Left: Cumulative peak circular velocity function for all subhalos within Via Lactea's r_{vir} (upper curve) and for the subpopulation within the inner $0.1\,r_{vir}$ (lower curve). Solid line with points: observed number of dwarf galaxy satellites around the Milky Way. Right: Same plotted versus virial temperature T_{vir}.

this sense, galaxies in the Local Group can provide a crucial diagnostic link to the physical processes that govern structure formation and evolution in the early universe, an approach termed "near-field cosmology". It is now well established that the hierarchical mergers that form the halos surrounding galaxies are rather inefficient, leaving substantial amounts of stripped halo cores or "subhalos" orbiting within these systems. Small halos collapse at high redshift when the universe is very dense, so their central densities are correspondingly high. When these merge into larger hosts, their high densities allow them to resist the strong tidal forces that acts to destroy them. Gravitational interactions appear to unbind most of the mass associated with the merged progenitors, but a significant fraction of these small halos survives as distinct substructure.

In order to recognize the basic building blocks of galaxies in the near-field, we have recently completed "Via Lactea", the most detailed simulation of Milky Way CDM substructure to date (Diemand, Kuhlen & Madau 2006). The simulation resolves a Milky Way-size halo with 85 million particles within its virial radius r_{vir}, and was run for $320,000$ cpu-hours on NASA's SGI Altix supercomputer *Columbia*. Figure 3 shows a projected dark matter density squared map of a 800×600 kpc region of this simulation at the current epoch. The high resolution region was sampled with 234 million particles of mass $2.1 \times 10^4\,M_\odot$, evolved with a force resolution of 90 pc, and centered on an isolated halo that had no major merger after $z = 1.7$, making it a suitable host for a Milky Way-like disk galaxy.

Approximately 10,000 surviving satellite halos can be identified at the present epoch: this is more than an order of magnitude larger than found in previous ΛCDM simulations. Their cumulative mass function is well-fit by $N(> M_{sub}) \propto M_{sub}^{-1}$ down to $M_{sub} = 4 \times 10^6\,M_\odot$. Sub-substructure is apparent in all the larger satellites, and a few dark matter lumps are now resolved even in the solar vicinity. In Via Lactea, the number of dark satellites with peak circular velocities above $5\ \mathrm{km\,s^{-1}}$ ($10\ \mathrm{km\,s^{-1}}$) exceeds 800 (120). As shown in Figure 4, such finding appears to exacerbate the so-called "missing satellite problem", the large mismatch between the twenty or so dwarf satellite galaxies observed around the Milky Way and the predicted large number of CDM subhalos

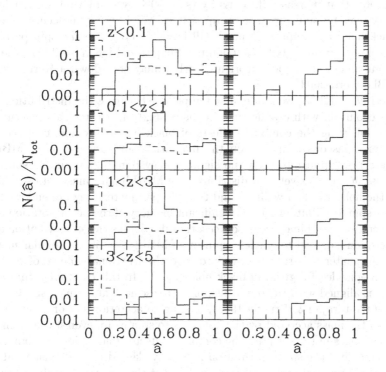

Figure 5. Distribution of MBH spins in different redshift intervals. Left panel: effect of black hole binary coalescences only. Solid histogram: seed holes are born with $\hat{a} \equiv a/m_{BH} = 0.6$. Dashed histogram: seed holes are born non-spinning. Right panel: spin distribution from binary coalescences and gas accretion. Seed holes are born with $\hat{a} = 0.6$, and are efficiently spun up by accretion via a thin disk.

(Moore *et al.* 1999; Klypin *et al.* 1999). Solutions involving feedback mechanisms that make halo substructure very inefficient in forming stars offer a possible way out (e.g. Bullock, Kravtsov & Weinberg 2000; Kravtsov, Gnedin & Klypin 2004; Moore *et al.* 2006). In this case seed black holes may not have grown efficiently in small minihalos just above the cosmological Jeans mass, and gas accretion may have had to await the buildup of more massive galaxies (with virial temperatures above the threshold for atomic cooling).

5. MBH spins

As discussed in the Introduction, growing very large MBHs at high redshift from small seeds requires low accretion efficiencies, or, equivalently, modest black hole spins. The spin of a MBH is also expected to have implications for the direction of jets in active nuclei and to determine the innermost flow pattern of gas accreting onto Kerr holes (Bardeen & Petterson 1975). The coalescence of two spinning black holes in a radio galaxy may cause a sudden reorientation of the jet direction, perhaps leading to the so-called "winged" or "X-type" radio sources (Merritt & Ekers 2002). MBH spins are determined by the competition between a number of physical processes. Black holes forming from the gravitational collapse of very massive stars endowed with rotation will in general be born with non-zero spin (e.g. Fryer, Woosley & Heger 2002). An initially non-rotating hole that increases its mass by (say) 50% by swallowing material from an accretion disk may be spun up to $a/m_{BH} = 0.84$. While the coalescence of two non-spinning black holes of comparable mass will immediately drive the spin parameter of the merged hole to $a/m_{BH} \gtrsim 0.8$ (e.g. Gammie, Shapiro & McKinney 2004), the capture of smaller companions in randomly-oriented orbits may spin down a Kerr hole instead (Hughes & Blandford 2003).

In Volonteri *et al.* (2005) we made an attempt at estimating the distribution of MBH spins and its evolution with cosmic time in the context of hierarchical structure formation theories, following the combined effects of black hole-black hole coalescences and accretion from a gaseous thin disk on the magnitude and orientation of MBH spins. Binary coalescences appear to cause no significant systematic spin-up or spin-down of MBHs: because of the relatively flat distribution of MBH binary mass ratios in hierarchical models, the holes random-walk around the spin parameter they are endowed with at birth, and the spin distribution retains significant memory of the initial rotation of "seed" holes. It is accretion, not binary coalescences, that dominates the spin evolution of MBHs (Fig. 5). Accretion can lead to efficient spin-up of MBHs even if the angular momentum of the inflowing material varies in time, provided the fractional change of mass during each accretion episode of a growing black hole is large. In this case, for a thin accretion disk, the hole is aligned with the outer disk on a timescale that is much shorter than the Salpeter time (Natarajan & Pringle 1998), leading to accretion via prograde equatorial orbits: most of the mass accreted by the hole acts to spin it up, even if the orientation of the spin axis changes in time. Under the combined effects of accretion and binary coalescences, we found that the spin distribution is heavily skewed towards fast-rotating Kerr holes, is already in place at early epochs, and does not change significantly below redshift 5. One way to avoid rapid rotation and produce slowly rotating fast-growing holes is to assume "chaotic feeding" in which small amounts of material, with $\Delta m \ll m_{BH}$, are swallowed by the hole in successive accretion episodes with random orientations (e.g., Moderski and Sikora 1996; Volonteri *et al.* 2005; King & Pringle 2006).

Acknowledgements

I would like to thank all my collaborators for their contributions to the ideas presented here. Support for this work was provided by NASA grant NNG04GK85G.

References

Abel T., Bryan G. & Norman, M. L. 2002, Science, 295, 93
Bardeen J. M. & Petterson J. A. 1975, ApJ, 195, L65
Barkana R. & Loeb A. 1999, ApJ, 523, 54
Bond J. R., Arnett W. D. & Carr B. J. 1984, ApJ, 280, 825
Bromm V., Coppi P. S. & Larson R. B. 2002, ApJ, 564, 23
Bullock J. S., Kravtsov A. V. & Weinberg D. H. 2000, ApJ, 539, 517
Diemand J., Kuhlen M. & Madau P. 2006, ApJ, in press (astro-ph/0611370)
Ferrarese L. & Merritt D. 2000, ApJ, 539, L9
Fryer C. L., Woosley S. E. & Heger A. 2001, ApJ, 550, 372
Gammie C. F., Shapiro S. L. & McKinney J. C. 2004, ApJ, 602, 312
Gebhardt K. *et al.* 2000, ApJ, 543, L5
Haiman Z., Abel T. & Rees M. J. 2000, ApJ, 534, 11
Heger A. & Woosley S. E. 2002, ApJ, 567, 532
Hughes S. A. & Blandford R. D. 2003, ApJ, 585, L101
King A. R. & Pringle J. E. 2006, MNRAS, 373, L90

Klypin A. A., Kravtsov A. V., Valenzuela O. & Prada F. 1999, ApJ, 522, 82
Kravtsov A. V., Gnedin O. Y. & Klypin A. A. 2004, ApJ, 609, 482
Kuhlen M. & Madau P. 2005, MNRAS, 363, 1069
Menou K., Haiman Z. & Narayanan V. K. 2001, ApJ, 558, 535
Machacek M. M., Bryan G. L. & Abel T. 2003, MNRAS, 338, 273
Madau P. & Rees M. J. 2001, ApJ, 551, L27
Madau P., Rees M. J., Volonteri M., Haardt F. & Oh S. P. 2004, ApJ, 606, 484
Merritt D. & Ekers R. D. 2002, Science, 297, 1310
Moderski R. & Sikora M. 1996, A&AS, 120, 591
Moore B., Diemand J., Madau P., Zemp M. & Stadel J. 2006, MNRAS, 368, 563
Moore B., Quinn T., Governato F., Stadel J. & Lake G. 1999, MNRAS, 310, 1147
Natarajan P. & Pringle J. E. 1998, ApJ, 506, L97
Oh S. P. & Haiman Z. 2003, MNRAS, 346, 456
Oh S. P., Nollett K. M., Madau P. & Wasserburg G. J. 2001, ApJ, 562, L1
Ricotti M., Ostriker J. P. & Gnedin N. Y. 2005, MNRAS, 357, 207
Scannapieco E., Madau P., Woosley S., Heger A. & Ferrara A. 2005, ApJ, 633, 1031
Schaerer D. 2002, A&A, 382, 28
Spergel D. N., *et al.* 2006, ApJ, submitted (astro-ph/0603449)
Tegmark M., Silk J., Rees M. J., Blanchard A., Abel T. & Palla F. 1997, ApJ, 474, 1
Volonteri M., Madau P., Quataert E. & Rees M. J. 2005, ApJ, 620, 69
Volonteri M. & Rees M. J. 2005, ApJ, 633, 624

MITCHELL BEGELMAN: Comment: If your are able to surround a small seed black hole with a sufficiently optically thick envelope that is much more massive than the seed black hole then the relevant Eddington limit is actually the Eddington limit for that envelope, not for the mass of the black hole. You could in principle accrete about 100 time the Eddington limit or even more, up to about 10^4–10^5 solar masses.

PIERO MADAU: Thanks for pointing this out. Such a supercritical accretion phase would help in relaxing some of the constraints associated with the early growth of billion solar mass holes by redshift 6.

MARTIN REES: Just a less radical comment. The "classic" efficiency arguments based on the binding energy of the innermost stable orbit are of course not strictly valid when the accretion is at the critical rate – the disc is then thick and the thickness is of the order of the Schwarzschild radius. As has been known since the 1980s, even with the Schwarzschild case the efficiency can be either lower or higher than the standard thin-disc value.

PIERO MADAU: Thanks Martin. I thought about thick radiation-supported tori – I was working on those configurations long time ago with my PhD supervisor, Marek Abramowicz. We have run a few models and they do not seem to help much with the "spin problem": as the radiation efficiency drops, the holes accrete material with angular momentum and spin up rapidly.

GÜNTHER HASINGER: The Supernova that produces the black hole in the first place could wipe out its birth halo because of the high explosion velocity. Could this be the reason for subhalos below the atomic cooling mass to be dark?

PIERO MADAU: Population III very massive stars above 260 solar masses are predicted to collapse to black holes without ejecting material. In general, feedback from supernova explosion is a plausible mechanism for suppressing further star formation in young small galaxies. Of course, and contrary to reionization from the "outside", it must be at work in every minihalo at high redshifts.

FUKUN LIU: One comment on the impact of recoil velocity on the formation of massive black holes in dwarf galaxies. In your simulations, you have assumed that black hole binary coalesces instantly when two galaxies merge. However, we know that dynamical friction becomes inefficient before gravitational wave radiation starts to dominate the loss of binary orbital angular momentum, so we do not know how black hole binaries become harder. Up to now the observational evidence for black hole binary merges, e.g. double radio galaxies proposed by us (Liu *et al.* 2003, MNRAS) or X-shaped radio sources suggested by Merritt & Ekers (2002, Science), are relevant for massive galaxies. Therefore, it is quite possible that two black holes in dwarf galaxies do not coalesce but grow separately until the next galaxy merger occurs and the host galaxy becomes massive.

PIERO MADAU: I agree. In this work we just simply assumed that two black holes coalesce on short timescales in order to investigate the effect of MBH coalescences on the spin distribution.

Black Holes from Stars to Galaxies – Across the Range of Masses
Proceedings IAU Symposium No. 238, 2006
V. Karas & G. Matt, eds.

© 2007 International Astronomical Union
doi:10.1017/S1743921307004735

Quasar evolution: black hole mass and accretion rate determination

Deborah Dultzin-Hacyan,[1] Paola Marziani,[2] C. Alenka Negrete[1] and Jack W. Sulentic[3]

[1] Instituto de Astronomía, UNAM, Apdo. Postal 70-264, 04510 Mexico D.F., Mexico

[2] INAF, Osservatorio Astronomico di Padova, Vicolo dell'Osservatorio 5, I-35122 Padova, Italy

[3] Department of Physics and Astronomy, University of Alabama, Tuscaloosa, AL 35487, USA

email: deborah@astroscu.unam.mx, paola.marziani@oapd.inaf.it, giacomo@merlot.astr.ua.edu

Abstract. Accurate measurements of emission line properties are crucial to understand the physics of the broad line region in quasars. This region consists of warm gas that is closest to the quasar central engine and has not been spatially resolved for almost all sources. We present here an analysis of optical and IR data for a large sample of quasars, covering the HI Hβ spectral region in the redshift range $0 \lesssim z \lesssim 2.5$. Spectra were interpreted within the framework of the the so-called "eigenvector 1" parameter space, which can be viewed as a tentative H-R diagram for quasars. We stress the lack of spectral evolution in the low ionization lines of quasars, with prominent FeII emission also at $z \gtrsim 2$. We also show how selection effects influence the ability to find quasars radiating at low Eddington ratio in flux-limited surveys. The quasar similarity at different redshift is probably due to the absence of super-Eddington radiators (at least within the caveats of black hole mass and Eddington ratio determination discussed in this paper) as well as to the limited Eddington ratio range within which quasars seem to radiate.

Keywords. Quasars: general – quasars: emission lines

1. Introduction

The current inability to spatially resolve the central regions of quasars is being circumvented by several alternative approaches. Results from reverberation mapping of NGC 5548 (Peterson *et al.* 2002) with HST and ground monitoring of broad line response to UV continuum variations suggest that high ionization lines (HILs) respond within a few days, while low ionization lines (LILs) respond within a month. By now more than 30 Seyfert nuclei have been studied in this way (Peterson *et al.* 2004). An important result is that the response time (proportional to the size of the emitting region) is correlated to luminosity (Kaspi *et al.* 2005): $r_{\rm BLR} \propto L^\alpha$. This relationship has been exploited to compute the central black hole mass $M_{\rm BH}$ in the absence of reverberation mapping data, under the assumption that the bulk line-emitting gas dynamics is virial. In this case one can apply the relationship $M_{\rm BH} = f \cdot G \cdot {\rm FWHM}^2/r_{\rm BLR}$, where G is the gravitational constant, FWHM the full-width at half maximum of a suitable emission line, and f a factor that depends on the broad line emitting region (BLR) structure and viewing angle. There are several caveats with this approach (reviewed by (Marziani *et al.* 2006; see also Sulentic *et al.* 2006). First, the $r_{\rm BLR} - L$ relationship is not free from uncertainties. Reverberation mapping-based determinations of $r_{\rm BLR}$ are affected by the non-negligible radial extent of the optically thick BLR. The derived $r_{\rm BLR}$ is not a very well defined quantity. The luminosity index α is also somewhat uncertain because of the intrinsic scatter in the correlation and lack of sampling at high luminosity. However, the main issue is that a good understanding of the structure and dynamics of BLR is required to justify the virial assumption. This knowledge is still lacking, and Principal Component Analysis (PCA) of quasars samples raise sources of concerns.

2. The eigenvector 1 of AGN

The low-ionization line properties of quasars can be organized along an "eigenvector 1" sequence, mainly governed by Eddington ratio, but also affected (to a second-order degree) by the angle at which the BLR are observed, and M_{BH} (still more speculative sources of scatter are iron abundances and black hole spin). On an optical plane defined by the intensity ratio R_{Fe} between the FeII blend centered at 4570 Å and the broad component of Hβ (Hβ_{BC}), and by the FWHM of Hβ_{BC} itself, sources are distributed along an elbow sequence: almost all objects show $R_{Fe} \lesssim 0.5$ if FWHM(Hβ_{BC}) $\gtrsim 4000$ km s^{-1}. We find most radio-loud quasars toward the distribution extreme with lowest R_{Fe} and broadest Hβ, where we also observe a rather high ionization spectrum, with prominent [OIII]$\lambda\lambda$4959,5007 lines. The median Hβ_{BC} profiles are deformed Gaussians with red-ward asymmetries. Toward the highest R_{Fe}, narrowest Hβ_{BC} end of the diagram, sources show a rather low-ionization spectrum, and less prominent [OIII]$\lambda\lambda$4959,5007 emission. The Hβ_{BC} profiles are most often of Lorentzian shape, sometimes with a blue-ward asymmetry. The prototypical HIL CIVλ1549 shows large blue-shifts with respect to the quasar reference frame. These observational constraints suggest that in the latter sources (which include Narrow Line Seyfert 1s but also objects with FWHM(Hβ_{BC}) $\lesssim$ 4000 kms^{-1}, collectively referred to as "Population A"). the LIL and HIL emitting regions obey to different kinematics: while the mostly symmetric profiles of Hβ_{BC} are naïvely consistent with virial motion, most CIVλ1549 profiles are not.

3. High-z quasars M_{BH} and Eddington ratio determination

The HIL CIVλ1549 has been the line of choice for black hole mass estimation in high-z quasars (see the careful analysis by Vestergaard & Peterson(2006)), although it is in principle unsuitable for at least two reasons: (1) blue-ward asymmetric profiles, or even centroid blue-shifts (up to several thousands km s^{-1}) with respect to the best estimates of the rest frame of the source and (2) FWHM CIVλ1549 does not correlate strongly with FWHM(Hβ_{BC}) which is the line of choice for low-redshift M_{BH} estimation. We made several attempts to deduce a correction for the CIVλ1549-derived M_{BH} from properties intrinsic to the CIVλ1549 profile shape (i.e., width, asymmetry and kurtosis), but we were unable to find an effective relationship. Blue-shifted CIVλ1549 profiles are thought to arise in a high ionization wind resulting in a radial flow velocity that is not negligible relative to any rotational component. Our fears about CIVλ1549–derived M_{BH} motivated us to pursue Hβ observations to the highest possible redshift.

The Balmer lines do not usually show very large line shifts although profiles can be very asymmetric. Both red and blue asymmetries are observed for Hβ_{BC} which is the most studied line because it is relatively un-blended and is observable with optical spectrometers up to $z \approx 1$. The most ambiguous sources from the point of view of M_{BH} are those with FWHM(Hβ_{BC}) $\gtrsim 4000$ km s^{-1} (Population B), and redshifted profiles and/or red asymmetries. In this case, not all parts of the Hβ profile respond to continuum changes in the same way implying that some of the line emitting gas may be optically thin or, less likely, not exposed to the variable HI ionizing radiation. A strong BLR response to continuum fluctuations, coupled with a weak or absent response of the broader optically thin component, can lead to an overestimate of FWHM for the virialized BLR component resulting in an overestimate of M_{BH}. For sources with FWHM(Hβ_{BC}) $\lesssim 4000$ km s^{-1},

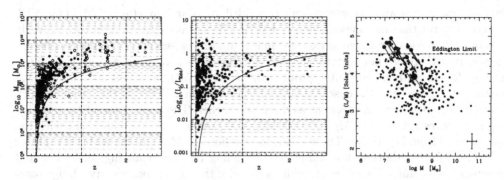

Figure 1. Left: Behavior of black hole mass as a function of z. Filled circles represent $M_{\rm BH}$ computed from corrected FWHM($H\beta_{\rm BC}$) values, while open circles from FWHM(FeIIλ 4570. Center: Eddington ratio as a function of z for calculations based on corrected FWHM($H\beta_{\rm BC}$) values. Filled lines indicate minimum $M_{\rm BH}$ (for Eddington ratio equal 1) and minimum Eddington ratio (for a "maximum" $M_{\rm BH} \approx 4 \cdot 10^9$ $M_\odot$) for a flux limited survey with $m_{\rm B} \approx 16.5$. Right: Super Eddington radiators may be due mainly to orientation effects. The larger circles indicate the so-called blue "outliers" which are sources believed to be viewed almost along the accretion disk axis.

the $H\beta_{\rm BC}$ profile is usually well fit with a symmetric function and the BLR emission is thought to arise from a Keplerian disk. The most unreliable Pop. A sources, in a disk emission scenario, should be those observed near face-on where the rotational (i.e. virial) contribution to FWHM($H\beta_{\rm BC}$) is minimal. At least some of the face-on sources may be identified as the so-called "blue outliers" which show a weak and significantly blue-shifted [OIII]$\lambda\lambda$4959,5007 lines (Collin *et al.* 2006).

A physical approach to $M_{\rm BH}$ estimation uses the FWHM of the variable part of the $H\beta_{\rm BC}$ profile Peterson *et al.* (2004). LILs like MgII and FeII may yield more reliable results than $H\beta_{\rm BC}$. HI Balmer line emission can be substantial from gas in a variety of physical conditions. A correction can be derived only at low-z, from objects for which a proper observation of the variable part (rms) of the $H\beta_{\rm BC}$ is possible. The optically thick BLR gas is responding to continuum changes shows a velocity dispersion correlated with – but slightly lower than that of the integrated profile at all FWHM. The reverberating part of $H\beta_{\rm BC}$ and FeII provides width estimates which are consistent, as expected since they may be emitted in a similar sub-region within the BLR. Similar considerations apply to MgIIλ2800. Given measurement difficulties, and doubts about the virial assumption for most other lines, corrected measures for $H\beta_{\rm BC}$, FeII, and MgIIλ2800 may offer the best hope for reliable $M_{\rm BH}$ and Eddington ratio estimates out to $z \approx 2.5$.

4. Quasar evolution: $M_{\rm BH}$ and Eddington ratio

The E1 sequence has been interpreted as an evolutionary sequence. Young AGNs still possess a compact NLR; they may have plenty of fuel and radiate close to the Eddington limit and even show spectral evidence of circum-nuclear Starbursts. More massive, evolved systems may have less fuel and have had all the time to ionize interstellar clouds in the bulge of their host galaxies as well as to produce radio lobes exceeding the host galaxy size. These conclusions, however preliminary they may be, are inferred from the diversity of properties of low-z quasar samples. To extend our observations to higher red-shift, we observed 50 quasars in the range $1 \lesssim z \lesssim 2.5$ with the IR spectrometer ISAAC on the ESO VLT (30 published in Sulentic *et al.* 2004, 2006). The passable S/N (≈ 30) and resolution allowed us to apply the same data analysis technique employed for about

300 spectra from optical spectrometers (provided by Marziani *et al.* 2003). Our VLT spectra confirm that any dependence of FWHM(Hβ_{BC}) on source luminosity is weak. Low redshift trends for asymmetries and line shifts are preserved in the intermediate z sample. It is therefore not certain that FWHM(Hβ_{BC}) is a valid estimator of the virial velocity for all sources even after proper [O$_{III}$]λ, λ4959,5007 and Fe$_{II}$ subtraction. We attempted to reconduct the observed Hβ_{BC} FWHM to the FWHM of the optically thick reverberating Hβ_{BC} sub-component as discussed earlier. The application of a correction largely removes the need of $M_{BH} \gtrsim 10^{10}$ M$_\odot$, and produces a trend consistent with the one obtained from Fe$_{II}$ width. There seems to be a limit to M_{BH} growth with z for masses determined from virialization assumption for Balmer and Fe$_{II}$ emitting gas at low and intermediate redshifts. This is reasonable because, if the dispersion velocity - bulge mass - M_{BH} relation holds for intermediate and high z, $M_{BH} \gtrsim 5 \cdot 10^9$ M$_\odot$ would imply bulge velocity dispersions in excess of 700 km s^{-1} and bulge masses $\sim 10^{13}$ M$_\odot$ which are not observed at low z (Bernardi *et al.* 2005).

Perhaps surprisingly, apparent super-Eddington radiators seem to be confined at low z. If our interpretation is correct, these sources are viewed with their accretion disk almost pole-on; their M_{BH} may be severely underestimated. If a tentative orientation correction is applied, even these sources could radiate close but below the Eddington Limit (Fig. 1, right panel). It is interesting to note that the curve shown in Fig. 1 (center panel) seem to envelop nicely the observations, suggesting that, in a flux limited sample (here a limiting magnitude $m_B \approx 16.5$ is assumed), we are starting to increasingly miss radiators above 0.01 the Eddington limit at $z \approx 0.3$, and that at $z \approx 1$ we may miss a significant population of low Eddington ratio radiators well sampled at low z (mainly Population B sources which, in general, show smaller C$_{IV}\lambda$1549 blue-shifts).

5. Conclusions

Unfortunately, observing Hβ in the IR and even obtaining a reliable recessional velocity for the rest frame remain a challenge. However, it is important to increase the sample of observed objects to confirm the findings about the lack of very large masses $M_{BH} \gtrsim 5 \cdot 10^9$ M$_\odot$ and super-Eddington radiators, as well as to recover information on the evolution of Eddington ratio with z.

References

Bernardi M., Sheth R. K., Nichol R. C., Schneider D. P. & Brinkmann J. 2005, AJ, 129, 61
Collin S., Kawaguchi T., Peterson B. M. & Vestergaard M. 2006, A&A, 456, 75
Kaspi S., Maoz D., Netzer H., Peterson B. M., Vestergaard M. & Jannuzi B. T. 2005, ApJ, 629, 61
Marziani P., Dultzin-Hacyan D. & Sulentic J. W. 2006, astro-ph/0606678
Marziani P., Zamanov R. K., Sulentic J. W. & Calvani M. 2003, MNRAS, 345, 1133
Peterson B. M. *et al.* 2002, ApJ, 581, 197
Peterson B. M. *et al.* 2004, ApJ, 613, 682
Sulentic J. W., Repetto P., Stirpe G. M., Marziani P., Dultzin-Hacyan D. & Calvani, M. 2006, A&A, 456, 929
Sulentic J. W., Stirpe G. M., Marziani P., Zamanov R., Calvani M. & Braito V. 2004, A&A, 423, 121
Vestergaard M., Peterson B. M. 2006, ApJ, 641, 689

Black Holes from Stars to Galaxies – Across the Range of Masses
Proceedings IAU Symposium No. 238, 2006
V. Karas & G. Matt, eds.

© 2007 International Astronomical Union
doi:10.1017/S1743921307004747

Local active black hole mass functions

Jenny E. Greene[1] and Luis C. Ho[2]

[1] Department of Astronomy, Princeton University, Princeton, NJ 08544, USA
[2] The Observatories of the Carnegie Institution of Washington, Pasadena, CA 91101, USA
email: jgreene@astro.princeton.edu, lho@ociw.edu

Abstract. While black holes (BHs) are apparently a ubiquitous component of the nuclei of local spheroids, their role in galaxy evolution remains largely unknown. The tight correlations between galaxy spheroid properties and BH mass provide an important boundary condition for models of the coevolution of BHs and galaxies. Here we consider another important boundary condition: the local mass function of broad-line active galaxies. We use standard virial mass estimation techniques to examine the distribution of BH masses and accretion rates for active galaxies in the local universe, and we also compare the distribution of BH masses in local broad and narrow-line objects, and find that both populations have a characteristic mass of $\sim 10^7$ $M_\odot$. Most importantly, this is the first BH mass function to consider BH with masses $< 10^6$ $M_\odot$. The space density of this important population allows us to place constraints on potential mechanisms for the creation of seed BHs in the early Universe.

Keywords. galaxies: active – galaxies: Seyfert – galaxies: evolution

1. Introduction

The distribution of black hole (BH) mass density in the local Universe provides an important boundary condition on the cosmological growth of BHs. Because of tight scaling relations between BH mass ($M_{\rm BH}$) and spheroid properties, including luminosity (e.g., Marconi & Hunt 2003) and stellar velocity dispersion (the $M_{\rm BH} - \sigma_*$ relation; Tremaine et al. 2002) it is possible to convert observed luminosity or σ_* functions into BH mass density, and thus constrain the primary growth modes of supermassive BHs (e.g., Yu & Tremaine 2002; Marconi et al. 2004). There is also useful information to be gleaned by looking at the masses and Eddington ratios of currently growing (i.e., active) BHs. Heckman et al. (2004) showed that while the characteristic BH mass for all BHs locally is $\sim 10^8$ $M_\odot$, narrow-line active galactic nuclei (AGNs) from the SDSS have significantly lower masses, $\sim 3 \times 10^7$ $M_\odot$, where $M_{\rm BH}$ is inferred from σ_*, assuming the $M_{\rm BH} - \sigma_*$ relation holds. Here, we present a complementary study based on broad-line AGNs, for which it is possible to derive BH masses using scaling relations between AGN luminosity and broad-line region size (e.g., Kaspi et al. 2005; Greene & Ho 2005). Amongst other things, this sample allows us to look at $M_{\rm BH}$ without direct dependence on the $M_{\rm BH} - \sigma_*$ relation, directly compare the mass distribution of broad and narrow-line AGNs, and, in particular, probe the mass regime $M_{\rm BH} \lesssim 10^6$ $M_\odot$. Although such low-mass BHs contribute a negligible fraction of both the total BH mass density and the integrated AGN luminosity density, their space density provide important constraints on the mass distribution of primordial seed BHs (e.g., Haehnelt 2004) and source counts for gravitational wave experiments such as *LISA* (e.g., Hughes 2002). Because of spectral resolution limitations, current mass distributions inferred from σ_* do not probe below 3×10^6 $M_\odot$, while we do not know how to convert galaxy luminosity into BH mass for low-mass galaxies.

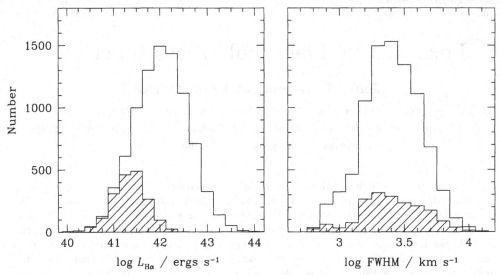

Figure 1. Distributions of broad $L_{\mathrm{H}\alpha}$ (*left*) and broad $\mathrm{FWHM}_{\mathrm{H}\alpha}$ (*right*) for the entire sample of broad-line AGNs. The shaded histograms indicate the distribution of the sample targeted by the SDSS main galaxy sample. Note the dominance of objects with $\mathrm{FWHM}_{\mathrm{H}\alpha}$ below the canonical division at $\mathrm{FWHM}_{\mathrm{H}\alpha} = 2000$ km s^{-1}, and that while the galaxy selected AGNs are uniformly at low luminosities, they span the entire range of M_{BH} in the sample. Massive BHs are predominantly in low accretion states at the present time.

2. Results

We follow the procedure outlined in Greene & Ho (2004) using the Fourth Data Release of SDSS (Adelman-McCarthy *et al.* 2006). Briefly, galaxy continuum is modeled and removed using the principal component analysis method of Hao *et al.* (2005), narrow Hα+[N$_{\mathrm{II}}$] $\lambda\lambda$6548, 6583 are modeled and removed using a model of the [SII] $\lambda\lambda$6716, 6731 lines, and finally, when necessary, broad Hα is modeled as the sum of Gaussian components. The resulting sample of ~ 8300 AGNs have the distribution of $\mathrm{FWHM}_{\mathrm{H}\alpha}$ and $L_{\mathrm{H}\alpha}$ shown in Figure 1. Using the scaling relations presented in Greene & Ho (2005), we find a raw distribution of M_{BH} and $L_{\mathrm{bol}}/L_{\mathrm{Edd}}$ shown in Figure 2.

We use the classic V/V_{max} method to compute the broad Hα luminosity function. The maximum volume to which we would detect a source depends both on its total luminosity (it must be spectroscopically targeted by the SDSS) and its Hα luminosity and signal-to-noise ratio (S/N; our algorithms must detect the object as a broad-line AGN). [Note that we include objects in our sample that were targeted both as galaxies (limiting magnitude $r < 17.77$) and quasars ($i < 19.1$), which slightly complicates the analysis.] To estimate the limit of our ability to recognize broad Hα we create model spectra at a range of redshifts and perform our entire selection procedure on each object. The final luminosity function is shown as filled circles in Figure 3, while the solid line shows our maximal luminosity function, including objects with AGN-like line ratios, but weak Hα emission for which simulations show we may not reliably measure a BH mass. We are clearly incomplete below a few $\times 10^{40}$ ergs s^{-1}; the Palomar survey, which by virtue of its smaller aperture and much brighter magnitude limit is sensitive to much lower Hα luminosities, continues to rise toward lower luminosity than probed by this study (Ho 2004). Nevertheless, we convert the observed luminosity function into a BH mass function using the V_{max} weights derived above. We find, in agreement with Heckman *et al.* (2004), that the characteristic mass for this sample is $\sim 10^7\ M_\odot$, which is significantly lower than

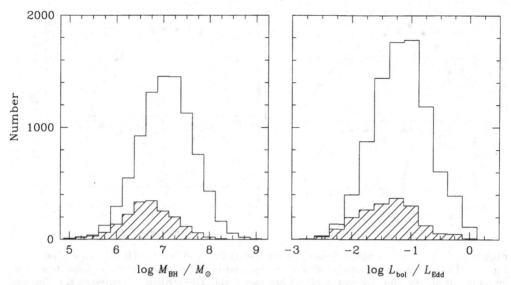

Figure 2. Distributions of BH mass (*left*) and L_{bol}/L_{Edd} (*right*) for the entire sample of broad–line AGNs. M_{BH} is calculated from the $L_{H\alpha}$ and $FWHM_{H\alpha}$ shown above, using the formalism of Greene & Ho (2005). The Eddington ratio is derived assuming an average bolometric correction of $L_{bol} = 9L_{5100}$Å. As in the previous figure, the galaxies targeted in the SDSS main sample are indicated with the shaded histogram.

that of inactive BHs. However, because we are sensitive to lower masses than Heckman *et al.* we find a flat mass function between $10^6 - 10^7$ $M_\odot$, before the mass function turns over. Our goal is to determine the extent to which the mass function truly turns over below 10^6 $M_\odot$, and the extent to which we are simply incomplete.

To investigate the impact of our incompleteness, we run Monte Carlo simulations of BHs with masses from $10^4 - 10^{9.5}$ $M_\odot$, Eddington ratios from 0.01–1, and redshifts up to $z = 0.352$. Each BH is assigned a galaxy luminosity using either the M_{BH}-L_{bulge} relation of Marconi & Hunt (2003) at high BH mass, or the $M_{BH} - \sigma_*$ relation combined with the σ_*-v_c relation of Pizzella *et al.* (2005) and the Tully-Fisher relation of Masters *et al.* (2006) at low BH mass; observed scatter is included for all relations. Artificial spectra are created for each BH, using a range of galaxy types, AGN spectral shapes and narrow-line strengths. Realistic S/N ratios are assigned based on empirical relations derived from the SDSS spectra. We then attempt to detect these artificial galaxies, and derive completeness fractions as a function of mass, Eddington ratio and redshift. At low redshift ($z \lesssim 0.02$) and high Eddington ratio ($0.3 < L_{bol}/L_{Edd} < 1$) we expect to be $\sim 75\%$ complete for BHs with masses of 10^5 $M_\odot$. Therefore, we tentatively conclude that we have detected a real decrease in the space density of BHs with masses $< 10^6$ $M_\odot$, unless BHs in this mass range are preferentially inactive compared to their more massive cousins.

3. Summary

We have constructed a BH mass function for local broad-line AGNs, sensitive, for the first time, to BHs with masses $< 10^6$ $M_\odot$. We find that active BHs have generally lower masses than inactive BHs at the present day, in agreement with results for narrow-line objects. Furthermore, we find tentative evidence for a decline in space density or activity below 10^6 $M_\odot$.

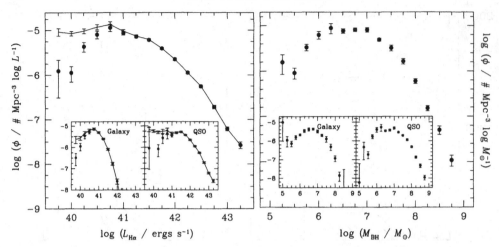

Figure 3. *Left*: Volume-weighted broad Hα luminosity function in bins of 0.25 dex (# Mpc^{-3} log $L_{H\alpha}^{-1}$). The maximum volume is calculated based on both the photometric and spectroscopic limits of the survey and our search algorithm (see text). The error bars represent the Poisson errors in each bin. The inset panels show the luminosity functions for objects targeted as galaxies or quasars, respectively, primarily based on a color selection. *Right*: Volume-weighted BH mass function in bins of 0.25 dex (# Mpc^{-3} log M_{BH}^{-1}). The weights used are identical to those for the luminosity function.

Acknowledgements

J. E. G. acknowledges an international travel grant from the AAS.

References

Adelman-McCarthy J. K. *et al.* 2006, ApJS, 162, 38

Greene J. E., Ho L. C. 2005, ApJ, 630, 122

——. 2004, ApJ, 610, 722

Haehnelt M. G. 2004, in Carnegie Observatories Astrophysics Series, Vol. 1: Coevolution of Black Holes and Galaxies, ed. L. C. Ho (Cambridge: Cambridge Univ. Press), 405

Hao L. *et al.* 2005, AJ, 129, 1783

Heckman T. M., Kauffmann G., Brinchmann J., Charlot S., Tremonti C. & White, S. D. M. 2004, ApJ, 613, 109

Ho L. C. 2004, in Carnegie Observatories Astrophysics Series, Vol. 1: Coevolution of Black Holes and Galaxies, ed. L. C. Ho (Cambridge: Cambridge Univ. Press), 292

Hughes S. A. 2002, MNRAS, 331, 805

Kaspi S., Maoz D., Netzer H., Peterson B. M., Vestergaard M. & Jannuzi B. T. 2005, ApJ, 629, 61

Marconi A., Hunt L. K. 2003, ApJ, 589, L21

Marconi A., Risaliti G., Gilli R., Hunt L. K., Maiolino R. & Salvati M. 2004, MNRAS, 351, 169

Masters K. L., Springob, C. M., Haynes M. P. & Giovanelli R. 2006, ApJ, in press (astro-ph/0609249)

Pizzella A., Corsini E. M., Dalla Bontà E., Sarzi M., Coccato L. & Bertola F. 2005, ApJ, 631, 785

Tremaine S. *et al.* 2002, ApJ, 574, 740

Yu Q., Tremaine S. 2002, MNRAS, 335, 965

Active Galactic Nuclei I

Active galactic nuclei I.

M. C. Begelman
Supermassive black holes: accretion and outflows
(Invited Review)

H. Kunieda
Hard X-ray spectra of AGN observed with Suzaku

K. R. W. Tristram
Mapping the circumnuclear dust in nearby AGN with the mid-infrared interferometric instrument MIDI

R. W. Goosmann
X-ray variability in AGN: implications of magnetic flares

Detail of the complicated exterior structure of the Church of Holy Saviour and the entrance into premises of Clementinum.

Kepler's House in Karlova Str. He lived in four places during his Prague period and wrote several books, including *Astronomia Nova*.

Black Holes from Stars to Galaxies – Across the Range of Masses
Proceedings IAU Symposium No. 238, 2006
V. Karas & G. Matt, eds.
© 2007 International Astronomical Union
doi:10.1017/S1743921307004760

Mapping the circumnuclear dust in nearby AGN with MIDI

Konrad R. W. Tristram,[1] Klaus Meisenheimer,[1] Walter Jaffe[2] and William D. Cotton[3]

[1] Max-Planck-Institut für Astronomie, Königstuhl 17, 69117 Heidelberg, Germany

[2] Sterrewacht Leiden, Postbus 9513, NL-2300 RA Leiden, The Netherlands

[3] NRAO, 520 Edgemont Road, Charlottesville, VA 22903-2475, USA

email: tristram@mpia.de

Abstract. We observed four nearby AGN with MIDI at the VLTI to investigate the mid-infrared emission from these sources. With our measurements we resolve the dusty structure around the nucleus of the Circinus galaxy. We find two dust components: a hot, small elongated structure with a size of 0.4 pc and a cooler, almost round component with a size of 1.9 pc. We interpret the emission to be originating in a geometrically thick dusty torus oriented perpendicular to the ionisation cone. Hence our finding nicely confirms the unified picture. We also observed the nucleus of Centaurus A and find that 70% of the mid infrared flux originates from an unresolved source with a size of less than 0.2 pc. In this case, the majority of the emission comes from a synchrotron source at the base of the radio jet. Two further galaxies, Mrk 1239 (Seyfert 1) and MCG -05-23-016 (Seyfert 2), also show unresolved mid infrared sources limiting the size of the dust distribution to less than 5 and 2 pc respectively.

Keywords. Galaxies: active – galaxies: nuclei – galaxies: Seyfert – radiation mechanisms: thermal – techniques: interferometric

1. Introduction

In the "unified scheme", the standard model for Active Galactic Nuclei (AGN), a hot accretion disk around a supermassive black hole is assumed to be surrounded by a geometrically thick distribution of gas and dust, the dusty torus. Originally, all arguments for the existence of such tori were based on theoretical considerations and on indirect observational evidence such as the spectral energy distributions (SED) of AGN. The torus is thought to give a simple explanation for the observed dichotomy between Seyfert 1 and Seyfert 2 galaxies: type 1 objects are seen face on allowing us to directly see the central engine, while type 2 objects are oriented edge on so that the dust in the torus blocks the line of sight towards the very centre.

There is hardly any direct evidence for this scenario due to the lack of angular resolution: even for the nearest AGN the expected structures are too small (on the order of a parsec and hence smaller than 100 mas) to be resolved by single dish telescopes in the mid-infrared (MIR). Only interferometry provides the necessary angular resolution. Indeed first observations of NGC 1068 with the MID-infrared Instrument (MIDI) at the Very Large Telescope Interferometer (VLTI) of the European Southern Observatory (ESO) on Cerro Paranal were successful to resolve the circumnuclear MIR emission of this prototype Seyfert 2 galaxy, confirming the picture of the unified scheme (Jaffe *et al.* 2004). Since this first success, we have continued our high-resolution study of nearby AGN with MIDI, combining the light from two of the 8 m unit telescopes (UTs) of the VLTI. All observations were performed using guaranteed time observations (GTO).

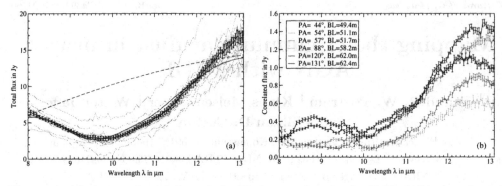

Figure 1. (a) Averaged total flux of the Circinus nucleus in the N band (blue) and total flux of the model (red) with silicate absorption (continuous) and without silicate absorption (dashed). (b) Correlated fluxes as observed during the night of 2005-02-28. Mainly the baseline angle changed, while the projected baseline length stayed roughly the same. From the dramatic change in the correlated fluxes, direct implications on the morphology can be derived: the source is considerably more extended in the direction of $PA \sim 50°$ than in the direction of $PA \sim 130°$.

2. Circinus

The Circinus galaxy is a highly inclined ($\sim 65°$) SA(s)b galaxy harbouring a Seyfert type 2 active nucleus as well as a circumnuclear starburst. At about 4 Mpc distance ($1'' \sim 20$ pc, Freeman *et al.* 1977), the galaxy is among the nearest AGN and hence it is an ideal object to study the nuclear region of an active galaxy.

The interferometric data obtained on Circinus constitute the most extensive infrared interferometric data of any extragalactic source collected so far. We were able to reconstruct a total of 21 visibility points from the data. Figure 1a shows the calibrated total flux F_{tot} of Circinus (thick blue line). The broad absorption feature dominating most of the spectrum is due to silicate absorption in Circinus. We see no evidence for any line emission or emission of Polycyclic Aromatic Hydrocarbons (PAH), as observed at greater distances from the nucleus (e.g. by Roche *et al.* 2006).

For all baselines observed, the correlated flux F_{cor} is much smaller than the total flux ($F_{cor} \ll F_{tot}$), indicating that the emission region is well resolved with our interferometric resolution of $\lambda/2B \leqslant 40$ mas. As an example, the correlated fluxes obtained on February 28th, 2005, are shown in figure 1b. During this night, the projected baseline length only increased by a minor factor from 50 to 60 m, while the position angle underwent a major change from 44 to 131°. An increase of the correlated flux can be clearly observed while the position angle increases. We directly interpret this as a change of size of the emitting source: at position angles of 44 to 57° the correlated flux is low, i.e. the emission is extended, while at angles of 120 and 131° the correlated flux is higher, i.e. the emission is less resolved. This is a direct and completely model independent evidence for an elongated dust structure perpendicular to the ionisation cone and outflow in Circinus (position angle $-50°$ and opening angle $\sim 90°$).

Even though the coverage is high for infrared interferometry, it is insufficient to attempt any image reconstruction. Therefore we used a simple model consisting of two black body Gaussian emitters with silicate absorption to explain the data. We find the silicate absorption profile for interstellar dust from Kemper *et al.* (2004) to be sufficient to describe the absorption feature we observe with MIDI, both for the total flux as well as for the correlated fluxes. The two component black body model has 10 free parameters, which are listed in figure 2 with their best fit values. The figure also contains a sketch of the best fit model.

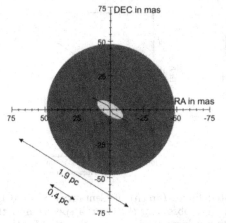

parameter	best fit value
$FWHM_1$	22.3
a_1/b_1	0.39
T_1	387.9
f_1	1.0
$FWHM_2$	98.4
a_2/b_2	0.96
T_2	293.6
f_2	0.5
τ_{SiO}	1.2
α	58.8
χ^2 (431 d.o.f.)	44.3

Figure 2. Sketch and best fit parameters of the model for Circinus. A highly elongated hot emission region (yellow) is surrounded by an extended, almost round and cooler emission region (brown). At the centre, the locus of H_2O maser emitters in a disk from Greenhill *et al.* (2003) is over-plotted. The parameters fitted are: *FWHM* – full width at half maximum (in mas), *a/b* – axis ratio, *T* – temperature (in K) and *f* – filling factor for each of the two components; τ_{SiO} – optical depth of the silicate absorption and α – position angle for both components.

We find a small hot component with a FWHM of 22 mas (~ 0.4 pc at the distance of Circinus), which is highly ellipsoidal and has a temperature of 390 K. This component has the same size and position angle as the rotating maser disc found by Greenhill *et al.* (2003) (see figure 2). The second component is significantly larger, reflecting the fact that a large part of the flux ($\sim 90\%$) is resolved with our interferometric setup. The component has a FWHM of 98 mas, which corresponds to 1.9 pc. The component only has a very small ellipticity and a temperature of less than 300 K. It is a grey body with a filling factor of 0.5. We consider the 390 K for the inner component as a lower limit for the highest dust temperature and the 300 K as an upper limit for the cool component. Clearly there is a temperature gradient from the inside to the outside, indicating that the dust is heated by the nuclear source.

3. Centaurus A

Centaurus A (NGC 5128) is the closest active galaxy at a distance of only 3.8 Mpc (1 pc $\sim$ 50 mas, Rejkuba 2004). It is an elliptical galaxy undergoing late stages of a merger event with a spiral galaxy. Radio interferometry reveals a structure composed of an unresolved core and a jet within the central parsec.

The results of our MIDI observations are summarised in figure 3a, which shows the total flux and the correlated fluxes of the two baseline sets observed. Similar to Circinus, most of the spectral region is affected by the broad silicate absorption feature. Its depth is identical in the correlated and in the total flux, indicating that both the core and the extended component suffer the same extinction. The [NeII] emission line at 12.9 μm is clearly detected at the edge of the total spectra, but not in any of the correlated fluxes. The interferometric observations reveal the existence of two components in the inner parsec of Centaurus A: a resolved component, the "disc", and an unresolved "core".

The disc which is most extended along a position angle of $\sim 110°$ is roughly oriented perpendicular to the parsec scale radio jet. The decrease in visibility towards longer wavelengths indicates that the extended emission has a spectrum which rises between 9 and 13 μm as expected for emission from thermal dust at temperatures $T < 300$ K. The

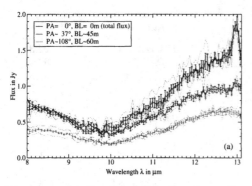 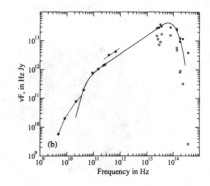

Figure 3. (a) Total flux (blue) and correlated fluxes (green) of Centaurus A. Two baseline sets (consisting of two visibility points each) were observed. (b) Overall spectrum of the core of Centaurus A. Open circles show observed flux values, filled dots are corrected for foreground extinction. The synchrotron spectrum (solid line) shows an optically thin power-law $F\nu \sim \nu^{-0.36}$.

size of the disk is poorly confined by the present observations but needs to be $> 30\,$mas ($\sim 0.57\,$pc) at $13\,\mu$m, in order to be consistent with our simple two-component model.

In order to understand the nature of the unresolved core emission, we supplemented our mid-infrared photometry by measurements at lower and higher frequencies, using VLBA, millimeter and near infrared data (see figure 3b). From the fact that both the mm-data and the core flux in the MIDI range can be nicely fitted by an optically thin power-law spectrum, we conclude that the core emission is dominated by non-thermal synchrotron radiation. This means that, in Centaurus A, the MIR emission is dominated by an unresolved synchrotron core making up between 60% (at $13\,\mu$m) and 80% (at $8\,\mu$m) of the nuclear MIR flux. The size of this core most likely is $R_c \sim 0.01\,$pc, as derived from VLBI observations at $43\,$GHz.

4. Other objects

The sensitivity of MIDI is better than initially predicted, so that two fainter AGN, MCG -05-23-016 and Mrk 1239, could be successfully observed. Only one UV point for each object was obtained so far. MCG -05-23-016 is a Seyfert 2 galaxy at a distance of approximately $40\,$Mpc. It also shows a silicate absorption feature. Mrk 1239, is the first Seyfert 1 galaxy observed with MIDI. It is located at a distance of about $90\,$Mpc and has a flat spectrum in the MIR. The correlated fluxes are consistent with marginally or unresolved objects ($F_{cor} \lesssim F_{tot}$). The flux level of Mrk 1239 is $400\,$mJy and that of MCG -05-23-016 only $300\,$mJy.

Considering this success of MIDI, we are confident to observe more AGN soon and to continue our investigations of the already observed targets.

References

Freeman, K. C. *et al.* 1977, A&A, 55, 445
Greenhill, L. J. *et al.* 2003, ApJ, 590, 162
Jaffe, W. *et al.* 2004, Nature, 429, 47
Kemper, F., Vriend, W. J. & Tielens, A. G. G. M. 2004, ApJ, 609, 826
Rejkuba, M. 2004, A&A, 413, 903
Roche, Patrick, F. *et al.* 2006, MNRAS, 367, 1689

THAISA STORCHI-BERGMANN: How do you infer that the dust structure you detect in Circinus is rotating?

KONRAD TRISTRAM: The H_2O maser disc observed by Greenhill et at. (2003) has the same orientation and size as the distribution in our model. The maser disc shows a clear Keplerian rotation velocity profile. We conclude hence that the H_2O masers are embedded in the dust structure and hence the dust is rotating as well.

FELIX MIRABEL: How does compare your extended dust emission at parsec scales with the very inner image obtained at 2μm with NICMOS on the HST?

KONRAD TRISTRAM: The high resolution HST images are dominated by the galactical dust-lane on kiloparsec scales in Centaurus A, which is between us and the nucleus. I am not aware of any detection of an extended dust emission in HST images at the nucleus of Centaurus A which would correspond to the emission we detect.

SEMICIRCVLVS MAGNVS
AZIMVTHALIS.

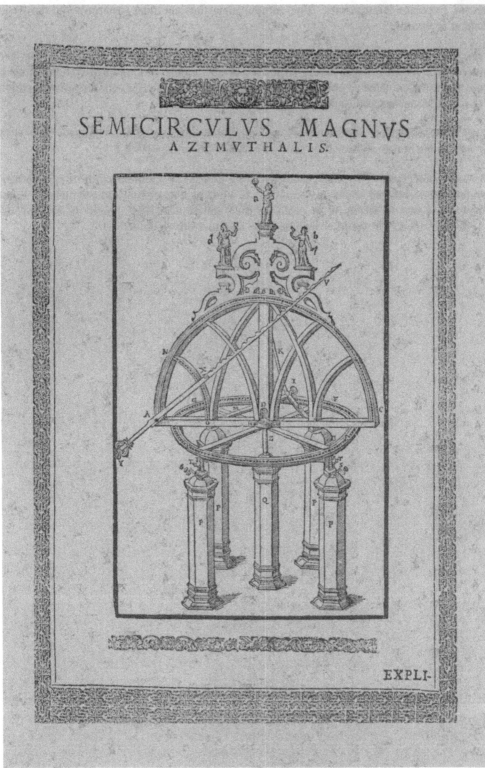

EXPLI-

Black Holes from Stars to Galaxies – Across the Range of Masses
Proceedings IAU Symposium No. 238, 2006
V. Karas & G. Matt, eds.
© 2007 International Astronomical Union
doi:10.1017/S1743921307004772

Constraints on a strong X-ray flare in the Seyfert galaxy MCG–6-30-15

R. W. Goosmann,[1] B. Czerny,[2] V. Karas,[1] M. Dovčiak,[1]
G. Ponti[3] and M. Mouchet[4]

[1] Astronomical Institute, Academy of Sciences, Boční II 1401, 14131 Prague, Czech Republic

[2] Copernicus Astronomical Center, Bartycka 18, 00-716 Warsaw, Poland

[3] Dipartimento di Astronomia, Università di Bologna, Via Ranzani 1, 40127, Bologna, Italy

[4] Laboratoire ApC, Université Denis Diderot, 2 place Jussieu, 75251 Paris Cedex 05, France

email: goosmann@astro.cas.cz

Abstract. We discuss implications of a strong flare event observed in the Seyfert galaxy MCG–6-30-15 assuming that the emission is due to localized magnetic reconnection. We conduct detailed radiative transfer modeling of the reprocessed radiation for a primary source that is elevated above the disk. The model includes relativistic effects and Keplerian motion around the black hole. We show that for such a model setup the observed time-modulation must be intrinsic to the primary source. Using a simple analytical model we then investigate time delays between hard and soft X-rays during the flare. The model considers an intrinsic delay between primary and reprocessed radiation, which measures the geometrical distance of the flare source to the reprocessing sites. The observed time delays are well reproduced if one assumes that the reprocessing happens in magnetically confined, cold clouds.

Keywords. Galaxies: active – galaxies: Seyfert – X-rays: individual (MCG–6-30-15)

1. Introduction

The Seyfert galaxy MCG–6-30-15 was observed for 95 ksec with *XMM-Newton* in the year 2000 (Wilms *et al.* 2001). The X-ray lightcurve of this observation reveals a strong flare event of which Ponti *et al.* (2004) conducted a detailed analysis. In this note we discuss a possibility that the flare is produced by localized magnetic reconnection and we constrain some details of such a flare setup.

We imagine that the strong flare in MCG–6-30-15 originates in a compact reconnection site elevated to a height H above the surface of the accretion disk. The radiation from this primary source partly shines toward the disk and creates a hot spot. The distance of the spot's center to the disk center is denoted by r. The flare is supposed to be in Keplerian co-rotation with the disk and we assume that the primary illumination sets on and fades out instantaneously. The irradiation of the disk then evolves across the hot spot, starting from the spot center and progressing toward the border. Therefore we expect the lightcurve of the reprocessed radiation to be curved even if the time evolution of the primary is box-shaped.

We want to model the exact shape of the lightcurve expected from such a flare setup at different orbital phases of the disk. We first conduct detailed local radiative transfer computations. The varying intensity of the irradiation across the hot spot is taken into account as well as the angular dependence of the reprocessed emission. The vertical structure of the disk is assumed to remain in the same hydrostatic equilibrium as before the onset of the flare. The profile is computed with an extended version of the code

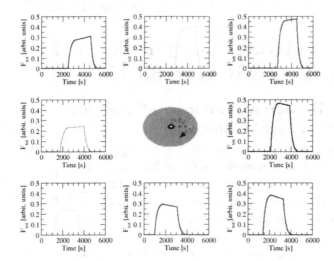

Figure 1. Lightcurves of the reprocessed spectrum integrated between 2 keV and 10 keV for a flare lasting 2000 sec occurring at different orbital phases. The position of the distant observer is toward the bottom.

described in Różańska *et al.* (2002). The local spectra across the spot are then computed by the codes TITAN and NOAR (Dumont *et al.* 2000, 2003). We assume a black hole mass $M = 10^7 M_\odot$, a disk accretion rate $\dot{m} = 0.02$ (in units of the Eddington accretion rate), and $r = 18\,R_g$ (with $R_g \equiv GM/c^2$).

Based on the local spectra, the relativistic ray-tracing code KY (Dovčiak *et al.* 2004) computes the time evolution of the spectrum seen by a distant observer. We set the disk inclination to $i = 30°$ and assume a Kerr parameter $a/M = 0.998$ having known characteristics of MCG–6-30-15 in mind. In KY we define the time-dependent disk emission according to the subsequent illumination and fade-out of the spot from its center toward the border. The duration of this time sequence is normalized by H, which we constrain from the observed soft-to-hard time delay of ~ 600 s (Ponti *et al.* 2004). We assume that this delay is entirely due to the light traveling time between the source and the disk leading to $H = 7\,R_g$. The flare lasts for 2000 sec, which at $r = 18\,R_g$ corresponds to 1/12 of the Keplerian time scale. More details about this type of flare model are given in Goosmann (2006).

In figure 1 we show the model lightcurves obtained for flares at 8 different orbital positions. They show significant differences in shape and in normalization when comparing different orbital phases. All curves have a broadened maximum which does not match the observed, peaked shape (see Fig. 4 in Ponti *et al.* 2004). Thus, even if the flare is reflection-dominated, neither the geometrically evolving irradiation across the spot nor relativistic and Doppler modifications can account for the observed shape of the lightcurve. An intrinsic time evolution of the primary source is therefore required.

2. Modeling spectral time delays for a flare

Ponti *et al.* (2004) conducted a timing analysis for the observed flare and computed cross-correlation functions and time delays between six different energy bands ΔE_i (see their Fig. 6). We attempt to reproduce the observed time delays using the following, simple mathematical parameterization for the primary radiation I_p and the reprocessed component I_r:

$$I_p(E,t) = \mathcal{L}_p(t)E^{-\alpha_p}, \quad I_r(E,t) = N\mathcal{L}_r(t)E^{-\alpha_r} \tag{2.1}$$

The spectral shapes are thus represented by power laws with indexes α_p and α_r. The time modulations $\mathcal{L}_p(t)$ and $\mathcal{L}_r(t)$ are defined by

$$\mathcal{L}_p(t) = \frac{T_f^2}{(t - t_0)^2 + T_f^2}, \quad \mathcal{L}_r(t) = \frac{b^2 T_f^2}{[t - (t_0 - \delta)]^2 + b^2 T_f^2}. \tag{2.2}$$

These terms account for the intrinsic change of the primary. The overall factor N normalizes I_r against the primary and δ denotes an intrinsic delay between the two components. The time evolution of the reprocessing can be broadened by choosing $b > 1$. This represents a spread of the light travel-time between the source and the reprocessing site. For further details on this model see Goosmann *et al.* (2007).

A distant observer detects the sum I_{obs} of I_p and I_r,

$$I_{obs}(E, t) = I_p(E, t) + I_r(E, t). \tag{2.3}$$

To investigate the time delay between the spectral response of the energy bands ΔE_i and ΔE_j we compute cross-correlation functions

$$F_{CCF}^{ij}(\tau) = \frac{\int\limits_{-\tau_{max}}^{+\tau_{max}} L_i(t) L_j(t - \tau) dt}{\sqrt{\int\limits_{-\tau_{max}}^{+\tau_{max}} L_i^2(t') \, dt'} \times \sqrt{\int\limits_{-\tau_{max}}^{+\tau_{max}} L_j^2(t'') \, dt''}}, \tag{2.4}$$

where the lightcurve L_i belongs to the energy bin ΔE_i. We adopt here the very same method as was used in the analysis of Ponti *et al.* (2004). The values for $t_0 = 500$ sec and $T_f = 2200$ sec are adjusted to the flare lightcurve plotted in Fig. 4 of Ponti *et al.* (2004).

The spectral slope of the reprocessed component has a major impact on the obtained energy-dependent time delays. We choose a rather hard slope of I_r and set $\alpha_r = 0.1$. Such a spectral shape corresponds to the reprocessing in a medium of magnetically confined, cold clouds as analyzed by Kuncic, Celotti & Rees (1997). Their Fig. 2 shows reprocessed model spectra that are much harder than the primary due to multiple soft X-ray absorption by the individual clouds. For the primary slope we set $\alpha_p = 1.3$ having in mind that for the higher flux state during the flare the spectrum should steepen (Shih, Iwasawa & Fabian 2002).

The model can reproduce the observed time lags for the observed flare of MCG–6-30-15. In Fig. 2 we show two examples of satisfactory data representation. The results account for the slight flattening of the delay curve toward higher energies. It turns out that different choices of parameters can lead to similar delay curves, as there are three free parameters involved (N, b, and δ) and the error bars of the data are large. To distinguish between different parameter sets some fine tuning of the delay curve's shape would be necessary. This is not meaningful with the current data.

The observed time-delay cannot be reproduced when the spectral slope of I_r is softer, i.e. adjusted to an ionized reflection scenario as proposed by Ballantyne, Vaughan & Fabian (2003). If the primary source is located within $\sim 10\,R_g$ of the black hole, the relative variability of the primary and the reflection component can be reproduced (Miniutti & Fabian 2004). Such a setup also explains the high equivalent width of the iron Kα line, which requires a relatively large amount of reflection. While these ionized reflection and light-bending models are certainly important to explain the behavior of MCG–6-30-15 on longer time-scales, the data obtained during the 2000 sec flare period does not allow to conclude on the time evolution of the iron Kα line.

From the results of our delay modeling, we therefore suggest the possibility that the strong individual flare does not originate in the inner accretion flow, where most of the

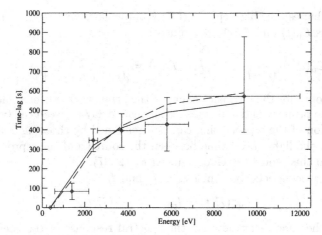

Figure 2. Best representation of the measured time-lags (red diamonds with error bars) for the flare/clumps scenario. The two curves rely on the following parameterization: $\delta = 1000$ sec, $b = 4$ and $N = 0.9$ (solid black line), and $\delta = 900$ sec, $b = 3$ and $N = 0.6$ (dashed blue line).

X-ray energy is dissipated. We rather imagine that the flare might occur farther away from the black hole, in a region where the disk is fragmented and replaced by magnetically confined, cold clumps.

Acknowledgements

We are grateful to A.-M. Dumont and A. Różańska for their help with computing the local spectra. VK and RG thank the ESA PECS Programme (ref. 98040) for support.

References

Ballantyne D. R., Vaughan S. & Fabian A. C. 2003, MNRAS, 342, 239
Dovčiak, M., Karas, V. & Yaqoob, T. 2004, ApJS, 153, 205
Dumont A.-M., Abrassart A. & Collin S. 2000, A&A, 357, 823
Dumont A.-M., Collin S., Paletou F., Coupé S., Godet O. & Pelat D. 2003, A&A, 407, 13
Goosmann R. W., Czerny B., Karas V., Ponti G., A&A in press (astro-ph 0702685)
Goosmann R. W. 2006, Accretion and Emission Close to Supermassive Black Holes in Quasars
        and AGN: Modeling the UV–X Spectrum, PhD thesis (Universität Hamburg, Germany)
Kuncic Z., Celotti A. & Rees M. J. 1997, MNRAS, 284, 717
Miniutti G., Fabian A. C. 2004, MNRAS, 349, 1435
Ponti G., Cappi M., Dadina M. & Malaguti G. 2004, A&A, 417, 451
Różańska A., Dumont A.-M., Czerny B. & Collin S. 2002, MNRAS, 332, 799
Shih D. C., Iwasawa K. & Fabian A. C. 2002, MNRAS, 333, 687
Wilms J., Reynolds C. S., Begelman M. C., Reeves J., Molendi S. *et al.* 2001, MNRAS, 328, L27

ELISABETE DE GOUVEIA DAL PINO: Is there any radio counterpart of these X-ray flares just as it is observed in the microquasars (or even in the solar corona)? If so, how does it correlate with the X-ray emission?

RENÉ GOOSMANN: I am not aware of any simultaneous radio and X-ray observations of MGG–6-30-15. It is a radio-quiet object. In 1984, Ulvestad & Wilson measured the radio flux at the VLA to only 1–2 mJy. I wonder if it is possible to find a radio/X-ray correlation at such a low radio flux.

Active Galactic Nuclei II

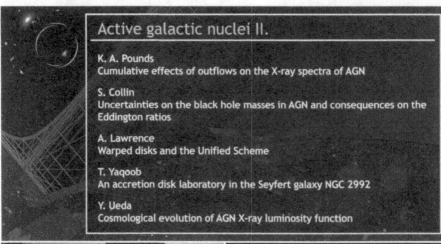

Active galactic nuclei II.

K. A. Pounds
Cumulative effects of outflows on the X-ray spectra of AGN

S. Collin
Uncertainties on the black hole masses in AGN and consequences on the Eddington ratios

A. Lawrence
Warped disks and the Unified Scheme

T. Yaqoob
An accretion disk laboratory in the Seyfert galaxy NGC 2992

Y. Ueda
Cosmological evolution of AGN X-ray luminosity function

Na Kampě – an admirable part of Lesser Quarter that spreads from the slopes below the Prague Castle down to Charles Bridge.

Black Holes from Stars to Galaxies – Across the Range of Masses
Proceedings IAU Symposium No. 238, 2006
V. Karas & G. Matt, eds.
© 2007 International Astronomical Union
doi:10.1017/S1743921307004796

The soft X-ray spectrum of PG1211+143

K. A. Pounds,[1] K. L. Page[1] and J. N. Reeves[2]

[1] Department of Physics and Astronomy, University of Leicester, UK

[2] NASA Goddard Space Flight Center, Code 663, Greenbelt, MD 20771, USA

Abstract. The narrow line QSO PG1211+143 has been a focus of recent attempts to understand the soft excess in AGN, while the 2001 *XMM-Newton* observation of this luminous AGN also provided evidence for a massive and energetic outflow. Here we consider a physical link between the energetic outflow and the variable soft excess.

Keywords. Galaxies: active – galaxies: nuclei

1. Introduction

One of the most important new results emerging from *XMM-Newton* and *Chandra* observations of AGN is the evidence for sub-relativistic outflows from a number of luminous quasars. For two of these, APM08279+5255 (Chartas *et al.* 2002) and PG1115+080 (Chartas *et al.* 2003), both BAL quasars, the immediate inference was to associate the X-ray columns with the high velocity gas seen in optical and UV spectra of these unusual objects. In such cases the high columns might result from viewing the continuum source through a wind being blown tangentially off the accretion disc (Proga *et al.* 2000).

However, in 2 further cases, PG1211+143 (Pounds *et al.* 2003) and PDS456 (Reeves *et al.* 2003), a moderate and a high luminosity QSO, respectively, evidence of a high velocity – and hence energetic – outflow suggested this could be a common property of luminous, or perhaps more specifically, of high accretion rate quasars (King and Pounds 2003). Unless highly collimated, such outflows will form a substantial component of the mass and energy budgets in luminous AGN, while having important implications for metal enrichment of the intergalactic medium and for the feedback mechanism implied by the correlation of black hole and galactic bulge masses (King 2005).

The *XMM-Newton* observation of PG1211+143 in 2001 has also been used recently to explore the physical origin of the 'soft excess' in the X-ray spectra of many AGN, following doubts raised about Comptonisation models (Gierlinski and Done 2004). Alternative descriptions have been proposed, invoking strong reflection from a highly ionised accretion disc (Crummy *et al.* 2006) and absorption in high velocity ionised gas (Schurch and Done 2006, Chevallier *et al.* 2006). Here we also examine EPIC data from a second observation of PG1211+143 in 2004, where a conventional plot shows a weaker but 'hotter' soft excess (Figure 1L).

A direct comparison of the EPIC data from 2001 and 2004 (Figure 1R) suggests the spectral difference is primarily due to a variable 'deficit' of flux at ~ 0.7–2 keV, supporting the contention that the spectral curvature near ~ 1 keV is indeed an artefact of absorption on a steeper underlying continuum. However, if the flux deficit is due to photoionised absorption, it is interesting that the hard X-ray flux – which should be a good proxy for the ionising luminosity – is essentially unchanged, suggesting that a reduced covering factor (or column density) is the cause of the weaker 'soft excess' in 2004. Here we suggest an alternative possibility (explored more fully in Pounds and Reeves 2006), that a variable continuum component partly 'fills in' the absorption flux deficit in the 2004

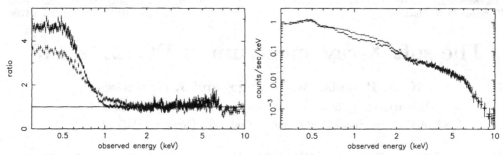

Figure 1. Left panel: conventional plot of the 2001 (black) and 2004 (grey) EPIC pn spectra showing a strong soft excess sitting above the hard power law. On this view it appears that the soft excess in 2004 is weaker but 'hotter'. Right panel: direct comparison of the background–subtracted EPIC data showing the spectral difference to be due to an increase in flux at ~ 0.7–2 keV and a (less obvious, but significant) decrease at ~ 0.4–0.7 keV in 2004 (grey).

spectrum. While contributing directly to the strong soft X-ray flux in PG1211+143, the additional soft continuum component also explains the lack of strong 'narrow' absorption lines in the soft X-ray band (Kaspi and Behar 2006), in conflict with the relatively strong, blue-shifted absorption lines in the EPIC data (Pounds and Page 2006).

We speculate that the additional, steep power law continuum is powered by the energetic high velocity outflow.

2. Confirming an energetic outflow in PG1211+143

PG1211+143 (at $z = 0.0809$) is a prototypical 'Big Blue Bump' quasar, and one of the brightest AGN in the soft X-ray band (Elvis *et al.* 1991). It is also classified (Kaspi *et al.* 2000) as a Narrow Line Seyfert 1 galaxy (FWHM Hβ 1800 km s^{-1}), with a small central black hole mass ($\sim 4 \times 10^7 M_\odot$) for an object of bolometric luminosity 4×10^{45} erg s^{-1}, implying that the accretion rate is close to the Eddington limit.

The initial ~ 50 ks observation of PG1211+143 with XMM-Newton in 2001 revealed several blue-shifted absorption lines, interpreted by Pounds *et al.* (2003) as evidence for a highly ionised outflow with a velocity v$\sim 0.09c$, a conclusion primarily based on identifying the strongest ~ 7 keV feature with FeXXVI Lyα. As Kaspi and Behar (2006) were unable to confirm this result in a careful ion-by-ion modelling of the RGS data we have re-examined the higher signal-to-noise EPIC data. That re-analysis resolved additional absorption lines in the intermediate (~ 1–4 keV) energy band (Pounds and Page 2006), strengthening the evidence for a high velocity outflow. In total, 7 statistically significant absorption lines are identified, with the strong ~ 7 keV line re-assigned to Heα of FeXXV, yielding an *increased* outflow velocity v$\sim 0.14 \pm 0.01c$.

Assuming a spherically symmetric flow, the mass loss rate is of then of order $\dot{M} \sim 35bM_\odot$ yr^{-1}, where $b \leqslant 1$ allows for the collimation of the flow. Modelling of the broadband spectrum of PG1211+143 (Pounds and Reeves 2006) has quantified both the absorbed and re-emitted fluxes for the ionised outflow, yielding a covering factor CF~ 0.1 for the high velocity, highly ionised outflow, and $\dot{M} \sim 3.5M_\odot$ yr^{-1}. This outflow rate compares with $\dot{M}_{\mathrm{Edd}} = 1.6M_\odot$ yr^{-1} for a non-rotating SMBH of mass $\sim 4 \times 10^7 M_\odot$ (Kaspi *et al.* 2000) accreting at an efficiency of 10%. The mechanical energy of the highly ionised outflow is then $\sim 10^{45}$ erg s^{-1}, easily sufficient to power a substantial component of the X-ray emission of PG1211+143.

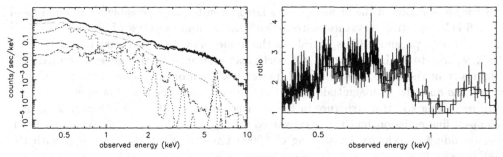

Figure 2. Left panel: deconvolution of the EPIC MOS spectrum of PG1211+143 with the primary (dark grey) and secondary (light grey) continua plus line emission from the highly and moderately ionised outflow (Pounds and Reeves 2006). Right panel: structure in the soft excess in the 2001 RGS data. XSTAR model fitting shows this structure is dominated by velocity-broadened resonance emission lines of N,O,Ne and Mg, together with a group of Fe-L lines at ~ 0.7 keV.

3. A second continuum component

For most AGN the energetically dominant emission component can be modelled by a power law of photon index $\Gamma \sim 1.9$ (Nandra and Pounds 1994), visible in Type 1 objects over the ~ 2–100 keV band. This 'primary' continuum component is widely believed to arise by Comptonisation of UV and optical accretion disc photons in a high temperature electron corona (Haardt and Maraschi 1991). However, recent studies of spectral variability in MCG–6-30-15 (Vaughan and Fabian 2004) and 1H0419-57 (Pounds *et al.* 2004) have provided clear evidence of a variable power law component of slope significantly steeper than the 'canonical value' of $\Gamma \sim 1.9$.

We now suggest this softer continuum component is physically distinct from the 'primary' continuum in PG1211+143, having established that the mechanical energy in the fast, highly ionised outflow is ample to power that additional continuum component, perhaps via internal shocks (analogous to the process suggested in Gamma Ray Bursts), or shock heating of slower moving clouds providing the bulk of the continuum opacity. The most likely X-ray emission mechanism would again be Comptonisation of optical-UV disc photons, with the steep power law resulting from a lower equilibrium temperature in the dispersed 'corona' compared with that responsible for the primary power law continuum.

A separate continuum component is attractive in the present context since, if absorption is to create the impression of a strong 'soft excess', then the underlying continuum must be steeper than observed above ~ 2 keV.

4. A new spectral model

The evidence of absorption in highly ionised (narrow absorption lines) and moderately ionised gas (low energy spectral curvature) suggests a new spectral model including 2 power law continua, both modified by absorption, and re-emission from a 2-component photoionised gas. Applying this model to the 2001 data from PG1211+143 produced a statistically excellent fit (Pounds and Reeves 2006).

Figure 2L illustrates the model components, showing that above ~ 2 keV the spectrum is dominated by the 'primary' power law, with a photon index $\Gamma \sim 2.2$, at the upper end of – but consistent with – the 'accepted' range for type 1 AGN spectra. At lower energies the secondary power law component ($\Gamma \sim 3.1$) dominates, while below ~ 1 keV re-emission, particularly from the moderately ionised absorber, becomes significant.

Strong absorption of the primary power law is seen to be responsible for the mid-band ($\sim$ 2–6 keV) spectral curvature, with ionised gas column densities of $N_H \sim 3$–5×10^{22} cm^{-2}. In both pn and MOS spectral fits the model shows a broad trough near ~ 0.75–0.8 keV, due to Fe-L absorption. Of particular relevance in the present context, the luminosity of the steep, less absorbed, power law component was $\sim 6 \times 10^{43}$ erg s^{-1}, more than an order of magnitude lower than the estimated fast outflow energy.

In modelling spectral structure in the higher resolution RGS data (Figure 2R), photoionised line emission had to be convolved with a gaussian ($\sigma \sim 25$ eV at 0.6 keV), corresponding to velocity broadening of 29000 km s^{-1} (FWHM), consistent with the high velocity flow.

Applying the 2001 model to the 2004 data, a good fit was obtained with element abundances and ionisation parameters fixed, the spectral change being largely modelled by an increase in the steep, weakly absorbed continuum component. In contrast, a change in the covering factor (modelled by a lower column) of ionised absorber on the primary power law was not able – alone – to fit both data sets.

5. Broader application of the new spectral model

The Black Hole Winds model (King and Pounds 2003) provides a simple physical basis whereby massive, high velocity outflows can be expected in AGN accreting at or above the Eddington limit. In addition to PG1211+143, this might apply to both PDS456 (Reeves *et al.* 2003) and to the bright Seyfert 1 IC4329A, recently shown (Markowitz *et al.* 2006) to exhibit a strongly blue-shifted Fe K absorption line indicating a highly ionised outflow at v$\sim$ 0.1c. While relatively rare in the local universe such X-ray spectra could be common for luminous, higher redshift AGN.

If the outflow velocity in PG1211+143 equates to the escape velocity at the launch radius R_{launch} (from an optically thick photosphere or radiatively extended inner disc), v$\sim$ 0.14c corresponds to $R_{\text{launch}} \sim 50R_s$ (where $R_s = 2GM/c^2$ is the Schwarzschild radius), or 3×10^{14} cm for PG1211+143. The EPIC data show significant flux variability in the harder (2–10 keV) band on timescales of 2–3 hours (fig 1 in Pounds *et al.* 2003)), compatible with the above scale size relating to the primary (disc/corona) X-ray emission region.

Observations of PG1211+143 over several days are needed to test the new spectral model. The variability timescale of the secondary power law will constrain the scale of the region where we predict the fast outflow undergoes internal shocks. Detecting hard X-ray emission from PG1211+143 above $\sim$ 20 keV would also support the need for a continuum component with photon index less steep than that in the single power law/ absorption models, while deeper RGS spectra are needed to resolve the broad emission line profiles indicated in the model fits.

In summary, it seems clear that further studies of high accretion rate AGN, such as PG1211+143, offer great potential for understanding outflows and their potential effects on the intergalactic medium and host galaxy growth.

References

Chartas G., Brandt W. N., Gallagher S. C. & Garmire G. P. 2002, ApJ, 579, 169

Chartas G., Brandt W. N. & Gallagher S. C. 2003, ApJ, 595, 85

Chevallier L., Collin S., Dumont A-M., Czerny B., Mouchet M., Goncalves A. C. & Goosmann R. 2006, A&A, 449, 493

Crummy J., Fabian A. C., Gallo L. & Ross R. R. 2006, MNRAS, 365, 1067

Elvis M., Giommi P., Wilkes B. J. & McDowell J. 1991, ApJ, 378, 537

Gierlinski M. & Done C. 2004, MNRAS, 349, L7

Haardt F. & Maraschi L. 1991, ApJ, 350, L81

Kaspi S., Smith P. S., Netzer H., Maoz D., Jannuzi B. T. & Giveon U. 2000, ApJ, 533, 631

Kaspi S. & Behar E. 2006, ApJ, 636, 674

King A. R. & Pounds K. A. 2003, MNRAS, 345, 657

King A. R. 2005, ApJ, 635, L121

Markowitz A., Reeves J. N. & Braito V. 2006, ApJ, 646, 783

Nandra K. & Pounds K. A. 1994, MNRAS, 268, 405

Pounds K. A., Reeves J. N., King A. R., Page K. L., O'Brien P. T. & Turner M. J. L. 2003, MNRAS, 345, 705

Pounds K. A., Reeves J. N., Page K. L. & O'Brien P. T. 2004, ApJ, 616, 696

Pounds K. A. & Page K. L. 2006, MNRAS, 372, 1275

Pounds K. A. & Reeves J. N. 2006, MNRAS, accepted (astro-ph/0610112)

Proga D., Stone J. M. & Kallman T. R. 2000, ApJ, 543, 686

Reeves J. N., O'Brien P. T. & Ward M. J. 2003, ApJ, 593, L65

Schurch N. J. & Done C. 2006, MNRAS, 371, 81

Vaughan S. & Fabian A. C. 2004, MNRAS, 348, 1415

KAZUO MAKISHIMA: Comment: Steep power law you are finding is extremely interesting and reminds me of similar phenomenon that occurs in Galactic black hole binaries when they get in the very high state. In that case this component appears between 10–50 keV, connecting the hard tail and the soft disc component.

GIORGIO MATT: Can you exclude that the 0.7–2 keV spectrum is due to ionized reflection rather than absorption?

KENNETH POUNDS: The spectral curvature (soft excess) in PG1211 is too strong to be produced by photoionized reflection from the accretion disc unless the direct continuum is somehow suppressed, for example by light bending in strong gravity near the black hole. Given the detection of a large column density outflow in this luminous AGN, absorption plus re-emission provides a more likely explanation.

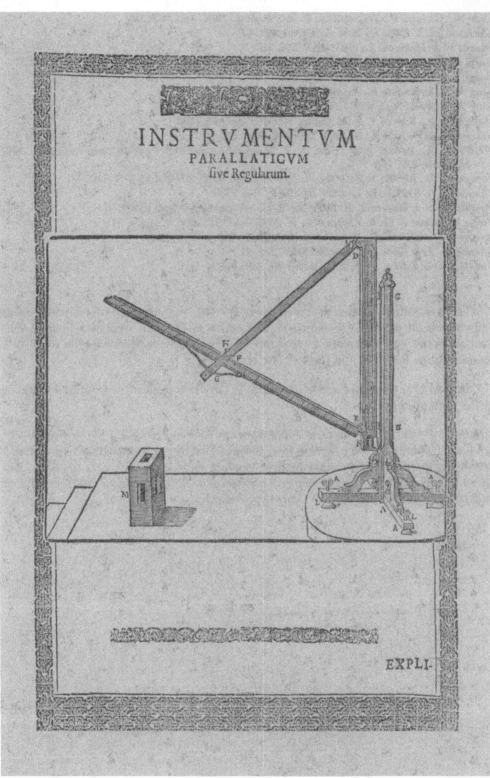

Black Holes from Stars to Galaxies – Across the Range of Masses
Proceedings IAU Symposium No. 238, 2006   © 2007 International Astronomical Union
V. Karas & G. Matt, eds.   doi:10.1017/S1743921307004802

Uncertainties of the masses of black holes and Eddington ratios in AGN

Suzy Collin

LUTH, Observatoire de Paris-Meudon, 5 Place Janssen, 92140 Meudon, France
email: suzy.collin@obspm.fr

Abstract. Black hole masses in Active Galactic Nuclei have been determined in 35 objects through reverberation mapping of the emission line region. I mention some uncertainties of the method, such as the "scale factor" relating the Virial Product to the mass, which depends on the unknown structure and dynamics of the Broad Line Region.

When the black hole masses are estimated indirectly using the empirical size-luminosity relation deduced from this method, the uncertainties can be larger, especially when the relation is extrapolated to high and low masses and/or luminosities. In particular they lead to Eddington ratios of the order of unity in samples of Narrow Line Seyfert 1. As the optical-UV luminosity is provided by the accretion disk, the accretion rates can be determined and are found to be much larger than the Eddington rates.

So, accretion must be performed at a super-critical rate through a slim disk, resulting in rapid growth of the black holes. The alternative is that the mass determination is wrong at this limit.

Keywords. Quasars: general – galaxies: active

---

1. Introduction

It is a paradox that the determination of Black Hole (BH) masses in Active Galactic Nuclei (AGN) is more difficult than in quiescent galaxies. Nevertheless, it is absolutely necessary to know the BH masses at high redshifts, and to understand how BHs grow. BH masses in AGN are determined through the 'virial technique', so-called because it assumes that the line emission region is gravitationally bound to the BH. Since it has become an industry and was used in several dozens of papers to determine the masses of thousands of BHs, I thought that it was necessary to recall some basic uncertainties plaguing this method. Also the question of the accretion rates is very important in the context of BH growth, and it is almost always confused with the question of the luminosity.

2. Virial masses

In 35 Seyfert and low redshift quasars, the BH masses have been determined directly from "reverberation mapping" (in the following these AGN are called "Reverberation Mapped", or RM objects). As a by-product of this method, an empirical relation was found between the size of the Broad Line Region (BLR) and the optical luminosity, $L_{\rm opt}$. In all other AGN except one (NGC 4258), the BH masses are determined indirectly using this empirical relation.

The direct method consists in measuring the time delay τ between the continuum and the line variations which respond to them; it gives a characteristic size of the BLR. Assuming that the BLR is gravitationally bound (which is certainly true for the Balmer line emitting region, cf. Peterson & Wandel 2000), the mass of the BH, $M_{\rm BH}$, is then equal to $M_{\rm BH} = fc\tau\,\delta V^2/G$, where δV is the dispersion velocity, and f a scale factor.

$c\tau\delta V^2/G$ is called the "Virial Factor". The usual way is to identify δV with the FWHM, and to assume $f = 3/4$, which correspond to an isotropic BLR with a random distribution of orbits.

There are strongly debated questions about the method. To quote only a few:

1. What is the best choice for measuring the dispersion velocity? Is it the FWHM (which is used in all works) or σ_{line}, i.e. the second moment of the line profile? There is a large scatter of the FWHM/σ_{line} ratio, and Peterson *et al.* (2004) showed that σ_{line} seems more reliable for the intrinsic dispersion of measurements, but it is generally not measured.

2. The scale factor f varies among the objects by a factor three: it is larger than two for RM objects with narrow peaked lines and smaller than unity in RM objects with broad flat topped lines. Collin *et al.* (2006) showed that it depends probably on the Eddington rate and on the inclination of the BLR (which is most likely not spherical but axially symmetric), and that some narrow line objects must be seen at low inclination and consequently have their masses underestimated by up to one order of magnitude.

3. Is it better to use the RMS or the mean spectrum? Peterson *et al.* (2004) reanalysed the data of all RM objects, used σ_{line} instead of the FWHM, RMS spectrum instead of mean spectrum, and changed the factor f, as scaled on the bulge masses by Onken *et al.* (2004). As a result, they obtained masses *systematically larger* by typically a factor two than those of Kaspi *et al.* (2000), some of them by factors up to one order of magnitude.

Everybody in this audience is aware that there is an empirical relation discovered with the RM objects between the size of the BLR and the optical luminosity (Peterson & Wandel 1999, Kaspi *et al.* 2000). This relation was revised by Kaspi *et al.* in 2005 and is given now with a good precision. It is not well understood theoretically, but it allows to determine M_{BH} for single epoch observations, by simply measuring $L(opt)$ ($=\nu L_\nu$ at 5100Å) and the FWHM. This is very useful since the reverberation method requires at least several months of monitoring of a given object to lead to a mass determination. Once the mass is known, it is also possible to determine the Eddington ratio $R_{Edd} = L_{bol}/L_{Edd}$, assuming a bolometric correction of the order of 10. It has been used for large samples of quasars, like the SDSS.

Some studies have shown that the most luminous quasars correspond to very massive black holes, up to $10^{10} M_\odot$ radiating at their Eddington luminosity. At the other extreme, it led to very small masses, of the order of a few $10^5 M_\odot$, in a sample of AGN with small host galaxies (Barth *et al.* 2005). It is not clear whether the extrapolation of the relation to large or small luminosities is valid (cf. Wang & Zhang 2003 for the low mass side). Moreover one must not forget that the indirect method eliminates the intrinsic dispersion of the size-luminosity relation, and that the scale factor is still not well known and certainly varies among objects.

3. The Eddington ratio versus the accretion rate

Using the RM objects and assuming that the optical luminosity was due to the accretion disk, Collin *et al.* (2002) showed that a fraction of them were accreting at super-Eddington rate. Actually the rates were overestimated, because the Hubble constant was assumed to be 50 and not 70 km s^{-1} Mpc^{-3}, and the masses taken from Kaspi *et al.* (2000) were probably underestimated by a factor two to three. In total the Eddington ratios were overestimated by a factor of about six, and the accretion rates in Eddington units had to be also reduced by larger factors (see below), so almost none of the RM objects are now found to accrete at super-Eddington rates. Nevertheless, applying the

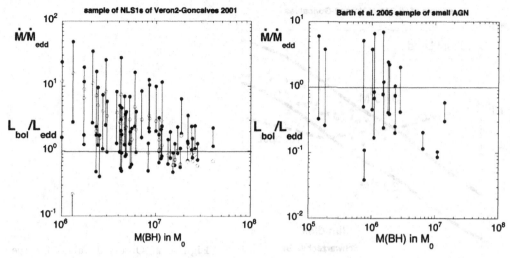

Figure 1. Accretion rates and luminosities in Eddington units for two samples (see the text).

same technique to samples of Narrow Line Seyfert 1 nuclei (NLS1s) where the BH masses were determined by the indirect Virial method, Collin & Kawaguchi (2004) showed that a large fraction of them are accreting at super-Eddington rate.

It is easy to understand why the accretion rate in Eddington units can be much larger than the Eddington ratio. In the optical range, the disk radiates locally like a black body (this is not true in the UV range, cf. Hubeny *et al.* 2001). Thus L_{opt} can be expressed as $L_{opt} = A \cos(\theta)(M\dot{M})^{2/3}$, where θ is the inclination of the normal of the disk with the line of sight, and $\dot{M}$ is the accretion rate. A is a constant, actually of the order of unity. It leads to the efficiency η for mass-energy conversion:

$$\eta \sim 0.006 \frac{\cos(\theta)^{3/2} C_{bol10} M_6^{3/2}}{\sqrt{R_{Edd}}} \tag{3.1}$$

where $C_{bol10} \approx 1$ is the observed bolometric correction divided by 10, and M_6 the mass in $10^6 M_\odot$. Thus η could be very small for large luminosities and small BH masses. It means that *the accretion rate expressed in Eddington units with a standard efficiency of 0.1 ($\dot{M}_{Edd} = L_{Edd}/0.1c^2$) is much larger than R_{Edd}.* This is illustrated in Fig. 1. The figure on the left (from Collin & Kawaguchi 2004) shows the application of this law to the NLS1 sample of Veron-Cetti *et al.* (2001), and that on the right to the sample of Barth *et al.* (2005) consisting of small mass BHs. The red points correspond to $\dot{M}/\dot{M}_{Edd}$, and the blue ones to R_{Edd}.

One can see that $\dot{M}/\dot{M}_{Edd}$ is typically one order of magnitude larger than R_{Edd} for the NLS1 sample, while it is slightly smaller for the Barth sample whose luminosities are smaller. On the left figure are also shown the values of $\dot{M}/\dot{M}_{Edd}$ computed assuming that a large fraction of the optical luminosity is provided by an external non-gravitational heating of the disk (cf. Collin & Kawaguchi 2004 for a detailed explanation). Though the values of $\dot{M}/\dot{M}_{Edd}$ are smaller than previously, they remain nevertheless larger than R_{Edd}.

Is this conclusion inescapable? Other explanations can be proposed:

1. Super-Eddington accretion rate can be relatively large at a distance of 100 to 1000 Schwarzschild radii corresponding to the emission of the optical band, and relativistic

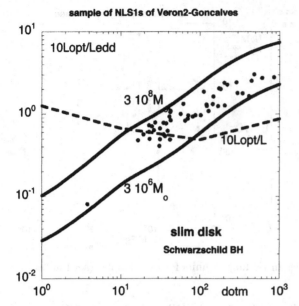

Figure 2. Observed values for the Veron *et al.* sample compared to slim disk parameters.

outflows can be created close to the BH. Note however that such outflows, if they exist as proposed by Gierlinski & Done (2004), are not super-Eddington.

2. The optical-UV emission is not due to the accretion disk, even when taking into account a non-gravitational external heating: but then to what else (see the discussion in Collin *et al.* 2002)?

3. The empirical size-luminosity relation may not be valid at large Eddington ratios and small masses, or the scale factor is much larger for these objects.

4. Alternatively, we observe really super-Eddington accretion rates, due to "slim" super-Eddington advective disks (Abramowicz *et al.* 1988).

Indeed the SED of such disks is in agreement with the relation between $\dot{M}/\dot{M}_{\rm Edd}$ and $R_{\rm Edd}$, and with the constancy of the ratio $L_{\rm bol}/L_{\rm opt}$, as shown on Fig. 2 (cf. the description of the slim disk model in Collin & Kawaguchi 2004). If these super-Eddington accretion rates are real and concern not only nearby quasars, they would have important cosmological consequences. During their low mass phase, the growth time of the BHs would not be Eddington but mass supply limited, and can be much smaller than the Eddington time. Super-Eddington accretion can thus account for the rapid early growth of BHs. It implies also that the BH/bulge mass relationship for NLS1s would be more dispersed than in other objects.

Acknowledgements

I am grateful to T. Kawaguchi, B. Peterson, and M. Vestergaard for many enlightening discussions on the subject.

References

Abramowicz, M. A., Czerny, B., Lasota, J.-P. & Szuszkiewicz, E. 1988, ApJ, 332, 646
Barth, A. J., Greene, J. E. & Ho, L. C. 2005, ApJL, 619, L151
Collin, S., Boisson, C., Mouchet, M., Dumont, A.-M., Coupé, S., Porquet, D. & Rokaki, E. 2002, A&A, 388, 771
Collin, S. & Kawaguchi, T. 2004, A&A, 426, 797
Collin, S., Kawaguchi, T., Peterson, B. M. & Vestergaard, M. 2006, A&A, 456, 75

Gierliński, M. & Done, C. 2004, MNRAS, 349, L7

Hubeny, I., Blaes, O., Krolik, J. H. & Agol, E. 2001, ApJ, 559, 680

Kaspi, S., Smith, P. S., Netzer, H., Maoz, D., Jannuzi, B. T. & Giveon, U. 2000, ApJ, 533, 631

Kaspi, S., Maoz, D., Netzer, H., Peterson, B. M., Vestergaard, M. & Jannuzi, B. T. 2005, ApJ, 629, 61

Onken, C. A., Ferrarese, L., Merritt, D., Peterson, B. M., Pogge, R. W., Vestergaard, M. & Wandel, A. 2004, ApJ, 615, 645

Peterson, B. M. & Wandel, A. 1999, ApJL, 521, L95

Peterson, B. M. & Wandel, A. 2000, ApJL, 540, L13

Peterson, B. M. *et al.* 2004, ApJ, 613, 682

Véron-Cetty, M.-P., Véron, P. & Gonçalves, A.C. 2001, A&A, 372, 730

Wang, T.-G., Zhang & X.-G. 2003, MNRAS, 340, 793

DEBORAH DULTZIN-HACYAN: Comment: It has been shown by Zhang & Win that the R–L relation is not valid for dwarf (low L) AGN.

SUZY COLLIN: You are perfectly right. If the size is larger that that given by R–L relation for small masses, it means that we underestimate them. On the other hand, as we have shown (Collin *et al.* 2006) that the scale factor is probably underestimated for the NLS1 class, this could be another reason of underestimating the mass.

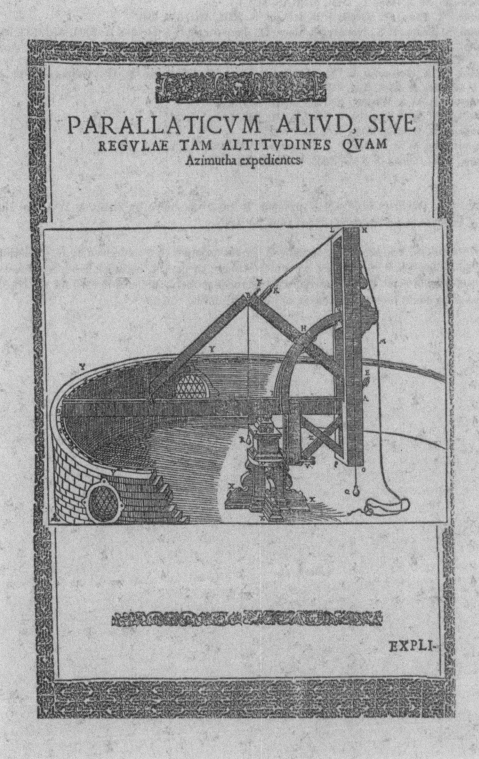

PARALLATICVM ALIVD, SIVE
REGVLAE TAM ALTITVDINES QVAM
Azimutha expedientes.

EXPLI-

Black Holes from stars to galaxies - Across the range of masses
Proceedings IAU Symposium No. 238, 2006
V. Karas, G. Matt, eds.
© 2007 International Astronomical Union
doi:10.1017/S1743921307004814

Warped discs and the Unified Scheme

Andrew Lawrence

Institute for Astronomy, University of Edinburgh Royal Observatory, Blackford Hill,
Edinburgh EH9 3HJ, UK
email: al@roe.ac.uk

Abstract. The standard torus picture for explaining the difference between Type I and Type II AGN is physically unlikely and provides no natural explanation for a number of simple facts - the average covering factor, the broad range of covering factors, and the characteristic reprocessing distance. Parsec scale warped discs are a strong alternative. A very simple "misaligned disc" model produces good agreement with covering factor statistics, but predicts too many off-axis emission line cones. I discuss possible ways to improve such a model.

1. Introduction

The standard Unified Scheme for AGN includes a central continuum source assumed to be an accretion disc around a black hole; a region somewhat further out emitting broad emission lines ; a dusty rotating "torus" beyond this; and gas emitting narrow emission lines on large scales, ionised through the open cone of the torus. (Antonucci and Miller 1985; Krolik and Begelman 1988). Type II AGN are those seen sideways through the obscuring torus, so that one sees only the narrow lines and the IR emission produced by reprocessing of the continuum source by the torus. The torus is assumed to be geometrically thick ($H/R \sim 1$) and to produce both optical extinction and X-ray absorption. In order to explain the X-ray background, it is popular to assume that most AGN ($\sim 80\%$) are obscured. The Unified Scheme has good general observational evidence (polarisation mirrors, and emission line cones, seen in some objects; the relative radio lobe sizes of broad and narrow line radio galaxies; the different relative orientations of polarisation and radio axis in broad and narrow line radio galaxies). However, in detail, the simple scheme has several problems.

2. Unsatisfactory aspects of the simple torus picture

(i) The idea of a geometrically thick rotating torus is problematic because vertical motions would need to be of the same order as rotational motions, which on the scales concerned would be around 1000 km/s. As the expected temperature is $\sim 200K$, the material would be highly supersonic. Discrete gas clouds could be in macroscopic motion, but many papers have argued that such a structure would rapidly dissipate to a flat disc.

(ii) Some statistical differences between Type I and II AGN populations cannot be explained by seeing the same structure at different angles. Circumnuclear imaging shows that on average Type II AGN are dustier and have more disturbed morphologies (Malkan, Gorjian and Tam 1998; Hunt and Malkan 2004). For a given radio power Type II AGN have weaker narrow-line OIII emission (Whittle 1985; Jackson and Browne 1990; Lawrence 1991; Grimes *et al.* 2004). Assuming radio power indicates true pre-obscured nuclear power, and relative OIII strength is proportional to the uncovered solid angle, there must be a wide range of covering factors; objects with large covering factors are more likely to appear as Type II and possibly also to appear dusty. The mean ratio of

OIII/radio for Type I vs Type II is $R_{OIII} \sim 4$ (Grimes *et al.* 2004 for radio galaxies; by inspection from the data of Whittle 1985 for Seyfert galaxies). The standard torus model has no natural explanation for the distribution of covering factors.

(iii) The simple torus model makes no simple prediction for the mean covering factor, let alone the distribution of covering factors. The correct 'quasar fraction' f_Q is in fact still a contentious issue. Radio, X-ray, and OIII selected samples show f_Q varying with luminosity from ~ 0.5 at high power to ~ 0.2 at low power (e.g. Lawrence 1991, Ueda *et al.* 2003, Simpson 2005). However, IRAS selected mid-IR samples show Type I and II AGN in equal numbers over all luminosities (e.g. Rush, Malkan and Spinoglio 1993), and early Spitzer analyses seem to be giving a similar answer (Lacy *et al.* 2004). Martinez-Sansigre (2005) have argued from observations of faint Spitzer sources that *most* high redshift quasars are obscured, but depending on various assumptions their analysis is consistent with anything in the range $f_Q \sim 0.2 - 0.8$. Willott *et al.* (2000) noted that considering only high excitation narrow line radio galaxies removes the luminosity dependence, giving $f_Q \sim 0.4$, and suggested that low excitation objects, which constitute the majority of the low luminosity narrow line radio galaxies, are a different phenomenon - possibly 'switched off' quasars rather than obscured quasars. (A similar analysis of X-ray selected samples would be of great interest.) Another clue to the covering factor is the observation that the luminosity of the IR-bump in quasars is on average 30% of the luminosity of the UV bump (Sanders *et al.* 1989). Note however that if there is a range of covering factors, the averaged reprocessed fraction f_R for objects seen as Type I is is not the same as the average covering factor for *all* AGN.

(iv) The simple torus model makes no simple prediction for the distance at which most extinction/reprocessing takes place. The mean quasar SED (Sanders *et al.* 1989; Elvis *et al.* 1994) shows a clear peak at $\sim 10 \mu m$ corresponding to $T \sim 200K$, much cooler than the dust sublimation temperature, and corresponding to around 1pc in Seyferts, and 10pc in quasars. It has been noted that this is roughly the 'sphere of influence' of the black hole (Krolik and Begelman 1988) but it is not clear why a 'wall of dust' should occur at this distance. Although there is a fairly well defined typical distance/dust temperature, the breadth of the IR bump shows that there is dust at a wide range of temperatures.

In conclusion, we ideally want a model that naturally produces (a) a geometrically thick structure, (b) mean values of $f_Q \sim 0.4$ and $f_R \sim 0.3$, (c) a distribution of covering factors $N(C)$ such that $R_{OIII} \sim 4$, and (d) a wide range of dust temperatures, but peaking at $T \sim 200K$.

3. A simple re-aligning disc model

It has been proposed in the past that warped dust discs may be a good way to explain the IR emission in quasars (Phinney 1989; Sanders *et al.* 1989), as they can cover a significant fraction of the sky, and naturally produce a wide range of temperatures - the inner radii do not obscure the outer radii. Can a warped disc model produce the required distribution of covering factors $N(C)$? Warped discs clearly occur frequently on kpc scales, but the degree of warping is relatively modest, and it is not clear what can happen on pc scales, although there are plausible mechanisms such as the Pringle instability (Pringle 1996, 1997), tumbling bars (Tohline and Osterbrock 1982) or perhaps torques from large scale magnetic fields.

Rather than trying to produce a physical model, I consider here a very general suggestion: that the angular momenta of the incoming disc and the nuclear disc are *completely unconnected*. The radio axes of AGN seemed to be aligned at random with respect to their host galaxies (Ulvestad and Wilson 1984; Clarke, Kinney and Pringle 1998; Nagar and

Wilson 1999; Kinney *et al.* 2000) so this is a quite reasonable hypothesis. The probability density for the difference θ of the angular momentum vectors is then $dP = \frac{1}{2}\sin\theta d\theta$. (This assumes θ can be anywhere from 0 to π, i.e. the incoming disc can even be counter-rotating with respect to the nuclear disc.) At some critical radius some unspecified torque causes the disc to *re-align*. The covering factor C then depends on the degree of precession during the re-alignment. Complete precession will effectively make an equatorial wall of height θ giving $C = \sin\theta$. A warp which tilts but doesn't twist at all gives $C = \theta/\pi$ for small θ (Phinney 1989; the original paper states $C = \theta/3$ but Phinney has confirmed privately that θ/π is correct). For either of these cases (precessed disc and tilted disc) one can then calculate $N(C)$, for all objects or just for those seen as Type I or II. Likewise, calculating $N(1 - C)$ for each Type gives the distribution of relative OIII strengths.

For precessed discs, any value $\theta > \pi/2$ (i.e. counter-rotating incoming discs) gives $C = 1$, predicting that half of all AGN are *completely* obscured and possibly haven't even been recognised. Even for objects with $\theta < \pi/2$, the distribution is peaked towards $C = 1$. The quasar fraction comes from integrating under $N_I(C)$ for Type I only and gives $f_Q = 0.22$ or $f_Q = 0.11$ including the completely obscured objects, significantly in disagreement with observations. The average reprocessed fraction for quasars is $f_R \sim 0.5$, which is somewhat too large. Taking the median values of $N(1 - C)$ for Types I and II we get $R_{OIII} \sim 5$, but the distribution of relative OIII values, piled up towards low OIII for Type IIs, does not resemble the observations. This very simplest model therefore does not work.

For *tilted discs* $N(C)$ peaks around $C = 0.5$. The quasar fraction found from integrating under $N_I(C)$ for Type I only gives $f_Q = 0.48$. The typical reprocessed fraction for quasars comes from the mean of $N_I(C)$ which is $f_R \sim 0.35$. The ratio of typical relative OIII strengths comes from comparing the peaks of $N_I(1 - C)$ and $N_{II}(1 - C)$ from which we find $R_{OIII} \sim 2$. So a tilted re-aligned disc, on these statistical grounds, works fairly well - it gives good agreement with observations for f_Q and f_R and somewhat too small a difference between Type I and II for R_{OIII}. The covering factor equation $C = \theta/\pi$ is strictly only correct for small θ, so this agreement may be somewhat fortuitous. However there are other reasons why a tilted disc with no twist cannot be the correct model. On a given side of the inner disc, the warped outer disc is not azimuthally symmetric, and so does not produce simple shadow-cones. Seen in projection, it will produce bipolar cone-like ionisation structures, but they will be strongly one-sided, and will not usually be aligned with the radio axis, whereas observations do seem to show such an alignment (Schmitt *et al.* 2003).

So, for different reasons, simple versions of both twisted and untwisted warped disc models don't quite work. The conclusion could be that warped discs are not the answer; or it could be that the observed cones are not 'shadow cones' but rather are 'matter cones' formed by an outflow. Finally of course, one could relax the strict assumptions of the simplest model. The degree of twist could be between the extremes of no twist at all and complete precession. The radiation instability model of Pringle (1997) produces such a partial twist, but still has a problem with jets and cones not aligning. Alternatively, it could be that the axis difference θ between the outer and inner disc is not random. Analysis by Schmitt *et al.* (2002) finds that for radio galaxies the difference between the radio axis and the axis of the kpc scale dust disc is not random over all solid angle, but over a large polar cap. This is probably the most obvious variant to explore next.

4. Conclusions

The standard torus model has some significant problems, and lacks any natural prediction for the average covering factor or the distribution of covering factors. A simple re-aligning disc model, where the incoming disc and the nuclear disc are unrelated, makes quite simple and definite predictions, depending on whether the incoming disc twists as it re-aligns.

Based on the simple numerical tests of the previous section, the precessed disc does not fit the facts, whereas the tilted disc fits quite well. However, the tilted disc model predicts ionisation structures that are not in general aligned with the nuclear axis, and so can almost certainly be ruled out.

Perhaps the most promising possibility is that there is a fully precessed re-aligned disc, but that the difference between incoming and nuclear discs is not completely random. This possibility will be explored in a fuller version of this work.

Acknowledgements

Thanks to Martin Elvis and Andrew King, with whom I have many related discussions over several years.

References

Antonucci R. R. J. & Miller A. S. 1985, ApJ, 297, 621
Clarke C. J., Kinney A. L. & Pringle J. E. 1998, ApJ, 495, 189
Elvis M., Wilkes B. J., McDowell J. C., Green R. F., Bechtold J., Willner S. P. *et al.* 1994, ApJSS, 95, 1
Grimes J. A., Rawlings S. & Willott C. J. 2004, MNRAS, 349, 503
Hunt L. K. & Malkan M. A. 2004, ApJ, 616, 707
Jackson N. & Browne I. W. A. 1990, Nature, 343, 43
Kinney A. L., Schmitt H. R., Clarke C. J., Pringle J. E., Ulvestad J. S. & Antonucci R. R. J. 2000, ApJ, 537, 152
Krolik J. H. & Begelman M. C. 1988, ApJ, 329, 702
Lawrence A. 1991, MNRAS, 252, 586
Lacy M., Storrie-Lombardi L. J., Sajina A. *et al.* 2004, ApJ, 154, 166
Malkan M. A. Gorjian V. & Tam R. 1998, ApJSS, 117, 25
Martinez-Sansigre A., Rawlings S., Lacy M., Fadda D., Marline F. R., Simpson C. *et al.* 2005, Nature, 436, 666
Nagar N. M. & Wilson A. 1999, ApJ, 516, 97
Phinney E. S. 1989, in Theory of Accretion Disks, p. 457, eds F.Meyer *et al.* (Dordrecht: Kluwer)
Pringle J. E. 1996, MNRAS, 281, 357
Pringle J. E. 1997, MNRAS, 292, 136
Rush B., Malkan M. A. & Spinoglio L. 1993, ApJSupp 89, 1
Sanders D. B., Phinney E. S., Neugebauer G., Soifer B. T. & Mathews K. 1989, ApJ, 347, 29
Schmitt H. R., Pringle J. E., Clarke C. J. & Kinney A. L. 2002, ApJ, 575, 150
Schmitt H. R., Donley J. L., Antonucci R. R. J., Hutchings J. B. & Kinney A. L. 2003, ApJSS, 148, 327
Simpson C. 2005, MNRAS, 360, 565
Tohline J. E. & Osterbrock D. E. 1982, ApJ, 252, L49
Ulvestad J. S. & Wilson A. S. 1984, ApJ, 285, 439
Ueda Y., Akiyama M., Ohta K. & Miyaji T. 2003, ApJ, 598, 886
Whittle M. 1985, MNRAS, 213, 33
Willott C. J., Rawlings S., Blundell K. & Lacy M. 2000, MNRAS, 316, 449

FUKUN LIU: Comment: The warped disc model has problems caused by radiation instability of Pringle, as you mentioned. My comment is that the disc warps are not due to radiation, but due to interaction of an inclined supermassive black hole binary with a standard accretion disc, which is the natural consequence of hierarchical galaxy formation model and has been investigated and introduced to explain the observations of radio jet morphology (Liu 2004, MNRAS). If this is the case, the difficulties with warped disc model may be overcome.

GIORGIO MATT: Can the warped disc model explain the mismatch between optical and X-ray classification occurring in $\sim 10\%$ of AGN in the local universe?

ANDREW LAWRENCE: The warped disc model per se is irrelevant to this problem – I was simply pointing out that many papers draw naive conclusions because they are not careful about distinguishing between X-ray absorption and optical extinction, which are often unconnected.

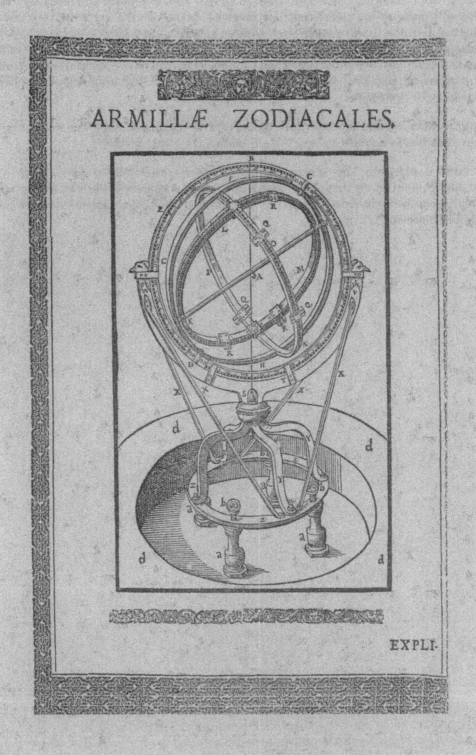

EXPLI-

Black Holes from Stars to Galaxies – Across the Range of Masses
Proceedings IAU Symposium No. 238, 2006
V. Karas & G. Matt, eds.
© 2007 International Astronomical Union
doi:10.1017/S1743921307004826

An accretion disk laboratory in the Seyfert 1.9 galaxy NGC 2992

Tahir Yaqoob,[1,2] Kendrah D. Murphy,[1,2] and Yuichi Terashima[3]

[1] Department of Physics and Astronomy, Johns Hopkins University, 3400 N. Charles St., Baltimore, MD 21218, USA

[2] Astrophysics Science Division, NASA Goddard Space Flight Center, Greenbelt Rd., Greenbelt, MD 20771, USA

[3] Department of Physics, Ehime University, Bunkyo-cho, Matsuyama, Ehime 790-8577, Japan

Abstract. Over twenty five years of X-ray observations of the Seyfert 1.9 galaxy NGC 2992 show that it is a promising test-bed for severely constraining accretion disk models. The previous interpretation of the historical activity of NGC 2992 in terms of the accretion disk slowly becoming dormant over many years and then 're-building' itself is not supported by new data. A recent year-long monitoring campaign with *RXTE* showed that the X-ray continuum varied by more than an order of magnitude on a timescale of weeks. During the large-amplitude flares the centroid energy of the Fe K emission-line complex became significantly redshifted, indicating that the violent activity was occurring close to the putative central black hole where gravitational energy shifts can be sufficiently large. For the continuum, the Compton-y parameter remains roughly constant despite the large-amplitude luminosity variability, with $(kT)\tau \sim 20–50$.

Keywords. Accretion, accretion disks – galaxies: active – black hole physics

---

1. Introduction

In this paper we discuss some new results from X-ray observations of the nearby ($z = 0.00771$, Keel 1996) Seyfert 1.9 galaxy NGC 2992, which has been observed by every X-ray astronomy mission since the time it was discovered by HEAO-1 to be one of the brightest hard X-ray AGN in the sky, with a 2–10 keV flux of $\sim 7.2 – 8.6 \times 10^{-11}$ erg cm^{-2} s^{-1} (Piccinotti *et al.* 1982). In more than a quarter of a century of X-ray observations, the hard X-ray flux has varied by over a factor of 20 (see Figure 1), corresponding to a range in the intrinsic 2–10 keV luminosity of $\sim 0.55 – 11.8 \times 10^{42}$ erg s^{-1} (assuming $H_0 = 70$ km s^{-1} Mpc^{-1}, $\Lambda = 0.73$, $\Omega = 1$). The mass of the central black hole in NGC 2992 has been estimated to be 5.2×10^7 $M_\odot$ from stellar velocity dispersions (Woo & Urry 2002, and references therein). The large dynamic range in luminosity corresponds approximately to a range $\sim (8 – 180) \times 10^{-4}$ in $L/L_{\rm Edd}$. This wide coverage in $L/L_{\rm Edd}$ attained by variability that is slow enough to obtain high signal-to-noise X-ray spectra in a particular luminosity state is also a key factor that makes NGC 2992 a good test-bed for accretion disk models.

2. Historical behavior

During an *ASCA* observation in 1994, NGC 2992 was in its lowest continuum flux state thus far observed, with a 2–10 keV flux of $\sim 4 \times 10^{-12}$ erg cm^{-2} s^{-1} (Weaver *et al.* 1996). In that observation the Fe K line equivalent width (EW) was very high ($\sim 500 – 700$ eV), indicating that the line intensity had not responded to the declining continuum. In 1997 a *BeppoSAX* observation (see Gilli *et al.* 2000 for details) found NGC 2992 to be still

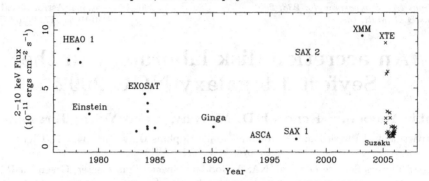

Figure 1. Historical 2–10 keV X-ray flux of NGC 2992.

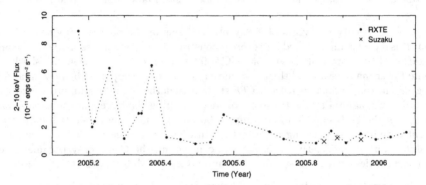

Figure 2. Flux versus time from the *RXTE* monitoring campaign.

in a low continuum state, and the Fe K line EW was even higher, with an additional component due to highly ionized Fe making an appearance (although the interpretation of the Fe K complex was ambiguous because of the limited spectral resolution). Then a second *BeppoSAX* observation in 1998 revealed that the source continuum had fully recovered to its bright HEAO-1 state (see Gilli *et al.* 2000). In this state, the EW of the Fe K line was very low (< 100 eV) since the line intensity again did not respond to the change in continuum level. These measurements were mostly sensitive to the Fe K line core, which from the lack of variability, likely arises in a distant, parsec-scale reprocessor, as suggested by Weaver *et al.* (1996). Gilli *et al.* (2000) interpreted the *apparently* slow decline in the continuum luminosity of NGC 2992 over many years, and the rise back to a high state over a period of ~ 1 year (see Figure 1) in terms of diminishing accretion activity followed by a 're-building' of the accretion disk during 1997–1998. This interpretation found support in the fact that in the high state NGC 2992 behaved more like a type 1 Seyfert galaxy, showing broad optical lines and rapid X-ray variability in the high state (see Gilli *et al.* 2000).

NGC 2992 was observed by *Suzaku* in 2005 and the data showed that the intensities of broad and narrow components to the Fe K line could be decoupled for the first time in this source. A lower limit of $\sim 30°$ on the inclination angle of the putative accretion disk was obtained from fitting disk-line profiles to the broad Fe K line, but this lower limit was not sensitive to other parameters in the disk-line model, including the black-hole angular momentum.

NGC 2992 was the subject of an *RXTE* monitoring campaign consisting of 24 observations in the period 2005 March to 2006 January. Full details will be presented by K. D. Murphy *et al.* (2007, in preparation) but here we summarize some key results.

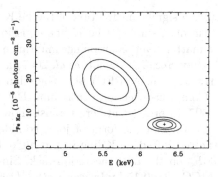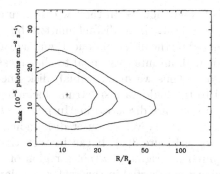

Figure 3. Joint 68%, 90%, and 99% *RXTE* confidence contours from (a) *(left)* the low-state and high-state spectra for the Fe Kα line intensity versus the line centroid energy, and (b) *(right)* the high-state Fe Kα disk-line intensity versus the outer disk radius when the the line is modeled with a dual, Gaussian and relativistic disk line.

3. Was Accretion Activity Quenched for Years?

We extracted spectra from the 24 *RXTE* observations and Figure 2 shows the 2–10 keV flux measured from spectral fitting to each of the spectra (details in Murphy *et al.* 2007, in press). Prior to this campaign, NGC 2992 had never been observed by *RXTE*. The *RXTE* fluxes are also shown in Figure 1. It can be seen that the flux varied by nearly an order of magnitude, showing large amplitude excursions on the sampling timescale (i.e. days–weeks). In particular, three large flares occurred in the first part of the campaign. The fact that in one year alone the *RXTE* flux and luminosity covered such a large range suggests that this is typical behavior and that sparse sampling in the historical lightcurve (Figure 1) gave the illusion that NGC 2992 slowly switched off over many years followed by a 'recovery'. Continuous large-amplitude variability is common in Seyfert 1 galaxies, yet 'switching off' only occasionally invoked in those AGN. There is little evidence to suggest that the accretion process is quenched in the low X-ray flux states of NGC 2992 and therefore it is unlikely that the two *BeppoSAX* observations (Figure 1) were revealing a re-building of the accretion disk.

To study the spectral variability of NGC 2992 during the *RXTE* monitoring campaign, we constructed averaged spectra and response matrices from 15 of the lowest-flux spectra (low-state spectrum), and the 3 highest-flux spectra (high-state spectrum). The spectra were fitted in the 3–15 keV band with a simple model consisting of a power-law continuum (with a photon index Γ), Galactic absorption, and a single Gaussian emission-line component. The latter, which represents Fe K line emission, had three free parameters, namely the centroid energy, E_c, the intrinsic width, σ, and the line intensity, I. We note that the 3–15 keV *RXTE* spectra are not sensitive to absorption at the level measured by *Suzaku* and previous missions.

4. Gravitationally redshifted Fe K line emission

In the low-state spectrum the Fe K emission line was consistent with a centroid energy of ~ 6.4 keV ($E_c = 6.3 \pm 0.2$ keV), had an EW of 550 ± 120 eV, and was unresolved ($\sigma = 0.3 \pm 0.3$). All errors quoted here are 90% confidence for one parameter. In the high state the Fe K line centroid energy became significantly redshifted ($E_c = 5.6 \pm 0.4$ keV), and the line became broader ($\sigma = 0.8^{+0.4}_{-0.3}$ keV). The EW of the line in the high state was 190^{+70}_{-60} eV. Joint 99% confidence contours of the line intensity versus the centroid energy for the low and high states are shown in Figure 3(a). We interpret this behavior as an indication that the violent activity during the continuum flaring illuminates the inner-most regions of the accretion disk, where strong gravity effects significantly redshift the Fe K line photons. In the low continuum state a significant fraction of the Fe K line emission may be from distant matter, but in the high state the Fe K line emission is dominated by the disk component, and a distant-matter component being simultaneously present with an intensity equal to that measured by *Suzaku* (Yaqoob *et al.* 2007) is not ruled out. Thus we fitted a dual, disk line (`diskline` in XSPEC) plus Gaussian line model to the high state spectrum, including a Compton-reflection continuum. Detailed fitting results are discussed in Murphy *et al.* (2007, in press) and here we just show the disk-line intensity versus outer disk radius constraints in the form of joint confidence contours in Figure 3(b). It can be seen that the redshifted Fe K line in the high state is required to originate within a region of $\sim 60 R_g$ on the inner accretion disk. Similar behavior was reported in the Seyfert 1 galaxy MCG −6-30-15 by Iwasawa *et al.* (1999).

5. Comptonization models of the X-ray continuum

We also fitted the low-state and high-state spectra using a simple thermal Comptonization model (the `comptt` model in XSPEC – see Titarchuk 1994), plus a Gaussian component to model the Fe K emission line). The model has only two free parameters, namely the temperature of the Comptonizing plasma, kT, and its Thomson depth, τ. Detailed fitting results are discussed in Murphy *et al.* (2007, in press) but here we simply note that the 99% confidence contour of kT versus τ for the high state lies inside that of the low state contour, and both contours are roughly diagonal (and open), satisfying $(kT)\tau \sim 20$–50. Essentially, this means that the Compton-y parameter remains approximately constant (around a few tenths) as the continuum luminosity varies by an order of magnitude. This is consistent with our finding that the hard X-ray slope ($\Gamma \sim 1.6$–1.8) from the power-law fits did not show significant variability. In order to achieve an approximately constant product of kT and τ in the face of such large changes in the energy of the Comptonizing plasma there must be a mechanism that decreases the Thomson depth of the plasma as its temperature is raised. Alternatively, if kT and τ individually remain approximately constant then the only way to increase the energy in the plasma is to increase the number of Comptonizing electrons. More sensitive spectral measurements (than currently available) in the hard X-ray band will be able to distinguish between different scenarios because the kT can then be directly measured from the high-energy spectral turnover.

In conclusion, nearly three decades of X-ray observations of NGC 2992 have shown that it has desirable properties for testing accretion disk models and future multi-wavelength observations should take advantage of this.

Acknowledgements

The authors thank the *Suzaku* team for their work in making the *Suzaku* observations of NGC 2992 possible and acknowledge partial support from NASA grants NNG05GM34G, and NNG0GB78A.

References

Gilli R., Maiolino R., Marconi A. *et al.* 2000, A&A, 355, 485
Iwasawa K., Fabian A. C., Young A. J. *et al.* 1999, MNRAS, 306, L19
Keel W. C. 1996, ApJS, 106, 27
Piccinotti G., Mushotzky R. F., Boldt E. A. *et al.* 1982 ApJ, 253, 485
Titarchuk L. 1994, ApJ, 434, 570
Weaver K. A., Nousek J., Yaqoob T. *et al.* 1996, ApJ, 458, 160
Woo J.-H. & Urry C. M. 2002, ApJ, 579, 530
Yaqoob T., Murphy K. D., Griffiths R. E. *et al.* 2007, PASJ, 59S, 283

Physical Processes Near Black Holes I

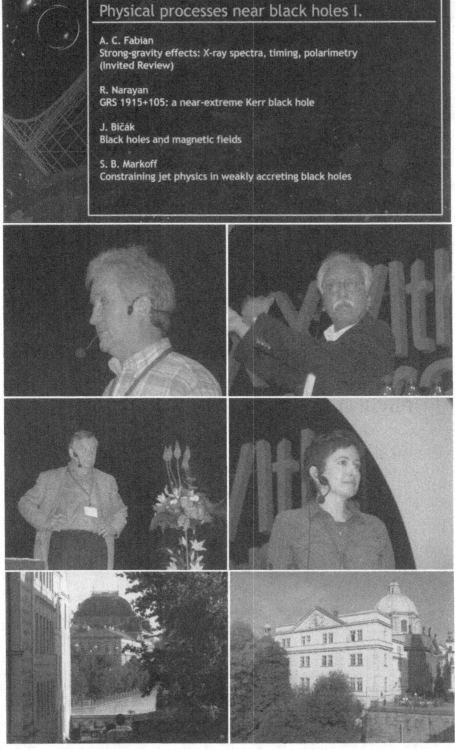

Physical processes near black holes I.

A. C. Fabian
Strong-gravity effects: X-ray spectra, timing, polarimetry
(Invited Review)

R. Narayan
GRS 1915+105: a near-extreme Kerr black hole

J. Bičák
Black holes and magnetic fields

S. B. Markoff
Constraining jet physics in weakly accreting black holes

A golden roof of the neo-renaissance building of the National Theatre. Vista from the Old Town Mills on the bank of Vltava River.

The Church of St Francis of Assisi (its sanctuary hides an admirable organ) and an adjacent Gallery of the Knights of the Cross.

Black Holes from Stars to Galaxies – Across the Range of Masses
Proceedings IAU Symposium No. 238, 2006   © 2007 International Astronomical Union
V. Karas & G. Matt, eds.   doi:10.1017/S174392130700484X

Strong gravity effects: X-ray spectra, variability and polarimetry

Andrew C. Fabian

Institute of Astronomy, Madingley Road, Cambridge CB3 0HA, UK
email: acf@ast.cam.ac.uk

Abstract. Accreting black holes often show iron line emission in their X-ray spectra. When this line emission is very broad or variable then it is likely to originate from close to the black hole. The theory and observations of such broad and variable iron lines are briefly reviewed here. In order for a clear broad line to be found, one or more of the following have to occur: high iron abundance, dense disk surface and minimal complex absorption.

Several excellent examples are found from observations of Seyfert galaxies and Galactic Black Holes. In some cases there is strong evidence that the black hole is rapidly spinning. Further examples are expected as more long observations are made with XMM-Newton, Chandra and Suzaku. The X-ray spectra show evidence for the strong gravitational redshifts and light bending expected around black holes.

Keywords. Black hole physics – X-rays – line: formation – accretion, accretion disks

---

1. Introduction

Much of the radiation from luminous accreting black holes is released within the innermost 10-20 gravitational radii (i.e. $10–20r_{\rm g} \equiv 10–20GM/c^2$). In such an energetic environment, iron is a major source of line emission, with strong emission lines in the 6.4–6.9 keV band. Observations of such line emission then provides us with a diagnostic of the accretion flow and the behaviour of matter and radiation in the strong gravity regime very close to the black hole (Fabian *et al.* 2000; Reynolds & Nowak 2003; Fabian & Miniutti 2005).

The rapid X-ray variability found in many Seyfert galaxies is strong evidence for the emission originating at small radii. The high frequency break in their power spectra, for example, corresponds to orbital periods at $\sim 20r_{\rm g}$ and variability is seen at still higher frequencies (Uttley & McHardy 2004; Vaughan *et al.* 2004). Key evidence that the very innermost radii are involved comes from Soltan's (1982) argument relating the energy density in radiation from active galactic nuclei (AGN) to the local mean mass density in massive black holes, which are presumed to have grown by accretion which liberated that radiation. The agreement found between these quantities requires that the radiative efficiency of accretion be 10 per cent or more (Yu & Tremaine 2002; Marconi *et al.* 2005). This exceeds the 6 per cent for accretion onto a non-spinning Schwarzschild black hole and inevitably implies that most massive black holes are rapidly spinning with accretion flows extending down to a just few $r_{\rm g}$. Moreover, this is where most of the radiation in such accretion flows originates.

The X-ray spectra of AGN are characterized by several components: a hard power-law which may turnover at a few hundred keV, a soft excess and a reflection component (Fig. 1). This last component is produced from surrounding material by irradiation by the power-law. It consists of backscattered X-rays, fluorescence and other line photons, bremsstrahlung and other continua from the irradiated surfaces. Examples of reflection

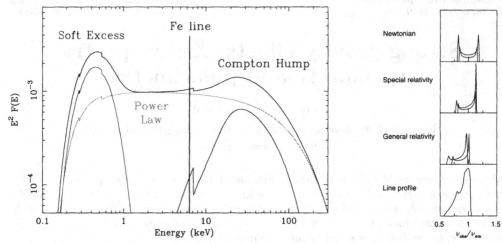

Figure 1. Left: The main components of the X-ray spectra of unobscured accreting BH are shown: soft X-ray emission from the accretion disc (red); power law from Comptonization of the soft X-rays in a corona above the disc (green); reflection continuum and narrow Fe line due to reflection of the hard X-ray emission from dense gas (blue). Right: The profile of an intrinsically narrow emission line is modified by the interplay of Doppler/gravitational energy shifts, relativistic beaming, and gravitational light bending occurring in the accretion disc (from Fabian *et al.* 2000). The upper panel shows the symmetric double–peaked profile from two annuli on a non–relativistic Newtonian disc. In the second panel, the effects of transverse Doppler shifts (making the profiles extend to lower energies) and of relativistic beaming (enhancing the blue peak with respect to the red) are included. In the third panel, gravitational redshift is turned on, shifting the overall profile to the red side and reducing the blue peak strength. The disc inclination fixes the maximum energy at which the line can still be seen, mainly because of the angular dependence of relativistic beaming and of gravitational light bending effects. All these effects combined give rise to a broad, skewed line profile which is shown in the last panel, after integrating over the contributions from all the different annuli on the accretion disc. Detailed computations are given by Fabian *et al.* (1989), Laor (1991), Dovčiak *et al.* (2004) and Beckwith & Done (2004).

spectra from photoionized slabs are shown in Fig. 2. At moderate ionization parameters ($\xi = F/n \sim 100$ erg cm s^{-1}, where F is the ionizing flux and n the density of the surface) the main components of the reflection spectrum are the Compton hump peaking at ~ 30 keV, the iron line at 6.4–6.9 keV (depending on ionization state) and a collection of lines and reradiated continuum below 1 keV. When such a spectrum is produced from the innermost parts of an accretion disk around a spinning black hole, the outside observer sees it smeared and redshifted (Fig. 2) due to doppler and gravitational redshifts.

Another, potentially powerful, diagnostic of the strong gravity regime is the study of timing and Quasi-Periodic Oscillations (QPOs). Despite the richness of the data (e.g. Strohmayer 2001), there is no consensus on how to interpret them and they will not be discussed further here.

2. Observations

All three main parts of the reflection spectrum have now been seen from AGN and Galactic Black Holes (GBH). The broad iron line and reflection hump are clearly seen in the Seyfert galaxy MCG–6-30-15 and in the GBH J1650-400. More recently it has been realised that the soft excess in many AGN can be well explained by smeared reflection (Crummy *et al.* 2006). It had been noted by Czerny *et al.* (2003) and by Gierlinski &

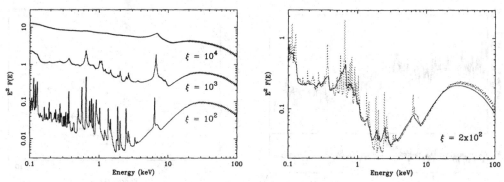

Figure 2. Left: Computed X-ray reflection spectra as a function of the ionization parameter ξ (from the code by Ross & Fabian 2005). The illuminating continuum has a photon index of $\Gamma = 2$ and the reflector is assumed to have cosmic (solar) abundances. Right: Relativistic effects on the observed X-ray reflection spectrum (solid line). We assume that the intrinsic rest-frame spectrum (dotted) is emitted in an accretion disc and suffers all the relativistic effects shown in Fig. 1.

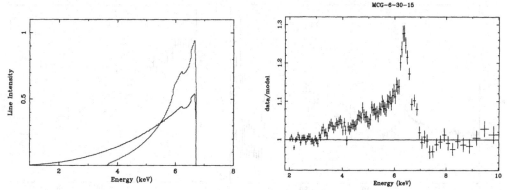

Figure 3. Left: The line profile dependence on the inner disc radius is shown for the two extremal cases of a Schwarzschild BH (red, with inner disc radius at $6\ r_g$) and of a Maximal Kerr BH (blue, with inner disc radius at $\simeq 1.24\ r_g$). Right: The broad iron line in MCG-6-30-15 from the XMM observation in 2001 (Fabian *et al.* 2002a) is shown as a ratio to the continuum model.

Done (2004) that the soft excess posed a major puzzle if thermal since the required temperature was always about 150 eV, irrespective of black hole mass, luminosity etc. Explaining it as a feature due to smeared atomic lines resolves this puzzle. An alternative interpretation involves smeared absorption lines (Gierlinski & Done 2004).

The extent of the blurring of the reflection spectrum is determined by the innermost radius of the disk (Fig. 3). Assuming that this is the radius of marginal stability enables the spin parameter a of the hole to be measured. Objects with a very broad iron line like MCG-6-30-15 are inferred to have high spin $a > 0.95$ (Dabrowski *et al.* 1997; Brennemann *et al.* 2006). Some (Krolik & Hawley 2002) have argued that magnetic fields in the disk can blur the separation between innermost edge of the disk and the inner plunge region so that the above assumption is invalid. This probably makes little difference for the iron line however since the low ionization parameter of most observed reflection reflection requires that the disc matter is very dense. The density of matter in the plunge region drops very rapidly to a low values (Reynolds & Begelman 1997) and only very strong magnetic fields, much larger than are inferred in disks, can stop this steep decline in

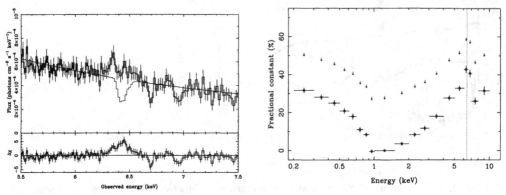

Figure 4. Left: The Chandra HEG spectrum overlaid with an ionized absorber model for the red wing (Young *et al.* 2005). Note that the absorption between 6.4 and 6.5 keV predicted by the absorber model is not seen. Right: The fractional spectrum of the constant component of MCG–6-30-15, constructed from the intercept in flux–flux plots (Vaughan & Fabian 2004). This component strongly resembles reflection.

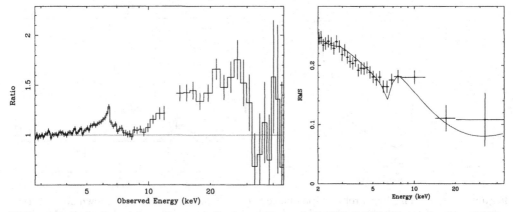

Figure 5. Left: Broad iron line and Compton hump in MCG–6-30-15 from Suzaku. Right: Fractional r.m.s. variability as a function of energy compared with model (solid line) in which the power-law is assumed to change in normalization whilst the reflection component is fixed.

density. Any reflection from the plunge region will be very highly ionized and so produce little iron emission.

Broad iron lines and reflection components are seen in both AGN (e.g. Tanaka *et al.* 1995; Nandra *et al.* 1997) and GBH (Martocchia & Matt 2002; Miller *et al.* 2002abc, 2003ab, 2004ab). A recent exciting development are the reports that broad iron lines are present in 50 per cent of all XMM AGN observations where the data are of high quality (more than 150,000 counts; Guainazzi *et al.* 2006; Nandra *et al.* 2006). They are not found in all objects or in all accretion states. There are many possible reasons for this, including over-ionization of the surface, low iron abundance, and beaming of the primary power-law away from the disk (e.g. if the power-law originates from the mildly-relativistic base of the jet).

In many cases there is a narrow iron line component due to reflection from distant matter. Absorption due to intervening gas, warm absorbers and outflows from the AGN, as well as the interstellar medium in both our Milky Way galaxy and the host galaxy must be accounted for. Moreover, if most of the emission emerges from within a few gravitational radii and the abundance is not high, then the extreme blurring can render the blurred reflection undetectable (Fabian & Miniutti 2005).

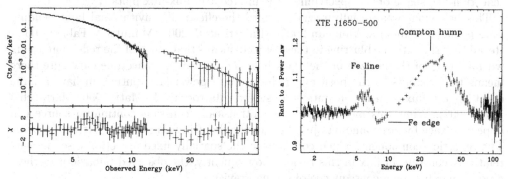

Figure 6. The broadband BeppoSAX spectrum of XTE J1650–500 (plotted as a ratio to the continuum). The signatures of relativistically-blurred reflection are clearly seen (Miniutti *et al.* 2004).

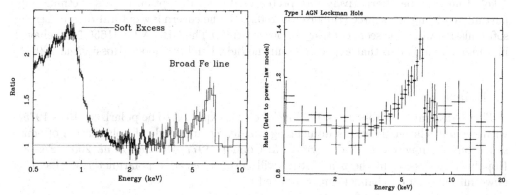

Figure 7. Left: Ratio of the spectrum of the NLS1 galaxy 1H0707 to a power-law. Spectral fits with either a very broad iron line or a partial covering with a steep edge are equally good for this object (Fabian *et al.* 2004; Boller *et al.* 2002). Right: Ratio plot of the mean unfolded spectrum for type–1 AGN in the Lockman Hole with respect to a power law (Streblyanska *et al.* 2005).

In order to distinguish between the various spectral components, both emission and absorption, we can use higher spectral resolution, broader bandwidth and variability. An example of the use of higher spectral resolution is the work of Young *et al.* (2005) with the Chandra high energy gratings (Fig. 4). Observations of MCG–6-30-15 fail to show absorption lines or feature associated with iron of intermediate ionization. Such gas could cause some curvature of the apparent continuum mimicking a very broad line. A broader bandwidth is very useful in determining the slope of the underlying continuum. This has been shown using BeppoSAX (e.g. Guainazzi *et al.* 1999) and now in several sources with Suzaku (Fig. 5, Miniutti *et al.* 2007; Reeves *et al.* 2007).

3. Variability

In the best objects where a very broad line is seen (e.g. MCG–6-30-15 – Fabian *et al.* 2002; NGC4051 – Ponti *et al.* 2006) the reflection appears to change little despite large variations in the continuum. The spectral variability can be decomposed into a highly variable power-law and a quasi-constant reflection component. This behaviour is also borne out by a difference (high-low) spectrum which is power-law in shape and the reflection-like shape of the spectrum of the intercept in flux-flux plots (Figs. 6–7).

This behaviour was initially puzzling, until the effects of gravitational light bending were included (Fabian & Vaughan 2003; Miniutti et al. 2004; Miniutti & Fabian 2005). Recall that the extreme blurring in these objects means that much of the reflection occurs with a few r_g of the horizon of the black hole. The enormous spacetime curvature there means that changes in the position of the primary power-law continuum have a large effect on the flux seen by an outside observer (Martocchia & Matt 1996; Martocchia et al. 2002). What this means is that an intrinsically constant continuum source can appear to vary by large amounts just be moving about in this region of extreme gravity. The reflection component, which comes from the spatially fixed accretion disc, appears relatively constant in flux in this region. Consequently, the observed behaviour of these objects may just be a consequence of strong gravity.

Some of the Narrow-Line Seyfert 1 galaxies such as 1H0707, IRAS13224 and 1H0439 appear to share this behaviour (Fabian et al. 2002b, 2004, 2005) and can be interpreted in terms of extreme light bending. Some of these objects can show sharp drops around 7 keV that may be alternatively interpreted as due to absorption from something only partially covering the source (Boller et al 2002). (If the covering was total then no strong soft emission would be seen, contrary to observation.) The GBH XTE J1650-500 behaved in a manner similar to that expected from the light bending model (Rossi et al. 2005).

4. Polarization

The X-ray emission from accreting black holes is expected to be polarized (Rees 1975), with general relativistic effects influencing the degree and angle of polarization of emission from the innermost regions (Stark & Connors 1977; Dovčiak et al. 2004, Fig. 8). Hopefully, a mission in the near future will carry a sensitive polarimeter so that this powerful information channel can be opened up.

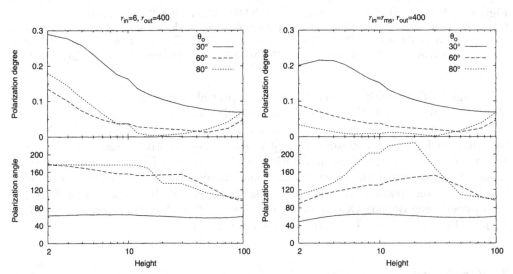

Figure 8. Polarization properties in the 9–12 keV band expected from a disc around a rapidly spinning black hole as a function of the height of the source above the disc in gravitational radii (Dovčiak, Karas & Matt 2004). The different lines are for different inclination angles. The inner radius of the disc is $6r_g$ on the left and $1.2r_g$ on the right.

5. Discussion

Clear examples of relativistically-broadened iron lines are seen in some AGN and GBH in some states. Such objects must have dense inner accretion disks in order that the gas is not over-ionized. Detection of a line is helped greatly if the iron abundance is super-Solar and if there is little extra absorption due to very strong warm absorbers or winds. Where broad lines are seen and can be modelled satisfactorilly then the spin of the central black hole can be reliably determined.

The study of absorption and emission variability of iron-K lines is in its infancy, with some interesting and tantalising results produced so far. The inner regions of accretion flows are bound to be structured and so give rise to variations. Some may be due to motion or transience in the corona or primary power-law source while others may reflect structure, e.g. spiral waves, on the disk itself, intercepting primary radiation from much smaller radii.

The number of broad lines detected is increasing and will continue to expand with the improved broad band coverage from Suzaku. Hints that broad lines are common in fainter objects such as in the Lockman Hole (Streblyanska *et al.* 2005) and Chandra Deep Fields (Brusa *et al.* 2005) could indicate that the conditions necessary for strong line production, perhaps high metalicity, are common in typical AGN at redshifts 0.5–1.

X-ray astronomers have an excellent tool with which to observe the innermost regions of accretion disks immediately around spinning black holes. The effects of redshifts and light bending expected from strong gravity in this regime are clearly evident. This can and should be exploited by future X-ray missions. To make significant progress we need large collecting areas. The count rate in the broad iron line of MCG–6-30-15 is about 2 ph m^{-2} s^{-1}, which means that square metres of collecting area are required around 6 keV in order to look for reverberation effects. GBH are much brighter but the orbital periods of matter close to the black hole are much, much smaller so reverberation is difficult here. Instead, variations with mass accretion rate and source state are accessible.

Acknowledgements

I am grateful to Vladimír Karas and Giorgio Matt for proposing and organising an interesting and successful Symposium. I am also grateful to many colleagues for work on broad iron lines, including Giovanni Miniutti, Jon Miller, Chris Reynolds, Andy Young, Kazushi Iwasawa, Randy Ross, Simon Vaughan, Luigi Gallo, Thomas Boller, Jamie Crummy, Josefin Larsson, and to The Royal Society for continued support. This brief review is an update of one I presented at the ESAC meeting on Variable and Broad Iron Lines.

References

Beckwith, K. & Done, C. 2004, MNRAS, 352, 353
Boller, Th. *et al.* 2002, MNRAS, 329, L1
Brennemann, L. & Reynolds, C. S. 2006, ApJ 652, 1028
Brusa, M., Gilli, R. & Comastri, A. 2005, ApJ, 621, L5
Crummy, J., Fabian, A. C., Gallo, L. & Ross, R. R. 2006, MNRAS, 365, 1067
Dabrowski, Y., Fabian, A. C., Iwasawa, K., Lasenby, A. N. & Rynolds, C. S. 1997, MNRAS, 288, L11
Dovčiak, M., Karas, V. & Yaqoob, T. 2004, ApJSS, 153, 205
Dovčiak, M., Karas, V. & Matt, G. 2004, MNRAS, 355, 1005
Fabian, A. C., Rees, M. J., Stella, L. & White, N. E. 1989, MNRAS, 238, 729
Fabian, A. C., Iwasawa, K., Reynolds, C. S. & Young, A. J. 2000, PASP, 112, 1145
Fabian, A. C. *et al.* 2002a, MNRAS, 335, L1

Fabian, A. C., Ballantyne, D. R., Merloni, A., Vaughan, S., Iwasawa, K. & Boller, Th. 2002b, MNRAS, 331, L35

Fabian, A. C. & Vaughan, S. 2003, MNRAS, 340, L28

Fabian, A. C., Miniutti, G., Gallo, L., Boller, Th., Tanaka, Y., Vaughan, S. & Ross, R. R. 2004, MNRAS, 353, 1071

Fabian, A. C., Miniutti, G., Iwasawa, K. & Ross, R. R. 2005, MNRAS, 361, 795

Gierliński, M., Done, C. 2004, MNRAS, 349, L7

Guainazzi, M. et al. 1999, A&A, 341, L27

Guainazzi, M. et al. 2006, AN, 327, 1032 (artro-ph/0610151)

Krolik, J. H. & Hawley, J. F. 2002, ApJ, 573, 754

Laor, A. 1991, ApJ, 376, 90

Marconi, A. et al. 2005, MNRAS, 351, 169

Martocchia, A., Matt, G. & Karas, V. 2002a, A&A, 383, L23

Martocchia, A., Matt, G., Karas, V., Belloni, T. & Feroci, M. 2002b, A&A, 387, 215

Miller, J. M. et al. 2002a, ApJ, 570, L69

Miller, J. M. et al. 2002b, ApJ, 577, L15

Miller, J. M. et al. 2002c, ApJ, 578, 348

Miller, J. M. et al. 2003a, MNRAS, 338, 7

Miller, J. M. et al. 2004a, ApJ, 601, 450

Miller, J. M. et al. 2004b, ApJ, 606, L131

Miller, J. M., Fabian, A. C., Nowak, M. A. & Lewin, W. H. G. 2003b, Proc. of the 10th Marcel Grossman Meeting, astro-ph/0402101

Miniutti, G., Fabian, A. C., Goyder, R. & Lasenby, A. N. 2003, MNRAS, 344, L22

Miniutti, G. & Fabian, A. C. 2004, MNRAS, 349, 1435

Miniutti, G., Fabian, A. C. & Miller, J. M. 2004, MNRAS, 351, 466

Miniutti, G. et al. 2006, PASJ, 59, S315 (astro-ph/0609521)

Nandra, K., George, I. M., Mushotzky, R. F., Turner, T. J. & Yaqoob, T. 1997a, ApJ, 476, 70

Nandra, K. et al. 2006, AN, 327, 1039 (astro-ph/0610585)

Ponti, G. et al. 2006, MNRAS, 282, L53

Rees, M. J. 1975, MNRAS, 171 457

Reeves, J. et al. 2006, PASJ, 59, S301 (astro-ph/0610434)

Reynolds, C. S. & Begelman, M. C. 1997, ApJ, 488, 109

Reynolds, C. S. & Nowak, M. A. 2003, PhR, 377, 389

Ross, R. R. & Fabian, A. C. 2005, MNRAS, 358, 211

Rossi, S., Homan, J., Miller, J. M. & Belloni, T. 2005, MNRAS, 360, 763

Soltan, A. 1982, MNRAS, 200, 115

Stark, R. F. & Connors, P. A. 1977, Nature, 266, 429

Streblyanska, A., Hasinger, G., Finoguenov, A., Barcons, X., Mateos, S. & Fabian, A. C. 2005, A&A, 432, 395

Strohmayer, T. 2001, ApJ, 552, L49

Tanaka, Y. et al. 1995, Nature, 375, 659

Uttley, P. & McHardy, I. M. 2004, Progr. Th. Phys. S155, 170

Vaughan, S. & Fabian, A. C. 2004, MNRAS, 348, 1415

Vaughan, S., Iwasawa, K., Fabian, A. C. & Hayashida, K. 2004, MNRAS, 356, 524

Young, A. J., Lee, J. C., Fabian, A. C., Reynolds, C. S., Gibson, R. R. & Canizares, C. R. 2005, ApJ, 631, 733

Yu, Q. & Tremaine, S. 2002, MNRAS, 335, 965

FELIX MIRABEL: Could you comment on the possibility of probing empirically the connection between black-hole spin derived from the skewed iron lines and the observed jets power?

ANDY FABIAN: Not at the present time. Such correlation is possible, but it cannot be currently probed with, say, MCG–6-30-15 which is the one I have worked on most. This galaxy does have a radio source – it's virtually unresolved and I know of no study of its variability, but I have also learned that radiative efficiency of jets can be as low as $\sim 10^{-4}$. So the radio emission does not clearly tell us about power of the jet. In terms of Galactic black holes there has been very little work on correlating broad iron line properties with radio properties.

MAREK ABRAMOWICZ: Comment: I would like to remark that there is another example of the effect described by the speaker – namely, the QPO variability in microquasars may also be caused or influenced by motion of the source in cooperation with the strong-gravity light bending (and *not* by the intrinsic variations of the flux). This is part of the model described by Michal Bursa who is also present here in the room.

THAISA STORCHI-BERGMANN: When you and others discuss observations of the Fe $K\alpha$ line, always the same source MCG–6-30-15 is presented. Why is that – what about other sources with broad iron K lines?

ANDY FABIAN: Broad lines are seen in other sources, but MCG–6-30-15 is the clearest. The fraction of sources (AGN) with broad lines is discussed in the work on XMM–Newton data by Paul Nandra *et al.* Also recently at a workshop on broad iron lines at VILSPA, Matteo Guainazzi *et al.* showed the fraction of sources with broad iron lines is similar to that found with ASCA ($\sim 30\%$). It turns out that MCG–6-30-15 happens to have, I believe, high iron abundance – we can see that also from the reflection hump in Suzaku data. Perhaps one has to be somewhat lucky with the objects that show high iron abundance.

ALEXANDER ZAKHAROV: Are the images of accretion discs – those images which were presented in the first part of your talk – an artist's view? If computed, the images should exhibit a bended part of the image from behind the black hole, especially in case of Kerr black hole.

ANDY FABIAN: Yes, these were artist's impressions apart from the simulation by Armitage & Reynolds.

GREGORY BESKIN: What about magnetic field influence on the iron lines, especially in binary systems?

ANDY FABIAN: I don't think there is any direct influence, although we do believe that the power-law continuum is powered by magnetic energy from the disc.

BORIS KOMBERG: I would like to know your opinion about the nature of X-ray sources that appear very high above of the accretion disc (primary sources of X-rays).

ANDY FABIAN: I believe they are connected with magnetic field twisting and reconnection, but we still do not understand details of that process. Perhaps the primary X-ray emission is generated at the base of the jet, but I would stress that's still a speculation.

MARGARITA SAFONOVA: Is it possible to quantify the dependence of the spin of the black hole with the iron line broadening?

ANDY FABIAN: Yes; for this the entire shape of the line and the continuum have to be fitted.

Black Holes from Stars to Galaxies – Across the Range of Masses
Proceedings IAU Symposium No. 238, 2006
V. Karas & G. Matt, eds.
© 2007 International Astronomical Union
doi:10.1017/S1743921307004851

Black holes and magnetic fields

Jiří Bičák,[1] Vladimír Karas[2] and Tomáš Ledvinka[1]

[1]Institute of Theoretical Physics, Faculty of Mathematics and Physics, Charles University,
V Holešovičkách 2, CZ-18000 Prague, Czech Republic

[2]Astronomical Institute, Academy of Sciences, Boční II, CZ-14131 Prague, Czech Republic

email: bicak@mbox.troja.mff.cuni.cz, vladimir.karas@cuni.cz, ledvinka@mbox.troja.mff.cuni.cz

Abstract. Stationary axisymmetric magnetic fields are expelled from outer horizons of black holes as they become extremal. Extreme black holes exhibit Meissner effect also within exact Einstein–Maxwell theory and in string theories in higher dimensions. Since maximally rotating black holes are expected to be astrophysically most important, the expulsion of the magnetic flux from their horizons represents a potential threat to an electromagnetic mechanism launching the jets at the account of black-hole rotation.

Keywords. Black hole physics – magnetic fields – galaxies: jets

---

1. Introduction

The exact mechanism of formation of highly relativistic jets from galactic nuclei and microquasars remains unknown. Four ways by which a black hole or its accretion disk could power two opposite jets are indicated in figure 1 (taken from the *Czech* edition of Kip Thorne's popular book to indicate how, as compared with the last 1967 IAU meeting in Prague, black holes domesticated even in central Europe): (a) wind from the disk may blow a bubble in a spinning gas cloud and hot gas makes the orifices through which jets are shot out; (b) the surface of the puffed rotating disk forms funnels which collimate the wind; (c) magnetic field lines anchored in the disk are spinning due to disk's rotation and push plasma to form jets; (d) magnetic lines threading through the hole are forced to spin by the "rotating geometry" and push plasma outwards along the rotation axis.

The last way, the Blandford–Znajek mechanism, is considered to be the most relevant. The field brought into the innermost region and onto the black hole from the outside has clean field structure while the field in/around the disk is expected to be quite chaotic (Thorne *et al.* 1986). The estimated power radiated out from the "load" regions farther away from the hole is

$$\Delta L_{\max} \simeq \left[10^{45}\frac{\text{erg}}{\text{sec}}\right]\left[\frac{a}{M}\right]^2\left[\frac{M}{10^9 M_\odot}\right]^2\left[\frac{B_n}{10^4 G}\right]^2. \tag{1.1}$$

Here $a \equiv J/M$ is the hole's angular momentum per unit mass (velocity of light $c = 1$, gravitational constant $G = 1$), B_n is the normal magnetic field at the horizon. Hence, the highest power is achieved when the hole is "extreme", i.e., rotating with maximal angular momentum $a = M$, and B_n, which directly determines magnetic flux across the horizon, is as high as possible.

2. Meissner effect

The main purpose of this contribution is to point out that these two aspects go one against the other: black holes approaching extremal states exhibit a "Meissner effect" – they expel external vacuum stationary (electro)magnetic fields.

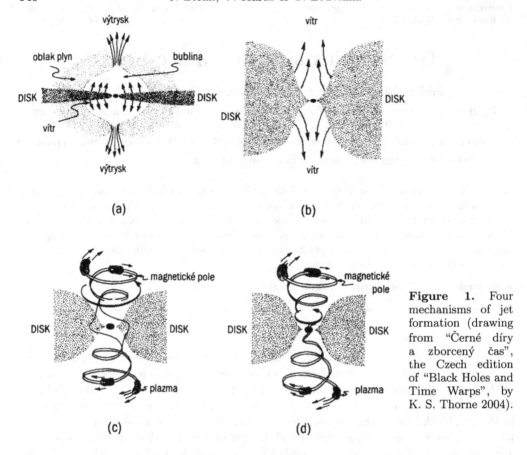

Figure 1. Four mechanisms of jet formation (drawing from "Černé díry a zborcený čas", the Czech edition of "Black Holes and Time Warps", by K. S. Thorne 2004).

To see this effect in the simplest situation, consider the magnetic test field B_0 which is uniform at infinity and aligned with hole's rotation axis. Solution of Maxwell's equations on the background geometry of a rotating (Kerr) black hole with boundary condition of uniformity at infinity and finiteness at the horizon yields the field components; from these the lines of force are defined as lines tangent to the Lorentz force experienced by test magnetic/electric charges at rest with respect to locally non-rotating frames (preferred by the Kerr background field). The field lines are plotted in figure 2 for $a = 0.5M$ and in extreme case $a = M$. Notice that only weak expulsion occurs in the former case. There is a simple analytic formula for the flux across the hemisphere of the horizon: $\Phi = B_0\pi r_+^2(1 - a^4/r_+^4)$, where $r_+ = M + (M^2 - a^2)^{1/2}$ (King et al. 1975; Bičák & Janiš 1985).

As a consequence of the coupling of magnetic field to frame-dragging effects of the Kerr geometry the electric field of a quadrupolar nature arises. Its field lines are shown in figure 3. Again the flux expulsion takes place. While even with $a = 0.95M$ it is still not very distinct, the expulsion becomes complete in the extreme case.

One can demonstrate that total flux expansion takes place for all axisymmetric stationary fields around a rotating black hole (Bičák & Janiš 1985; Bičák & Ledvinka 2000). In figure 4 the field lines of a current loop in the equatorial plane are shown.

The Meissner-type effect arises also for charged (Reissner–Nordström) black holes. Although extremely charged black holes ($e^2 = M^2$) are probably not important astrophysically they may be significant in fundamental physics (as very special supersymmetric BPS states mass of which does not get any quantum corrections). Since electromagnetic

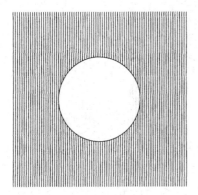 

Figure 2. Field lines of the test magnetic field uniform at infinity and aligned with hole's rotation axis. Two cases with $a = 0.5M$ (left) and $a = M$ (right) are shown.

 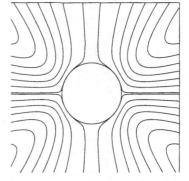

Figure 3. Field lines of the electric field induced by the "rotating geometry" of Kerr black hole in asymptotically uniform test magnetic field. $a = 0.95M$ (left), and $a = M$ (right).

perturbations are in general coupled to gravitational perturbations, the resulting formalism is involved. Nevertheless, one may construct explicit solutions, at least in stationary cases. From these the magnetic field lines follow as in the Kerr case. The magnetic field lines of a dipole far away from the hole look like in a flat space (Fig. 5a), however, when the dipole is close to the horizon, the expulsion in the extreme case is evident (Fig. 5b). Due to the coupling of perturbations closed field lines appear without any electric current inside; see Bičák & Dvořák (1980) for details.

There exist exact models (exact solutions of the Einstein–Maxwell equations) representing in general rotating, charged black holes immersed in an axisymmetric magnetic field. The expulsion takes place also within this exact framework – see Bičák & Karas (1989), Karas & Vokrouhlický (1991), Karas & Budínová (2000).

Very recently the Meissner effect was demonstrated for extremal black-hole solutions in higher dimensions in string theory and Kaluza–Klein theory. The question of the flux expulsion from the horizons of extreme black holes in more general frameworks is not yet understood properly. The authors of Chamblin, Emparan & Gibbons (1998) "believe this to be a generic phenomenon for black holes in theories with more complicated field content, although a precise specification of the dynamical situations where this effect is present seems to be out of reach."

The flux expulsion does *not* occur for the fields which are not axisymmetric. In figure 6a,b the field lines are constructed for fields asymptotically uniform and perpendicular to

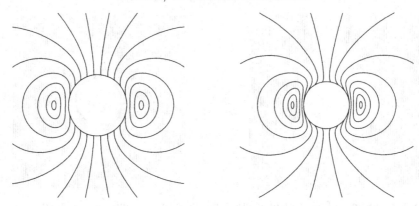

Figure 4. Field lines of the magnetic field of a current loop in the equatorial plane of Kerr black hole located at $r = 1.5\,r_+$. Two cases with $a = 0.9M$ (left) and $a = 0.995M$ (right) are shown.

Figure 5. Field lines of the magnetic dipole placed far away from the extreme Reissner–Nordström black hole ($e = M$, left panel), and close to the hole (right panel).

the axis of an extreme Kerr black hole. There is an angle $\delta_{\mathrm{max}} \sim -63°$ for which the flux of B_1 component across a hemisphere is maximal, $\Phi_{\mathrm{max}} \sim 2.25 B_1 \pi r_+^2$. This is the effect of the rotating geometry (for detailed description, see Bičák & Janiš 1985; Dovčiak *et al.* 2000). In these "misaligned" situations, there is an angular momentum flux to infinity, so such conditions are not stationary.

Naturally, the shape of magnetic and electric field lines depends on the choice of observers, however, the whole discussion can be cast in equivalent and invariant form by employing surfaces of constant flux in which field lines reside. In this way, figure 7 demonstrates that the cross-section of the black hole for the capture of non-aligned magnetic fields indeed does not vanish as the hole rotation aproaches the extreme value.

3. Recent progress, open problems

Can these important properties of electromagnetic fields in neighborhood of black holes be relevant in astrophysical conditions? Recently remarkable progress in axisymmetric simulations of rotating black holes surrounded by magnetized plasma based on general-relativistic MHD has been achieved by various authors, in particular by C. Gammie, S. Komissarov, J. McKinney, J. Krolik, D. Uzdensky, H. Kim, A. Aliev and others (see the E-print archive for references). Such simulations could bring an answer although some new analytic insight may also be required. For example, McKinney & Gammie (2004) remark "we see no sign of expulsion of flux from the horizon... It is possible that we have

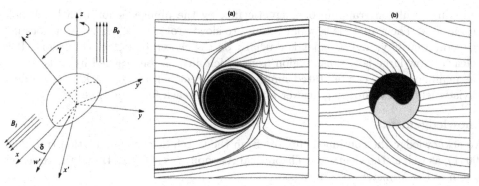

Figure 6. Lines of the magnetic field which is asymptotically uniform and perpendicular to the rotation axis. The equatorial plane is shown as viewed from top, i.e. along the rotation axis, (a) in the frame of zero angular momentum observers orbiting at constant radius; (b) in the frame of freely falling observers. In the panel (b), two regions of ingoing/outgoing lines are distinguished by different levels of shading of the horizon (the hole rotates counter-clockwise, $a = M$).

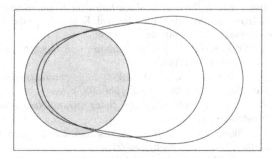

Figure 7. The cross-sectional area for the capture of magnetic field lines (asymptotically uniform magnetic field perpendicular to the rotation axis). The three curves correspond to different values of the black-hole angular momentum: $a = 0$ (the black hole cross-section is a perfect circle and its projection coincides with the black-hole horizon of radius $2M$, indicated here by shading), $a = 0.95M$, and $a = M$ (the most deformed shape refers to the maximally rotating case). Non-vanishing cross-sectional area for the extreme case demonstrates that the Meissner-type expulsion of the magnetic field does not operate on non-axisymmetric fields. The hole's rotation axis is vertical and the magnetic field lines are pointing "towards us" (see Dovčiak *et al.* 2000 for details).

not gone close enough to $a/M = 1$." They have $a/M = 0.938$ and, indeed, in view of simplest (though just vacuum) situations illustrated in figures 2 and 3 this value may still be far from 1 to see the effect.

An interesting issue arises in connection with the "black-hole membrane paradigm" of Thorne *et al.* (1986). For an extreme black hole the proper distance from any $r > r_+$ to the horizon is logarithmically infinite, so one might tend to explain the vanishing of magnetic flux by this fact. However, although the flux across a "stretched horizon" (at a finite distance) is non-vanishing even in the extreme case, it depends on where the stretched horizon is located. It turns out, (see section 4 of Bičák & Ledvinka 2000) that for any $\epsilon > 0$ one can find such a stretched horizon that the flux is less then ϵ. This suggests the following question: does the power in the Blandford–Znajek model arise from regions with "relatively large $\sqrt{-g_{00}}$" in near extreme cases?

Another potential obstacle for Blandford–Znajek mechanism to operate efficiently is a low value of magnetic field brought onto the black hole from an accretion disk in realistic situations. Most recently Reynolds, Garofalo & Begelman (2006) obtained an encouraging result on trapping of magnetic flux by the plunge region of a black hole accretion disk. Their analysis is so far limited to slowly rotating black holes; it does not use general relativity; and it depends crucially on the chosen boundary conditions. It is not clear of how large is the disk region from which flux can be dragged inwards.

The main open issue can be stated simply: "Do extremely rotating black holes produce relativistic jets?" A compelling answer may be out of reach for some time yet.

Acknowledgements

We acknowledge the continued support from the Czech Science Foundation (ref. 202/06/0041), as well as the Centre for Theoretical Astrophysics (LC06014) and the Research Program of the Ministery of Education.

References

Bičák, J. & Dvořák, L. 1980, *Stationary electromagnetic fields around black holes III. General solutions and the fields of current loops near the Reissner–Nordström black hole*, Phys. Rev. D, 22, 2933

Bičák, J. & Janiš, V. 1985, *Magnetic fluxes across black holes*, MNRAS, 212, 899

Bičák, J. & Karas, V. 1989, *The influence of black holes on uniform magnetic fields*, in Proc. of the 5th Marcel Grossman Meeting on General Relativity, eds. D. G. Blair & M. J. Buckingham, (World Scientific: Singapore), p. 1199

Bičák, J. & Ledvinka, T. 2000, *Electromagnetic fields around black holes and Meissner effect*, Nuovo Cim., B115, 739 (gr-qc/0012006)

Chamblin, A., Emparan, R. & Gibbons, G. W. 1998, *Superconducting p-branes and extremal black holes*, Phys. Rev. D, 58, 084009 (hep-th/9806017)

Dovčiak, M., Karas, V. & Lanza, A. 2000, *Magnetic fields around black holes*, European Journal of Physics, 21, 303 (astro-ph/0005216)

Karas, V. & Budínová, Z. 2000, *Magnetic fluxes across black holes in a strong magnetic field regime* Physica Scripta, 61, 25 (astro-ph/0005216)

Karas, V. & Vokrouhlický, D. 1991, *On interpretation of the magnetized Kerr-Newman black hole*, J. Math. Phys., 32, 714

King, A. R., Lasota, J. P. & Kundt, W. 1975, *Black holes and magnetic fields*, Phys. Rev. D, 12, 3037

McKinney, J. C. & Gammie, C. F. 2004, *A measurement of the electromagnetic luminosity of Kerr black hole*, ApJ, 611, 977 (astro-ph/0404512)

Reynolds, C. S., Garofalo, D. & Begelman, M. C. 2006, *Trapping of magnetic flux by the plunge region for a black hole accretion disk*, ApJ, 651, 1023 (astro-ph/0607381)

Thorne, K. S. 2004, *Černé díry a zborcený čas: Pozoruhodná dědictví Einsteinova génia*, the Czech edition of *Black Holes and Time Warps: Einstein's Outrageous Legacy*, translation by J. Langer *et al.* (Mladá Fronta: Prague)

Thorne, K. S., Price, R. M. & MacDonald, D. A. 1986, *Black Holes: the Membrane Paradigm* (Yale Univ. Press: New Haven)

ANDREW KING: Comment: A misaligned magnetic field tries to reach axisymmetric configuration. It is interesting to notice that the timescale for that is comparable to the Blandford-Znajek timescale.

JIŘÍ BIČÁK: Yes indeed, the tendency to align the magnetic field with the black hole rotation axis is natural here, since in a misaligned case there is the angular momentum flux out of the system. Of course, even in case of the misaligned geometry the magnetic field component parallel to the axis becomes expelled out of the extreme Kerr black hole.

Black Holes from Stars to Galaxies – Across the Range of Masses
Proceedings IAU Symposium No. 238, 2006
V. Karas & G. Matt, eds.
© 2007 International Astronomical Union
doi:10.1017/S1743921307004863

Constraining jet physics in weakly accreting black holes

Sera Markoff

Astronomical Institute "Anton Pannekoek", University of Amsterdam, Kruislaan 403,
1098SJ Amsterdam, The Netherlands
email: sera@science.uva.nl

Abstract. Outflowing jets are observed in a variety of astronomical objects such as accreting compact objects from X-ray binaries (XRBs) to active galactic nuclei (AGN), as well as at stellar birth and death. Yet we still do not know exactly what they are comprised of, why and how they form, or their exact relationship with the accretion flow. In this talk I focus on jets in black hole systems, which provide the ideal test population for studying the relationship between inflow and outflow over an extreme range in mass and accretion rate.

I present several recent results from coordinated multi-wavelength studies of low-luminosity sources. These results not only support similar trends in weakly accreting black hole behavior across the mass scale, but also suggest that the same underlying physical model can explain their broadband spectra. I discuss how comparisons between small- and large-scale systems are revealing new information about the regions nearest the black hole, providing clues about the creation of these weakest of jets. Furthermore, comparisons between our Galactic center nucleus Sgr A* and other sources at slightly higher accretion rates can elucidate the processes which drive central activity, and pave the way for new tests with upcoming instruments.

Keywords. Black hole physics – accretion, accretion disks – radiation mechanisms: nonthermal – X-rays: binaries – galaxies: active – galaxies: jets

---

1. Introduction

One of the most basic predictions of general relativity is that black hole physics should be the same regardless of black hole mass. However, black holes at different ends of the stellar-to-galactic mass range are accreting from different environments; i.e., from one star versus an entire stellar cluster. This difference, together with the huge range in dynamical timescales which go roughly linearly with the mass, has made it challenging to discover if the observed physical processes are indeed similar between accreting systems such as X-ray binaries (XRBs) and Active Galactic Nuclei (AGN).

Recent advances in simultaneous broadband observing techniques have resulted in the first confirmation of scaling physics across the black hole masses, in particular for the case of weakly accreting black holes. For XRBs this means the Low/Hard state (for definitions see, e.g. McClintock & Remillard 2006), and for AGN this includes the general class of low-luminosity AGN (LLAGN; Ho 1999), and likely FR Is and BL Lacs. All of these classes are believed to be accreting at rates significantly below the Eddington limit, and thus in similar state to each other. This state is of particular interest for studying jet physics because it is the only XRB state associated with steady jet production, and jets are observed to increasingly dominate the power output of the accreting system as $\dot{M}$ decreases (Fender *et al.* 2003).

In brief summary, simultaneous observations in the radio and X-ray bands of the XRB GX 339-4 in its Low/Hard state established a tight correlation between the respective

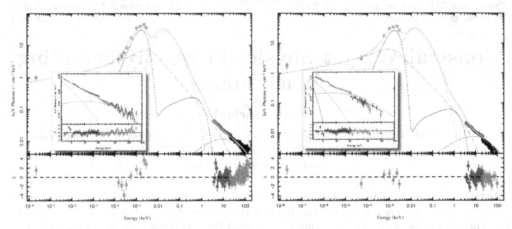

Figure 1. Model fits with residuals to broadband simultaneous observations of X-ray binary GRO J1655-40 in the Low/Hard state on 24 Sep (left) and 29 Sep (right), 2005. The data are from the VLA, Spitzer, SMARTS and RXTE, see Migliari *et al.* (2007) for details. Inset shows the X-ray band only. Emission components indicated are jet base/corona synchrotron (blue-green), post-acceleration outer jet synchrotron (green), jet base/corona Comptonization (external and SSC; red), and the single blackbody for the star plus the multicolor blackbody thermal accretion disk (magenta). The reduced χ^2 for these two fits are 1.7 and 0.9, respectively.

luminosities holding over orders of magnitude in power (Corbel *et al.* 2000, 2003). Gallo *et al.* (2003) then found that this correlation appears in most Low/Hard state XRBs for which we have good simultaneous data. Markoff *et al.* (2003) and Heinz & Sunyaev (2003) explored the physical mechanisms responsible for the correlation, and two independent groups then showed that the same correlation is present in samples of weakly accreting AGN (Merloni *et al.* 2003; Falcke *et al.* 2004). The theoretically predicted and empirically confirmed relationship between radio luminosity, X-ray luminosity and mass has been named the "fundamental plane of black hole accretion". Some recent papers have further explored and improved upon the original statistics (Merloni *et al.* 2006; Körding *et al.* 2006).

These results support the picture that similar processes are at work in these classes of sources. In order to further test this idea, as well as to learn more about the physics near the black holes and especially jet formation, my colleagues and I have been exploring how well the same model can apply to objects at extreme ends of the mass scale. Not only would this provide independent checks on the "fundamental plane", but also by comparing stellar and supermassive black hole accretion we can better understand which physical mechanisms act exactly the same across the mass scale.

2. Scaling physics across the mass scale

The model used in the fits presented here is described in detail in Markoff *et al.* (2005). It is an outflow-dominated model in the sense that we include the corona as a compact, weakly beamed region comprising the base of the jets. Beyond arguments why such outflowing, magnetized coronae are useful for reducing the amount of reflected emission (e.g. Beloborodov 1999), we also wanted to explore the relationship between the corona and jets for this most extreme case. A cooler, thermal accretion disk is also included both as a weak spectral component, as well as a source of photons for Compton upscattering within the jets/coronae. In Markoff *et al.* (2005), we showed that statistical fits of this model to the X-ray region alone (including a Gaussian iron line, and convolved with a

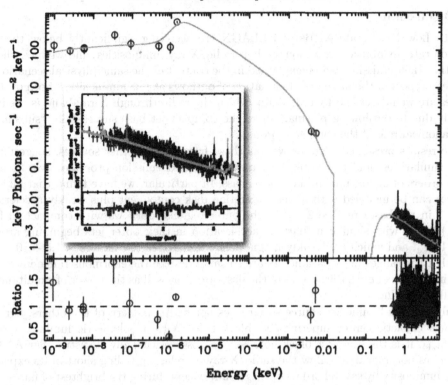

Figure 2. The same exact model used for GRO J1655-40 shown here applied to data from a simultaneous broadband campaign of the low-luminosity AGN M81*, with mass $\sim 10^7$ times greater. The data are from the GMRT, VLA, IRAM, SMA, HST (not simultaneous) and Chandra with the grating spectrometer (see Markoff *et al.*, in prep. for details). The reduced χ^2 of this fit is 1.37. Several lines were detected in the X-ray band as well, see Young *et al.* (2007).

reflection model) could describe data from two Galactic XRBs, GX 339-4 and Cyg X-1, as well as thermal corona models. Furthermore this model also naturally addresses the simultaneous radio emission and thus the correlations. Based on the fitting of several epochs of Low/Hard state data for both sources, we concluded that several physical parameters which we had previously left free to vary could be frozen for future fits. These parameters included the ratio between the height and width of the jet base/corona, which always remained quite compact ($\sim 1 \div 1.5$), and the fraction of particles accelerated in the jets, which we found to be consistently high ($\sim 75\%$).

More recently we have obtained new broadband spectra for other sources which also include the infrared/optical (IRO) frequencies, providing even further constraints on this outflow-dominated model. In Fig. 1 we show the result of two examples of such fits for the Low/Hard state of XRB GRO J1655-40, with an inset detailing the X-ray fit (from Migliari *et al.* 2007). The IRO range is mostly dominated by the stellar companion, but a NIR excess indicates a significant contribution from jet synchrotron, consistent with the conclusions of Russell *et al.* (2006), who found $\sim 90\%$ of the NIR is contributed by the jets in bright Low/Hard state sources.

The exact same model has also been successfully applied to several weakly-accreting supermassive black holes. In Fig. 2 we show the results for the nearby galaxy M81.

3. Discussion

Both Low/Hard state XRBs and LLAGN are accreting significantly below the Eddington rate, exhibit a correlation in their radio/X-ray luminosities, and we have shown here that their emission can be explained in the context of the same physical scenario. As further support of the increased domination of outflows at low-luminosities, both of these classes are well described by a model in which the radio through X-ray band is predominantly due to outflowing plasma, where the compact jet base successfully "subsumes" the canonical role of the compact corona.

The results presented here, as well as those from several other sources, suggest that basic similarities hold in the physical parameters and emission processes over at least seven orders of magnitude in black hole mass. In particular, we have found that all such sources can be modeled with a weak accretion disk component plus weakly accelerated ($\Gamma \sim 1$ in the bases to $\Gamma \sim 2 \div 3$ in the outer regions) outflows with compact (a few- $10\ r_g$) bases, with significant particle acceleration in their outer jets beginning around $10 \div 100 r_g$, and which are weakly magnetically dominated (by factors $\sim 1.5 \div 10$). The fact that the ranges in fit parameters are so small despite the enormous range in masses is an indication of both the success of the description as well as the power of simultaneous broadband fitting.

Only one low-luminosity source so far does not fit the pattern of the others: Sgr A*, our own Galactic center supermassive black hole. With a bolometric luminosity of $\sim 10^{-9} L_{\mathrm{Edd}}$, it is the dimmest black hole we can observe with any statistics. Sgr A* first of all does not appear to follow the radio/X-ray correlation, falling short of its expected X-ray luminosity by several orders of magnitude except during the brightest of flares (see, e.g. Markoff 2005). Interestingly, we can use the results of the above spectral modeling in combination with morphological constraints from VLBI (Bower *et al.* 2004; Shen *et al.* 2005; Bower *et al.* 2006) to probe the reasons behind such differences. Spectrally, Sgr A* can be explained with the same context of other weakly accreting black holes *only* if particle acceleration is very weak in its jets, either with a very low fraction or a very steep ($p > 4$) electron distribution spectrum. The lack of a strong optically thin power component in its spectrum makes Sgr A* quite different from other LLAGN, and can also account for the difficulty in resolving its jets through the intervening electron scattering screen. We demonstrate this explicitly in Markoff *et al.* (2007). Our results suggest that particle acceleration in the outflows may break down at extremely low accretion rates, hinting at at least one of the many mechanisms that must build up in order to create signs of black hole "activity". This also suggests that classes/states of active objects which do not show resolved jets may still have outflow structures which lack the mechanism to accelerate.

References

Beloborodov, A. M. 1999, ApJL, 510, L123
Bower, G. C., Falcke, H., Herrnstein, R. M., *et al.* 2004, Science, 304, 704
Bower, G. C., Goss, W. M., Falcke, H., *et al.* 2006, ApJL, 648, L127
Corbel, S., Fender, R. P., Tzioumis, A. K., *et al.* 2000, A&A, 359, 251
Corbel, S., Nowak, M., Fender, R. P., *et al.* 2003, A&A, 400, 1007
Falcke, H., Körding, E. & Markoff, S. 2004, A&A, 414, 895
Fender, R. P., Gallo, E. & Jonker, P. G. 2003, MNRAS, 343, L99
Gallo, E., Fender, R. P. & Pooley, G. G. 2003, MNRAS, 344, 60
Heinz, S., Sunyaev, R. A. 2003, MNRAS, 343, L59
Ho, L. C. 1999, ApJ, 516, 672

Körding, E., Falcke, H. & Corbel, S. 2006, A&A, 456, 439

Markoff, S. 2005, ApJL, 618, L103

Markoff, S., Bower, G. C. & Falcke, H. 2007, MNRAS, submitted

Markoff, S., Nowak, M., Corbel, S., Fender, R. & Falcke, H. 2003, A&A, 397, 645

Markoff, S., Nowak, M. A. & Wilms, J. 2005, ApJ, submitted

McClintock, J. E., Remillard, R. A. 2006, in: W. H. G. Lewin & M. van der Klis (eds.), Compact Stellar X-ray Sources (Cambridge University Press: Cambridge), p. 157

Merloni, A., Heinz, S. & di Matteo, T. 2003, MNRAS, 345, 1057

Merloni, A., Körding, E., Heinz, S., Markoff, S., *et al.* 2006, New Astronomy, 11, 567

Migliari, S., Tomsick, J. A., Markoff, S., *et al.* 2007, ApJ, submitted

Russell, D. M., Fender, R. P., Hynes, R. I., *et al.* 2006, MNRAS, 371, 1334

Shen, Z.-Q., Lo, K. Y., Liang, M.-C., Ho, P. T. P. & Zhao, J.-H. 2005, Nature, 438, 62

Young, A. J., Nowak, M. A., Markoff, S., Marshall, H. L., *et al.* 2007, ApJ, submitted

GREGORY BESKIN: Some years ago we detected several very short optical flashes of duration around one millisecond and brightness temperature above 10^{10} K in A0620–00. In some sense this is a proof of magnetized corona.

SERA MARKOFF: Yes, this is a very interesting and lively topic which I could not cover in detail. There is a lot of variability that cannot be coming from the star.

Physical Processes Near Black Holes II

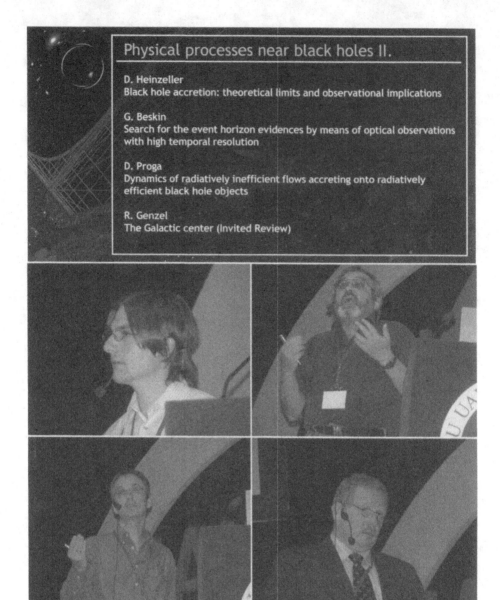

Physical processes near black holes II.

D. Heinzeller
Black hole accretion: theoretical limits and observational implications

G. Beskin
Search for the event horizon evidences by means of optical observations with high temporal resolution

D. Proga
Dynamics of radiatively inefficient flows accreting onto radiatively efficient black hole objects

R. Genzel
The Galactic center (Invited Review)

The Stone Bell House, the origins of which date back to the 13th century. The richly decorated front facade in Gothic style.

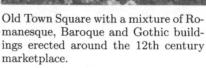

Old Town Square with a mixture of Romanesque, Baroque and Gothic buildings erected around the 12th century marketplace.

Black Holes from Stars to Galaxies – Across the Range of Masses
Proceedings IAU Symposium No. 238, 2006
V. Karas & G. Matt, eds.

Black hole accretion: theoretical limits and observational implications

Dominikus Heinzeller,[1,2] **Wolfgang J. Duschl,**[1,2,3] **Shin Mineshige**[4] **and Ken Ohsuga**[5]

[1]Zentrum für Astronomie Heidelberg, Institut für Theoretische Astrophysik, Heidelberg, Germany

[2]Institut für Theoretische Physik und Astrophysik, Universität Kiel, Kiel, Germany

[3]Steward Observatory, The University of Arizona, Tucson, AZ 85721, USA

[4]Yukawa Institute for Theoretical Physics, Kyoto University, Sakyo-ku, Kyoto, Japan

[5]Department of Physics, Rikkyo University, Toshimaku, Tokyo, Japan

email: hd, wjd@astrophysik.uni-kiel.de, minesige@yukawa.kyoto-u.ac.jp, k_ohsuga@rikkyo.ac.jp

Abstract. Recently, the issue of the role of the Eddington limit in accretion discs became a matter of debate. While the classical (spherical) Eddington limit is certainly an over-simplification, it is not really how to treat it in a flattened structure like an accretion disc. We calculate the critical accretion rates and resulting disc luminosities for various disc models corresponding to the classical Eddington limit by equating the attractive and repulsive forces locally. We also discuss the observational appearance of such highly accreting systems by analyzing their spectral energy distributions. Our calculations indicate that the allowed mass accretion rates differ considerably from what one expects by applying the Eddington limit in its classical form, while the luminosities only weakly exceed their classical equivalent. Depending on the orientation of the disc relative to the observer, mild relativistic beaming turns out to have an important influence on the disc spectra. Thus, possible super-Eddington accretion, combined with mild relativistic beaming, supports the idea that ultraluminous X-ray sources host stellar mass black holes and accounts partially for the observed high temperatures of these objects.

Keywords. accretion, accretion discs – radiative transfer – galaxies: active – galaxies: nuclei

1. Introduction

With the constant improvement of observational techniques in the past decades, more and more detailed information about accretion disc systems could be gained. In particular, a large number of ultraluminous X-ray sources (ULXs) have been discovered (Fabbiano 1989), imposing a severe problem upon the existing general idea of accretion disc systems: With a bolometric luminosity exceeding $10^{39} \, \mathrm{erg \, s^{-1}}$, at least some of them show relatively low radiation temperatures ($\sim 0.1 \, \mathrm{keV}$). These systems have been suggested to be intermediate mass black hole (IMBH), sub-Eddington accretion disc systems (Roberts, Warwick, Ward et al. (2005)). However, large samples of ULXs from recent observational data reveal that a distinct class exists, showing higher temperatures – sometimes exceeding $1 \, \mathrm{keV}$ – than can be explained by IMBHs (Mizuno, Ohnishi, Kubota et al. (1999)). In contrast, stellar mass black holes accreting above their Eddington limit can account for these sources (Watarai, Mizuno & Mineshige (2001)).

In particular, the applicability of the classical Eddington limit is an important point in this discussion: Similar to the stellar case, disc accretion may be limited by radiation pressure, counteracting gravity and viscous dissipation. While in the stellar case, we are dealing with an approximately spherically symmetric, i. e., 1-dimensional situation, discs require an – at least – 2-dimensional treatment. The stellar Eddington limit implies several assumptions, which do not apply for the disc case, though it is not clear *a priori* to what degree a proper treatment will alter the resulting numbers.

In our work, we calculate the *local* analog to the Eddington limit for a classical thin disc model of the Shakura & Sunyaev (1973) variety and for a slim disc model of the Abramowicz, Czerny, Lasota *et al.* (1988) type (Sect. 2). In both descriptions, a standard α-viscosity is applied, therefore restricting the models to the non-selfgravitating case. We also investigate the spectral energy distribution of supercritical accretion flows in the stellar mass black hole case, based on the radiation hydrodynamic (RHD) simulations computed by Ohsuga, Mori, Nakamoto *et al.* (2005) (Sect. 3).

2. Black hole accretion and the Eddington limit

The physical reasoning for the Eddington limit is an equilibrium of the attractive force F_g and the repelling force F_r. The original stellar Eddington limit relies on several assumptions: (1) spherical symmetry of the system; (2) isotropic radiation; (3) homogeneous degree of ionisation; (4) Thomson scattering as the sole source of opacity; (5) negligible gas pressure; (6) no relativistic effects; and (7) no time dependence (stationarity). To investigate if these assumptions are an oversimplification in the disc case, we relinquish the approximations (1)–(5), but keep the approximation of non-relativistic stationarity for our calculation.

We assume azimuthal symmetry in the following. Therefore, one can define two different "Eddington limits", corresponding to the equilibria in the vertical (z) and radial (s) direction. (We use a cylindrical coordinate system $\{s, \varphi, z\}$ with the distance r to the origin, $r^2 = s^2 + z^2$.) For a rotating viscous disc, further contributions to the total force have to be included in the radial direction:

$$F_{\text{tot}}^{(z)}(s) = F_r^{(z)}(s) - F_g^{(z)}(s) = 0, \qquad F_{\text{tot}}^{(s)}(s) = F_r^{(s)}(s) - F_g^{(s)}(s) + \ldots = 0.$$

In the following, as an example, we calculate the maximum amount of matter $\dot{M}_{\text{crit}}$ that can be accreted towards the central object in the case of a stellar mass black hole, $M = 10\,M_\odot$, for non-selfgravitating discs with an α-parameter of 0.1. For both the thin and the slim disc case, we use two descriptions for the opacity: Firstly, simple Thomson scattering is applied: $\kappa_{\text{es}} = 0.4\,\text{cm}^2\text{g}^{-1}$. Secondly, we use an interpolation formula, $\kappa = \kappa_{\text{in}}$ (Gail, priv. comm.), which accounts for further contributions like bound-free and free-free absorption. For a detailed description of κ_{in} see Heinzeller & Duschl (2006).

We refer our results to the classical Eddington-limit, which in the disc case with a torque-free boundary condition at the disc's inner radius $s_i = 3r_g = 6GM/c^2$ is given by $L_E = 4\pi c G M/\kappa_{\text{es}} = 1.2 \cdot 10^{39}\,\text{erg/s}$, and $\dot{M}_E = 8\pi c s_i/\kappa_{\text{es}} = 2.6 \cdot 10^{-7}\,M_\odot/\text{a}$. Detailed calculations reveal that the vertical limit sets severer conditions on the critical accretion rates and corresponding luminosities than the radial limit in both the thin and the slim disc model. We therefore concentrate on the results for the vertical limit in the following.

Figure 1 shows the results for the critical accretion rates $\dot{M}_{\text{crit}}$ in the vertical Eddington limit as a function of the radial distance s from the central object. In the slim disc case, one free parameter remains to be determined by hand: the ratio of the local height h of the disc to the local radial distance s. Since the slim disc model is valid for $h/s \leqslant 1$, we plot the results for the two cases $(h/s)_{\text{max}} = 1.0$ and $(h/s)_{\text{max}} = 0.5$.

Obviously, the critical accretion rates are no longer given by a global quantity like $\dot{M}_E$: they depend on the radius $\propto s^{1.2\ldots1.9}$. While the opacity has a strong influence on the results ($\dot{M}_{\text{crit}} \propto \kappa^{-1}$) for low accretion rates, it becomes more and more unimportant

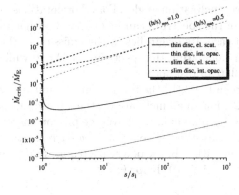

Figure 1. Critical local accretion rates in the vertical Eddington limit as a function of the radial distance s from the central object. For slim discs, $(h/s)_{\mathrm{max}} = 1.0$ in the upper and $(h/s)_{\mathrm{max}} = 0.5$ in the lower case. Black lines correspond to Thomson scattering, $\kappa = \kappa_{\mathrm{es}}$, gray lines to interpolated opacities, $\kappa = \kappa_{\mathrm{in}}$.

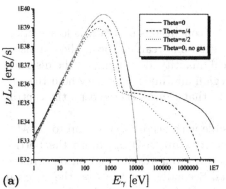

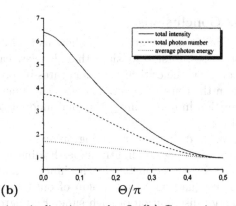

Figure 2. (a) Spectral energy distribution for various inclination angles Θ; (b) Comparison of total intensity, average photon energy and total photon number for $\Theta = 0 \ldots \pi/2$, normalized to the corresponding values at $\pi/2$.

in the slim disc model for high accretion rates. This is due to the fact that advection dominates the energy transport in this regime. As can be seen from the results for thin discs, the inner boundary condition plays a crucial role in determining the maximum amount of matter which finally reaches the central black hole. For slim discs, the critical accretion rates may reach up to $10^3 \dot{M}_{\mathrm{E}}$, while at the same time the luminosity stays close to the classical limit, $L \approx 20 L_E$ (Watarai, Fukue, Takeuchi *et al.* (2000)). Thus, highly super-Eddington accretion comes along with only mildly super-Eddington luminosities.

3. Spectral energy distribution of super-Eddington flows

In the second part, we investigate the observational appearance of super-Eddington accretion discs. This work is based on 2D RHD simulations, which are presented in detail in Ohsuga, Mori, Nakamoto *et al.* (2005). In a spherical computational box of $500 R_{\mathrm{G}}$ radius, matter is injected continuously in the disc plane $x - y$ around a $10 M_\odot$ black hole at high rates ($\dot{M} = 10^3 \dot{M}_{\mathrm{E}}$). Again, the viscosity is parameterized by $\alpha = 0.1$. For the energy transport equation, radiation and advection are taken into account, as well as photon trapping effects (Begelman (1978)). We calculate the spectral energy distribution of this system for various inclination angles Θ between 0 (face-on) and $\pi/2$ (edge-on). For details about the computation, we refer the reader to Heinzeller, Mineshige & Ohsuga (2006). The resulting SEDs for inclination angles 0, $\pi/4$ and $\pi/2$ are shown in Fig. 2(a).

While in the Rayleigh-Jeans part of the spectrum no difference between the orientations of the system can be seen, the peak intensity region around 1 keV is shifted towards higher photon energies and higher luminosities for lower inclination angles. For all orientations, a plateau-like structure in the high-energy part of the spectrum can be observed, although this feature is more prominent for the face-on case. Its origin lies in the emission of the hot gas in the vicinity of the black hole: when calculating the spectrum for $\Theta = 0$ while neglecting gas emission and absorption, this plateau disappears.

To outline the orientation effects, we compare in Fig. 2(b) the total intensity emitted by the system for inclination angles $\Theta = \{0 \ldots \pi/2\}$. For small inclinations, the intensity rises by a factor of up to 6.4 for $\Theta = 0$; this is due to both an enhanced average photon energy and an increased total photon number. We identify this behavior with *mild relativistic beaming*. When fitting the peak intensity region with a black body spectrum, we derive temperatures between $1.1 \cdot 10^6$ K for $\Theta = \pi/2$ and $1.7 \cdot 10^6$ K for $\Theta = 0$.

4. Conclusions

The above investigations allow us to draw several conclusions on black hole accretion processes. Firstly, we show that the classical Eddington limit can not be applied in discs. Contrary, the critical accretion rates become a local quantity, depending on the distance from the central object. Thereby, the inner disc region and inner boundary hold the key position in determining the final amount of matter that can be accreted by the central object.

We find that super-Eddington accretion is possible with accretion rates up to $10^3 \dot{M}_E$, corresponding to slightly super-Eddington luminosities up to $20 L_E$. Since the bottleneck for accretion lies in the inner region of the disc, our results may also correlate to the question on the origin of outflows and jets. Secondly, our work on the spectral energy distribution of such super-Eddington accreting systems shows that for a proper interpretation of the spectrum, the combined system of the disc and its surroundings has to be taken into account.

Features like the high-energy plateau can not be reproduced by considering the accretion disc only. Also, we find that mild relativistic beaming becomes important for small inclination angles, leading to higher luminosities and also higher temperatures. However, these temperatures are still too low to account for the observed ULX sources which sometimes reach temperatures up to 10^7 K. This may be due to the lack of Comptonization effects in our calculations: Socrates, Davis & Blaes (2004), for example, showed that turbulent Comptonization produces a significant contribution to the far-UV and X-ray emission of black hole accretion discs. Thus, the inclusion of Comptonization effects will be a primary goal in our future investigations.

References

Abramowicz, M. A., Czerny, B., Lasota, J.-P. & Szuzkiewicz, E. 1988, ApJ, 332, 646
Begelman, M. C. 1978, MNRAS, 184, 53
Fabbiano, G. 1989, ARA&A, 27, 87
Heinzeller, D. & Duschl, W. J. 2006, to appear in MNRAS
Heinzeller, D., Mineshige, S. & Ohsuga, K. 2006, to appear in MNRAS
Mizuno, T., Ohnishi, T., Kubota, A., Makishima, K. & Tashiro, M. 1999, PASJ, 51, 663
Ohsuga, K., Mori, M., Nakamoto, T. & Mineshige, S. 2005, ApJ, 628, 368
Roberts, T. P., Warwick, R. S., Ward, M. J., Goad, M. R. & Jenkins, L. P. 2005, MNRAS, 357, 1363
Shakura, N. I. & Sunyaev, R. A. 1973, A&A, 24, 337
Socrates, A., Davis, S. W. & Blaes, O. 2004, ApJ, 601, 405
Watarai, K.-Y., Fukue, J., Takeuchi, M. & Mineshige, S. 2000, PASJ, 52, 133
Watarai, K.-Y., Mizuno, T. & Mineshige, S. 2001, ApJ, 549, L77

RICHARD MUSHOTZKY: Two questions: first, have you done a stability analysis? And second, temperatures seem to be low compared to ULX...

DOMINIKUS HEINZELLER: 1. No, we have not performed a stability analysis yet. Also the time-dependency of these highly accreting systems has not been studied to date, but it is one of primary goals for the future work. 2. Indeed, our temperatures are too low to explain the highest observed temperatures of ULXs. However, Comptonization effects have not been included in the radiative transfer calculations. We expect that Comptonization will increase the effective temperatures, mostly for low inclination angles (i.e. a face-on view).

ANDREW LAWRENCE: Isn't the simplest way to explain super-Eddington sources to abandon the assumption that they are in steady state? In other words, what if they are in outburst?

DOMINIKUS HEINZELLER: May be – or may not. I can't answer this at the present, but investigations have to reveal if the duration of outburst lasts sufficiently long to explain the lifetime of super-Eddington states in these sources.

Black Holes from Stars to Galaxies – Across the Range of Masses
Proceedings IAU Symposium No. 238, 2006   © 2007 International Astronomical Union
V. Karas & G. Matt, eds.   doi:10.1017/S1743921307004899

Search for the event horizon by means of optical observations with high temporal resolution

G. Beskin,[1] V. Debur,[1] S. Karpov,[1] V. Plokhotnichenko[1] and A. Biryukov[2]

[1]Special Astrophysical Observatory of Russian Academy of Sciences, Nizhniy Arkhyz, Karachaevo-Cherkessia, Russia

[2]Sternberg Astronomical Institute of Moscow State University, Moscow, Russia

Abstract. The critical property of the black hole is the presence of the event horizon. It may be detected only by means of the detailed study of the emission features of its surroundings. The temporal resolution of such observations has to be better than $\sim rg/c$, and it lies in the 10^{-6}–10 s range depending on the black hole mass. In SAO RAS we have developed the MANIA hardware and software complex based on the panoramic photon counter and use it in observations on 6m telescope for the search and investigation of the optical variability on the time scales of 10^{-6}–10^3 s of various astronomical objects. We present the hardware and methods used for these photometrical, spectroscopic and polarimetrical observations, the principles and criteria of the object selection. The list of the latter includes objects with featureless optical spectra (DC white dwarfs, blazars) and long microlensing events.

We present the results of the observations of two objects-candidates – long MACHO event MACHO-1999-BLG-22 and radio-loud x-ray source with featureless optical spectrum J1942+10 – on the 6-m telescope in June-July 2006.

Keywords. Accretion – black hole physics – methods: data analysis – instrumentation: photometers – instrumentation: polarimeters – instrumentation: spectrographs

1. Introduction

Even though more than 60 years have passed since the theoretical prediction of black holes as an astrophysical objects (Oppenheimer & Snyder 1939) in some sense they have not been discovered yet. To identify an object as a black hole, one needs to show that its mass exceeds $3M_\odot$, its size is close to $r_g = 2GM/c^2$ and it has an event horizon instead of a normal surface – the distinguishing property of black holes which separates them from massive compact objects of finite size in some theories of gravity (Will 1998). However, only the two former criteria are used now for selection of black hole candidates of two types: a) with masses of 5-18 $M_\odot$, in X-ray binaries (see, for example, Greiner *et al.* 2001); and b) supermassive black holes in galaxy nuclei with masses of $10^6 - 10^{10} M_\odot$ (Shields 1999). Existence of the event horizon in such objects is usually implied by the absence of periodic pulsations of the X-ray emission from strong regular magnetic fields (the black hole "no-hair" theorem) and I type X-ray flares due to thermonuclear bursts of the accreted matter on the surface of the neutron star. High accretion rates in X-ray binaries and active galactic nuclei result in the screening of regions close to the event horizon, and the most luminous parts of accretion flow are situated at distances of 10–100r_g (Chakrabarti 1996, Cherepaschuk 2003) where general relativity effects are negligible.

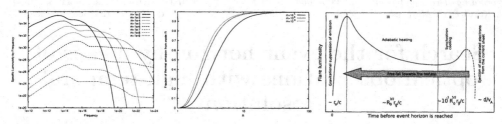

Figure 1. Properties of the isolated black hole emission. Left panel – spectra for various accretion rates, middle panel – the ratio of emission generated inside given radius, right panel – the temporal structure of the emission flare.

2. Search for event horizon in optics

At the same time, single stellar-mass black holes, which accrete interstellar medium of low density (10^{-2}–1 per cm^3), are the ideal case for detection and study of the event horizon. Shvartsman 1971 first demonstrated that an emitting halo of accreted matter forms around such objects and generates optical featureless emission. For the majority of Galaxy, filled with hot and warm ionized hydrogen, the regime of accretion onto the stellar mass isolated black holes is spherical, as the captured specific angular momentum is much smaller than one on the last stable orbit. The accretion rate is also small, $\dot{m} = \dot{M}/\dot{M}_{edd} \sim 10^{-10}$–$10^{-5}$ for the velocity of 50–100 km/s (Shvartsman 1971, Ipser & Price 1982). So, the emission of the accretion flow is dominated by the non-thermal synchrotron component due to accelerated particles(Beskin & Karpov 2005) (see Fig. 1). The total luminosity of such object is then $L = 9.6 \cdot 10^{33} M_{10}^3 n_1^2 (V^2 + c_s^2)_{16}^{-3} \sim 10^{29}$–$10^{34}$ erg/s. The majority of such emission comes from the regions near the horizon at $(2$–$5)r_g$ (see Fig. 1).

The most striking property of the accretion flow onto the single black hole is its inhomogeneity – the clots of plasma act as a probe testing the space-time properties near the horizon. The characteristic timescale of emission variability is $\tau_v \sim r_g/c \sim 10^{-4}$–$10^{-5}$ sec and such short stochastic variability may be considered as a distinctive property of black hole as the smallest possible physical object with a given mass. Its parameters – spectra, energy distribution and light curves – carry important information on space-time properties of the horizon (Beskin & Karpov 2005) (see Fig. 1).

The general observational appearance of a single stellar-mass black hole at typical interstellar medium densities is the same as other optical objects without spectral lines – DC-dwarfs and ROCOSes (Radio Objects with Continuous Optical Spectra, a subclass of blazars) (Pustilnik 1977). The suggestion that isolated BHs can be among them is the basis of the observational programme of search for isolated stellar-mass black holes – MANIA (Multichannel Analysis of Nanosecond Intensity Alterations). It uses photometric observations of candidate objects with high time resolution, special hardware and data analysis methods (Shvartsman 1977, Beskin *et al.* 1997), and is based on the fact that the fast variability is the critical property of isolated black hole emission.

In observations using the 6-meter telescope of the Special Astrophysical Observatory and the standard high time resolution photometer based on photomultipliers of 40 DC-dwarfs and ROCOSes, only upper limits for variability levels of 20% – 5% on the timescales of 10^{-6}–10 sec, respectively, were obtained, i.e. BHs were not detected (Shvartsman *et al.* 1989a, Shvartsman *et al.* 1989b, Beskin *et al.* 1997).

Recently, some evidences appeared that single stellar-mass black holes may be found among the stationary unidentified gamma-ray sources (Gehrels *et al.* 2002), gravitational lenses causing long-lasting MACHO events (Bennett *et al.* 2001), white dwarf – black

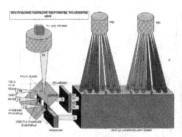

Figure 2. The scheme and the image of multichannel panoramic spectropolarimeter, and the sample star field observed in one-color photopolarimetric mode.

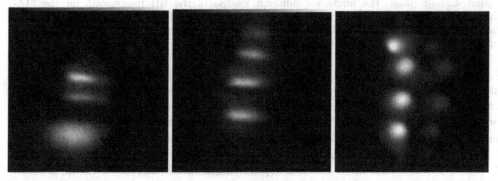

Figure 3. The different modes of MPSP operation – spectroscopic (left panel), spectropolarimetric (middle panel) and multicolor photopolarimetric (right panel).

hole binaries detected by means of self-microlensing flashes (Beskin & Tuntsov 2002), and radio-loud x-ray sources with featureless optical spectra (Tsarevsky 2005).

To study such a faint objects (down to 22^m) we have developed the multichannel panoramic spectro-polarimeter (MPSP) based on position-sensitive detector (PSD) with 1 μs time resolution (Debur *et al.* 2003, Plokhotnichenko *et al.* 2003) (see Fig. 2). Such detectors use the set of multichannel plates (MCP) for electron multiplication, and multi-electrode collector to determine its position. PSD used in our observations has the following parameters: quantum efficiency of 10% in the 3700-7500 A range, (S20 photocathode), MCP stack gain of 10^6, spatial resolution of 70 μm ($0.21''$ for the 6-m telescope), 700 ns time resolution, $7 \cdot 10^4$ pixels with 22 mm working diameter, and the 200-500 counts/s detector noise. The acquisition system used is the "Quantochron 4-480" spectral time-code convertor with 30 ns time resolution and 10^6 counts/s maximal count rate.

MPSP has the following modes of operation (see Fig. 3): one-color (U, B, V, R) photometry and polarimetry in the 1 arcmin field of view, four-color photometry and polarimetry of object and comparison star simultaneously in the 15 arcsec field, spectroscopy and the spectropolarimetry with 1-5 arcsec slit. It allows to measure the 3 Stokes parameters simultaneously, and is able to study 20-22^m objects for 1-hour exposure under good weather conditions.

In summer 2006 we performed the set of observations of two objects-candidates to isolated stellar-mass black holes, the MACHO-99-BLG-22 and the J1942+10.

The MACHO-99-BLG-22 is the longest microlensing event with 1120±90 days duration, projected velocity of 75±8 km/s, and the baseline magnitude $I = 19.2^m$ (Bennett *et al.* 2002). The analysis of microlensing data provided three possible models of the lens, with masses of 130 $M_\odot$ (at 0.5 kpc with $V_t = 70$ km/s), 30 $M_\odot$ (at 2 kpc with

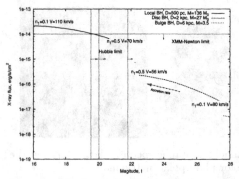 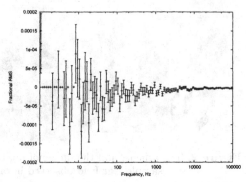

Figure 4. Left panel: the limits for different models of the BH parameters for the MA-CHO-99-BLG-22 object from the x-ray and optical observations. Right panel: the optical Fourier power spectrum for the radio-loud x-ray source with featureless optical spectrum J1942+10. The upper limit for the variable emission component is 10% in 10^{-5}–1 s range.

$V_t = 56$ km/s) and 3.5 $M_\odot$ (at 6 kpc, $V_t = 19$ km/s). We placed the upper limit on the variability of the emission from the position of the object on the level of $B < 21.5^\mathrm{m}$ (3σ) over the 10^{-5}–1 s time scales. The combination of the I-band Hubble data ($I < 20^\mathrm{m}$) and XMM-Newton (flux ¡ 10^{-14} erg/s/cm^2) with our theoretical estimations for possible models (Beskin & Karpov 2005) allow us to rule out the first model (see Fig. 4).

The J1942+10 is the radio-loud x-ray object with featureless optical spectrum with $B \sim 18^\mathrm{m}$. We observed it for 40 minutes in photometric mode and placed the upper limits on the B-band variability on the 10% level over the 10^{-5} - 1 s time scale (see Fig. 4).

3. Conclusions

In the near future we are planning to continue the investigation of the long MACHO events, radio-loud x-ray sources and ROCOSes.

For the increase of the efficiency of search for fast variability we are now developing the new generation of position-sensitive detector, with GaAs photocathode, and 16-electrode collector. It will have the quantum efficiency up to 30% in the 4000-8500 A range, and spatial resolution of 20 μm for the 10^6 number of pixels.

Acknowledgements

This work has been supported by the INTAS (grant No 04-78-7366), Russian Foundation for Basic Research (grant No 04-02-17555) and by the Russian Science Support Foundation.

References

Bennett, D. P. *et al.* 2002, ApJ, 579, 639
Beskin, G. M., Komarova, V. N., Neizvestny, S. I. *et al.* 1997, ExA, 7, 413
Beskin, G. M. & Tuntsov, A. V. 2002, A&A, 394, 489
Beskin, G. M. & Karpov, S. V. 2005, A&A, 440, 223
Chakrabarti, S. K. 1996, PhysRep, 266, 229
Cherepashchuk, A. M. 2003, Usp.Fiz.Nauk, 173, 345
Debur, V. *et al.* 2003, Nuclear Instruments and Methods in Physics Research, A 513, 127.
Greiner, J., Cuby, J.-G. & Mc Caughrean, M. J. 2001, Nature, 414, 522
Ipser, J. R. & Price, R. H. 1982, ApJ, 255, 654
Oppenheimer, J.& Snyder, H. 1939, Phys. Rev., 56, 455

Pustilnik, S. A. 1977, Soobsch. SAO, 18, 3

Plokhotnichenko, V. *et al.* 2003, Nuclear Instruments and Methods in Phys. Res., A 513, 167.

Shields, G. A. 1999, PASP, 111, 661

Shvartsman, V. F. 1971, AZh, 48, 479

Shvartsman, V. F. 1977 Soobsch. SAO, 19, 3

Shvartsman, V. F., Beskin, G. M. & Pustilnik, S. A. 1989, AFz, 31, 457

Shvartsman, V. F., Beskin, G. M. & Mitronova, S. N. 1989, Astron. Report Letters, 15, 145

Tsarevsky, G. *et al.* 2005, A&A, 438, 949

Will, C. M. 1998, gr-qc/9811036

MARGARITA SAFONOVA: When did you observe the reported MACHO event, 99-BLG-22, and how did you look for the lens?

GREGORY BESKIN: We have searched the field around the reported MACHO event two years after its occurrence and found three candidates for source stars. We also found an unidentified object that is quite possibly the moving lens for that event.

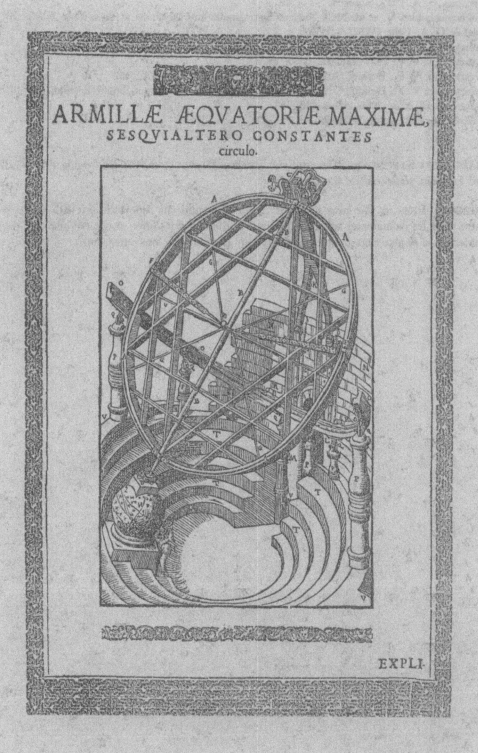

Black Holes from Stars to Galaxies – Across the Range of Masses
Proceedings IAU Symposium No. 238, 2006   © 2007 International Astronomical Union
V. Karas & G. Matt, eds.   doi:10.1017/S1743921307004905

Dynamics of radiatively inefficient flows accreting onto radiatively efficient black hole objects

Daniel Proga

Department Physics, University of Nevada, Las Vegas, Las Vegas, NV 89154, USA

email: dproga@physics.unlv.edu

Abstract. I present results from numerical simulations of gas dynamics outside luminous accretion disks in active galactic nuclei. The gas, gravitationally captured by a super massive black hole, can be driven away by the energy and momentum of the radiation emitted during black hole accretion. Assuming axisymmetry, I study how the mass accretion and outflow rates, and the flow dynamics respond to changes in radiation heating relative to radiation pressure.

I find that for a 10^8 M$_\odot$ black hole with the accretion luminosity of 0.6 of the Eddington luminosity the flow settles into a steady state and has two components: (1) an equatorial inflow and (2) a bipolar inflow/outflow with the outflow leaving the system along the disk rotational axis. The inflow is a realization of a Bondi–like accretion flow. The second component is an example of a non-radial accretion flow which becomes an outflow once it is pushed close to the rotational axis where thermal expansion and radiation pressure accelerate it outward.

The main result of this preliminary work is that although the above two-component solution is robust, its properties are sensitive to the geometry and spectral energy distribution of the radiation field.

Keywords. Accretion, accretion disks – methods: numerical – hydrodynamics

1. Introduction

The radiation properties of active galactic nuclei (AGN) and the AGN central location in their host galaxies imply that they play a very important role in determining the ionization structure and dynamics of matter not only in their vicinity but also on larger, galactic and even intergalactic scales (Cotti & Ostriker, 1997, 2001; King 2003; Murray, Quataert, & Thompson 2005; Sazonov *et al.* 2005; Springel, Di Matteo & Hernquist 2005; Hopkins *et al.* 2005, and references therin). Many observational results support this suggestion, in particular the presence of broad emission and absorption lines in AGN spectra. The ionization structure and dynamics of the gas responsible for these lines can be driven by radiation, even for sub-Eddington sources. The driving can be due to radiation pressure or radiation heating, or both (e.g., Begelman, McKee and Shields, 1982; Shlosman, Vitello & Shaviv 1985; Ostriker, McKee, & Klein 1991; Arav & Li 1994; Murray *et al.* 1995; Proga, Stone & Kallman 2000; Proga & Kallman 2002, 2004).

In this paper, I present results from hydrodynamical simulations of a non-rotating gas on sub-parsec- and parsec-scales in AGNs. I use a simplified version of the numerical method developed by Proga, Stone & Kallman (2000) to study a related problem of radiation driven disk winds in AGN (for details see Proga *et al.* 2006 in preparation). I consider an axisymmetric flow accreting onto a supermassive black hole (BH). The flow is non-spherical because it is irradiated by an accretion disk. The disk radiation flux is the highest along the disk rotational axis and is gradually decreasing with increasing polar angle, θ as $\cos\theta$. The flow is also irradiated by an isotropic corona (see eq. 2.1). I take into account the radiation heating and cooling, radiation pressure due to the electron scattering and spectral lines. I adopt a simplified treatment of photoionization, and radiative cooling and heating allowing for a self-consistent calculation of the ionization state, and therefore the line force, in the flow.

2. Results

I assume the mass of the non-rotating BH, $M_{BH} = 10^8$ M$_\odot$ and the disk inner radius, $r_* = 3$ $r_S = 8.8 \times 10^{13}$ cm throughout this paper. I consider the case with the rest mass conversion efficient $\eta = 0.0833$ and the mass accretion rate, $\dot{M}_a = 10^{26}$ g s^{-1} (=1.6 M$_\odot$ yr^{-1}). These system parameters yield the accretion luminosity, $L = 7.5 \times 10^{45}$ erg s^{-1}, corresponding to 0.6 of the Eddington luminosity. To determine the radiation field, I specify the fraction of L in the UV and X-ray band, as f_{UV} and f_X, respectively.

Although, a quasar is powered by accretion, the disk accretion rate, that determines radiation, is not coupled to the rate, $\dot{M}_{in}(r_i)$ at which the non-rotating gas leaves the computational domain of these simulations through the inner boundary. This $\dot{M}_a - \dot{M}_{in}(r_i)$ decoupling is physically motivated because an accretion disk is build from rotating gas but non-rotating gas will not significantly contribute to the disk mass and to the system luminosity. In this sense, I study radiatively inefficient flows accreting onto an object with a radiatively efficient accretion disk.

I present here preliminary results from simulations where all model parameters are fixed except for f_{UV} and f_X. I consider three cases: case A with $f_{UV} = 0.5$ and $f_X = 0.5$, case B with $f_{UV} = 0.8$ and $f_X = 0.2$, and case C with $f_{UV} = 0.95$ and $f_X = 0.05$ (see Table 1 for summary of the runs).

At large radii, the radial radiation force from the disk and spherical corona can be approximated as

$$F_r^{rad}(r, \theta) = \frac{\sigma_e L}{4\pi r^2 c}[2\cos\theta f_{UV}(1 + M(t)) + f_X], \qquad (2.1)$$

where r and θ are the radius and polar angle in the spherical polar coordinate system, respectively while the terms in the brackets with f_{UV} and f_X correspond to the disk and corona contribution, respectively. For simplicity, I assume that all UV photons are emitted by the disk whereas all X-rays are emitted by the corona. The $M(t)$ is the so-called force multiplier - the numerical factor which parameterizes by how much spectral lines increase the scattering coefficient compared to the electron scattering coefficient (Castor, Abbott & Klein 1975).

Figure 1 compares the results from run A and run C. I specified the outer boundary in the following way. The density and temperature at the outer radius, r_o was set to $\rho_0 = 10^{-21}$ g cm^{-3} and $T_0 = 2 \times 10^7$ K, respectively. The velocity was set to zero. At the outer radial boundary, during the evolution of each model I continue to apply these constraints that the density and temperature are fixed at constant values at all times. For the initial conditions, I set all variables constant and equal to their values at the outer boundary, as listed above. The figure shows the instantaneous density, temperature, and distributions and the poloidal velocity field of the models. Additionally, it also shows the so-called Compton radius corrected for the effects of radiation pressure due to electron scattering

$$\bar{R}_C \equiv R_C[1 - \Gamma(2\cos\theta f_{UV} + f_X)], \qquad (2.2)$$

where $R_C \equiv GM_{BH}\mu m_p/kT_C = 8.03 \times 10^{18}$ cm $= 9.1 \times 10^4$ r_* is the uncorrected Compton radius for the Compton temperature, $T_C = 2 \times 10^7$ K.

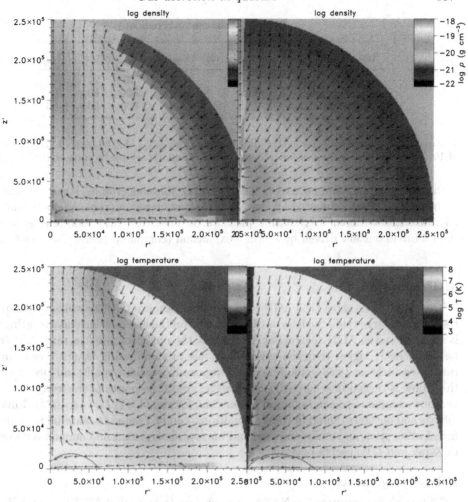

Figure 1. Comparison of the results for run A and C (left and and right column, respectively). *Top row of panels:* Maps of logarithmic density over-plotted by the direction of the poloidal velocity. *Bottom row of panels:* Maps of logarithmic temperature over-plotted by the direction of the poloidal velocity. The solid curve in the bottom left corner marks the position of the Compton radius corrected for the effects of radiation pressure due to electron scattering (see eq. 2.2 in the main text). The length scale is in units of the inner disk radius (i.e., $r' = r/r_*$ and $z' = z/r_*$). The computational domain is defined to occupy the angular range $0^o \leqslant \theta \leqslant 90^o$ and the radial range $r_i = 500\, r_* \leqslant r \leqslant r_o = 2.5 \times 10^5\, r_*$.

In all runs, the flow settles quickly into a steady state (within $\sim$ a few $\times 10^{12}$ s which correspond to a few dynamical time scales at r_o, $\tau = (r_o^3/GM_{BH})^{1/2} = 9 \times 10^{11}$ s). The steady state consists of two flow components (1) an equatorial inflow and (2) a bipolar inflow/outflow with the outflow leaving the system along the pole. The outflow is collimated by the inflow. The calculations capture the subsonic and supersonic parts of both the inflow and outflow. Although the same components can be identified in all runs, their size, density and temperature, and the degree of outflow collimation depend on f_{UV} and f_X (the spectral energy distribution and geometry of the radiation field). In particular, the outflow power (e.g., measured as the kinetic energy, P_k, and thermal energy P_{th} carried by the outflow) and the degree of collimation is higher for the model

Table 1. Summary of results

Run	f_{UV}	f_X	$\dot{M}_{in}(r_o)$	$\dot{M}_{in}(r_i)$	$\dot{M}_{out}(r_o)$	v_r	$P_k(r_o)$	$P_t(r_o)$
A	0.5	0.5	4	1	3	700	2	4
B	0.8	0.2	8	3	5	4000	100	1
C	0.95	0.05	9	1	8	6700	300	0.03

The quantities in the table are in the following units: $\dot{M}_{in}(r_o)$, $\dot{M}_{in}(r_i)$, and $\dot{M}_{out}(r_o)$ are in units of 10^{25} g s^{-1}, v_r is in units of km s^{-1}, and $P_k(r_o)$ and $P_t(r_o)$ are in units of 10^{40} erg s^{-1}.

with the radiation dominated by the UV/disk emission (run C) than for the model with the radiation dominated by the X-ray/corona emission (run A). I note that a very narrow outflow driven by radiation pressure on lines can carry more energy and mass than a broad outflow driven by thermal expansion (compared results for runs C and A).

3. Conclusions

The simulations show that AGN can have a substantial outflow which originates from the inflow at large radii. Such an outflow can control the rate at which non-rotating matter is supplied to the AGN central engine because the outflow mass loss rate, $\dot{M}_{out}(r_i)$, can be significantly higher than the mass inflow rate at small radii, $\dot{M}_{in}(r_i)$. For example, in run C, as little as 10% of the inflow at large radii reaches small radii because 90% of the inflow is turned into an outflow. However, even the power of the strongest outflow is very low compared to the radiation power (i.e., for run C, $P_k/L = 4 \times 10^{-4}$). Finally, the inflows and outflows, found in these simulations, can be related to material responsible for broad absorption and emission lines, and narrow absorption and emission lines observed in X-ray and UV spectra of AGN.

Acknowledgements

This work is supported by NASA through grants and HST-AR-10305 and HST-AR-10680 from the Space Telescope Science Institute, which is operated by the Association of Universities for Research in Astronomy, Inc., under NASA contract NAS5-26555.

References

Arav, N. & Li, Z. Y. 1994, ApJ, 427, 700
Begelman, M. C., McKee, C. F. & Shields, G. A. 1983, 271, 30
Castor, J. I., Abbott, D. C., & Klein, R. I. 1975, ApJ, 195, 157
Ciotti, L. & Ostriker, J. P. 1997, ApJ, 487, L105
Ciotti, L. & Ostriker, J. P. 2001, ApJ, 551, 131
Hopkins, P. F., Hernquist, L., Cox, T. J., Di Matteo, T., Martini, P., Robertson, B. & Springel, V. 2005, ApJ, 630, 705
King, A. 2003, ApJ, 596, L27
Murray, N., Quataert, E. & Thompson, T. A. 2005, ApJ, 618, 569
Ostriker, E. C., McKee, C. F. & Klein, R. I. 1991, ApJ, 377, 5930
Proga, D. & Kallman, T. R. 2002, ApJ, 565, 455
Proga, D. & Kallman, T. R. 2004, ApJ, 616, 688
Proga, D., Stone, J. M. & Kallman, T. R. 2000, ApJ, 543, 686
Sazonov, S. Y., Ostriker, J. P., Ciotti, L. & Sunyaev, R. A. 2005, MNRAS, 358, 168
Shlosman, I., Vitello, P. A. & Shaviv, G. 1985, ApJ, 294, 96
Springel, V., Di Matteo, T. & Hernquist, L. 2005, ApJ, 620, L79

MITCHELL BEGELMAN: Have you looked at what happens when you adjust heating versus radiation pressure? One would like to see to what extent heating inhibits accretion. That is, you want to see how much pre-heating is inhibiting the accretion flow, but you also have radiation pressure there... How close are you to the Eddington limit with this system?

DANIEL PROGA: The calculation was for 0.6 Eddington limit. The last example was actually a manifestation of unclear interplay between the two effects. I am currently in the middle of the parameter survey.

MAREK ABRAMOWICZ: What the gas does after it reaches the boundary of your computational domain?

DANIEL PROGA: We do not consider that the gas will be interacting with the medium on larger scales, so in this model the gas will leave the system. I look at velocities at the outer radius, and they can be quite high – order of 10^4 km/s. This exceeds the escape velocity from the system.

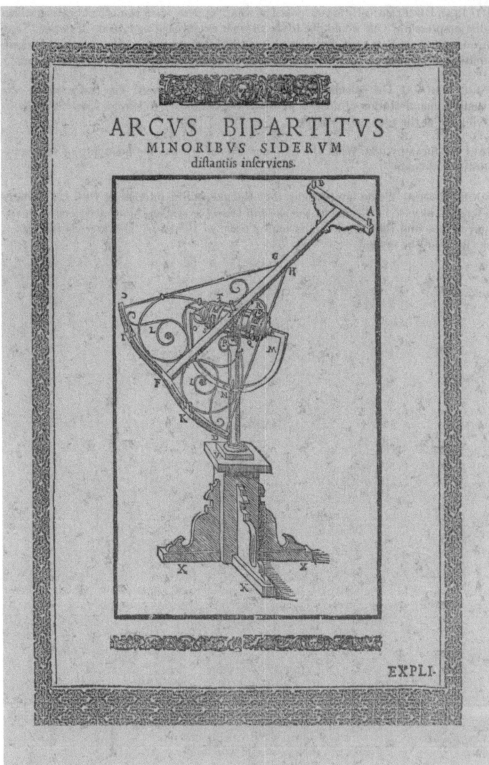

ARCVS BIPARTITVS
MINORIBVS SIDERVM
diftantiis inferviens.

EXPLI.

The Galactic Center

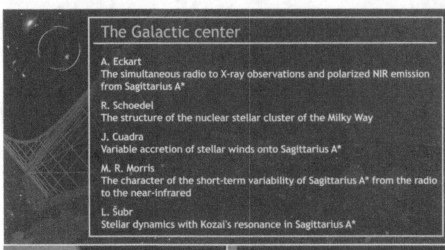

The Galactic center

A. Eckart
The simultaneous radio to X-ray observations and polarized NIR emission from Sagittarius A*

R. Schoedel
The structure of the nuclear stellar cluster of the Milky Way

J. Cuadra
Variable accretion of stellar winds onto Sagittarius A*

M. R. Morris
The character of the short-term variability of Sagittarius A* from the radio to the near-infrared

L. Šubr
Stellar dynamics with Kozai's resonance in Sagittarius A*

The Theatre of Estates. Two world premieres of Mozart's operas were staged here: Don Giovanni and La Clemenza di Tito.

Black Holes from Stars to Galaxies – Across the Range of Masses
Proceedings IAU Symposium No. 238, 2006
V. Karas & G. Matt, eds.

© 2007 International Astronomical Union
doi:10.1017/S1743921307004929

The Galactic Center

Reinhard Genzel[1,2] and Vladimír Karas[3]

[1]Max-Planck Institut für Extraterrestrische Physik, Garching, Germany

[2]Department of Physics, University of California, Berkeley, USA

[3]Astronomical Institute, Academy of Sciences, Prague, Czech Republic

Abstract. In the past decade high resolution measurements in the infrared employing adaptive optics imaging on 10m telescopes have allowed determining the three dimensional orbits stars within ten light hours of the compact radio source at the center of the Milky Way. These observations show the presence of a three million solar mass black hole in Sagittarius A* beyond any reasonable doubt. The Galactic Center thus constitutes the best astrophysical evidence for the existence of black holes which have long been postulated, and is also an ideal 'lab' for studying the physics in the vicinity of such an object. Remarkably, young massive stars are present there and probably have formed in the innermost stellar cusp. Variable infrared and X-ray emission from Sagittarius A* are a new probe of the physical processes and space-time curvature just outside the event horizon.

Keywords. Galaxy: center – black hole physics

1. Introduction – Sagittarius A*

The central light years of our Galaxy contain a dense and luminous star cluster, as well as several components of neutral, ionized and extremely hot gas (Genzel, Hollenbach & Townes 1994). The Galactic Center also contains a very compact radio source, Sagittarius A* (Sgr A*; Balick & Brown 1974) which is located at the center of the nuclear star cluster and ionized gas environment. Short-wavelength centimeter and millimeter VLBI observations have established that its intrinsic radio size is a mere 10 light minutes (Bower *et al.* 2004; Shen *et al.* 2005). Sgr A* is also an X-ray emission source, albeit of only modest luminosity (Baganoff *et al.* 2001). Most recently, Aharonian *et al.* (2004) have discovered a source of TeV γ-ray emission within 10 arcsec of Sgr A*. It is not yet clear whether these most energetic γ-rays come from Sgr A* itself or whether they are associated with the nearby supernova remnant, Sgr A East.

Sgr A* thus may be a supermassive black hole analogous to QSOs, albeit of much lower mass and luminosity. Because of its proximity – the distance to the Galactic Center is about 10^5 times closer than the nearest quasars – high resolution observations of the Milky Way nucleus offer the unique opportunity of stringently testing the black hole paradigm and of studying stars and gas in the immediate vicinity of a black hole, at a level of detail that will not be accessible in any other galactic nucleus in the foreseeable future.

Since the center of the Milky Way is highly obscured by interstellar dust particles in the plane of the Galactic disk, observations in the visible light are not possible. Investigations require measurements at longer wavelengths – the infrared and microwave bands, or at shorter wavelengths – hard X-rays and γ-rays, where the veil of dust is transparent. The dramatic progress in our knowledge of the Galactic Center over the past two decades is a direct consequence of the development of novel facilities, instruments and techniques across the whole range of the electromagnetic spectrum.

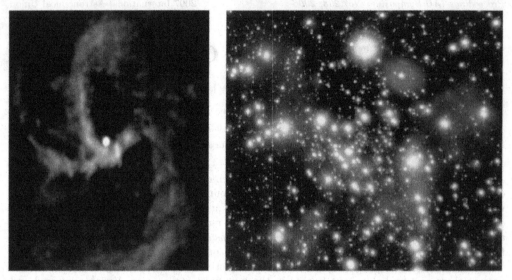

Figure 1. Left: VLA radio continuum map of the central parsec (Roberts & Goss 1993). The radio emission delineates ionized gaseous streams orbiting the compact radio source Sgr A*. Spectroscopic measurements in the radio band (Wollman *et al.* 1977) provided the first dynamical evidence from large gas velocities that there might be a hidden mass of 3–4 million solar masses located near Sgr A*. Right: A diffraction limited image of Sgr A* (~ 0.05 arcsec resolution) from the 8m ESO VLT, taken with the NACO AO-camera and an infrared wavefront sensor at $1.6/2.2/3.7$ μm (Genzel *et al.* 2003b). The central black hole is located in the centre of the box. NACO is a collaboration between ONERA (Paris), Observatoire de Paris, Observatoire Grenoble, MPE (Garching), and MPIA (Heidelberg) (Lenzen *et al.* 1998; Rousset *et al.* 1998).

2. High angular resolution astronomy

The key to the nature of Sgr A* obviously lies in very high angular resolution measurements. The Schwarzschild radius of a 3.6 million solar mass black hole at the Galactic Center subtends a mere 10^{-5} arcsec. For the high-resolution imaging from the ground it is necessary to correct for the distortions of an incoming electromagnetic wave by the refractive and dynamic Earth atmosphere. VLBI overcomes this hurdle by phase-referencing to nearby QSOs; sub-milliarcsecond resolution can now be routinely achieved.

In the optical/near-infrared wavebands the atmosphere smears out long-exposure images to a diameter at least ten times greater than the diffraction limited resolution of large ground-based telescopes (Fig. 1). From the early 1990s onward initially speckle imaging (recording short exposure images, which are subsequently processed and co-added to retrieve the diffraction limited resolution) and then later adaptive optics (AO, correcting the wave distortions on-line) became available. With these techniques it is possible to achieve diffraction limited resolution on large ground-based telescopes. The diffraction limited images are much sharper and also much deeper than the seeing limited images. In the case of AO (Beckers 1993) the incoming wavefront of a bright star near the source of interest is analyzed, the necessary corrections for undoing the aberrations of the atmosphere are computed (on time scales shorter than the atmospheric coherence time of a few milli-seconds) and these corrections are then applied to a deformable optical element (e.g. a mirror) in the light path.

The requirements on the brightness of the AO star and on the maximum allowable separation between star and source are quite stringent, resulting in a very small sky coverage of natural star AO. Fortunately, in the Galactic Center there is a bright infrared star only 6 arcsec away from Sgr A*, such that good AO correction can be achieved with an infrared wavefront sensor system. Artificial laser beacons can overcome the sky coverage problem to a considerable extent. For this purpose, a laser beam is projected from the telescope into the upper atmosphere and the backscattered laser light can then be used for AO correction. The Keck telescope team has already begun successfully exploiting the new laser guide star technique for Galactic Center research (Ghez *et al.* 2005a). After AO correction, the images are an order of magnitude sharper and also much deeper than in conventional seeing limited measurements. The combination of AO techniques with advanced imaging and spectroscopic instruments (e.g. integral field imaging spectroscopy) have resulted in a major breakthrough in high resolution studies of the Galactic Center.

3. Nuclear star cluster and the paradox of youth

One of the big surprises is a fairly large number of bright stars in Sgr A*, a number of which were already apparent on the discovery infrared images of Becklin & Neugebauer (1975, 1978). High-resolution infrared spectroscopy reveals that many of these bright stars are actually somewhat older, late-type supergiants and AGB stars. Starting with the discovery of the AF-star (Allen *et al.* 1990; Forrest *et al.* 1987), however, an ever increasing number of the bright stars have been identified as being young, massive and early type. The most recent counts from the deep SINFONI integral-field spectroscopy yields about one hundred OB stars, including various luminous blue supergiants and Wolf-Rayet stars, but also normal main-sequence OB stars (Paumard *et al.* 2006a). The nuclear star cluster is one of the richest concentrations of young massive stars in the Milky Way.

The deep adaptive optics images also trace the surface density distribution of the fainter stars, to about K 17–18 mag, corresponding to late B or early A stars (masses of 3–6 solar masses), which are a better probe of the density distribution of the overall mass density of the star cluster. While the surface brightness distribution of the star cluster is not centered on Sgr A*, the surface density distribution is. There is clearly a cusp of stars centered on the compact radio source (Genzel *et al.* 2003b; Schödel *et al.* 2006). The inferred volume density of the cusp is a power-law $\propto R^{-1.4\pm0.1}$, consistent with the expectation for a stellar cusp around a massive black hole (Alexander 2005).

If there is indeed a central black hole associated with Sgr A* the presence of so many young stars in its immediate vicinity constitutes a significant puzzle (Allen & Sanders 1986; Morris 1993; Alexander 2005). For gravitational collapse to occur in the presence of the tidal shear from the central mass, gas clouds have to be denser than $\sim 10^9 (R/(10''))^{-3}$ hydrogen atoms per cm^{-3}. This 'Roche' limit exceeds the density of any gas currently observed in the central region. Recent near-diffraction limited AO spectroscopy with both the Keck and VLT shows that almost all of the cusp stars brighter than K $\sim$ 16 mag appear to be normal, main sequence B stars (Ghez *et al.* 2003; Eisenhauer *et al.* 2005a). If these stars formed in situ, the required cloud densities approach the conditions in outer stellar atmospheres.

Several scenarios have been proposed to account for this paradox of youth. In spite of this effort the origin of central stars (S-stars) is not well understood: models have difficulties in reconciling different aspects of the Galaxy Centre – on one side it is a low level of present activity, indicating a very small accretion rate, and on the other side it is the spectral classification that suggests these stars have been formed relatively recently; see Alexander (2005) for a detailed discussion and references. The most prominent ideas to resolve the apparent problem are in situ formation in a dense gas accretion disk that can overcome the tidal limits, re-juvenation of older stars by collisions or stripping, and

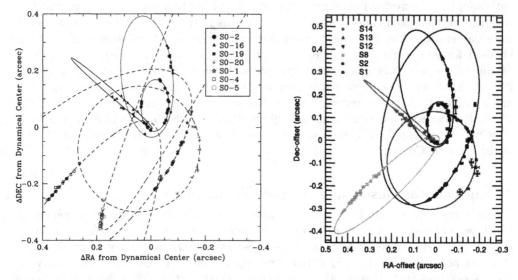

Figure 2. Positions on the sky as a function of time for the central stars orbiting the compact radio source Sgr A*. Left: the data from the UCLA group working with the Keck telescope (Ghez *et al.* 2005b). Right: the data from the MPE–Cologne group at the ESO-VLT (Schödel *et al.* 2003; Eisenhauer *et al.* 2005a; Gillessen *et al.*, in preparation).

rapid in-spiral of a compact, massive star cluster that formed outside the central region and various scattering a three body interaction mechanisms, including resonant relaxation (Alexander 2005). Several other mechanisms have been proposed that could set stars on highly eccentric orbits and bring them to the neighbourhood of the central black hole (e.g., Hansen & Milosavljević 2003; McMillan & Portegies Zwart 2003; Alexander & Livio 2004; Šubr & Karas 2005), but the problem of the S-stars remains open.

4. Compelling evidence for a central massive black hole

With diffraction limited imagery starting in 1991 on the 3.5m ESO New Technology Telescope and continuing since 2002 on the VLT, a group at MPE was able to determine proper motions of stars as close as ∼ 0.1 arcsec from Sgr A* (Eckart & Genzel 1996, 1997). In 1995 a group at the University of California, Los Angeles started a similar program with the 10m diameter Keck telescope (Ghez *et al.* 1998). Both groups independently found that the stellar velocities follow Kepler laws and exceed 10^3 km/s within the central light month.

Only a few years later both groups achieved the next and crucial step: they were able to determine individual stellar orbits for several stars very close to the compact radio source (Fig. 2; Schödel *et al.* 2002, 2003; Ghez *et al.* 2003, 2005b; Eisenhauer *et al.* 2005a). In addition to the astrometric imaging they obtained near-diffraction limited Doppler spectroscopy of the same stars (Ghez *et al.* 2003; Eisenhauer *et al.* 2003a,b), yielding precision measurements of the three dimensional structure of several orbits, as well as the distance to the Galactic Center. At the time of writing, the orbits have been determined for about a dozen stars in the central light month. The central mass and stellar orbital parameters derived by the two teams agree mostly very well. The orbits show that the gravitational potential indeed is that of a point mass centered on Sgr A* within the relative astrometric uncertainties of ∼ 10 milliarcsec. Most of the mass must be concentrated well within the peri-approaches of the innermost stars, ∼ 10–20 light hours, or 70 times the Earth orbit radius and about 1000 times the event horizon of a 3.6 million solar mass black hole. There is presently no indication for an extended mass greater than about 5% of the point mass.

Simulations indicate that current measurement accuracies are sufficient to reveal the first and second order effects of Special and General Relativity in a few years time (Zucker *et al.* 2006). Observations with future 30m+ diameter telescopes will be able to measure the mass and distance to the Galactic Center to $\sim 0.1\%$ precision. They should detect radial precession of stellar orbits due to General Relativity and constrain the extended mass to $< 10^{-3}$ of the massive black hole (Weinberg, Milosavljevic & Ghez 2005). At that level a positive detection of a halo of stellar remnants (stellar black holes and neutron stars) and perhaps dark matter would appear to be likely. Future interferometric techniques will push capabilities yet further.

Long-term VLBA observations have set 2σ upper limits of about 20 km/s and 2 km/s (or 50 micro-arcsec per year) to the motion of Sgr A* itself, along and perpendicular to the plane of the Milky Way, respectively (Reid & Brunthaler 2004; see also Backer & Sramek 1999). This precision measurement demonstrates very clearly that the radio source itself must indeed be massive, with simulations indicating a lower limit to the mass of Sgr A* of $\sim 10^5$ solar masses. The intrinsic size of the radio source at millimeter wavelengths is less than 5 to 20 times the event horizon diameter (Bower *et al.* 2004; Shen *et al.* 2005). Combining the radio size and proper motion limit of Sgr A* with the dynamical measurements of the nearby orbiting stars leads to the conclusion that Sgr A* can only be a massive black hole, beyond any reasonable doubt. An astrophysical dark cluster fulfilling the observational constraints would have a life-time less than a few 10^4 years and thus can be safely rejected, as can be a possible fermion ball of hypothetical heavy neutrinos. In fact all non-black hole configurations can be excluded by the available measurements (Schödel *et al.* 2003; Ghez *et al.* 2005b) – except for a hypothetical boson star and the gravastar hypothesis, but it appears that the two mentioned alternatives have difficulties of their own, and they are less likely and certainly much less understood than black holes (e.g. Maoz 1998; Miller *et al.* 1998). We thus conclude that, under the assumption of the validity of General Relativity, the Galactic Center provides the best quantitative evidence for the actual existence of (massive) black holes that contemporary astrophysics can offer.

5. Zooming in on the accretion zone and event horizon

Recent millimeter, infrared and X-ray observations have detected irregular, and sometimes intense outbursts of emission from Sgr A* lasting anywhere between 30 minutes and a number of hours and occurring at least once per day (Baganoff *et al.* 2001; Genzel *et al.* 2003a; Marrone *et al.* 2006). These flares originate from within a few milli-arcseconds of the radio position of Sgr A*. They probably occur when relativistic electrons in the innermost accretion zone of the black hole are significantly accelerated, so that they are able to produce infrared synchrotron emission and X-ray synchrotron or inverse Compton radiation (Markoff *et al.* 2001; Yuan *et al.* 2003; Liu *et al.* 2005). This interpretation is also supported by the detection of significant polarization of the infrared flares (Eckart *et al.* 2006b), by the simultaneous occurrence of X- and IR-flaring activity (Eckart *et al.* 2006a; Yusef-Zadeh *et al.* 2006) and by variability in the infrared spectral properties (Ghez *et al.* 2005b; Gillessen *et al.* 2006a; Krabbe *et al.* 2006). There are indications for quasi-periodicities in the light curves of some of these flares, perhaps due to orbital motion of hot gas spots near the last circular orbit around the event horizon (Genzel *et al.* 2003a; Aschenbach *et al.* 2004; Bélanger *et al.* 2006).

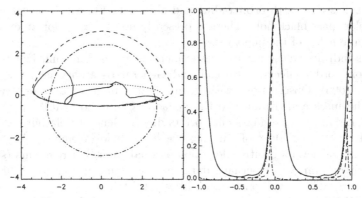

Figure 3. Photo-center wobbling (left) and light curve (right) of a hot spot on the innermost
stable orbit around Schwarzschild black hole (inclination of 80 deg), as derived from ray-tracing
computations. Dotted curve: 'true' path of the hot spot; dashed curves: apparent path and a predicted light curve of the primary image; dash-dotted curves: the same for secondary image; solid
curves: path of centroid and integrated light curve. Axes on the left panel are in Schwarzschild
radii of a 3 million solar-mass black hole, roughly equal to the astrometric accuracy of 10 arcsec;
the abscissa axis of the right panel is in cycles. The loop in the centroids track is due to the secondary image, which is strongly sensitive to the space-time curvature. The overall motion can be
detected at good significance at the anticipated accuracy of GRAVITY. Details can be obtained
by analyzing several flares simultaneously (Gillessen *et al.* 2006b; Paumard *et al.* 2005).

The infrared flares as well as the steady microwave emission from Sgr A* may be important probes of the gas dynamics and space-time metric around the black hole (Broderick
& Loeb 2006; Meyer *et al.* 2006a,b; Paumard *et al.* 2006b). Future long-baseline interferometry at short millimeter or sub-millimeter wavelengths may be able to map out the
strong light-bending effects around the photon orbit of the black hole. It is interesting
to realize that the angular size of the "shadow" of black hole (Bardeen 1973) is not
very far from the anticipated resolution of interferometric techniques and it may thus be
accessible to observations in near future (Falcke, Melia & Agol 2000).

Polarization measurements will help us to set further constraints on the emission processes responsible for the flares. Especially the time-resolved lightcurves of the polarized
signal carry specific information about the interplay between the gravitational and magnetic fields near Sgr A* horizon, because the propagation of the polarization vector is
sensitive to the presence and properties of these fields along the light trajectories (Bromley, Melia & Liu 2001; Horák & Karas 2006; Paumard *et al.* 2006b). Polarization is also
very sensitive also to intrinsic properties of the source – its geometry and details of
radiation mechanisms responsible for the emission.

Synthesis of different techniques will be a promising way for the future: the astrometry
of central stars gives very robust results because the stellar motion is almost unaffected
by poorly known processes of non-gravitational origin, while the flaring gas occurs much
closer to the black hole horizon and hence it directly probes the innermost regions of
Sgr A*. Eventually the two components – gas and stars of the Galaxy Center – are
interconnected and form the unique environment in which the flaring gas is influenced
by intense stellar winds whereas the long-term motion and the 'non-standard' evolution
of the central stars bears imprints of the gaseous medium though which the stars pass.

Eisenhauer *et al.* (2005b) are developing GRAVITY (an instrument for 'General Relativity Analysis via VLT Interferometry'), which will provide dual-beam, 10 micro-
arcsecond precision infrared astrometric imaging of faint sources. GRAVITY may be
able to map out the motion on the sky of hot spots during flares with a high enough

resolution and precision to determine the size of the emission region and possibly detect the imprint of multiple gravitational images (see Fig. 3). In addition to studies of the flares, it will also be able to image the orbits of stars very close to the black hole, which should then exhibit the orbital radial oscillations and Lense-Thirring precession due to General Relativity. Both the microwave shadows as well as the infrared hot spots are sensitive to the space-time metric in the strong gravity regime. As such, these ambitious future experiments can potentially test the validity of the black hole model near the event horizon and perhaps even the validity of General Relativity in the strong field limit.

Acknowledgements

An extended version of this lecture was presented by RG as Invited Discourse during the 26th General Assembly of the International Astronomical Union in Prague, 22nd August 2006, (*Highlights of Astronomy*, Volume 14, in preparation). VK thanks the Czech Science Foundation for continued support (ref. 205/07/0052).

References

Aharonian, F., Akhperjanian, A. G., Aye, K.-M *et al.* 2004, A&A, 425, L13

Alexander, T. 2005, Phys. Rep., 419, 65

Alexander, T. & Livio, M. 2004, 606, L21

Allen, D. A. & Sanders, R. 1986, Nature, 319, 191

Allen, D. A., Hyland, A. R. & Hillier, D. J. 1990, MNRAS, 244, 706

Aschenbach, B., Grosso, N., Porquet, D. & Predehl, P. 2004, A&A, 417, 71

Backer, D. C. & Sramek, R. A. 1999, ApJ, 524, 805

Baganoff, F., Bautz, M. W., Brandt, W. N. *et al.* 2001, Nature, 413, 45

Balick, B., Brown, R. 1974, ApJ, 194, 265

Bardeen, J. M. 1973, in Black Holes, eds. C. DeWitt & B. S. DeWitt (New York: Gordon & Breach), p. 215

Beckers, J. M. 1993, ARAA, 31, 13

Becklin, E. E. & Neugebauer, G. 1975, ApJ, 200, L71

Becklin, E. E., Matthews, K., Neugebauer, G. & Willner, S. P. 1978, ApJ, 219, 121

Bélanger, G., Terrier, R., De Jager, O. C., Goldwurm, A. & Melia, F. 2006, J. Phys., 54, 420

Bower, G. C. Falcke, H., Herrnstein, R. M., Zhao, Jun-Hui, Goss, W. M. & Backer, D. C. 2004, Science, 304, 704

Broderick, A., Loeb, A. 2006, MNRAS, 367, 905

Bromley, B. C., Melia, F. & Liu Siming, 2001, ApJ, 555, L83

Eckart, A. & Genzel, R. 1996, Nature, 383, 415

Eckart, A. & Genzel, R. 1997, MNRAS, 284, 576

Eckart, A., Baganoff, F. K., Schödel R. *et al.* 2006a, A&A, 450, 535

Eckart, A., Schödel, R., Meyer, L., Trippe, S., Ott, T. & Genzel, R. 2006b, A&A, 455, 1

Eisenhauer, F., Abuter, R., Bickert, K. *et al.* 2003b, Proc. SPIE, 4841, 1548

Eisenhauer, F., Genzel, R., Alexander, T. *et al.* 2005a, ApJ, 628, 246

Eisenhauer, F., Perrin, G., Rabien, S., *et al.* 2005b, AN, 326, 561

Eisenhauer, F., Schödel, R., Genzel, R. *et al.* 2003a, ApJ, 597, L121

Falcke, H., Melia, F. & Agol, E. 2000, ApJ, 528, L13

Forrest, W. J., Shure, M. A., Pipher, J. L. & Woodward, C. A. 1987, in The Galactic Center, AIP Conf. 155, ed. D. C. Backer (New York: AIP), 153

Genzel, R., Hollenbach, D. & Townes, C. H. 1994, Rep. Prog. Phys., 57, 417

Genzel, R., Schödel, R., Ott, T. *et al.* 2003a, Nature, 425, 934

Genzel, R., Schödel, R., Ott, T. *et al.* 2003b, ApJ, 594, 812

Ghez, A. M., Duchêne, G., Matthews, K. *et al.* 2003, ApJ, 586, L127

Ghez, A. M., Klein, B. L., Morris, M. & Becklin, E. E. 1998, ApJ, 509, 678

Ghez, A. M., Hornstein, S. D., Lu, J. R. *et al.* 2005a, ApJ, 635, 1087

Ghez, A. M., Salim, S., Hornstein, S. D. *et al.* 2005b, ApJ, 620, 744

Gillessen, S., Eisenhauer, F., Quataert, E. *et al.* 2006a, ApJ, 640, L163

Gillessen, S., Perrin, G., Brandner, W. *et al.* 2006b, in Advances of Stellar Interferometry, eds.
    J. D. Monnier *et al.*, Proc. SPIE, vol. 6268, 626811

Hansen, B. M. S. & Milosavljević, M. 2003, ApJ, 593, L80

Horák, J. & Karas, V. 2006, MNRAS, 365, 813

Krabbe, A., Iserlohe, C., Larkin, J. E. *et al.* 2006, ApJ, 642, L145

Lenzen, R., Hofmann, R., Bizenberger, P. & Tusche, A. 1998, Proc. SPIE, 3354, 606

Liu, S., Melia, F. & Petrosian, V. 2005, ApJ, 636, 798

Maoz, E. 1998, ApJ, 494, L181

Markoff, S., Falcke, H., Yuan, F. & Biermann, P. L. 2001, A&A, 379, L13

Marrone, D., Moran, J. M., Zhao, J.-H. & Rao, R. 2006, ApJ, 640, 308

McMillan, S. L. W. & Portegies, Zwart, S. F. 2003, ApJ, 596, 314

Meyer, L., Eckart, A., Schödel, R., Duschl, W. J., Muzic, K., Dovčiak, M. & Karas, V. 2006a,
    A&A, 460, 15

Meyer L., Schödel R., Eckart A., Karas V., Dovčiak M. & Duschl W. J. 2006b, A&A, 458, L25

Miller, J. C., Shahbaz, T. & Nolan, L. A. 1998, MNRAS, 29

Morris, M. 1993, ApJ, 408, 496

Paumard, T., Genzel, R., Martins, F. *et al.* 2006a, ApJ, 643, 1011

Paumard, T., Mueller, T., Genzel, R., Eisenhauer, F. & Gillesen, S. 2006, in preparation

Paumard, T., Perrin, G., Eckart, A. *et al.* 2005, AN, 326, 568

Reid, M. J. & Brunthaler, A. 2004, ApJ, 616, 872

Roberts, D. A. & Goss, W. M. 1993, ApJSS, 86, 133

Rousset, G., Lacombe, F., Puget, P. *et al.* 1998, in Adaptive Optical System Technologies, eds.
    D. Bonaccini & R. K. Tyson, Proc. SPIE, vol. 3255, 508

Schödel, R., Ott, T., Genzel, R. *et al.* 2002, Nature, 419, 694

Schödel, R., Ott, T., Genzel, R. *et al.* 2003, ApJ, 596, 1015

Schödel, R., Eckart, A., Alexander, T. *et al.* 2006, A&A, in press (astro-ph/0703178)

Shen, Z. Q., Lo, K. Y., Liang, M. C., Ho, P. T. P. & Zhao, J. H. 2005, Nature, 438, 62

Šubr, L. & Karas, V. 2005, A&A, 433, 405

Weinberg, N. N., Milosavljevic, M. & Ghez, A. M. 2005, ApJ, 622, 878

Wollman, E. R., Geballe, T. R., Lacy, J. H., Townes, C. H. & Rank, D. M. 1977, ApJ, 218, L103

Yuan, F., Quataert, E. & Narayan, R. 2003, ApJ, 598, 301

Yusef-Zadeh, F., Bushouse, H., Dowell, C. D *et al.* 2006, ApJ, 644, 198

Zucker, S., Alexander, T., Gillessen, S., Eisenhauer, F. & Genzel, R. 2006, ApJ, 639, L21

ZDENĚK STUCHLÍK: Do you observe any signs of interaction between a gaseous disc and stars near the massive central black hole in Sagittarius A*?

REINHARD GENZEL: We have no observational evidence for such gaseous disc. It might be there to some level – Jorge Cuadra might discuss this matter in more detail – but we have been looking for it and it does not appear to be present.

ALEXANDER ZAKHAROV: Is there a cusp in the center, e.g. within the distance of S2-star orbit?

REINHARD GENZEL: We are of course running out of counting statistics, but a core is possible. Physical collisions should be likely in that region.

THAISA STORCHI-BERGMANN: Comment: In my presentation on Friday, I will show that, as in the Galactic center, nearby galaxies, when in inactive phase, exhibit nuclear stellar discs, while AGN matched to the inactive galaxies in host galaxy properties exhibit instead nuclear gas plus dust spirals.

Black Holes from Stars to Galaxies – Across the Range of Masses
Proceedings IAU Symposium No. 238, 2006
V. Karas & G. Matt, eds.

© 2007 International Astronomical Union
doi:10.1017/S1743921307004930

Variable and polarized emission from SgrA*†

A. Eckart,[1] R. Schödel,[1] L. Meyer,[1] C. Straubmeier,[1] M. Dovčiak,[2] V. Karas,[2] M. R. Morris[3] and F. K. Baganoff[4]

[1]I. Physikalisches Institut, University of Cologne, Zülpicher Str. 77; D-50937 Köln, Germany

[2]Astronomical Institute, Academy of Sciences, Boční II, CZ-14131 Prague, Czech Republic

[3]Department of Physics & Astronomy, UCLA, Los Angeles, CA 90095-1547, USA

[4]Kavli Institute for Astrophysics and Space Research, MIT, Cambridge, MA 02139-4307, USA

Abstract. The super-massive black hole in the Galactic Center (Sagittarius A*) is one of the most exciting targets in the sky. At a distance of ∼ 8 kpc it is about one hundred times closer than the second nearest nucleus of a similar galaxy, M31, and therefore the closest galactic nucleus that we can study. Here we report on the modeling of polarized near-infrared flare emission from SgrA* using a model in which a hot spot is moving on a relativistic orbit around the massive black hole. We also summarize the results from simultaneous radio/near-infrared/X-ray measurements of flare emission.

Keywords. Galaxy: center – Galaxy: nucleus – accretion, accretion disks – black hole physics – radiation mechanisms: non-thermal – relativity

1. Introduction

Compelling evidence for a massive black hole at the position of Sagittarius A* (SgrA*) is provided by observations of stellar dynamics and of the variable emission from that position, both in the X-ray and in near-infrared (NIR) domains (see also Eckart, Schödel & Straubmeier 2005 and references therein). NACO adaptive optics (AO) instrument at the ESO VLT provided the infrared data of the first simultaneous NIR/X-ray flare detections. Repeated measurements have shown that to within less than 10 minutes the brighter X-ray flare events occur simultaneously with the corresponding NIR flares.

Recent near-infrared polarimetric observations with NACO at the VLT UT4 (Yepun) have revealed that some of the 1–2 hour flares exhibit a surprising fine structure in the form of polarized sub-flares that have duration of only about 7–10 minutes and are spaced by about 20±3 minutes from peak to peak. These features can successfully be interpreted as emission from hot spots that are on relativistic orbits around the central black hole.

2. Polarized sub-flares from SgrA*

We have obtained new polarization data of the variable NIR emission of SgrA* using the NACO instrument in 2004 and 2005 (Eckart et al. 2006a,b). These new data reveal that some of the typically 60–100 minute long infrared flares are modulated by highly polarized sub-flares with durations of only about 10 minutes (Eckart et al. 2005) with an overall degree of polarization of the order of 10–20% (Figure 1). In 2005 the main underlying flare was long enough to record three consecutive sub-flares that are consistent with quasi-periodicity of 20 ± 3 minutes, similar to the value of 17 ± 2 minutes found in previous NACO observations (Genzel et al. 2003; see also Gillessen et al. 2005). This

† Based on observations with CHANDRA and ESO VLT observations 271.B-5019, 073.B-0249, 75.B-0093, 075.B-0113, 076.B-0863, and 077.B-0028.

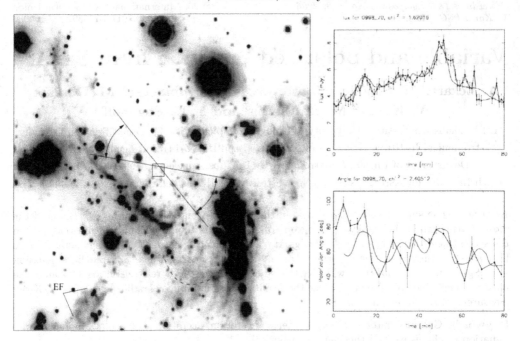

Figure 1. Left: An image of the central 0.5×0.5 square parsec (1 arcsec ~ 0.039 pc) of the Galactic Center at a wavelength of 3.8μm using the NACO adaptive optics system at the VLT, showing stars and dust emission from the mini-spiral. The two lines centered on the position of SgrA* cover the range over which the polarization angle observed in July 2005 and June 2004 varies on the sky. By dashed line we show the approximate location and shape of the mini-cavity. The source EF is described in Eckart *et al.* (2006b). Right: Total flux (top) and polarization angle (bottom) shown for a flare observed in 2004 (Eckart *et al.* 2005; Meyer *et al.* 2006a). In red we show a fit assuming a disk/spot relativistic model (see text for details).

resembles periodicity that has been reported recently for a bright X-ray flare (Bélanger *et al.* 2006).

The rapid variation of polarized emission is though to be indicative of synchrotron radiation by relativistic electrons and the intrinsic polarization of the sub-flares could be up to 60%. Polarization measurements have the potential to reveal strong-gravity effects of cosmic black holes at significantly better precision than what has been possible until now. By analyzing time-variable polarization degree and the polarization angle, we will be able to set tight constraints on the SgrA* black hole properties and relativistic models describing the motion and radiation of matter at distances of only a few gravitational radii from the Galactic Center.

Observed polarimetric properties are influenced by the gravitational field of the black hole as light passes near its horizon. Photons follow null geodesics and the Stokes vector is determined by the law of parallel propagation (Connors & Stark 1977). Therefore, unlike in Newtonian gravity, the polarization angle changes its direction if gravitational field is present and this relativistic phenomenon becomes strong near the black hole. Purely gravitational effects mix up with the intrinsic microphysical properties of gas emitting the light, hence, both gravitational physics and radiation mechanisms must be taken into account when interpreting the observations.

Our preferred model to explain the quasi-periodic polarized flux density fluctuations is that of an under-luminous transient disk, part of which is a spot or a blob orbiting the central black hole (see Figs. 2–3; more details of the relativistic calculations were given

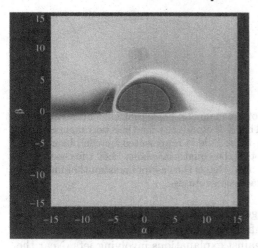 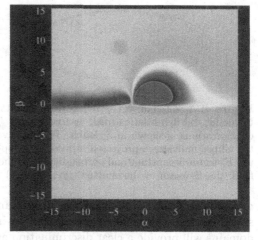

Figure 2. Color-coded levels of the energy shift for a source of light in Keplerian motion around a black hole, as projected onto observer's image plane (α, β) at inclination of $i = 85°$ with respect to the common rotation axis. Left: $a = 0$ (non-rotating black hole). Right: $a = 1$ (maximally rotating Kerr black hole). Projections of the the black hole horizon and the marginally stable orbit are also indicated.

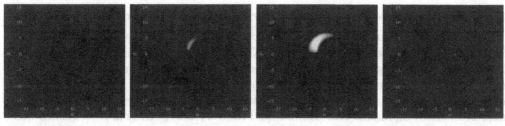

Figure 3. Photometric images of an orbiting hot-spot as seen by an observer at four different orbital phases. The time series covered by the images were calculated for a case of inclination $i = 85°$, $a = 1$ for a spot at a distance of $4R_g$ (four gravitational radii from the SgrA* black hole; Dovčiak et al. 2004). Models like this can successfully fit the data presented in Eckart et al. (2005, 2006a,b) and Meyer et al. (2006a,b).

in Dovčiak, Karas & Yaqoob 2004). In this model the temporal variations are explained by relativistic enhancement and subsequent dilution of the radiation flux when the spot is approaching/receding with respect to the observer. At the same time the formation of partial Einstein rings due to gravitational lensing affects the overall polarization of the hot spot.

All these effects are a function of the spot properties as well as the spin parameter of the black hole and the spin orientation with respect to the spot orbit. Another model parameter is the orientation of the magnetic field. A toroidal form of the B-field will result in apparent rotation of E-field vector. As an alternative, the B-field arrangement of the spot may be such that the apparent E-field is perpendicular to the disk (see Fig. 4). We found that the minimum spin parameter of $a \sim 0.5$ is consistent with the observed quasi-periodicity. We also found a tendency for high inclinations, and spot radii larger than the last stable orbit. Detailed calculations are presented by Meyer et al. (this volume, and refs. 2006a,b) and Eckart et al.(2005, 2006a).

As an unexpected surprise, for all the bright polarized flares observed between 2003 and 2006 the position angle of the mean E-field vector maintains a value of about 60 ± 30 degrees on the sky during sub-flares (Meyer et al. 2006a,b). This suggests a preferred

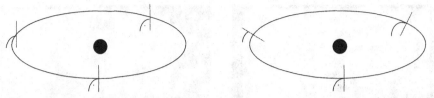

Figure 4. A sketch of two extreme arrangements of the magnetic fields that have been employed to model the hypothetical disk around SgrA*. The E-vector (red) for the two magnetic field configurations is shown in a sketch. The massive black hole is represented by the black circle; the ellipse indicates a projected orbit of a hot spot in the equatorial plane. Left: the case where the E-vector is constant and perpendicular to the disk. Right: the case of an azimuthal magnetic field; the E-vector of the emitted (synchrotron) radiation rotates.

orientation of the overall black hole/disk arrangement with respect to the observer. Future simultaneous observations covering the near-infrared, radio millimeter and sub-millimeter domains will provide a clear discrimination against explanations involving jets. Near the last stable orbit, the expected signatures of a short jet with a length of only a few Schwarzschild radii (the extent of the jet base, i.e., a nozzle) emerging from a disk would likely look almost indistinguishable from signal from a pure disk with orbiting spots.

3. Simultaneous observations of flares in the NIR and X-ray domains

Following the first successful experiment coordinated between the VLT and the Chandra satellite during which simultaneous NIR and X-ray flare emission was detected (Eckart *et al.* 2004; see also Yusef-Zadeh *et al.* 2006a), new simultaneous NIR/sub-millimeter/X-ray observations of the SgrA* counterpart were recently presented by Eckart *et al.* (2006b). In addition to NACO, the Chandra X-ray Observatory as well as the Sub-millimeter Array on Mauna Kea (Hawaii) and the Very Large Array in New Mexico were involved.

For a total of 4 near-IR flares we found an upper limit for a time lag between the X-ray and NIR flare of $\leqslant 10$ minutes. The NIR/X-ray flares from SgrA* can be explained with a synchrotron self-Compton (SSC) model involving up-scattered sub-millimeter photons from a compact source component. Inverse Compton scattering of the THz-peaked flare spectrum by the relativistic electrons then accounts for the X-ray emission. This model is in full agreement with the relativistic orbiting spot model described above. The excess flux densities detected in the radio and sub-millimeter may be linked with the NIR flare activity via cooling through adiabatic expansion of a synchrotron component (see also Marrone *et al.* 2006; Mauerhan *et al.* 2005). A similar behaviour was recently found in dual wavelength radio data by Yusef-Zadeh *et al.* (2006b).

Until now only a few hours of overlapping observations were gained between the NIR/X-ray data and the VLA and SMA data; extensive simultaneous data have not been achieved available so far. Future progress will depend on further successful simultaneous observing campaigns. Polarization data from the NIR to the radio as well as (sub-)mm-VLBI and NIR interferometric experiments are also highly desirable to study the details of the accretion process in SgrA*.

Acknowledgements

This work was supported in part by the Deutsche Forschungsgemeinschaft (DFG) via grant SFB 494. We are grateful to all members of the NAOS/CONICA and the ESO PARANAL team. The X-ray work was supported by NASA through Chandra award

G05-6093X. VK and MD acknowledge continued support from the Academy of Sciences (ref. 300030510) and the Czech Science Foundation (ref. 205/07/0052).

References

Bélanger, G., Terrier, R., De Jager, O., Goldwurm, A. & Melia, F. 2006, Journal of Physics: Conference Series, 54, 420 (astro-ph/0604337)

Connors, P. A. & Stark, R. F. 1977, Nature, 269, 128

Dovčiak, M., Karas, V. & Yaqoob, T. 2004, ApJS, 153, 205

Eckart, A., Baganoff, F. K., Morris, M., Bautz, M. W. *et al.* 2004, A&A, 427, 1

Eckart, A., Schödel, R. & Straubmeier, C. 2005, The Black Hole at the Center of the Milky Way (London: Imperial College Press)

Eckart, A., Schödel, R., Meyer, L., Trippe, S., Ott, T. & Genzel, G. 2006a, A&A, 455, 1

Eckart, A. Baganoff, F. K., Schdel, R., Morris, M. *et al.*, 2006b, A&A, 450, 535

Genzel, R., Schdel, R., Ott, T., Eckart, A., Alexander, T., Lacombe, F., Rouan, D. & Aschenbach, B. 2003, Nature, 425, 934

Gillessen, S., Davies, R., Kissler-Patig, M., Lehnert, M. *et al.* 2005, ESO Messenger, 120, 26

Marrone, D. P., Moran, J. M., Zhao, J.-H. & Rao, R. 2006, ApJ, 640, 308

Mauerhan, J. C., Morris, M., Walter, F. & Baganoff, F. K. 2005, ApJ, 623, L25

Meyer, L., Eckart, A., Schödel, R., Duschl, W. J., Muzic, K., Dovčiak, M. & Karas, V. 2006a, A&A, 460, 15

Meyer, L., Schödel, R., Eckart, A., Karas, V., Dovčiak, M. & Duschl, W. J. 2006b, A&A, 458, L25

Yusef-Zadeh, F., Bushouse, H., Dowell, C. D., Wardle, M., Roberts, D. *et al.* 2006a, ApJ, 644, 198

Yusef-Zadeh, F., Roberts, D., Wardle, M., Heinke, C. O., Bower, G. C. *et al.* 2006b, ApJ, 650, 189

ALEXANDER ZAKHAROV: What would be the appropriate size of the emitting region which produces the observed signal in your model?

ANDREAS ECKART: The actual size of the emitting region should be between $\sim$ 20–30 μarcsec. The spot would then be only a fraction of that size.

FELIX MIRABEL: Do you favor the model of a rotating hotspot rather than an ejected blob and, if this is the case, for what reason?

ANDREAS ECKART: Quasi-periodicity is a slight argument in favor of the orbiting spot model. On the other hand a short jet has not been excluded and indeed we also consider such possibility in our work. The question whether there is a jet has not yet been solved.

FELIX MIRABEL: My question was motivated by the fact that we find some kind of similar quasi-periodicity also in GRS1915+105. And if we are dealing with an ejected blob than the mass scalling of timescales does not apply – although at the beginning we also considered a hotspot model, such interpretation was then excluded by imaging.

ANDREAS ECKART: As I mentioned above we do not exclude the possibility of the ejected matter in SgrA* and actually we discuss more details on this idea in our A&A paper published recently.

QVADRANS MAXIMVS QVALEM
OLIM PROPE AVGVSTAM VINDELICORVM
exstruximus.

EXPLI-

Black Holes from Stars to Galaxies – Across the Range of Masses
Proceedings IAU Symposium No. 238, 2006
V. Karas & G. Matt, eds.
© 2007 International Astronomical Union
doi:10.1017/S1743921307004942

The structure of the nuclear stellar cluster of the Milky Way

Rainer Schödel[1] and Andreas Eckart[1,2]

[1]I. Physikalisches Institut, Universität zu Köln, Zülpicher Str. 77, 50937 Köln, Germany

[2]Max-Planck-Institut für Radioastronomie, Auf dem Hügel 69, 53121 Bonn, Germany

email: rainer@ph1.uni-koeln.de, eckart@ph1.uni-koeln.de

Abstract. High-resolution adaptive optics observations of the inner 0.5 pc of the Milky Way with multiple intermediate band filters are presented. From the images, stellar number counts and a detailed map of the interstellar extinction were extracted. The extinction map is consistent with a putative southwest-northeast aligned outflow from the central arcseconds.

An azimuthally averaged, crowding and extinction corrected stellar density profile presents clear evidence for the existence of a stellar cusp around Sgr A*. Several density peaks are found in the cluster that may indicate clumping of stars, possibly related to the last epoch of star formation in the Galactic Center. An analysis of stars in the brightness range $14.25 < mag_K < 15.75$ shows possible signs of mass segregation.

Keywords. Stellar dynamics – Galaxy: center – Galaxy: nucleus

1. Introduction

The center of our Milky Way is located at a distance of 7.6 ± 0.3 kpc and contains a $3.6 \pm 0.3 \times 10^6 M_\odot$ black hole (BH), Sagittarius A* (Sgr A*) (Schödel, Ott, Genzel *et al.* 2003; Ghez, Duchêne, Matthews *et al.* 2003; Eisenhauer, Genzel, Alexander *et al.* 2005). It is the only nucleus of a normal spiral galaxy that can be resolved in detail by observations. When the age of a galaxy is older than the relaxation time due to two-body encounters, the formation of a density excess around the central black hole, the so-called *stellar cusp*, is expected. The density in this cusp decreases from the black hole with a power law. The index of this power-law ranges between $-1/2$ and $-7/4$, depending on the stellar population and formation history of the cusp (Bahcall & Wolf 1976; Bahcall & Wolf 1977; Lightman & Shapiro 1977; Murphy, Cohn & Durisen 1991). Since the Galactic Center (GC) can be resolved into individual stars by observations, it is *the* template object, where to look for a stellar cusp. The first clear evidence for a density excess in the central arcseconds around Sgr A* was only obtained after adaptive optics (AO) became available at 8m-class telescopes (Genzel, Schödel, Ott *et al.* 2003). We present new AO observations of the central 0.5 pc of the GC. The images were used for a thorough analysis of the GC stellar cluster, determining not only the two-dimensional density structure of the cluster, but also taking into account interstellar extinction and analysing the distribution of stars in different brightness ranges. Here, we can only give a very brief introduction into the data, their analysis, and the results. The interested reader is referred to the article by Schödel *et al.* (2006, submitted to A&A).

2. Observations and data reduction

AO imaging observations were acquired with NACO/VLT im June and July 2004, using intermediate band (IB) filters of $0.06\,\mu m$ width, centered at 2.00, 2.06, 2.24, 2.27, and $2.30\,\mu m$ (see left panel in Fig. 1). *StarFinder* (Diolaiti, Bendinelli, Bonaccini *et al.* 2000) was used for point source detection and photometry via PSF fitting.

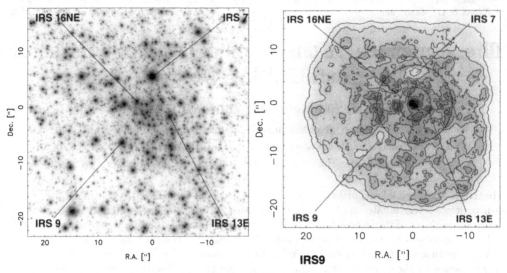

Figure 1. Left: Sum of the 2.27 + 2.30 μm IB filter NACO images. This image was used for the star counts. Offsets from Sgr A* in right ascension and declination. North is up and east is to the left. Right: Adaptively smoothed, crowding and extinction corrected map of the stellar surface density. The smoothing was performed such that 40 stars contributed to the density at each point in the map. The smoothing radius varies between 0.6″ near Sgr A* and 2″ near the edge of the field. IRS 7, IRS 9 and IRS 16NE are particularly bright stars. The detection of faint stars is suppressed in their environment. The low density near these stars is therefore an artefact. The circles around Sgr A* indicate radial distances of 3″ and 7″. At these distances several density peaks appear to be concentrated. Note that bumps in the azimuthally averaged surface density can be seen at these distances in Fig. 3.

3. Results

Extinction. The IB 2.00 to 2.27 images were used to construct a map of interstellar extinction. The near-infrared (NIR) spectra of stars are largely free from strong emission or absorption features at these wavelengths and A_K can be determined from differential reddening. An extinction law $A_\lambda \propto \lambda^{-1.75}$ was assumed (Draine 1989). The median of all measurements within 2″ of a given point in the FOV was taken. The resulting extinction map is shown in the left panel of Fig. 2. The 1σ uncertainty of the map is $\leqslant 0.1$ mag, except in the regions with highest extinction, where it is $\leqslant 0.2$ mag due to the reduced number of detected stars. It is intriguing to note that the so called mini-cavity can be distinguished in the extinction map. It has been suggested in previous work that the mini-cavity is caused by fast winds from the Helium stars in the central arcseconds or by an outflow from Sgr A* (Lutz, Krabbe & Genzel 1993; Melia, Coker & Yusef-Zadeh 1996; Yusef-Zadeh, Roberts & Biretta 1998). The southwest-northeast aligned channel-like feature that includes the mini-cavity in the extinction map is consistent with reduced extinction along a potential outflow along this direction (see Muzic *et al.*, in prep.).

Stellar number density. The right panel of Fig. 1 shows the extinction corrected two-dimensional stellar surface density in the central 0.5 pc of the GC. It can be seen that the GC stellar cluster appears fairly circularly symmetric. A strong density peak is visible centered on Sgr A*. Some regions of reduced density can be attributed to the presence of extremely bright stars, such as IRS 7, that hinder source detection in their immediate environment. It can be also seen that the stellar surface density does not decrease in a homogeneous way from the center, but that there are several density peaks present in the field, preferentially located at radial distances of 3″ and 7″ from Sgr A* (see circles

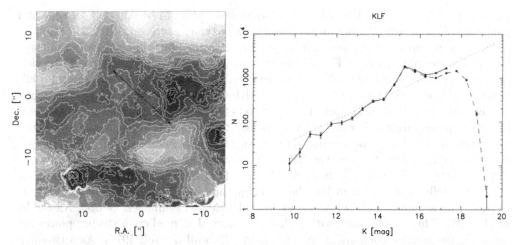

Figure 2. Left: Map of the interstellar extinction at $2.157\,\mu$m derived from NACO intermediate band imaging at 2.00, 2.06, 2.24, and 2.27 μm. The contours are spaced from $A_{2.157\mu m} = 2.0-3.5$ at intervals of 0.1. The cross indicates the position of Sgr A* and the dashed circle the approximate location of the mini-cavity. The arrows indicate a putative southwest-northeast aligned outflow from the central arcseconds. Right: KLF of the GC stellar cluster; dashed: raw counts, straight line: completeness corrected counts. The bump around $mag_K = 15$ is due to horizontal branch/red clump (HB/RC) stars. The dotted line is a simple power law fit, omitting the HB/RC bump.

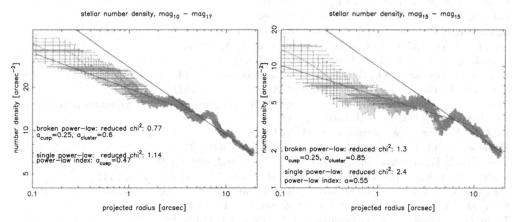

Figure 3. Left: Azimuthally averaged extinction and crowding corrected surface density vs. distance from Sgr A*. Fits to the data with a single (blue) and with a broken (red) power-law are indicated, along with the corresponding power-law indices and reduced χ^2 values. The χ^2-values refer to the *overall* fit. The $\chi^{(}2)$-value of the single power law fit is, of course, much worse in the central arcseconds. Right: The same as the left panel, but only for stars with $14.75 < mag_K < 15.75$.

in right panel of Fig. 1). One of these clumps is coincident with the dense co-moving group of stars in the IRS 13E complex (Maillard, Paumard, Stolovy *et al.* 2004; Schödel, Eckart, Iserlohe *et al.* 2005).

The azimuthally averaged extinction and crowding corrected surface density is presented in the left panel of Fig. 3. Power-laws were fitted to the plot of the stellar surface density. As can be seen in the Figure, a broken power law is necessary in order to provide a reasonable fit for the innermost arcseconds. The least-squares fit results in a broken power law with a break radius of $7 \pm 1''$, a power-law index of $\alpha = 0.6 \pm 0.1$ for the outer part, and of $\alpha_{cusp} = 0.25 \pm 0.1$ for the inner part of the cluster.

Evidence for mass segregation. The K-band luminosity function (KLF) of the cluster is shown in the right panel of Fig. 2. The prominent bump around $mag_K = 15$ is typical for observations toward the GC and galactic bulge (Tiede, Frogel & Terndrup 1995; Genzel, Schödel, Ott *et al.* 2003; Figer, Rich, Kim *et al.* 2004) and is due to horizontal branch/red clump (HB/RC) stars. These old stars of roughly $\geqslant 1\,M_{\odot}$ are the lightest ones in the examined magnitude range. Due to its age, this part of the stellar population may be relaxed by two-body encounters and bear signatures of mass segregation. In fact, as the right panel in Fig. 3 shows, the HB/RC stars display a flat or even slightly inverted slope at distances from $1 - 3''$ from Sgr A*. Inside of $1''$, the cusp slope is very steep. Here, the stellar population in the chosen magnitude range consists almost entirely of B-type main sequence stars with masses around $10\,M_{\odot}$ (Eisenhauer, Genzel, Alexander *et al.* 2005). This is consistent with stellar dynamics that predicts a steeper power law index for the heavier components of the cluster (Bahcall & Wolf 1977). As a caveat it must be said that it is not clear whether the young stars near Sgr A* had time to relax dynamically.

References

Schödel, R., Ott, T., Genzel, R. *et al.* 2003, ApJ, 596, 1015

Ghez, A. M., Duchêne, G., Matthews, K. *et al.* 2003, ApJ, 586, L127

Eisenhauer, F., Genzel, R., Alexander, T. *et al.* 2005, ApJ, 628, 246

Bahcall, J. N. & Wolf, R. A. 1976, ApJ, 209, 214

Bahcall, J. N. & Wolf, R. A. 1977, ApJ, 216, 883

Lightman, A. P. & Shapiro, S. L. 1977, ApJ, 211, 244

Murphy, B. W., Cohn, H. N. & Durisen, R. H. 1991, ApJ, 370, 60

Genzel, R., Schödel, R., Ott, T. *et al.* 2003, ApJ, 594, 812

Diolaiti, E., Bendinelli, O., Bonaccini, D. *et al.* 2000, A&AS, 147, 48

Paumard, T., Genzel, R., Martins, F. *et al.* 2006, ApJ, 643, 1011

Draine, B. T. 1989, Proceedings of the 22nd Eslab Symposium held in Salamanca, Spain, 7-9 December, 1988, ed. B.H. Kaldeich, 93

Lutz, D., Krabbe, A. & Genzel, R. 1993, ApJ, 418, 244

Melia, F., Coker, R. F. & Yusef-Zadeh, F. 1996, ApJ, 460, L33

Yusef-Zadeh, F., Roberts, D. A. & Biretta, J. 1998, ApJ, 499, L159

Schödel, R., Eckart, A., Iserlohe, C., Genzel, R. & Ott, T. 2005 ApJ, 625, L111

Maillard, J. P., Paumard, T., Stolovy, S. R. & Rigaut, V. 2004, A&A, 423, 155

Figer, D., Rich, V., Kim, S.S., Morris, M. & Serabyn, E. 2004, ApJ, 601, 319

Tiede, G. P., Frogel, J. A. & Terndrup, D. M. 1995, AJ, 110, 2788

EMMANUIL VILKOVISKIY: What is the total mass of the cusp and the stellar cluster?

RAINER SCHÖDEL: The cusp has mass of about $10^3 M_{\odot}$.

YIPING WANG: Do you see any metallicity gradient towards the central part of the Galactic center?

RAINER SCHÖDEL: No, we don't have enough information on the metallicity of the central star cluster, however, it is my understanding that solar metallicity is entirely consistent with available data.

Black Holes from Stars to Galaxies – Across the Range of Masses
Proceedings IAU Symposium No. 238, 2006
V. Karas & G. Matt, eds.
© 2007 International Astronomical Union
doi:10.1017/S1743921307004954

Accretion of stellar winds onto Sgr A*

Jorge Cuadra[1] and Sergei Nayakshin[2]

[1]Max-Planck-Institut für Astrophysik, D-85741 Garching, Germany
[2]Department of Physics and Astronomy, University of Leicester, LE1 7RH, UK
email: jcuadra@mpa-garching.mpg.de, Sergei.Nayakshin@astro.le.ac.uk

Abstract. We report a 3-dimensional numerical study of the accretion of stellar winds onto Sgr A*, the super-massive black hole at the centre of our Galaxy. Compared with previous investigations, we allow the stars to be on realistic orbits, include the recently discovered slow wind sources, and allow for optically thin radiative cooling. We first show the strong influence of the stellar dynamics on the accretion onto the central black hole. We then present more realistic simulations of Sgr A* accretion and find that the slow winds shock and rapidly cool, forming cold gas clumps and filaments that coexist with the hot X-ray emitting gas. The accretion rate in this case is highly variable on time-scales of tens to hundreds of years. Such variability can in principle lead to a strongly non-linear response through accretion flow physics not resolved here, making Sgr A* an important energy source for the Galactic centre.

Keywords. Galaxy: center – black hole physics

---

1. Introduction

Sgr A* is identified with the $M_{BH} \sim 3.5 \times 10^6 M_\odot$ super-massive black hole (SMBH) at the centre of our Galaxy (Schödel *et al.* 2002). By virtue of its proximity, Sgr A* plays a key role in the understanding of Active Galactic Nuclei (AGN). Indeed, this is the only SMBH where observations detail the origin of the gas in its vicinity. This information is absolutely necessary for the accretion problem to be modelled self-consistently.

One of Sgr A* puzzles is its low luminosity with respect to estimates of the accretion rate. Young massive stars in the inner parsec of the Galaxy fill this region with hot gas. From *Chandra* observations, one can measure the gas properties around the inner arcsecond (one arcsecond, $1''$, corresponds to ~ 0.04 pc, $\sim 10^{17}$ cm, or $\sim 10^5 R_S$ for Sgr A*) and then estimate the Bondi accretion rate (Baganoff *et al.* 2003). The expected luminosity is orders of magnitude higher than the measured $\sim 10^{36}$ erg s^{-1}.

The hot gas, however, is continuously created in shocked winds expelled by the stars near Sgr A*, and the stars themselves are distributed in discs (Paumard *et al.* 2006). The situation then is far more complex than in the idealised Bondi model. An alternative approach is to model the gas dynamics of stellar winds, assuming that the properties of the wind sources are known (Coker & Melia). Here we present our numerical modelling of wind accretion onto Sgr A*, the first to allow the wind-producing stars to be on Keplerian orbits.

2. Numerical approach and first results

We use the smooth particle hydrodynamics (SPH) code GADGET-2 (Springel 2005) to simulate the dynamics of stars and gas in the gravitational field of the SMBH. To model the stellar winds, new gas particles are continuously created around the stars. The SMBH is modelled as a 'sink' particle (Springel *et al.* 2005), with all the gas passing within $0.05''$ from it disappearing from the computational domain. In addition, we eliminate all the gas that goes beyond a 20″ outer boundary. More details of the numerical method along with validation tests have been presented elsewhere (Cuadra *et al.* 2006).

We first performed several simulations with different configurations for the stellar orbits. We found that the angular momentum of the gas has a strong dependence on the orbital motions. Moreover, only the fraction of the gas with the 'right' low angular momentum goes to the inner region and can be accreted. Therefore, when the stars are rotating in a disc, the gas has more angular momentum and the accretion rate is lower than in the case where stars orbit the central black hole isotropically (Cuadra *et al.* 2006). This result highlights the necessity of using realistic orbits to model the accretion onto Sgr A*.

3. Set-up of the new simulations

In our previous calculations (Cuadra *et al.* 2006), we used an ensemble of stars in circular orbits with roughly the same properties as the observed population. For the new simulations presented here, we use the latest available data for individual stars, going a step forward on the creation of more realistic models.

We include in our calculations the 30 stars identified as WRs or LBVs (Paumard *et al.* 2006). For the initial conditions of the stellar winds, we use the wind velocities and mass loss rates calculated by Martins *et al.* (in preparation). For the rest of the stars we use the wind velocities derived previously (Paumard *et al.* 2001) when available, otherwise we set $v_w = 1000$ km s^{-1}. The mass loss rate for the stars without data is set to a common value such that the total mass loss rate from all the stars is $10^{-3} M_\odot$ yr^{-1}. More tests changing the mass loss rate of the stars are discussed by Cuadra *et al.* (in prep.).

We use the published positions and velocities for the mass-losing stars (Paumard *et al.* 2006). For the stars without a z coordinate, this is set assuming that the star is in the corresponding disc. With the current 3d positions and velocities for the stars, we integrate the orbits back in time for ~ 1200 yr. The positions and velocities at this time were used as initial conditions for the SPH simulations.

4. Two-phase gas

Figure 1 shows the resulting morphology of the gas at the end of one simulation. Cool dense regions in the gas distribution are mainly produced by winds from LBVs. When shocked, these slow winds attain a temperature of only around 10^6 K, and, given the high pressure environment of the inner parsec of the GC, quickly cool radiatively (Cuadra *et al.* 2005). LBV winds form bound clouds of gas, often flattened into filaments due to the SMBH potential. On the other hand, the WRs by themselves do not produce much structure. The fast winds they emit have temperatures $> 10^7$ K after shocking, and do not cool fast enough to form clumps. This temperature is comparable to that producing X-ray emission detected by *Chandra*. Gas cooler than that would be invisible in X-rays due to the finite energy window of *Chandra* and the huge obscuration in the Galactic plane.

We find that it is harder to form a disc-like structure like the one found in our previous studies (Cuadra *et al.* 2006). The main reason for this is that in the present simulations there is not as much cold gas. There are less LBVs – from whose winds the cold clumps are mostly formed – and they have smaller mass loss rates than previously assumed. Moreover, while before we put most slow wind stars in the same plane, now we put only two out of the three innermost stars in one plane, so in this case the preference for one plane is lower.

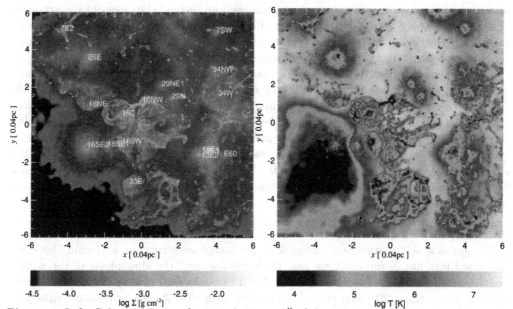

Figure 1. Left: Column density of gas in the inner $6''$ of the computational domain. Stars are shown as green symbols with labels. Right: Averaged temperature of the same region. Dense cold clumps form around the IRS 16 and 13 groups, where slow winds from different stars collide. Winds from the WR stars also remain cold but diffuse before interacting with other winds.

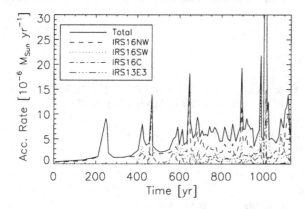

Figure 2. Accretion history from one of our simulations. Different lines show the contribution of material coming from a few selected stars. Notice that most of the accreted matter comes from the stars in the IRS 16 group.

5. Accretion onto Sgr A*

From any of our simulations, we can extract the accretion history onto Sgr A*. Accretion is here defined by the quantity of gas entering the inner boundary ($0.05''$) of our computational domain.

In all the simulations the accretion rate is quite variable, changing by factors ~ 10 on timescales as short as the chosen resolution of 10 yr. The variability is mainly caused by the accretion of clumps originating from material produced by the few innermost slow wind stars in the IRS 16 group (see Fig. 5). But even if the mass loss rates from these stars are reduced by more than a factor 10, a large degree of variability remains (see Cuadra *et al.*, in prep.).

In our previous work, we had found that the accretion onto the SMBH could be separated into a quasi-steady hot component and a variable cold one coming from the episodic accretion of clumps. In these new calculations, we find instead that most of the accreted material is actually hot ($T > 10^7$ K), regardless of which star it is coming from. This is most likely the effect of the eccentric orbits – they increase the total velocity of the wind, making it acquire a large kinetic energy which is later thermalized.

While we cannot resolve the inner accretion flow to predict the actual accretion rate onto Sgr A*, we estimate that variability by a factor of a few should still reach the black hole. In the extremely sub-Eddington regime of Sgr A*, a small change in the accretion rate produces a non-linear response on the luminosity (Yuan *et al.* 2002). The results from our simulations, the observational evidence for higher luminosity in the recent past (Revnivtsev *et al.* 2004), and the idea of star formation in an AGN-like accretion disc a few million yr ago (Nayakshin & Cuadra 2005), all suggest that on long time-scales Sgr A* is an important energy source for the inner Galaxy.

6. Conclusions

The gas at $r \sim 1''$ distances from Sgr A* has a two-phase structure, with cold filaments immersed into hot X-ray emitting gas. Both the fast and the slow phase of the winds contribute to the accretion flow onto Sgr A*. The accretion rates we obtain are of the order of a few times $10^{-6} M_\odot$ year^{-1}, consistent with the *Chandra* estimates, although very variable on time-scales as short as tens of years. This implies that the current very low luminosity state of Sgr A* may be the result of a relatively unusual quiescent state.

Acknowledgements

JC thanks the IAU and the ASCR for their financial support to attend the meeting.

References

Baganoff, F. K. *et al.* 2003, ApJ, 591, 891
Coker, R. F. & Melia, F. 1997, ApJL, 488, L149
Cuadra, J., Nayakshin, S., Springel, V. & Di Matteo, T. 2005, MNRAS, 360, L55
Cuadra, J., Nayakshin, S., Springel, V. & di Matteo, T. 2006, MNRAS, 366, 358
Nayakshin, S. & Cuadra, J. 2005, A&A, 437, 437
Paumard, T., Maillard, J. P., Morris, M. & Rigaut, F. 2001, å, 366, 466
Paumard, T. *et al.* 2006, ApJ, 643, 1011
Revnivtsev, M. G. *et al.* 2004, A&A, 425, L49
Schödel, R. *et al.* 2002, Nature, 419, 694
Springel, V. 2005, MNRAS, 364, 1105
Springel, V., Di Matteo, T. & Hernquist, L. 2005, MNRAS, 361, 776
Yuan, F., Markoff, S. & Falcke, H. 2002, A&A, 383, 854

THIERRY COURVOISIER: It seems very striking that there is no substantial net angular momentum. How does matter present in the center gets rid of angular momentum?

JORGE CUADRA: I guess you refer to the process that led the material that formed the stars to the inner parsec. This is an important issue but I do not think the answer to your question is known. It is possible that the gas originated in a collision between molecular clouds with opposite angular momentum for instance.

Black Holes from Stars to Galaxies – Across the Range of Masses
Proceedings IAU Symposium No. 238, 2006
V. Karas & G. Matt, eds.
© 2007 International Astronomical Union
doi:10.1017/S1743921307004966

Short-term variability of Sgr A*

M. R. Morris,[1] S. D. Hornstein,[1] A. M. Ghez,[1] J. R. Lu,[1]
K. Matthews[2] and F. K. Baganoff[3]

[1]Dept. of Physics & Astronomy, Univ. of California, Los Angeles, CA 90095-1547, USA

[2]Caltech Optical Observatories, California Institute of Technology, Mail Code 105-24,
Pasadena, CA 91125, USA

[3]Center for Space Research, Massachusetts Institute of Technology,
Cambridge, MA 02139, USA

email: morris@astro.ucla.edu

Abstract. Observations of Sgr A* over the past 4 years with the Keck Telescope in the near-infrared, coupled with millimeter and submillimeter observations, show that the $3.7 \times 10^6 M_\odot$ Galactic Black Hole, Sagittarius A*, displays continuous variability at all these wavelengths, with the variability power concentrated on characteristic time scales of a few hours, and with a variability fraction that increases with wavelength. We review the observations indicating that the few-hour time scale for variability is reproduced at all accessible wavelengths. Interpreted as a dynamical time, this time scale corresponds to a radial distance of 2 AU, or ~25 Schwarzschild radii. Searches for quasi-periodicities in the near-infrared data from the Keck Telescope have so far been negative. One interpretation of the character of these variations is that they result from a recurring disk instability, rather than from variations in the mass accretion rate flowing through the outer boundary of the emission region. However, neither a variable accretion rate nor a mechanism associated with a jet can presently be ruled out.

Keywords. Galaxy: nucleus – black hole physics – accretion disks

1. Radio variability

The Galactic black hole, Sgr A*, has been observed across the electromagnetic spectrum from radio to X-rays, except where interstellar dust blocks our view. Furthermore, it is variable at all accessible wavelengths. It has been known for some time that this source is variable in the radio on time scales from weeks to years (Brown & Lo 1982; Zhao *et al.* 1989, 1992; Herrnstein *et al.* 2004). The variability at the longer centimeter wavelengths is ascribable to interstellar scintillation, but at a few cm and shorter wavelengths, the variations are largely intrinsic (e.g., Macquart & Bower 2006). The amplitude of the intensity modulation depends on the time scale, but there is a general trend for the amplitude to decrease with increasing wavelength (Herrnstein *et al.* 2004). Relatively weak variability (~10%) of the 1.3 and 0.7-cm emission from Sgr A* on an hourly time scale (Yusef-Zadeh *et al.* 2006) is consistent with that trend; this result was interpreted in terms of an expanding plasmon model because of a half-hour time lag between the peaks of a flare at these two frequencies.

Variability on hourly time scales is particularly interesting, as such time scales correspond to light travel times across distances of <100 Schwarzschild radii (R_s). Intraday variability of 3-mm emission was reported by Miyazaki *et al.* (2004, 2006). They noted "flares" occurring on time scales of a few weeks to a month that have maximum amplitude increases of $100 \div 200\%$ at 100 GHz and $200 \div 400\%$ at 140 GHz, and on a number of occasions they noted significant factor-of-two variability on time scales of a few hours.

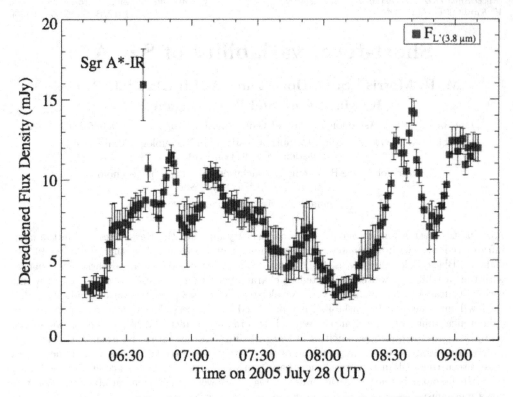

Figure 1. Light curve of Sgr A* at a wavelength of 3.8 μm on 28 July 2005. This observation with the NIRC2 camera on the Keck II telescope was simultaneous with an X-ray observation with the Chandra X-ray Observatory, but no X-ray activity was detected above level of the extended quiescent emission.

Variability on few-hour time scales was investigated in depth by Mauerhan *et al.* (2005), who observed fluctuations of only 20 ÷ 40% during an 8-day observing period. They carried out a red noise analysis of the variations, and concluded that there was excess power on a time scale of 2.5 hr.

Variations on hourly time scales have also been seen at submillimeter wavelengths. Marrone *et al.* (2006ab; see also Eckart *et al.* 2006) report observations of Sgr A* with the Submillimeter Array, in which they see hour-time-scale variability on the order of 10÷20% in total intensity at 230 and 345 GHz. The polarized sub-mm intensity, however, shows a far greater variation, in both position angle and percent polarization. While we do not discuss the implications of the polarization variations here, we note that they open up an exciting new dimension of investigation; the polarization variability, for example, may arise as a result of aspect changes associated with rotation in the inner disk.

2. Infrared variability

Sgr A* was found to be a variable infrared source at the same time that it was first convincingly detected independently by Genzel *et al.* (2003) at 1.65 and 2.16 μm and Ghez *et al.* (2004) at 3.8 μm. These studies reported so-called "flares" having 40 ÷ 80 minute durations, and showing substructure on smaller time scales, including a possible 17-minute periodicity (Genzel *et al.* 2003). Since then, a number of observations, all employing adaptive optics, have shown many further examples of near-IR events in Sgr A*

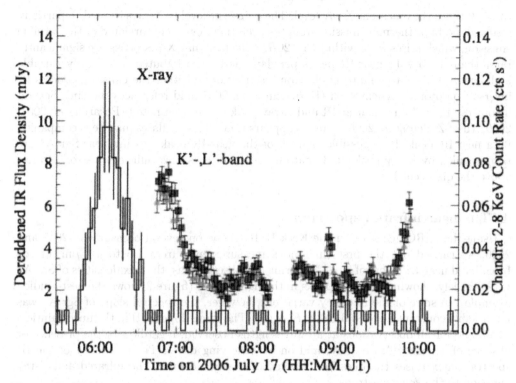

Figure 2. Light curves at 2.1μm (triangles), and 3.8μm (squares) measured alternately in quick succession on 17 July 2006; from Hornstein *et al.* (2007). When corrected for unresolved background starlight, the ratios of these fluxes are consistent with being constant. At bottom is the simultaneously obtained Chandra light curve, showing a substantial X-ray flare occurring just prior to the onset of the IR observations.

(Eckart *et al.* 2004, 2006, and in this volume; Hornstein *et al.* 2006; 2007). In the course of our Keck observations on several occasions over the past 3 years, we have not confirmed any quasi-periodicity on time scales comparable to 17 minutes, so the appearance of such quasi-periodicities in the total intensity does not appear to be commonplace.

The concept of a "flare" implies a temporary rise of the emission above some quiescent level, but it is not clear that a steady-state quiescent level exists in the near-IR. Indeed, the variability of Sgr A* appears at 3.8 μm to be continuous (figure 1), perhaps characterizable as a red noise light curve. At 2.1 μm, the flux density has been seen to range from <2 to 27 mJy (Hornstein *et al.* 2007). However, it appears likely that, at both wavelengths, there is excess power on time scales of a few hours, such that the broad peaks manifest themselves as flares. In recognition of the continuous variability, Eckart *et al.* (2004, 2006) refer the intensity of their near-IR flares to an inconstant "interim quiescent" level. As with submillimeter emission, the polarized near-IR emission is more variable than the total intensity (Eckart *et al.*, this volume).

3. X-ray variability

Since the first discovery of an X-ray flare from Sgr A* by Baganoff *et al.* (2001), many other X-ray flares having a broad range of intensities have been found with both Chandra and XMM (Baganoff 2005; Belanger *et al.* 2005, and references therein). The flares last from 30 ÷ 170 minutes, and occur about once per day, with intensities so far ranging up to ~150 times the quiescent X-ray level. The quiescent level in X-rays is owed to partially resolved, perhaps thermal emission from the accretion flow, so is unrelated to the variable emission, which arises from within $10 \div 20R_s$. The fact that X-ray peaks are significantly rarer than the $5 \div 12$ near-IR peaks per day has been attributed to a highly variable high-energy tail on the emitting electron distribution, whether the emission process be inverse Compton or synchrotron (Hornstein *et al.* 2007, and references therein). Several instances of simultaneous near-IR and X-ray peaks have been noted (Eckart *et al.* 2004, 2006; Yusef-Zadeh *et al.* 2006b), and it appears that all X-ray flares may be accompanied by a near-IR peak. It is possible that all of the near-IR peaks arising near Sgr A* are accompanied by X-ray peaks, but that only a few of these are sufficiently strong to rise above the quiescent level.

4. The near-infrared spectrum

Using the NIRC2 camera on the Keck II Telescope on several occasions in 2005 and 2006, we carried out the first time-series measurements, from 1.6 to 4.7 μm, of the broadband near-IR colors of Sgr A*. During our observations, the flux densities of Sgr A* varied widely, showing multiple peaks in the light curves (figure 2 shows the best-studied example). In spite of the intensity variations, however, the spectral slope of Sgr A* was remarkably constant: $\alpha = -0.6 \pm 0.2$ ($F_\nu \propto \nu^\alpha$). The same slope (within the uncertainties) was found for all observations at all wavelengths observed, regardless of the presence or absence of a nearby X-ray flare, and on all observing dates. The constancy of the IR spectral index raises the question of whether one can connect the infrared and X-ray emission to the *same* electrons.

We hypothesize that the electron acceleration mechanism produces a strikingly reproducible energy spectrum for the electrons responsible for the near-IR emission, while it occasionally generates a tail of electrons above 1 GeV that, when sufficiently populated, gives rise to X-ray outbursts that accompany some of the near-IR peaks. If the electrons are accelerated in a turbulent medium, perhaps induced by a disk instability, then the high-energy cutoff in the electron energy spectrum might be linked to the presumably variable outer scale length of the turbulence.

Acknowledgements

This research has been supported by NSF grant AST-0406816 to UCLA.

References

Baganoff, F. K. *et al.* 2001, Nature, 413, 45

Baganoff, F. K. 2005, KITP Conference: Paradoxes of Massive Black Holes: A Case Study in the Milky Way, http://online.kitp.ucsb.edu/online/galactic_c05/baganoff

Belanger, G. *et al.* 2005, ApJ, 635, 1095

Belanger, G., Terrier, R., de Jager, O. C., Goldwurm, A. & Melia, F. 2006, in: R. Schödel, G. C. Bower, M. P. Muno, S. Nayakshin & T. Ott (eds.), J. Phys. Conf. Ser., 54, 420

Brown, R. L. & Lo, K. Y. 1982, ApJ, 253, 225

Eckart, A. *et al.* 2004, A&A, 427, 1

Eckart, A. *et al.* 2006, A&A, 450, 535

Genzel, R. *et al.* 2003, Nature, 425, 943

Ghez, A. M. *et al.* 2004, ApJL, 601, L159

Herrnstein, R. M., Zhao, J.-H., Bower, G. C. & Goss, W. M. 2004, AJ, 127, 3399

Hornstein, S. *et al.* 2004, in: R. Schödel, G. C. Bower, M. P. Muno, S. Nayakshin & T. Ott (eds.), J. Phys. Conf. Ser., 54, 399

Hornstein, S. *et al.* 2007, ApJ, submitted

Macquart, J.-P. & Bower, G. C. 2006, ApJ, 641, 302

Marrone, D. P., Moran, J. M., Zhao, J.-H. & Rao, R. 2006a, ApJ, 640, 308

Marrone, D. P., Moran, J. M., Zhao, J.-H. & Rao, R. 2006b, in: R. Schödel, G. C. Bower, M. P. Muno, S. Nayakshin & T. Ott (eds.), J. Phys. Conf. Ser., 54, 354

Mauerhan, J. C., Morris, M., Walter, F. & Baganoff, F K. 2005, ApJL, 623, L25

Miyazaki, A., Tsutsumi, T. & Tsuboi, M. 2004, ApJL, 611, L97

Miyazaki, A., Shen, Z.-Q., Miyoshi, M., Tsuboi, M. & Tsutsumi, T. 2006, in: R. Schödel, G. C. Bower, M. P. Muno, S. Nayakshin & T. Ott (eds.), J. Phys. Conf. Ser., 54, 363

Yusef-Zadeh, F., Roberts, D., Wardle, M., Heinke, C. O. & Bower, G. C. 2006a, ApJ, 650, 189

Yusef-Zadeh, F. *et al.* 2006b, ApJ, 644, 198

Zhao, J.-H., Ekers, R. D., Goss, W. M., Lo, K. Y. & Narayan, R. 1989, in: M. Morris (ed.), The Center of the Galaxy (Kluwer: Dordrecht), p. 535

Zhao, J.-H., Goss, W. M., Lo, K. Y. & Ekers, R. D. 1992, in: A. Filippenko (ed.), Relationships Between Active Galactic Nuclei and Starburst Galaxies, ASP Conference Series, vol. 31, p. 295

FELIX MIRABEL: There are time-lags between 22 GHz and 43 GHz, K-band and X-rays etc., roughly up to order of one hour. This may suggest an adiabatic expansion. How would you accommodate this fact in a rotating medium of the disc?

MARK MORRIS: A sudden energy release can be responsible for sudden rise of intensity and can be accompanied with the expansion. This is consistent with recent results. However, it is baffling that the near-infrared spectral indices do not change through these strong variations of the intensity.

Black Holes from Stars to Galaxies – Across the Range of Masses
Proceedings IAU Symposium No. 238, 2006
V. Karas & G. Matt, eds.
© 2007 International Astronomical Union
doi:10.1017/S1743921307004978

Kozai resonance model for Sagittarius A* stellar orbits

Ladislav Šubr,[1,2] Vladimír Karas,[3] and Jaroslav Haas[2]

[1]Argelander Institut für Astronomie, University of Bonn, Bonn, Germany

[2]Charles University, Astronomical Institute, Prague, Czech Republic

[3]Astronomical Institute, Academy of Sciences, Prague, Czech Republic

email: subr@sirrah.troja.mff.cuni.cz

Abstract. We study a possibility of tidal disruptions of stars orbiting a supermassive black hole due to eccentricity oscillations driven by Kozai's mechanism. We apply the model to conditions relevant for the Galactic Centre where we consider two different sources of the perturbation to the central potential, which trigger the resonance mechanism. Firstly, it is a disc of young massive stars orbiting Sgr A* at $r \gtrsim 0.08\,\mathrm{pc}$, and, secondly, a molecular circumnuclear disc. Each of the two possibilities appears to be capable of exciting eccentricities to values sufficient for the tidal disruption of ~ 100 stars from the nuclear stellar cluster on a time-scale of 0.1–10 Myrs. Tidally disrupted stars may cause periods of increased accretion activity of Sgr A*.

Keywords. Galaxy: center – stellar dynamics – celestial mechanics

1. Introduction

Tidal disruptions of stars from a gravitationally bound dense cluster may represent one of the channels of the supermassive black holes (SMBH) growth. This process has a relatively small cross-section as the radius at which the tidal disruption of a solar type star occurs is of the order of several tens gravitational radii of the SMBH, while a characteristic radius of the stellar cluster under the dominance of the massive centre is of the order of $10^6 - 10^7 R_g$. Hence, only stars on nearly radial orbits, i.e. those with very small angular momenta, can be a subject of this process. In the case of a star cluster with semi-major axis distribution $\propto a^{1/4}$ (Bahcall & Wolf 1976) and eccentricity distribution $\propto e$ the loss-cone contains only a few ($\lesssim 10$) stars and it is emptied on a very short, orbital time-scale. Therefore, the tidal disruptions channel can be effective only if there are some physical processes that refill or enlarge the loss-cone.

Kozai resonance (Kozai 1962, Lidov 1961) is a process that leads to secular evolution of the orbital elements of a particle moving in the central Keplerian potential perturbed by an axisymmetric perturbation. In such a system a component L_z of the angular momentum vector parallel to the symmetry axis is conserved. The total angular momentum, evolves in time with the lower limit $\geqslant L_z$. According to the analysis of Kozai (1962) and Lidov (1961) there exists an additional integral of motion in the system, which poses further constraint on the angular momentum. This integral appears to be an average of the perturbing (non-Keplerian) part of the Hamiltonian over one orbital revolution, $\bar{V}_p$.

The maximum eccentricity attained during the orbital evolution depends on the form of the (axisymmetric) perturbing potential. An additional, spherical perturbation of the Keplerian potential decreases the eccentricity oscillations. In this paper we discuss two different kinds of perturbations that damp the Kozai oscillations, preventing the stars from being tidally disrupted: (i) smoothed potential of the stellar cluster gravitationally bound to the central BH and (ii) relativistic pericentre advance.

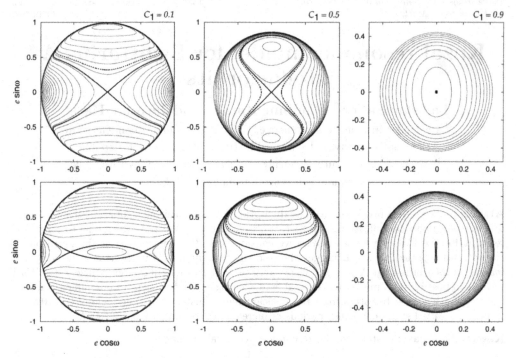

Figure 1. Iso-contours of time-averaged perturbing potential of a ring (top) and a disc of constant surface density (bottom) are plotted with solid lines; thick solid line emphasizes a separatrix. Dotted lines show tracks of numerically integrated orbits evolving in the respective potential from arbitrarily selected initial conditions. Both axisymmetric perturbations are characterised by radius $R_d = 5 \times 10^4 R_g$; value of semi-major axis $a = 2.4 \times 10^4 R_g$ is common for all panels. Value of z-component of angular momentum (expressed by means of C_1) differs in individual panels in order to demonstrate dependence of the topology of contours on this quantity.

2. Evolutionary diagrams

Motion in the axially symmetric static potential obeys two 'trivial' integrals of motion — semimajor axis a and z-component of angular momentum (expressed by means of $C_1 \equiv \sqrt{1 - e^2} \cos i$; $C_1 = 1$ corresponds to a circular orbit in the equatorial plane) — that reduce the number of free orbital parameters. Additional integral of motion, $\bar{V}_p$, further reduces the independent variables to three and enables us to analyse topology of orbital tracks in the space of eccentricity e and argument of pericentre ω.

Figure 1 shows examples of the iso-contours of the averaged perturbing potential for two different sources: (i) an infinitesimally narrow ring and (ii) a razor-thin disc of constant surface density. These cases were used as convenient basic examples with analytic form of potential (see e.g. Lass & Blitzer 1983). Both sources are characterised by mass M_d and (outer) radius R_d. We see that topology of the curves in the (e, ω) space qualitatively differs for the two sources. In both cases, increase of C_1 increases fraction of the plots covered with 'rotating' curves ($\omega \in \langle 0, \pi \rangle$) with decreasing amplitude of the eccentricity oscillations along them. In addition, in some panels we plotted examples of tracks of numerically integrated orbits evolving in the respective potential. These tracks fit perfectly in the congruences of iso-contours, confirming that the quantity $\bar{V}_p$ is an integral of motion. This allows us to interpret the plots as 'evolutionary diagrams'.

In Figure 2 we plot evolutionary diagrams for the case of ring-like axisymmetric perturbation accompanied by a smoothed potential of a spherical stellar cusp with radial

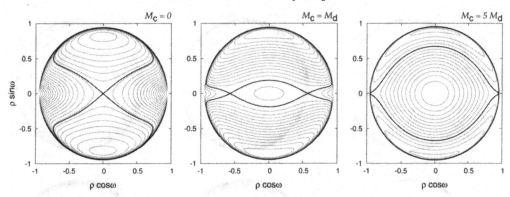

Figure 2. Evolutionary diagrams of orbits in a compound potential of a ring and a spherically symmetric cusp. Ratio of their masses is indicated above each panel. Value of the integral $C_1 = 0.1$; other parameters are identical to Fig. 1. Left panel corresponds to the upper left panel in Fig. 1; here we use variable ρ; $\rho^2 = 1 - (1 - e^2)^{1/2}$ as the radial coordinate, which allows us to resolve the structure of the contours at large eccentricities (i.e. on the perimeter).

density profile $\propto r^{-7/4}$ which implies potential of the form $V_c \propto r^{1/4}$. The stellar cusp is characterised by its mass M_c enclosed within the radius R_h (we set $R_h = R_d$ in this case). The ratio of M_c/M_d clearly plays an important role. While the topology of the evolutionary diagrams, in general, becomes more complex when the star cluster is considered, the inner zone of rotation gradually increases with M_c increasing and the eccentricity oscillations are damped in the whole parameter space.

General relativistic pericentre advance stands as another effect that damps the Kozai oscillations. In order to visualise it by means of the iso-contours of the perturbing potential we employed an additional spherically symmetric term, $V_{PW} = -2GM_\bullet r^{-1}(r-R_g)^{-1}$, which represents a non-Keplerian part of the Paczyński–Wiita (1980) pseudo-Newtonian potential. Examples of the evolutionary diagrams including this perturbation are shown in Figure 3. By comparison of the left and middle panels we clearly see that the relativistic pericentre advance tends to smear the structure of the curves and bring the libration point to lower eccentricities, i.e. it enlarges the outer rotation zone. On the other hand, the precession due to the stellar cluster pushes the libration point to higher eccentricities and enlarges inner rotation zone. The axisymmetric source in the form of the disc competes more successfully with the damping perturbations.

3. Capture rates

Due to eccentricity oscillations the stars may get close enough to the central BH to be tidally disrupted. In order to examine importance of this effect we evaluate the fraction of stars that reach the centre within the tidal radius $R_t = (M_\bullet/M_\star)^{1/3}R_\star$, i.e. we integrate the distribution function over appropriate loss-cone in the space (a, C_1, e, ω). We consider a distribution function in the form $D_f(a, C_1, e, \omega) = Ka^{1/4}e\eta^{-1}$, where $\eta \equiv (1 - e^2)^{1/2}$, which corresponds to spatially isotropic Bahcall & Wolf (1976) distribution in energy and linear distribution of eccentricities.

In a purely Newtonian regime the loss-cone would be defined as $a(1-e) < R_t$ and the appropriate fraction of stars with pericentre below R_t is then

$$\mathcal{F}(R_t) \simeq 10 \, R_t \, a_{\max}^{-1}, \tag{3.1}$$

where $a_{\max}$ is an upper limit on the semi-major axis. An upper estimate for the case when the Kozai mechanism drives the secular orbital evolution is determined by the loss-cone

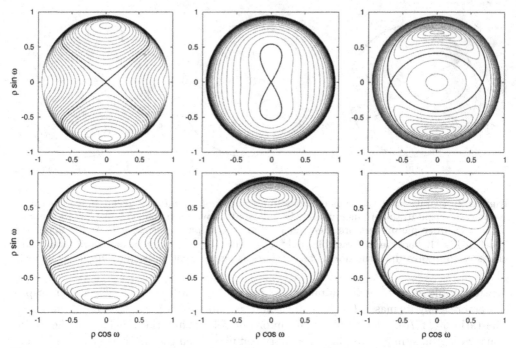

Figure 3. Evolutionary diagrams for different forms of the perturbing potential. Axisymmetric perturbation V_d due to the ring (top) or the disc (bottom) is considered in all panels. In the middle panels, the perturbing potential consists of $V_d + V_{PW}$ and in the right panels all terms $V_d + V_{PW} + V_c$ are included. Common parameters are: $C_1 = 0.1$, $a = 1.2 \times 10^4 R_g$, $M_\bullet = 3.5 \times 10^6 M_\odot$, $M_d = M_c = 0.01 M_\bullet$ and $R_d = R_h = 5 \times 10^4 R_g$.

defined as: $a(1 - (1 - C_1)^{1/2}) < R_t$ which gives:

$$\mathcal{F}(R_t) \simeq \tfrac{10}{3} \sqrt{2R_t/a_{\max}} \, . \tag{3.2}$$

The real values should lie somewhere between the two estimates, depending on the perturbing terms considered. Example of numerically determined fraction $\mathcal{F}(R_t)$ is shown in Figure 4. We employed a single-parameter model with $M_c = M_\bullet$, $M_d = 0.01 M_\bullet$ and $R_d = R_h = 2.25 \times 10^{10} (M_\bullet/M_\odot)^{-1/2} R_g$. In the left panel, $\mathcal{F}(R_t, a)$ represents a partial fraction of stars with given value of semi-major axis that reach the central BH within R_t. Again, the way how the two additional spherical perturbations damp the Kozai oscillations can be easily distinguished. The relativistic effect sharply drops $\mathcal{F}(R_t, a)$ to the lower estimate (3.1) at $a \simeq a_t$ which can be estimated (Karas & Šubr 2007) as:

$$a_t^7 \simeq \tfrac{32}{9} R_d^6 R_g^2 R_{\min}^{-1} \mu^{-2} \, . \tag{3.3}$$

In contrary, the gravity of the extended cluster damps amplitude of the Kozai oscillations over the whole range of semi-major axis, without any clear threshold. Fraction $\mathcal{F}(R_t)$ integrated over the semi-major axis as a function of $M_\bullet$ is plotted in the right panel of Figure 4. We see that the relativistic damping is becoming important for $M_\bullet \gtrsim 10^6 M_\odot$, while the damping due to the cluster potential influences $\mathcal{F}(R_t)$ in the whole mass spectrum. In general, we may conclude that the tidal disruption loss-cone increases with decreasing mass of the system.

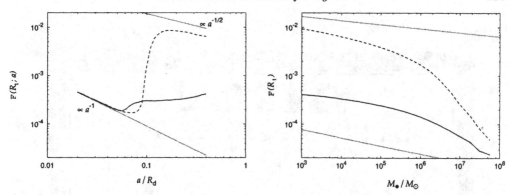

Figure 4. Fraction of stars from the cluster that reach R_t due to Kozai oscillations. Left: $\mathcal{F}$ as a function of semi-major axis for $M_\bullet = 10^4 M_\odot$; right: integrated fraction as a function of $M_\bullet$. Dashed and solid curves correspond to models omitting or including gravity of the stellar cluster, respectively. Dotted lines represent the lower and upper estimates (3.1) and (3.2). Vertical thin dotted line indicates the terminal value of semi-major axis determined by eq. (3.3).

4. Application to the Sgr A*

Recent observations of the Galactic Centre have revealed two structures that could be sources of the perturbation of the central potential required by our model: (i) At relatively large radii it is the circumnuclear disc of molecular gas ($M_d \approx 0.1 M_\bullet$) which extends to $r \gtrsim 1.5$pc (e.g. Christopher *et al.* 2005). Its influence on the stellar orbits of the bound cluster can be in the first approximation simulated with a ring of radius R_h. This configuration leads to $\mathcal{F}(R_t) \approx 3 \times 10^{-4}$, which can be translated to few$\times$100 tidally disrupted stars. (ii) Another structure is a coherently rotating disc of young stars, which can be modelled by a disc of surface density $\Sigma \propto R^{-2}$ (e.g. Paumard *et al.* 2006) with 0.04pc $\leqslant R \leqslant 0.4$pc and $M_d \approx 0.01 M_\bullet$. This latter source competes more successfully with the damping effects; therefore, the loss-cone gets enlarged, giving $\mathcal{F}(R_t) \approx 2 \times 10^{-2}$. Nevertheless, the part of the cluster that is affected by this source is small. Hence, the number of stars tidally disrupted due to this perturbation is again few$\times$100. These stars can be pushed to the tidal radius within one Kozai period (i.e. order of 1 Myr). Such events may lead to episodes of enhanced activity of the Galactic SMBH.

Acknowledgements

This work was supported by the grant of the Charles University Prague (LŠ, ref. 299/2004) and the Czech Science Foundation (VK, ref. 205/07/0052). We also acknowledge support from the Centre for Theoretical Astrophysics in Prague.

References

Bahcall, J. N. & Wolf, R. A. 1976, ApJ, 209, 214
Christopher, M. H., Scoville, N. Z., Stolovy, S. R. & Yun, M. S. 2005, ApJ, 622, 346
Karas, V. & Šubr, L. 2007, A&A, submitted
Kozai, Y. 1962, AJ, 67, 591
Lass, H. & Blitzer, L. 1983, Celest. Mech., 30, 225
Lidov, M. L. 1961, Isskust. Sputniky Zemly, 8, Acad. Sci. USSR
Paczyński, B. & Wiita, P. J. 1980, A&A, 88, 23
Paumard, T. *et al.* 2006, ApJ, 643, 1011

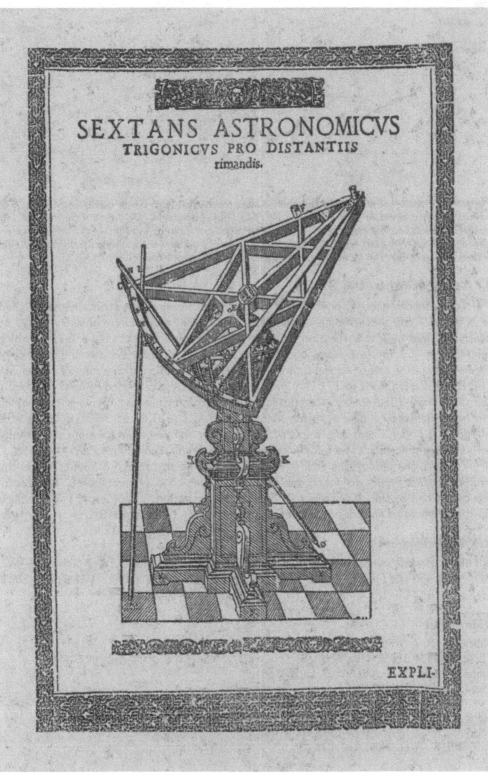

Ultraluminous X-Ray Sources I

Ultraluminous X-ray sources I.

K. Makishima
Observational evidence for intermediate-mass black holes
(Invited Review)

P. A. Charles
SS433-type X-ray binaries and the nature of ULXs

S. N. Fabrika
The supercritical accretion disk in SS433 and ultraluminous X-ray sources

P. Abolmasov
The optical counterpart of an ultraluminous X-ray source NGC 6946 X-1

The Church of St Ignatius of Loyola, with the wealth of gilding and stucco decoration.

New Town Hall, built between the years of 1377–1418 on the spacious Charles Square.

Black Holes from Stars to Galaxies – Across the Range of Masses
Proceedings IAU Symposium No. 238, 2006
V. Karas & G. Matt, eds.
© 2007 International Astronomical Union
doi:10.1017/S1743921307004991

Observational evidence for intermediate-mass black holes: ultra-luminous X-ray sources

Kazuo Makishima[1,2]

[1]Department of Physics, Graduate School of Science, The University of Tokyo,
7-3-1 Hongo, Bunkyo-ku, Tokyo, Japan 113-0033
email: maxima@phys.s.u-tokyo.ac.jp

[2]Cosmic Radiation Laboratory, The Institute of Physical and Chemical Research,
2-1 Hirosawa, Wako, Saitama, Japan 351-0198

Abstract. Incorporating early data from the *Suzaku* satellite launched in July 2005, properties of Ultra-Luminous compact X-ray sources (ULXs) were studied in close comparison with those of Galactic and Magellanic black-hole binaries. Based on an analogy between these two types of X-ray sources, ULXs showing power-law type spectra are considered to host Comptonized accretion disks, while those with multicolor-disk type spectra are interpreted to harbor "slim" disks. The analogy also suggests that ULXs are radiating near their Eddington limits, and hence their central black holes are significantly more massive than the ordinary stellar-mass black holes contained in Galactic and Magellanic black-hole binaries. In this sense, ULXs can be regarded as intermediate-mass black holes.

Keywords. Accretion, accretion disks – black hole physics – galaxies: spiral – X-rays: binaries

---

1. Introduction

X-ray images of nearby spiral galaxies, obtained with the *Einstein Observatory*, revealed very luminous point-like X-ray sources in their arm regions (e.g., Fabbiano 1989). Their X-ray luminosity reaches $10^{39.5-40.5}$ erg s^{-1}, corresponding to the Eddington limits, $L_{\rm E}$, for 20–200 $M_\odot$ objects. Their nature has long remained a big mystery, because such objects are absent in the local-group galaxies (with a possible exception of M33 X-8) including the Milky Way, and because none of them have optical/radio counterparts.

A breakthrough in the study of these enigmatic objects has been brought about by *ASCA*; Corbett & Mushotzky (1999) and Makishima *et al.* (2000) found that the 0.5–10 keV spectra of the most luminous of these objects resemble those of Galactic black-hole binaries (BHBs). This, together with their high luminosity, led Makishima *et al.* (2000) to name them Ultra-Luminous compact X-ray sources (ULXs), and to propose that they are binaries involving black holes (BHs) which are significantly more massive than ordinary stellar-mass BHs, to be called intermediate-mass black holes (IMBHs). If so, ULXs will provide a long-sought missing link between stellar-mass BHs and active galactic nuclei.

The *ASCA* view has been greatly expanded by *Chandra* and *XMM-Newton*. Nevertheless, the nature of ULXs has remained controversial, since the formation of such IMBHs cannot easily be explained by the contemporary scenario of stellar evolution. As a result, alternative explanations for ULXs have been proposed: that they are BHBs with "ordinary" masses (e.g., $\lesssim 20\ M_\odot$) under very high accretion rates, and their high luminosity is due either to strong X-ray beaming toward us (e.g., King 2002), or to genuine super-Eddington emission (e.g., Begelman 2002). To tell whether ULXs really host IMBHs or not is one of the most important issues of the current BH study.

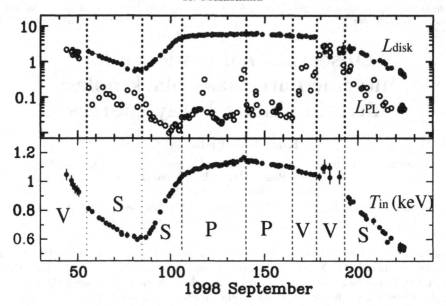

Figure 1. Spectral parameters of XTE 1550−564, obtained by fitting the *RXTE* PCA spectra with an MCD plus power-law model (Kubota & Makishima 2004). (*top*) The bolometric MCD luminosity (upper trace), and the 1–100 keV power-law luminosity (lower trace), both in the unit of 10^{38} ergs s^{-1}. (*bottom*) The innermost disk temperature $T_{\rm in}$ in keV. "S", "V", and "P" specify the High/Soft state, the Very-High state, and the Slim-Disk states, respectively.

2. Information from black hole binaries

2.1. *The transient source XTE 1550−564*

To understand ULXs which are though to involve BHs under high accretion rates, we may take lessons from BHBs (e.g., Remillard & McClintock 2006), including in particular transient sources (Tanaka & Shibazaki 1996) because of their large luminosity changes. An important role in our study has been played by the transient BHB, XTE 1550−564 (Kubota & Makishima 2004), which contains a BH of $8.4 - 11.2\ M_\odot$ (Orosz *et al.* 2002).

Figure 1 shows evolution of spectral parameters of XTE 1550−564 in an extended outburst. These were obtained by analyzing a large number of pointing data taken with the *RXTE* PCA, with a conventional model consisting of an MCD (multi-color disk) component and a PL (Kubota & Makishima 2004); as widely known, the MCD model (Mitsuda *et al.* 1984, Makishima *et al.* 1986) describes X-ray spectra emergent from standard accretion disks around BHs. As the luminosity varied by an order of magnitude, the source experienced three characteristic states, specified as "S", "V", and "P".

2.2. *Four spectral states*

Figure 2 exemplifies the PCA spectra of XTE 1550−564 for the above three states, all fitted with a conventional MCD plus power-law model. Thus, the state "S" is characterized by a canonical High/Soft (=standard-disk) state spectrum, with a luminosity of 5–40% of $L_{\rm E}$. Assuming an inclination of $i \sim 70°$ (Orosz *et al.* 2002), the obtained inner disk radius $r_{\rm in} \sim 60$ km (Kubota & Makishima 2004) agrees, within $\sim 20\%$, with $3R_{\rm S}$ ($R_{\rm S}$ being the Schwarzschild radius) predicted by the optically measured BH mass quoted above. Therefore, the state "S" can be securely identified with the High/Soft state.

Figure 3 is so-called luminosity-temperature diagram of accreting black holes (Makishima *et al.* 2000), where their innermost disk temperature $T_{\rm in}$, derived with the

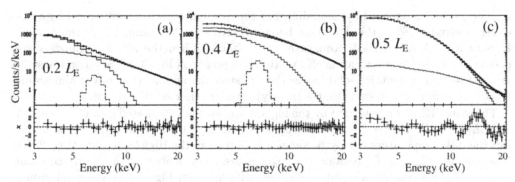

Figure 2. Response-inclusive PCA spectra of XTE 1550−565, in the three spectral states specified in Fig.1. An approximate 1–20 keV luminosity is also indicated in the unit of L_E. (a) High/Soft state ("S"). (b) Very-High state ("V"). (c) Slim-Disk state ("P").

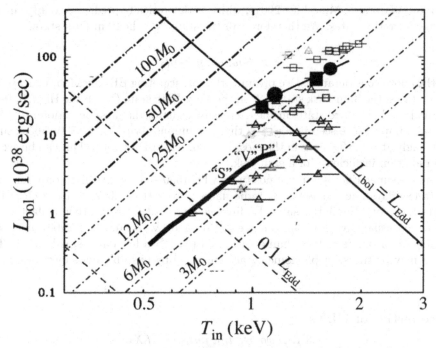

Figure 3. A luminosity-temperature diagram of accreting black holes (Makishima *et al.* 2000), where results on some CS-type ULXs are plotted. The family of dash-dotted lines show loci of standard accretion disks seen at an inclination of $i = 60°$, assuming non-rotating BHs with $r_{in} = 3R_S$. The thick solid line represents the results on XTE 1550−564 (Kubota & Makishima 2004). The filled squares and filled circles indicate NGC 1313 X-2, observed with *ASCA* (Makishima *et al.* 2000) and *Suzaku* (§ 4.3), respectively.

conventional MCD fit, are plotted against their bolometric disk luminosity L_{disk}. While XTE 1550−564 is in the state "S", its locus is indeed seen to follow the $L_{disk} \propto T_{in}^4$ relation (Makishima *et al.* 2000), drawing a straight section of the thick solid line.

When the source brightens to $\sim 0.4L_E$, there appears the state "V", which is identified with the previously reported Very-High state (Miyamoto *et al.* 1991) or the *anomalous regime* (Kubota & Makishima 2004). The fitted PL component dominates the whole energy range, so the overall spectrum assumes a power-law shape with a weak soft excess (Fig. 2b). Although the conventional fit yields unphysical values of T_{in} and r_{in}, these anomalies can be removed by assuming that the disk emission is Comptonized by a hot electron cloud to form the dominant hard tail (Kubota, Makishima & Ebisawa 2001, Kubota & Makishima 2004, Kobayashi et al. 2003). When the MCD parameters are corrected for the Comptonization effects, the data points on Fig. 3 still align on the same straight section of the thick solid line as the "S" state data points (Kubota & Makishima 2004). Therefore, we presume that a (nearly) standard disk is still present.

The state "P" is characterized by the highest and strongly saturated disk luminosity at $\sim 0.5L_{\mathrm{E}}$ (Fig. 1). The hard tail diminishes so that the spectrum resumes a thermal shape, but the disk temperature clearly becomes higher than in the High/Soft state (Fig. 2a,c). The canonical $L_{\mathrm{disk}} \propto T_{\mathrm{in}}^4$ relation no longer holds, as evidenced by the nearly constant L_{disk} as T_{in} changes. As a result, the slope of its locus on Fig. 3 becomes significantly flatter than in the state "S". These properties are explained (Kubota & Makishima 2004, Abe et al. 2005) by assuming that the standard accretion disk has changed into an optically-thick advection-dominated accretion disk, called a "slim" disk (Abramowicz et al. 1988, Watarai & Mineshige 2001), which is theoretically predicted to appear under very high accretion rates. We therefore call the state "P" the Slim-Disk state.

2.3. Eddington limits

From the above considerations, we presume that an accreting BH takes four characteristic spectral states; the Slim-Disk state ("P"), the Very-High state ("V"), the High/Soft state ("S"), and the Low/Hard state, in the decreasing order of the mass accretion rate. A fact of essential importance seen in Fig. 1 is that the luminosity of XTE 1550−565 saturates very strongly at $\sim 0.5L_{\mathrm{E}}$, even in the most luminous Slim-Disk state. This can be ascribed to the radiation inefficiency of a slim disk.

A genuine super-Eddington luminosity of XTE 1550−565 was realized only for a short while (a few days), accompanied by radio flares (Orosz et al. 2002) at day ~ 5 in Fig. 1 (outside the plot). Similarly, super-Eddington luminosities from other Galactic BHBs are observed usually as short episodes (e.g, Wijnands & van der Klis 2000), e.g., at the beginning of outbursts, rather than a sustained (e.g., $>$ a few days) steady state. These episodes may be transient phenomena, accompanied by unstable mass ejections.

3. Properties of ULXs

3.1. Two spectral types of ULXs

ULXs typically show two spectral types. One is to be called convex-spectrum (CS) type, in which their 1–10 keV spectra exhibit featureless convex shapes. The other is power-law (PL) type, wherein the spectra approximately take power-law shapes. As shown in Fig. 4, some ULXs make transitions between these two types (Kubota et al. 2001, La Parola et al. 2001), with factor 1.5–3 higher luminosity in the CS-type state. Thus, the two types of ULXs are thought to be the same class of objects in different spectral states.

It would be most straightforward to consider the CS-type and PL-type ULXs to be analogous to BHBs in the High/Soft and Low/Hard states, respectively. Then, the ULX transitions such as seen in Fig. 4 would be identified with those of BHBs between these two classical states (Kubota et al. 2001). However, if so, these ULXs would be inferred to be emitting at a few percent of L_{E} (a typical transition luminosity of BHBs; Remillard & McClintock 2006), and hence their true Eddington limit would have to be $\sim 10^{41}$ erg s^{-1} (Tsunoda et al. 2007). Added to this difficulty, other problems (see below) argue against the above simple-minded state assignments.

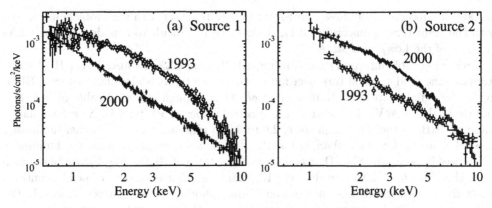

Figure 4. Response-removed spectral of two ULXs (X-1 and X-2) in IC 342 observed with *ASCA* (Kubota *et al.* 2001), revealing transitions between the CS and PL states.

3.2. *CS-type ULXs*

Spectra of CS-type ULXs are reminiscent of those of BHBs in the High/Soft state. In fact, the low-luminosity end of the CS-type ULX population may be confused with the high-luminosity end of the BHB population (Tanaka *et al.* 2005). However, compared with BHBs in the High/Soft state, the majority of CS-type ULXs exhibit a marked anomaly; their values of T_{in}, typically 1.3–1.8 keV, are too high for their luminosity (Makishima *et al.* 2000), to presume that they host sub-Eddington standard disks with $r_{in} = 3R_S$. This problem is visualized in Fig. 3, where CS-type ULXs are seen to fall mostly on the super-Eddington region.

The problem could be solved if BHs in these ULXs are rapidly spinning (Makishima *et al.* 2000); then, r_{in} would decrease down to $0.5R_S$. However, due to relativistic effects, T_{in} will not increase as much as hoped, so that the problem of too high T_{in} would not be solved unless the disk inclination is very high (Ebisawa *et al.* 2003). Furthermore, there remains another problem, that their values of r_{in} apparently varies roughly as $r_{in} \propto T_{in}$ (Mizuno *et al.* 2001), whereas that of a High/Soft state BHB stays constant at $\sim 3R_S$ (the last stable orbit). As a result, the loci of ULXs on Fig. 3 show a flatter slope than the canonical ones, just in the same way as XTE 1550−564 in the state "P" .

A most promising explanation for these apparent anomalies of the CS-type ULXs is to assign the state "P" (§ 2.2) to them, namely to assume that they host slim disks rather than standard disks. This can explain (Watarai *et al.* 2001, Mizuno *et al.* 2001, Tsunoda *et al.* 2007) both their too high values of T_{in}, and their deviation from the $L_{disk} \propto T_{in}^4$ relation in Fig. 3 .

If CS-type ULXs in fact harbor slim disks, we can no longer determine their BH masses reliably from the observed spectra (particularly r_{in}), since a slim disk would no longer be sharply truncated at $3R_S$, and the color hardening factor under such a condition is un-calibrated. Instead, we may get a crude estimate of the BH mass by equating the observed luminosity with L_E. This is based on the empirical knowledge from BHBs in § 2.3, rather than any rigorous belief in the Eddington limit. Thus, we consider that CS-type ULXs involve BHs with masses several tens to a few hundreds $M_\odot$.

3.3. *PL-type ULXs*

Compared with the photon index of of ~ 1.7 which is typical of BHBs in the Low/Hard state, the spectral slopes of PL-type ULXs scatter more, e.g., $1.5 \sim 2.8$. In addition, their spectra often show high-energy turn-over (Kubota, Done & Makishima 2002, Dewangan *et al.* 2005), in energies of 5–10 keV which is much lower than the thermal cutoff (at

~ 100 keV) observed from Low/Hard-state BHBs. Together with the problem associated with the transition luminosity (§ 3.1), these make it difficult to explain PL-type ULXs in terms of the Low/Hard state.

A more plausible interpretation is to regard PL-type ULXs as in the Very-High state. Actually, the mild spectral turn-over found with these ULXs is also observed from BHBs in the Very-High state (e.g., Kobayashi *et al.* 2003), though at considerably higher energies (several tens keV). In addition, the spectral slopes of PL-type ULXs are similar to those of BHBs in the Very-High state, 2.0 to 3.0 in photon index. The higher luminosity of a ULX in the CS state than in the PL state is also consistent with the luminosity differences between the Slim-Disk and Very-High states of BHBs (Fig. 2). We hence conclude that PL-type ULXs is in the Very-High state, so that their power-law like continua result from unsaturated Comptonization of disk photons by a hot electron cloud. The high-energy spectral curvature can be attributed to the electron temperature.

As another similarity to BHBs in Very-High state, PL-type ULXs often show a weak soft excess above the dominant power-law. The excess has been interpreted (e.g., Miller *et al.* 2003) as the emission from a cool ($T_{\rm in} \sim 0.1$ keV) standard accretion disk, of which $r_{\rm in}$ corresponds to $3R_{\rm S}$ of a $\sim 10^3\ M_\odot$ BH (e.g., Miller *et al.* 2003). Although this has been taken as evidence for genuine IMBHs, the disk parameters are derived with a conventional MCD fit without considering the possible Comptonized effects. Therefore, we should not use face values from the MCD modeling of the soft excess. Instead, our new state assignment implies that PL-type ULXs are radiating at $\sim L_{\rm E}$, and hence the required BH mass reduces typically to $\sim 10^2\ M_\odot$.

In some cases, however, the soft excess may be interpreted indeed as emission from a standard disk. One such candidate is the low-luminosity ($\sim 5 \times 10^{39}$ erg s^{-1}) state of Holmberg II X-1 (Dewangan *et al.* 2004); this could be a source excursion from the Very-High state into the High/Soft state. If this luminosity is identified with $0.5\ L_{\rm E}$ after XTE 1550$-$564, a BH mass of $\sim 60\ M_\odot$ is indicated.

3.4. *Other properties of ULXs*

A fair fraction of ULXs are surrounded by optical line-emitting nebulae (Pakull & Mirioni 2003), each several hundred parsecs in radius and several hundreds $M_\odot$ in mass. The lines are likely to be excited by the central ULXs themselves. The required excitation luminosity is estimated to amount to $\sim 10^{40}$ erg s^{-1}, assuming an isotropic radiation. This argues against the beaming interpretation of ULXs, because the excitation flux would become insufficient if the X-ray radiation were beamed toward us. The detections of slow quasi-periodic oscillations from some ULXs (Strohmayer & Mushotzky 2003, Dewangan *et al.* 2006) also point to significantly higher BH masses than in ordinary BHBs.

4. *Suzaku* observations of ULXs

4.1. *The Suzaku satellite*

The fifth Japanese cosmic X-ray satellite *Suzaku* (*Astro-E2*; Mitsida *et al.* 2007) was launched on 2005 July 10. It carries onboard X-ray CCD cameras called the X-ray Imaging Spectrometer (XIS; Koyama *et al.* 2007), placed at the focal plane of the X-ray Telescopes (XRT), and the Hard X-ray Detector (HXD; Takahashi *et al.* 2007, Kokubun *et al.* 2007). Unfortunately, the X-ray micro-calorimeter (XRS) stopped its function before the start of observations. Although the HXD is providing useful upper limits on the 10–50 keV emission from a few ULXs, here we concentrate on the XIS results.

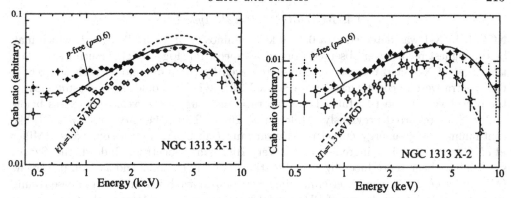

Figure 5. *Suzaku* XIS spectra (sum of the three FI sensors) of NGC 1313 X-1 (*left*) and X-2 (*right*), shown as ratios to the Crab Nebula spectrum. Filled and open circles indicate brighter and fainter phases of each source, respectively. Predictions by an MCD model (dashed curve) and a *p*-free model with $p = 0.6$ (solid curve) are illustrated.

4.2. *Observations of NGC 1313*

With *Suzaku*, we observed the nearby (3.7 Mpc) Sb galaxy NGC 1313 for 28 ksec on 2005 September 15. From previous X-ray studies (Petre *et al.* 1994, Makishima *et al.* 2000, Mizuno *et al.* 2001, Miller *et al.* 2003, Dewangan *et al.* 2005), this galaxy is known to host three luminous X-ray sources; two ULXs called X-1 and X-2, and SN1978k. During the observation, both X-1 and X-2 varied by $\sim 50\%$, with mild spectral changes.

Figure 5 shows Crab ratios of the spectra of these two ULXs, taken with the *Suzaku* XIS. These are roughly equal to response-removed $\nu F \nu$ forms of the spectra, except in energies below ~ 1.5 keV where the ratios are affected by interstellar absorption. A clear advantage of these *Suzaku* spectra is the very high signal-to-noise ratios in harder energies, thanks to the large effective area of the XRT and a low background of the XIS. Below, we first discuss X-2, and then X-1, after a detailed report by Mizuno *et al.* (2007).

4.3. *NGC 1313 X-2: a CS-type ULX*

This source was detected at a time-averaged 0.4–10 keV luminosity of 6×10^{39} erg s^{-1}. Because its spectra exhibit clearly convex shapes (Fig. 5 right), we confirm that the source was in an CS state like in most of the previous observations. Both the brighter-phase and fainter-phase spectra can be approximated by an MCD model, yielding the two data points (filled circles) in Fig. 3. Thus, the *Suzaku* results are roughly consistent with those from *ASCA* (Makishima *et al.* 2000). As mentioned § 3.2, these data points define a flatter slope (thin solid line) in Fig. 3 than the $L_{\rm disk} \propto T_{\rm in}^4$ relation.

In Fig. 5 (right), the MCD model gives a relatively good fit to the faint-phase spectrum (except below 1.5 keV). However, the model is not fully successful on the brighter-phase spectrum (Mizuno *et al.* 2007), since it is significantly less convex. Instead, a successful fit is obtained using a *p*-free model with $p = 0.6$, as illustrated in Fig. 5 (right). The *p*-free model (Hirano *et al.* 1995) generalizes the MCD model, by assuming that the local disk temperature depends on the radius r as $T(r) = T_{\rm in}(r/r_{\rm in})^{-p}$; the case of $p = 0.75$ reduces to the MCD model. The *p*-free model can approximate theoretical spectra calculated for slim disks (Watarai & Mineshige 2001). Furthermore, it can successfully reproduce the state "P" spectra of XTE 1550−564, and the observed spectra of some other CS-type ULXs (Tsunoda *et al.* 2007). These results support the view (§ 3.2) that X-2 was in the Slim-Disk state during the *Suzaku* observation (Mizuno *et al.* 2007).

4.4. *NGC 1313 X-1: a PL-type ULX*

NGC 1313 X-1 was detected at a 0.4–10 keV luminosity of 2×10^{40} erg s^{-1}, which may be the highest record ever observed from it. Its spectrum (Fig. 5 left) is much less convex than that of X-1, and cannot be represented even by the p-free model with $p = 0.6$. If the spectra were unconstrained above ~ 5 keV, a single PL would be successful; clearly, the object was in the PL state. The Crab ratio also suggests a weak soft excess below ~ 2 keV, as reported previously (Miller *et al.* 2003). Thus, the spectra show a PL-like continuum, a high-energy curvature (Dewangan *et al.* 2005), and a soft excess (Miller *et al.* 2003), all typical ingredients of a Very-High state spectrum. Indeed, the *Suzaku* spectra have been reproduced by a combination of an MCD model and a cutoff power-law (Mizuno *et al.* 2007), the latter emulating the Comptonized disk emission. These results strengthen our interpretation of PL-type ULXs in terms of the Very-High state.

Finally, we quote an important argument from Mizuno *et al.* (2007). As seen above, X-1 is about 4 times more luminous than X-2. However, if normalized to their Eddington values, X-1 should be less luminous than X-2, because X-1 is in the Very High state while X-2 in the Slim-Disk state. Then, X-1 must have at least 4 times higher Eddington luminosity, and hence 4 times higher mass, than X-2. Even employing an extreme assumption that X-2 has a rather ordinary stellar mass, e.g., $\sim 10\ M_\odot$ that requires a super-Eddington factor of 4, X-1 is inferred to have at least $\sim 40\ M_\odot$. This indicates that UXLs have a rather broad mass spectrum, making a clear contrast to the ordinary stellar-mass BHs.

5. Conclusions

Based on the close analogy with BHBs under high accretion rates, we conclude that ULXs are close binaries involving IMBHs, in the sense that they are significantly more massive (several tens to a few hundred solar masses) than ordinary stellar BHs. The two spectral types found with them can be assigned to the Very-High state and Slim-Disk state of BHBs.

Acknowledgements

The author would like to thank R. Miyawaki, A. Kubota, T. Mizuno, M. Miyamoto, N. Isobe, and K. Ebisawa for their help in the data analysis and helpful discussion.

References

Abe, Y., Fukazawa, Y., Kubota, A., Kasama, D. & Makishima, K. 2005, APSJ, 57, 629
Abramowicz, M. A., Czerny, B., Lasota, J. P. & Szuszkiewicz, E. 1988, ApJ, 332, 646
Begelman, M. 2002, ApJL, 568, L97
Colbert, E. J. M. & Mushotzky, R. F. 1999, ApJ, 519, 89
Dewangan, G. C., Griffiths, R. E. & Rao, A. R. 2005, astro-ph/0511112
Dewangan, G. C., Griffiths, R. E. & Rao, A. R. 2006, ApJL, 641, L125
Dewangan, G. C., Miyaji, T., Griffiths, R. E. & Lehmann, I. 2004, ApJL, 608, L57
Ebisawa, K., Zycki, P., Kubota, A., Mizuno, T. & Watarai, K. 2003, ApJ, 597, 780
Fabbiano, G. 1989, ARA&A, 27, 87
Hannikainen, D., Campbell-Wilson, et al. 2001, Astrophys. Space Sci. Suppl., 276, 45
Hirano, A., Kitamoto, S., Yamada, T. T., Mineshige, S. & Fukue, J. 1995, ApJ, 446, 350
King, A. 2002, MNRAS, 335, L13
Kobayashi, Y., Kubota, A., Nakazawa, K. & Takahashi, T. 2003, PASJ, 55, 273
Kokubun, K., Makishima, K., Takahashi, T., Murakami, T., et al. 2007, PASJ, 59, in press
Koyama, K., et al. 2007, PASJ, 59, in press

Kubota, A., Done, C. & Makishima, K. 2002, MNRAS, 337, L11

Kubota, A. & Makishima, K. 2004, ApJ, 601, 428

Kubota, A., Makishima, K. & Ebisawa, K. 2001, ApJL, 560, L147

Kubota, A., Mizuno, T., Makishima, K., Fukazawa, Y. *et al.* 2001, ApJL, 547, L119

La Parola, V., Peres, G., Fabbiano, G., Kim, D. & Bocchino, F. 2001, ApJ, 556, 47

Makishima, K., Kubota, A., Mizuno, T., Ohnishi, T., *et al.* 2000, ApJ, 535, 632

Makishima, K., Maejima, Y., Mitsuda, K., Bradt, H., Remillard, R., *et al.* 1986, ApJ, 308, 635

Miller, J. M., Fabbiano, G., Miller, M. C. & Fabian, A. C. 2003, ApJL, 585, L37

Mitsuda, K., *et al.* 2007, PASJ, 59, in press

Mitsuda, K., Inoue, H., Koyama, K., Makishima, K. *et al.* 1984, PASJ, 36, 741

Miyamoto, S., Kimura, K., Kitamoto, S., Dotani, T. & Ebisawa, K. 1991, ApJ, 383, 784

Mizuno, T., Kubota, A. & Makishima, K., 2001, ApJ, 554, 1282

Mizuno, T., Miyawaki, R., Ebisawa, *et al.* 2007, PASJ, **59**, in press (astro-ph/0610185)

Orosz, J. A., Groot, P. J., van der Klis, M., McClintock, J. E. *et al.* 2002, ApJ, 568, 845

Pakull, M. & Mirioni, L. 2003, Revista Mexicana de Astronomia y Astrofisica, 15, 197

Petre, R., Okada, K., Mihara, T., Makishima, K. & Colbert, E. J. M. 1994, PASJ, 46, L115

Remillard, R. A. & McClintock, J. E. 2006, ARA&A, 44, 49

Strohmayer, T. E. & Mushotzky, R. F. 2003, ApJL, 586, L61

Takahashi, T. Abe, K., Endo, M., Endo, Y., Ezoe, Y., *et al.* 2007, PASJ, 59, in press

Tanaka, T., Sugiho, M., Kubota, A., Makishima, K. & Takahashi, T. 2005, PASJ, 57, 507

Tanaka, Y. & Shibazaki, N. 1996, ARA&A, 34, 607

Tsunoda, N., Kubota, A., Namiki, M., *et al.* 2007, PASJ, 59, in press (astro-ph/0610495)

Watarai, K. & Mineshige, S. 2001, PASJ, 53, 915

Watarai, K., Mizuno, T. & Mineshige, K. 2001, ApJL, 549, L77

Wijnands, R., van der Klis, M. 2000, ApJ, 528, L93

RICHARD MUSHOTZKY: The absence of iron lines is puzzling. What is the nature of Fe K feature in galactic black holes in Very High, slim disc states?

KAZUO MAKISHIMA: Honestly, I do not know the answer to this question very well. What I can only say is that these objects do not emit very strong iron lines.

HORACIO DOTTORI: Could you say something on the environment of the intermediate-mass black-hole binaries?

KAZUO MAKISHIMA: First of all, these objects generally reside in an environment of active star formation. And secondly, they are often surrounded by huge optical nebulosity, possibly a remnant of the explosion which created the central black hole.

SEXTANS CHALYBEVS PRO
DISTANTIIS PER VNICVM OBSERVATOREM
dimetiendis.

EXPLI-

Black Holes from Stars to Galaxies – Across the Range of Masses
Proceedings IAU Symposium No. 238, 2006
V. Karas & G. Matt, eds.
© 2007 International Astronomical Union
doi:10.1017/S1743921307005005

SS433 and the nature of ultra-luminous X-ray sources

P. A. Charles,[1,2] A. D. Barnes,[2] J. Casares,[3] J. S. Clark,[4] R. Cornelisse,[2,3] C. Knigge[2] and D. Steeghs[5]

[1]South African Astronomical Observatory, P.O. Box 9, Observatory 7935, South Africa

[2]School of Physics and Astronomy, University of Southampton, Southampton SO17 1BJ, UK

[3]Instituto de Astrofísica de Canarias, 38200 – La Laguna, Tenerife, Spain

[4]Department of Physics and Astronomy, The Open University, Walton Hall, Milton Keynes MK7 6AA, UK

[5]Harvard-Smithsonian CfA, MS-67, 60 Garden Street, Cambridge, MA 01238, USA

email: pac@saao.ac.za

Abstract. The prototypical micro-quasar, SS433, one of the most bizarre objects in the Galaxy, is a weak X-ray source, yet the kinetic energy of its relativistic, precessing jets is vastly greater. In spite of its importance as the nearest example of directly observable relativistic phenomena, we know remarkably little about the nature of this binary system. There are ongoing arguments not only about the mass of the compact object, but even as to whether it is a black hole or a neutron star, an argument that recent high resolution optical spectroscopy has contributed to.

Combined with the INTEGRAL discovery of a new class of highly obscured galactic high-mass X-ray binaries, one of which has been found to precess on a similar timescale to SS433, we suggest that these would indeed be seen by external observers as ULXs, once additional effects such as beaming (either relativistic or geometrical) are included.

Keywords. Accretion, accretion disks – black hole physics – X-rays: binaries

1. Introduction

Over the last 25 years SS433 has become one of the most well-known and yet most enigmatic objects in the Galaxy. In spite of intensive studies at all wavelengths, some of its key features are little understood, there even remains controversy over the nature of the compact object. Yet SS433 is extremely important as the first relativistic jet source discovered in the Galaxy, and only continuously emitting micro-quasar. Its key property is the 162d precession period in the *moving* emission lines associated with the 0.27c jets which are ejected in opposite directions from the compact object (Margon 1984). Whilst their properties are well described by the Kinematic Model (Margon 1984; Fabrika 2004), the jet kinetic energy exceeds 10^{40} erg s^{-1}, and yet the observed X-ray flux (from A1909+04) is only $\sim 10^{36}$ erg s^{-1} (for its assumed 5.5kpc distance, Blundell *et al.* 2004). This has been explained by the presence of both optical and X-ray eclipses, indicating a high binary inclination (79°, Margon & Anderson 1989) and hence that SS433 may be an Accretion Disc Corona (ADC) source in which we only see a small fraction of the intrinsic X-ray flux which is scattered into our line-of-sight. If true, this would imply that the intrinsic L_X is indeed much greater, and that if observed at a lower inclination SS433 would be one of the most luminous objects in the Galaxy. Based on the jet kinetic energy, presumably powered by accretion, SS433 may well provide a link with the ultra-luminous X-ray (ULX) sources, whose luminosities require compact object masses $>10 M_\odot$ if they are Eddington limited (e.g. Fabbiano 2004).

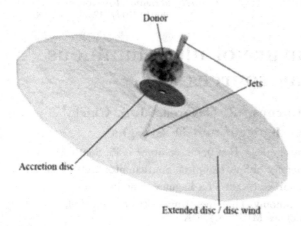

Figure 1. Schematic diagram of the SS433 system with a simplified portrayal of the impact an extended disc might have on the visibility of different components (from Barnes *et al.* 2006).

The eclipses occur on a 13d period, and modelling of INTEGRAL light-curves (Cherepashchuk *et al.* 2005) gives a mass ratio, $q(= M_X/M_2)$ ~0.2, implying that the mass donor is a large, early-type (~B) star and hence that SS433 is a high-mass X-ray binary (HMXB) with an extremely high mass transfer rate ($\sim 10^{-4} M_\odot yr^{-1}$, King *et al.* 2000). This is confirmed by optical spectra, which exhibit powerful P Cyg profiles in the "stationary" (presumably disc-based) lines of H, He (see e.g. Barnes *et al.* 2006), which are similar to those seen in Wolf-Rayet spectra with comparable mass-loss rates. Indeed ISO spectra (Fuchs *et al.* 2006) of SS433 have been compared with WN8 stars, and indicates that we are observing a dense outflow at the centre of which is embedded an ionising source, and it matters little whether this is driven by a very hot star or accretion onto a compact object. The eclipses establish 13d as P_{orb}, and hence the 162d precessional period is one of the *superorbital* periods seen in ~20 X-ray binaries (see e.g. Clarkson *et al.* 2003). Such behaviour makes SS433 an ideal laboratory for studying precessional disc properties, producing insights relevant in a wide variety of scenarios such as AGN, young stellar objects and galactic discs.

Crucially there is also a diffuse radio component that has been observed *perpendicular* to the main jets (Blundell *et al.* 2001) which is interpreted as an equatorial outflow from the accretion disc, occurring as a result of the extremely high $\dot{M}$ (see fig 1).

In spite of our knowledge of its binary orbit, SS433 is currently absent from the list of dynamically determined compact object masses (Casares, these proceedings). This is because it is well-known (e.g. Crampton & Hutchings 1981) that the donor's spectral type is poorly determined (a combination of high reddening and strong, broad disc lines that mask the principal stellar absorption features), and so previous mass determinations have been obtained from emission line radial velocity curves. These range from $M_X = 0.8$–$62 M_\odot$ (d'Odorico *et al.* 1991; Antokhina & Cherepashchuk 1985), thereby leaving the nature of the compact object completely indeterminate, a most regrettable situation for such an important system.

This is the first of two related papers on SS433 (see Fabrika, these proceedings), addressing recent observational and theoretical results that have attempted to provide better constraints on its fundamental parameters.

2. Direct observation of the donor?

The holy grail in making further progress clearly resides with the direct detection of the donor, ideally through stellar absorption features, or possibly via the recently developed technique of Bowen fluorescence emission from the donor's X-ray irradiated

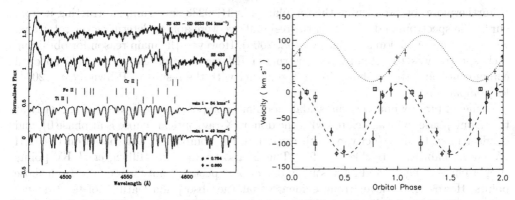

Figure 2. Left: WHT spectrum of SS433 obtained on 2004 June 29 at (Ψ, ϕ=0.66,0.78) together with the A4Iab standard HD9233 (below) shown broadened by different amounts. The top spectrum shows the SS433 residuals after subtracting the optimally broadened spectrum of HD9233. Right: Radial velocities from our SS433 campaign (squares are spectra discussed in the text) and best-fitting RV curve (dashed line; the upper dotted RV curve and velocities, pluses, are from Hillwig *et al.* 2004).

atmosphere (see Casares, these proceedings). However, the absence of any sharp features in the $\lambda\lambda$4640-50 blend precludes the latter approach. Furthermore, most previous spectroscopic studies of SS433 have concentrated on low resolution in the visual/red region (where SS433 is much brighter) so as to monitor the longer term behaviour of the jets using small telescopes. However, if the donor is an early-type star then the best spectral features are in the blue (λ <4800Å), but the combination of SS433's reddening ($A_V \sim$8) and the weakness of the features requires the use of large telescopes.

Several groups have attempted such studies, with apparently positive results. Gies *et al.* (2002) claimed the detection of A7I features during optical eclipse and at precessional phase $\Psi \sim$0 (i.e. disc open, which is why these features were only detected during eclipse). Their reason for observing at such Ψ was their assumption that the extended equatorial disc (fig 1) would otherwise obscure the photosphere of the donor, and that it needed to be observed "above" this disc. However, with velocity points only around orbital phase $\phi \sim$0, the sampling of the radial velocity (RV) curve is poor. Nevertheless, by combining these with the archival HeII emission line RV curve (Fabrika & Bychkova 1990), masses of 19 ± 7, $11 \pm 5 M_\odot$ were derived for M_2, M_X (Gies *et al.* 2002), with $M_2 > M_X$ as expected for very high $\dot{M}$(King *et al.* 2000). However, further observations (Hillwig *et al.* 2004) which included a reanalysis of the earlier data led to a compact object mass M_X of only $2.9 M_\odot$. This large uncertainty in M_X is inevitable when the RV curve amplitude, K_2, is poorly constrained, as this translates directly into very large uncertainties in the mass function $f(M) = P K_2^3/2\pi G$.

What is needed is high S/N, intermediate resolution blue spectra over a wide range of Ψ and ϕ, which we obtained using various telescopes and instruments, but mainly the 4.2m William Herschel Telescope on La Palma equipped with the ISIS spectrograph (see Barnes *et al.* 2006). In contrast to Gies *et al.*, we elected to concentrate on $\Psi \sim$0.3–0.7, when the disc would be closest to edge-on, as SS433 is then $\sim$0.5mag fainter (Fabrika & Irsmambetova 2002), and hence the disc contamination is reduced. This approach appeared to be vindicated when we obtained the spectrum shown in fig 2 at (Ψ, ϕ=0.66,0.78), an excellent match to a "normal" A4Iab stellar spectrum, and where most of the weak absorption features present are due to FeII. Interestingly the phasing of this spectrum is at a time when the donor is mostly *below* the extended disc outflow of fig 1.

Furthermore, by broadening the template star HD9233 spectrum, subtracting it from our SS433 spectrum and χ^2-testing the residuals we were able to determine the rotational velocity to be 84 ± 5 km s^{-1}(Barnes *et al.* 2006). However, the main reason for obtaining such spectra was to attempt to construct a RV curve from these "normal" spectral components, and this is shown in fig 3, together with the results of Hillwig *et al.* (2004) for comparison.

While an (approximately) sinusoidal modulation appears to be present, it is clearly at the wrong phase to be associated with the donor (X-ray/optical eclipse is at phase 0), and is offset in systemic velocity, γ, with respect to the compact object (based on the HeII RV curve, Fabrika & Bychkova 1990). This is also true for the Hillwig *et al.* RV points (upper curve in fig 3), and interestingly one of our spectra is not far removed from those points. However, we have strong evidence that the absorption features of fig 2 do *not* arise on the donor, in that (a) at (Ψ, ϕ=0.14,0.85) they are measured to be significantly *narrower* than in fig 2 (and hence the broadening measured is *not* due to the rotation of the donor, and (b) at (Ψ, ϕ=0.22,0.86) the features are *doubled*. Hence there are sites within SS433 that can clearly mimic the spectrum of a $\sim$mid-A supergiant, but that are not physically associated with the mass donor.

Additional observations of SS433 (Cherepashchuk *et al.* 2005) also find cooler spectral features during eclipse, but their $\sim$3Å resolution spectra give K_2=132 km s^{-1} which, when combined with the same HeII K_X, yields a q which is incompatible with the observed X-ray eclipse, and masses of $M_X, M_2 = 62, 206M_\odot$! Consequently they used irradiation effects to reduce K_2 to 85 km s^{-1}(which is then compatible with the q of 0.3 inferred from the X-ray eclipse), and gives M_X, M_2=17, 55$M_\odot$, although the effect of heating by the jets can reduce K_2 further to 70 km s^{-1}and the masses to M_X, M_2=9, 30$M_\odot$ (which is then in line with the black hole masses presented by Casares). However, it may simply be that these features are not clear indicators of the donor properties or even associated with the donor at all, as indicated by our higher resolution spectra above, and which would explain the difficulty in obtaining sensible mass constraints.

3. Implications for SS433

Our spectra raise uncertainties as to whether the mass donor has been identified at all. Furthermore there is now substantial observational evidence (e.g. Clark *et al.* 2006) for the presence of powerful outflows, possibly an extreme disc wind, which may well be the source of the $\sim10^4$K gas identified in our spectra. Hence they are of limited use in dynamical studies.

Such outflows require very large $\dot{M}$ from the donor, and hence an extremely high L_X, but observing this would depend very much on the viewing angle, i.e. there will be beaming, which may be a significant effect in the extragalactic ULX sources (King *et al.* 2001). However, at higher X-ray energies and γ-rays, it is possible for these objects to be seen even through high column densities of material, and so SS433 may also be related to the recently discovered class of highly obscured INTEGRAL sources (Revnivtsev *et al.* 2003). Interestingly, one of these, XTE J1716-389, has been found to exhibit a 99d modulation (Cornelisse *et al.* 2006).

Are there other sources in the Galaxy which may be related to SS433 and the ULXs? Perhaps the best candidate is GRS 1915+105, which is both a luminous X-ray source and a micro-quasar (see Casares, these proceedings). Indeed, while peaking at $L_X \sim 10^{39}$ erg s^{-1}, when corrected for the absorbing column this value could be 10 or even 100 times greater (Greiner *et al.* 1998). And many of the X-ray transients show relativistic jet ejections and extremely high, but non-steady, X-ray fluxes (McClintock & Remillard

2006, Fender 2006). However, it must also be noted that there are at least two well-known sources, A0538-66 (Charles *et al.* 1983) and Cir X-1 (Saz Parkinson *et al.* 2003), which have X-ray fluxes $>10^{38}$ erg s^{-1} and are clearly identified as neutron star systems, on the basis of pulsations and bursts respectively.

4. Conclusions

(*a*) determining the component masses of SS433 is extremely important, both for understanding the late stages of evolution of massive stars and the nature of ULXs, but the current observations are able to provide little in the way of sensible constraints;

(*b*) we have detected normal, early-type stellar absorption features, but their behaviour (variable $v\sin i$, line doubling, RV phasing, γvelocity) excludes them being associated with the donor star;

(*c*) the spectroscopic data are consistent with the presence of large mass outflows, possibly a form of extreme disc wind, as expected if $M_2 > M_X$ and SS433 is a HMXB with a black hole compact object;

(*d*) to make further progress in understanding SS433 we need more systematic coverage of *all* orbital and precessional phases with high resolution, high S/N blue spectroscopy. This is difficult with classically scheduled telescopes, but is ideal for Q-scheduled telescopes such as SALT;

(*e*) SS433 would likely be classed as a ULX if it were being observed closer to the jet axis, i.e. a result of the highly non-isotropic pattern of its radiation, which could be due to either relativistic beaming or a more simple geometric effect due to an extremely thick disc which obscures the X-rays except when viewing perpendicular to the disc. Interestingly, if this were true it implies that ULXs in external galaxies should be examined for regular X-ray variability on long timescales comparable to those of the SS433 precession period where the motion of the jet across the line-of-sight would produce a large modulation in the flux;

(*f*) it should nevertheless be noted that other candidates for galactic ULXs include at least two known neutron star systems, A0538-66 and Cir X-1;

(*g*) there are other black-hole candidates, such as the recently discovered INTEGRAL class of highly-obscured HMXBs, that may also be related to SS433.

References

Antokhina, E. A. & Cherepashchuk, A. M. 1985, SvAL, 11, 4A
Barnes, A. D. *et al.* 2006, MNRAS, 365, 296
Blundell, K. M. *et al.* 2001, ApJ, 562, L79
Blundell, K. M., Bowler, M. G. 2004, ApJ, 616, L159
Charles, P. A., Booth, L., Densham, R. H. *et al.* 1983, MNRAS, 202, 657
Cherepashchuk, A. M., Sunyaev, R. A., Fabrika, S. N. *et al.* 2005, A&A, 437, 561
Clark, J. S. *et al.* 2006, MNRAS, submitted
Clarkson, W. I. *et al.* 2003, MNRAS, 343, 1213
Cornelisse, R., Charles, P. A. & Robertson, C. 2006, MNRAS, 366, 918
Crampton, D. & Hutchings, J. B. 1981, ApJ, 251, 604
D'Odorico, S., Oosterloo, T., Zwitter, T. & Calvani, M. 1991, Nature, 353, 329
Fabbiano, G. 2004, RevMexAA, 20, 46
Fabrika, S. 2004, ApSpSci Rev, 12, 1
Fabrika, S. N. & Bychkova, L. V. 1990, A&A, 240, L5
Fabrika, S. N., Irsmambetova, T. R. 2002, in: P. Durouchoux, Y. Fuchs & J. Rodriguez (eds.),
    New Views on Microquasars (Centre for Space Physics: Kolkata), p. 276

Fender, R. 2006, in: W. H. G. Lewin & M. van der Klis (eds.), Compact Stellar X-ray Sources (Cambridge University Press: Cambridge), p. 381

Fuchs, Y., Koch Miramond, L. & Abraham, P. 2006, A&A, 445, 1041

Gies, D. R., Huang, W. & McSwain, M. V. 2002, ApJ, 578, L67

Greiner, J., Morgan, E. H. & Remillard, R. A. 1998, NewAstrRev, 42, 597

Hillwig, T. C. *et al.* 2004, ApJ, 615, 422

King, A. R., Taam, R. E. & Begelman, M. C. 2000, ApJ, 530, L25

King, A. R. *et al.* 2001, ApJ, 552, L109

Margon, B. 1984, ARA&A, 22, 507

Margon, B. & Anderson, S. F. 1989, ApJ, 347, 448

McClintock, J. E. & Remillard, R. A. 2006, in: W. H. G. Lewin & M. van der Klis (eds.), Compact Stellar X-ray Sources (Cambridge University Press: Cambridge), p. 157

Revnivtsev, M. G., Sazonov, S. Yu., Gilfanov, M. R. & Sunyaev, R. A. 2003, Ast. Lett., 29, 587

Saz Parkinson, P. M., Tournear, D. M., Bloom, E. D. *et al.* 2003, ApJ, 595, 333

FUKUN LIU: In the classical scenario for SS433, the accretion disc is flat but with an inclination angle with respect to the binary orbital plane, while in the configuration presented by Andrew King the accretion disc is warped. If we observe emission lines from accretion disc surface, the emission line profile in the latter configuration should be asymmetric and change periodically. Have you observed such signature?

PHIL CHARLES: The broad H_α emission line of SS433 is very complex. We should refer this question to Andrew King to ask him whether we could observe it; I doubt it is possible...

FELIX MIRABEL: Is the absence of an IMBH/ULX sources in our Galaxy consistent with the statistics on ULXs in nearby galaxies? Other way to put the question, why there is no IMBH in the Milky Way ULXs?

PHIL CHARLES: There are luminous X-ray sources in the Milky Way (e.g. GRS 1915+105, A0538–66) which might be viewed externally as ULXs, but we know that both are consistent with $M_X \sim 14 M_\odot$ (GRS 1915+105) and $\sim 1.5 M_\odot$ (A0538–66) based on dynamical information. This suggests that IMBH interpretations of ULXs have to be treated with caution.

Black Holes from Stars to Galaxies – Across the Range of Masses
Proceedings IAU Symposium No. 238, 2006
V. Karas & G. Matt, eds.
© 2007 International Astronomical Union
doi:10.1017/S1743921307005017

The supercritical accretion disk in SS 433 and ultraluminous X-ray sources

S. N. Fabrika, P. K. Abolmasov and S. Karpov

Special Astrophysical Observatory of the Russian AS, Nizhnij Arkhyz 369167, Russia
email: fabrika@sao.ru

Abstract. SS 433 is the only known persistent supercritical accretor, it may be very important for understanding ultraluminous X-ray sources (ULXs) located in external galaxies. We describe main properties of the SS 433 supercritical accretion disk and jets. Basing on observational data of SS 433 and published 2D simulations of supercritical accretion disks we estimate parameters of the funnel in the disk/wind of SS 433. We argue that the UV radiation of the SS 433 disk (~ 50000 K, $\sim 10^{40}$ erg/s) is roughly isotropic, but X-ray radiation ($\sim 10^7$ K, $\sim 10^{40}$ erg/s) of the funnel is mildly anisotropic. A face-on SS 433 object has to be ultraluminous in X-rays (10^{40-41} erg/s). Typical time-scales of the funnel flux variability are estimated. Shallow and very broad (0.1-0.3c) and blue-shifted absorption lines are expected in the funnel X-ray spectrum.

Keywords. X-rays: general – X-rays: individual (SS 433) – accretion – jets

1. Introduction

The main properties of the ultraluminous X-ray sources (ULXs) — the extremely high luminosities (10^{39-41} erg/s), diversity of X-ray spectra, strong variability, their connection with star-forming regions and surrounding nebulae. Here we continue to develop the idea that the galactic supercritical accretor SS 433 intrinsically is very bright X-ray source and it is a prototype of ULXs in external galaxies (Katz 1987, Fabrika & Mescheryakov 2001, King *et al.* 2001, Begelman *et al.* 2006, Poutanen *et al.* 2006). This means that ULXs are supercritical accretion disks in close binaries with stellar mass black holes or microquasars. We discuss possible properties of the funnel in the supercritical accretion disk of SS 433 to predict X-ray spectral features and temporal behaviour of the funnel in "face-on SS 433" stars in application to ULXs.

2. Properties of SS 433

The main difference between SS 433 and other known X-ray binaries is highly supercritical and persistent mass accretion rate ($\dot{M}_a \geqslant 10^{-4}\,M_\odot/y$) onto the relativistic star, a probable black hole ($M \sim 10M_\odot$), which has led to the formation of a supercritical accretion disk and the relativistic jets. SS 433 properties were reviewed recently (Fabrika 2004). A total observed luminosity of SS 433 is $L_{bol} \sim 10^{40}$ erg/s. Practically all the energy is realised near the black hole. Temperature of the source is $T = (5-7) \cdot 10^4$ K, when the accretion disk is the most open to the observer and it is $T \sim 2 \cdot 10^4$ K, when it is observed edge-on (Dolan *et al.* 1997). If the source SED is represented by a single blackbody source, its size is $\sim 10^{12}$ cm. The line of sight wind velocity varies from ~ 1500 km/s (the most open disk) to ~ 100 km/s (the disk is observed edge on). One may adopt the value ~ 2000 km/s for the wind velocity closer to the jets. The mass loss rate in the wind is $\dot{M}_w \sim 10^{-4}\,M_\odot/y$. The optical jets ($\sim 10^{15}$ cm) consist of dense small gas clouds, the X-ray jets ($\sim 10^{12}$ cm) consist of hot ($T \sim 10^8$ K) cooling plasma. Both the optical and X-ray jets are well collimated ($\sim 1°$). SS 433 X-ray luminosity (the cooling

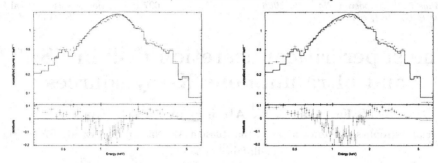

Figure 1. Simulated MCF spectra as seen with XMM MOS1 with total counts 10^5, obtained with $T_0 = 1$ keV, $\theta_f = 20°$, $V_w(r_0) = 6000$ km/s, $N_H = 0$ and $\xi = 10^4$ (left), $\xi = 10$ (right). Solid lines show the *diskbb* model fitted to the spectra. In the model spectra we found $T_{inn} = 1.55$ keV, $N_H = 1.5 \cdot 10^{21}$ cm^{-2}. The MCF spectrum is broader than the MCD (diskbb) spectrum.

X-ray jets) is $\sim 10^4$ times less than bolometric luminosity, however kinetic luminosity of the jets is very high, $L_k \sim 10^{39}$ erg/s, the jet mass loss rate is $\dot{M}_j \sim 5 \cdot 10^{-7} \, M_\odot/\text{y}$.

The jets have to be formed in a funnel in the disk or the disk wind. Supercritical accretion disks simulations (Eggum *et al.* 1985, Ohsuga *et al.* 2005, Okuda *et al.* 2005) show that a wide funnel ($\theta_f \approx 20° - 25°$) is formed close to the black hole. Convection is important in the inner accretion disk. Recent hydrodynamic simulations of super-Eddington radiation pressure-dominated disks with photon trapping (Ohsuga *et al.* 2005) have confirmed the importance of advective flows in the most inner parts of the disks. The mass accretion rate into the black hole a few times exceeds the mass loss rate in the disk wind. One may adopt a rough estimate of the accretion rate in the case of SS 433 (or the same for the gas supply rate in the accretion disk), $\dot{M}_a \sim 3 \cdot 10^{-4} \, M_\odot/\text{y}$.

3. Funnel in the supercritical accretion disk

A critical luminosity for a $10 \, M_\odot$ black hole is $L_{edd} \sim 1.5 \cdot 10^{39}$ erg/s and corresponding mass accretion rate $\dot{M}_{edd} \sim 3 \cdot 10^{-7} \, M_\odot/\text{y}$. There are three factors (Fabrika *et al.* 2006) increasing observed face-on luminosity of a supercritical accretion disk. (1) inside the spherization radius the disk is locally Eddington (Shakura & Sunyaev 1973), which gives logarithmic factor $(1 + \ln(\dot{M}_a/\dot{M}_{edd})) \sim 8$; (2) the doppler boosting factor is ($\beta = V_j/c = 0.26$ in SS 433) is $1/(1 - \beta)^{2+\alpha} \sim 2.5$, where α is spectral index; and (3) the geometrical funneling ($\theta_f \sim 25°$) is $\Omega_f/2\pi \sim 10$. Thus, one may expect an observed face-on luminosity of such supercritical disk $L_{edd} \sim (2 - 3) \cdot 10^{41}$ erg/s. On the other hand, if one adopts for the funnel luminosity, that it is about the same as SS 433 bolometric luminosity (Fabrika & Mescheryakov 2001) one obtains with the same opening angle of the funnel, the "observed" face-on luminosity of SS 433 is $L_x \sim 10^{41}$ erg/s and the expected frequency of such objects is ~ 0.1 per a galaxy like Milky Way.

The disk spherization radius (Shakura & Sunyaev 1973) in SS 433 is estimated $r_{sp} \approx 3\kappa\dot{M}_a/8\pi c \sim 2.6 \cdot 10^{10}$ cm, where the Thomson opacity for a gas with solar abundance is $\kappa = 0.35$ cm^2/g. Corresponding wind velocity is $V_w \sim (GM/r_{sp}) \sim 2300$ km/s, where we adopted the black hole mass $M = 10 \, M_\odot$. The size of the hot wind photosphere and the photosphere temperature (for not a face-on observer) are estimated $r_{ph,w} = \dot{M}_w\kappa/4\pi \cos\theta_f V_w \sim 1 \cdot 10^{12}$ cm, $T_{ph,w} = (L_{bol}/4\pi\sigma \cos\theta_f r_{ph,w}^2)^{1/4} \sim 6 \cdot 10^4$ K. Both the size of the hot body and the temperature are quite close to those we observe in SS 433.

We find the jet photosphere size $r_{ph,j} = \dot{M}_j\kappa/\Omega_f V_j \sim 4 \cdot 10^9$ cm, this value indicates the bottom of the funnel, below $r_{ph,j}$ the funnel walls can not be observed. A temperature of the inner funnel walls at a level of $r_{ph,j}$ is estimated as (Fabrika *et al.* 2006, Fabrika *et al.* 2006) $T_{ph,f}$ between $\sim 1.7 \cdot 10^7$ K and $\sim 1 \cdot 10^6$ K. Such temperatures provide a

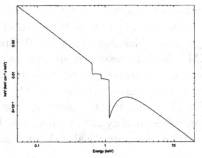 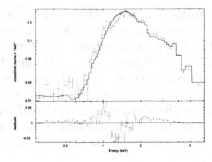

Figure 2. Possible fake broad emission/absorption features in ULXs spectral residuals. The model is a power law spectrum ($\Gamma = 2.5$) with $N_H = 1.0 \cdot 10^{22}\,\mathrm{cm}^{-2}$ and with introduced Lc edges of C VI, N VII and O VIII blue-shifted to the velocity $V_j = 0.26c$ (shown in left panel). The XMM MOS1 fit to the model ($\Gamma = 2.2$, $N_H = 1.05 \cdot 10^{22}\,\mathrm{cm}^{-2}$) and spectral residuals are shown by solid line.

high ionisation of elements needed for the line-locking mechanism (Shapiro *et al.* 1986) to operate for the jet acceleration. We conclude that the UV radiation of the SS 433 disk ($T \sim 50000\,\mathrm{K}$, $L \sim 10^{40}$ erg/s) is roughly isotropic, but its X-ray funnel radiation ($T \sim 10^6 - 10^7\,\mathrm{K}$, $L \sim 10^{40}$ erg/s) is mildly anisotropic. The same conclusions have to be done for face-on SS 433 stars (ULXs), they are expected to be very bright UV sources and their X-ray luminosities have to be even higher, up to $\sim 10^{40} - 10^{41}$ erg/s, due to the additional geometrical collimation.

Temporal variability of the SS 433 funnel is expected on time scales $r_{ph,w}/c \sim 30\,sec$ and $r_{ph,j}/c \sim 0.1\,sec$. A typical accretion disk variability power density spectrum at scales $\gg 0.1\,sec$ is expected. However the most rapid variability may be observed only for face-on SS 433 stars, in the case of SS 433 (the edge-on system), all the short-scale variability of the funnel ($\Delta t < 10 - 100\,sec$) has to be smoothed down in the funnel. Some of the expected X-ray variability patterns have been observed recently (Revnivtsev *et al.* 2005, 2006, Kotani *et al.* 2006).

4. The funnel spectrum and applications to ULXs

The multicolor funnel (MCF) model has been developed (Fabrika *et al.* 2006, Fabrika *et al.* 2006) to estimate the emerging X-ray spectrum in the funnel. The main parameters of the models are (i) the initial radius, the deepest visible level in the inner funnel walls ($r_0 = r_{ph,j}$), (ii) the walls temperature at $T_0(r_0)$, (iii) the ratio of radiation to gas pressure $\xi = aT_0^3/3k_b n_0$ at r_0, (iv) the wind velocity $V_w(r_0)$ and (v) the funnel opening angle θ_f. Photon trapping in the wind is considered by a simple comparison of advection time ($t_{mov} \sim r/v(r)$) and diffusion time ($t_{esc} \sim \beta_t r_{ph,w}/V(r)$), where β_t is a terminal wind velocity. These times are equal at the radius $r(t_{mov} = t_{esc}) = 2\dot{M}_w r_{sp}/3\cos\theta_f \dot{M}_a \sim r_{sp}$. There is a local advective radiation transfer at $r \ll r_{sp}$ and global diffusion transfer at $r \gg r_{sp}$.

In Fig. 1 we present simulated MCF spectra as seen with XMM MOS1 with total counts 10^5, obtained with $T_0 = 1\,keV$, $\xi = 10^4$ and 10, $\theta_f = 20°$, $V_w(r_0) = 6000$ km/s, $N_H = 0$ and their fittings using the *diskbb* model. This figure demonstrates that the MCF spectrum is broader than that of the MCD (diskbb). Shallow, very broad (0.1-0.3c) and blue-shifted absorption lines are expected in the funnel X-ray spectrum (Fabrika 2004, Fabrika *et al.* 2006). The absorption bands should belong to H- and He-like ions of the most abundant (O, Ne, Mg, Si, S, Fe) elements and should extend from the Kc to the Kα energies of the corresponding ions. Fig. 2 shows possible fake broad emission/absorption features which could be observed in ULXs spectral residuals. We have simulated a power law spectrum with introduced Lc edges of C VI, N VII and O VIII blue-shifted to the SS 433 jet velocity $V_j = 0.26c$. The optical thickness of these introduced edges corresponds to "effective" hydrogen thickness $\tau(L_c) = 20$. Spectral residuals in the figure were derived from the XMM MOS1 power law spectral model to the simulated spectrum. The fake emission/absorption features appear both due to the blue-shifts of the edges and the energetic differences between neutral and highly ionised absorptions. Residuals of such type are observed in ULXs spectra (for example, Dewangan et al. 2006, Roberts et al. 2006). The absorption edges, if they are confirmed in ULXs spectra give a possibility to measure the jet velocities in supercritical disk funnels.

Acknowledgements

The work is supported by RFBR under grants number 03-02-16341, 04-02-16349 and by joint RFBR/JSPS grant N 05-02-12710.

References

Begelman, M. C., King, A. R. & Pringle, J. E. 2006, MNRAS, 370, 399
Eggum, G. E., Coroniti, F. V. & Katz, J. I. 1985, ApJ, 298, L41
Dewangan, G. C., Griffiths, R. E. & Rao, A. R. ApJ, 641, L125
Dolan, J. F., et al. 1997, A&A, 327, 648
Fabrika, S. & Mescheryakov, A. 2001, in: R. T. Schilizzi (ed.), Galaxies and their Constituents at the Highest Angular Resolution, IAU Symposium No. 205, p. 268
Fabrika, S. 2004, Astrophysics and Space Physics Reviews, vol. 12, p. 1
Fabrika, S., Karpov, S., Abolmasov, P. & Sholukhova, O. 2006, in Meurs E. J. A., Fabbiano G., eds., IAU Symposium No. 230, Populations of High Energy Sources in Galaxies (Cambridge University Press, Cambridge), p. 278
Fabrika, S., Karpov, S. & Abolmasov, P. 2006, in preparation
Katz, J. J. 1987, ApJ, 317, 264
King, A. R., Davies, M. B., Ward, M. J., Fabbiano, G. & Elvis, M. 2001, ApJ, 552, L109
Kotani, T. et al. ApJ, 637, 486
Ohsuga, K., Mori, M., Nakamoto, T. & Mineshige, S. 2005, ApJ, 628, 368
Okuda, T., Teresi, V., Toscano, E. & Molteni, D. 2005, MNRAS, 357, 295
Poutanen, J., Fabrika, S., Butkevich, A. & Abolmasov, P. 2006, astro-ph/0609274
Revnivtsev, M. et al. 2005, A&A, 424, L5
Revnivtsev, M. et al. 2006, A&A, 447, 545
Roberts, T. P. et al. 2006, MNRAS, 371, 1877
Shakura, N. I. & Sunyaev, R. A. 1973, A&A, 24, 337
Shapiro, P. R., Milgrom, M. & Rees, M. J. 1986, ApJSS, 60, 393

KAZUO MAKISHIMA: If the funneling model of ULXs is correct, we should observe a variety of X-ray absorption, depending on the viewing angle. In reality, we do not find such a wide range of N_H. How about this point?

SERGEI FABRIKA: The X-ray absorption value, depending on the viewing angle, will also depend on the funnel structure and the gas ionization. Those funnels observed at larger angles may appear as super soft ULXs. Regarding the predicted broad smeared absorption features, I hope it is a consequence of signal-to-noise ratios in current data.

RICHARD MUSHOTZKY: Answer to question by Kazuo Makishima: The X-ray column densities to ULX's are very close to N_H derived from 21 cm data. There is just a paper accepted on this in ApJ.

Black Holes from Stars to Galaxies – Across the Range of Masses
Proceedings IAU Symposium No. 238, 2006
V. Karas & G. Matt, eds.
© 2007 International Astronomical Union
doi:10.1017/S1743921307005029

The optical counterpart of the ultraluminous X-ray source NGC6946 ULX-1

P. K. Abolmasov, S. N. Fabrika and O. N. Sholukhova

Special Astrophysical Observatory of the Russian AS, Nizhnij Arkhyz 369167, Russia

email: pasha@sao.ru

Abstract. We present a study of a peculiar nebula MF16 associated with an Ultraluminous X-ray Source NGC6946 ULX-1. We use integral-field and long-slit spectral data obtained with the 6-m telescope (Russia). The nebula was for a long time considered powered by strong shocks enhancing both high-excitation and low-excitation lines. However, kinematical properties point to rather moderate expansion rates ($V_S \sim 100 \div 200$ km s^{-1}). The total power of the emission-line source exceeds by one or two orders of magnitude the power observed expansion rate can provide, that points towards the existence of an additional source of excitation and ionization. Using CLOUDY96.01 photoionization code we derive the properties of the photoionizing source. Its total UV/EUV luminosity must be about 10^{40} erg/s.

Keywords. X-rays: individual (NGC6946 X8) – ISM: individual (MF16) – ISM: jets and outflows – ultraviolet: general

1. Introduction

Quite a large number of Ultraluminous X-ray Sources (ULXs) are associated with emission-line nebulae (ULX Nebulae, ULXNe), mostly large-scale bubbles powered by shock waves (Pakull & Mirioni 2003). However, several exceptions are known like the nebula associated with HoII X-1 (Lehmann *et al.* 2005), that is clearly a photoionized H II region. Another well-known example is the nebula MF16 coincident with the ULX NGC6946 ULX1.

The attention to MF16 was first drawn by Blair & Fesen (1994), who identified the object as a Supernova Remnant (SNR), according to the emission-line spectrum with bright collisionally-excited lines. It was long considered an unusually luminous SNR, because of its huge optical emission-line, $L_{H\alpha} = 1.9 \times 10^{39}$ erg s^{-1}, according to Blair & Fesen (1994), for the tangential size 20×34 pc – and in X-rays, $L_X = 2.5 \times 10^{39}$ erg s^{-1} in the $0.5 - 8$ keV range, according to the luminosities given by Roberts & Colbert (2003).

However, it was shown by Roberts & Colbert (2003), that the spectral, spatial and timing properties of the X-ray source do not agree with the suggestion of a bright SNR, but rather suppose a point source with a typical "ULX-like" X-ray spectrum: cool Multi-color Disk (MCD) and a Power Law (PL) component. So, apart from the physical nature of the object, MF16 should be considered a *ULX nebula*, one of a new class of objects, described by Pakull & Mirioni (2003).

2. Optical Spectroscopy

All the data were obtained on the SAO 6m telescope, Russia. Two spectrographs were used: panoramic MultiPupil Fiber Spectrograph MPFS (Afanasiev *et al.* 2001) and SCORPIO focal reducer (Afanasiev & Moiseev 2005) in long-slit mode. The details of data reduction processes and analysis technique will be presented in Abolmasov *et al.* (2007).

Panoramic spectroscopy has the advantage of providing unbiased flux estimates. However, SCORPIO results have much higher signal-to-noise ratio and reveal rich emission-line spectrum of [Fe III]. We also confirm the estimates of the total nebula emission-line luminosities by Blair et al. (2001). $H\beta$ line luminosity obtained from our MPFS data is $L(H\beta) = (7.2 \pm 0.2) \times 10^{37}$ erg s^{-1}.

Using line ratios for the integral spectrum we estimate the mean parameters of emitting gas as: $n_e \simeq 500 \pm 100$ cm^{-3}, $T_e \simeq (1.9 \pm 0.2) \times 10^4$ K. Interstellar absorption is estimated as $A_V \sim 1^m3$, close to the Galactic value ($A_V^{Gal} = 1^m14$, according to Schlegel et al. (1998)).

We confirm the estimate of the expansion rate obtained by Dunne et al. (2000), coming to the conclusion that the expansion velocity is $V_S \lesssim 200$ km s^{-1}. In this case the total emission-line luminosity can be estimated using for example the equations by Dopita & Sutherland (1996):

$$F_{H\beta} = 7.44 \times 10^{-6} \left(\frac{V_s}{100 \text{ km s}^{-1}}\right)^{2.41} \times \left(\frac{n_2}{\text{cm}^{-3}}\right) + 9.86 \times 10^{-6} \left(\frac{V_s}{100 \text{ km s}^{-1}}\right)^{2.28} \times \left(\frac{n_1}{\text{cm}^{-3}}\right) \text{ erg cm}^{-2}\text{s}^{-1}.$$

Here V_S is the shock velocity and n_1 the pre-shock hydrogen density. If the surface area is known, one can obtain the total luminosity in $H\beta$ from here. For $V_S = 200$ km/s and $n_1 = 1$ cm^{-3} it appears to be $L(H\beta) \simeq 1.6 \times 10^{36}$erg s^{-1}, that is too low compared to the observed value. So we suggest an additional source of power providing most of the energy of the optical nebula.

3. Photoionization Modelling

We have computed a grid of CLOUDY96.01 (Ferland et al. 1998) photoionization models in order to fit MF16 spectrum avoiding shock waves. We have fixed X-ray spectrum known from *Chandra* observations (Roberts & Colbert 2003), assuming all the plasma is situated at 10 pc from the central point source, and introduced a blackbody source with the temperature changing from 10^3 to 10^6K and integral flux densities from 0.01 to 100 erg cm^{-2} s^{-1}.

The best fit parameters are $\lg T(K) = 5.15 \pm 0.05$, $F = 0.6 \pm 0.1$ erg cm^{-2} s^{-1}, that suggests quite a luminous ultraviolet source: $L_{UV} = (7.5 \pm 0.5) \times 10^{39}$ erg s^{-1}. The UV source is more than 100 times brighter then what can be predicted by extrapolating the thermal component of the best-fit model for X-ray data (Roberts & Colbert 2003).

4. Ultraluminous UV sources?

At least for one source we have indications that its X-ray spectrum extends in the EUV region. It is interesting to analyse the implications in the frameworks of two most popular hypotheses explaining the ULX phenomenon.

For the standard disk of Shakura & Sunyaev (1973) the inner temperature scales as:

$$T_{in} \simeq 1 \text{ keV} \left(\frac{M}{M_\odot}\right)^{-1/4} \left(\frac{\dot{M}}{\dot{M}_{cr}}\right)^{1/4}.$$

In Figure 1 we present the reconstructed Spectral Energy Distribution of NGC6946 ULX-1 including optical identification by Blair et al. (2001) and the best-fit blackbody from our model. For comparison, a set of MCD SEDs for IMBHs accreting at 1% of critical rate is shown. To explain the high EUV luminosity and roughly flat SED in the EUV region, a rather high IMBH mass is needed, $M \gtrsim 10^4 M_\odot$.

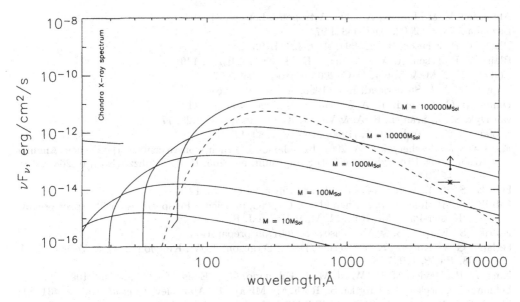

Figure 1. NGC6946 ULX1 SED reconstruction. Optical source d (Blair *et al.* 2000) is shown by an asterisk, and the upward arrow above indicates the unabsorbed optical luminosity: it is the lower estimate because only Galactic absorption was taken into account, $A_V = 1\overset{m}{.}14$ according to Schlegel *et al.* (1998). Dashed line represents the best-fit blackbody from our CLOUDY fitting. Thin solid lines are MCD models for accreting IMBHs with infinite outer disk radii. Mass accretion rate was set everywhere to $0.01\dot{M}_{cr}$.

For supercritical disk this relation breaks (Poutanen *et al.* 2006), and the outcoming radiation becomes much softer, except for the X-rays escaping along the disk axis (Fabrika *et al.* 2007). Most part of the luminosity is supposed to be reprocessed into EUV and UV quanta, creating the nearly-flat SED of NGC6946 ULX1. In optical/UV range contribution of the donor star may become significant.

In Abolmasov *et al.* (2007) we make estimates for the detectability of ULXs with GALEX, coming to the conclusion that at least some of them (the sources with lower Galactic absorption) may be bright enough targets even for low-resolution spectroscopy.

5. Conclusions

We conclude that MF16 is most likely a dense shell illuminated from inside. This can be a certain stage of the evolution of a ULXN, when the central source is bright and the shell itself rather compact. We suggest that ULXs must be luminous EUV sources as well in some cases, and may be also luminous UV sources.

Acknowledgements

This work was supported by the RFBR grants NN 05-02-19710, 04-02-16349, 06-02-16855.

References

Abolmasov, P., Fabrika, S., Sholukhova, O. & Afanasiev, V. 2005, in Science Perspectives for 3D Spectroscopy, ed. M. Kissler-Patig, M. M. Roth. & J. R. Walsh (Springer Berlin/Heidelberg); astro-ph/0602369

Abolmasov, P., Fabrika, S. & Sholukhova, O. 2007, in preparation

Afanasiev, V. L., Dodonov, S. N. & Moiseev, A. V. 2001, in Stellar Dynamics: from Classic to Modern, eds. Osipkov L.P., Nikiforov I.I., Saint Petersburg, 103

Afanasiev, V. & Moiseev, A. 2005, Astronomy Letters, 31, 194

Begelman, M. C. 2002, ApJ, 568, L97

Blair, W. P. & Fesen, R. A. 1994, ApJ, 424, L10371.8

Blair, W. P., Fesen, R. A. & Schlegel, E. M. 2001, AJ, 121, 1497

Colbert, E. J. M. & Miller, E. C. 2005, astro-ph/0402677

Dopita, M. A. & Sutherland, R. S. 1996, ApJSS, 102, 161

Dunne, B. C., Gruendl, R. A. & Chu, Y.-H. 2000, AJ, 119, 1172

van Dyk, S. D., Sramek, R. A. & Weiler, K. W. 1994, ApJ, 425, 77

Fabian, A. C. & Terlevich, R. 1996, MNRAS, 280, L5

Fabrika, S. & Mescheryakov, A. 2001, In Galaxies and their Constituents at the Highest Angular Resolutions, Proceedings of IAU Symposium No. 205, R. T. Schilizzi (Ed.), p. 268, astro-ph/0103070

Fabrika, S. 2004, Astrophysics and Space Physics Reviews, 12, 1

Fabrika, S., Abolmasov, P. & Sholukhova, O. 2005, in Science Perspectives for 3D Spectroscopy, eds. Kissler-Patig M., Roth M. M. & Walsh J. R.

Fabrika, S., Karpov, S. & Abolmasov, P. 2007, in preparation

Ferland, G. J., Korista, K. T., Verner, D. A., Ferguson, J. W., Kingdon, J. B. & Verner, E. M. 1998, PASP, 110, 761

King, A. R., Davies, M. B., Ward, M. J., Fabbiano, G. & Elvis, M. 2001, å, 552, 109

Lehmann, I., Becker, T., Fabrika, S., Roth, M., Miyaji, T., Afanasiev, V. et al. 2005, å, 431, 847

Liu, J.-F. & Bregman, N. 2005, ApJSS, 157, 59L

Matonick, D. M. & Fesen, R. A. 1997, ApJSS, 112, 49

Osterbrock, D. E. 1974, Astrophysics of Gaseous Nebulae, San Francisco, eds. W. H. Freeman and Company

Pakull, M. W. & Mirioni, L. 2003, RevMexAA (Serie de Conferencias), 15, 197

Poutanen, J., Fabrika, S., Butkevich, A. & Abolmasov, P. 2006, in press

Roberts, T. P. & Colbert, E. J. M. 2003, MNRAS, 341, 49

Schlegel, D. J., Finkbeiner, P. F. & Davis, M. 1998, ApJ, 500, 525

Shakura, N. I. & Sunyaev, R. A. 1973, A&A, 24, 337

Swartz, A. D., Ghosh, K. K., Tennant, A. F. & Wu, K. 2004, ApJSS, 154, 519

SUZY COLLIN: Perhaps there can be a similarity with LINERs, except that in LINERs the black holes are supermassive.

PAVEL ABOLMASOV: Yes indeed, objects discussed in my talk have line ratios resembling LINERs.

Ultraluminous X-Ray Sources II

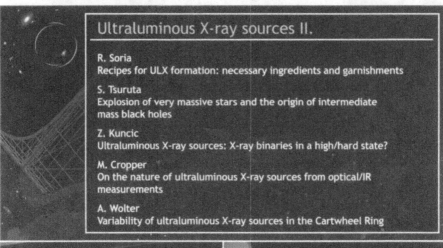

Ultraluminous X-ray sources II.

R. Soria
Recipes for ULX formation: necessary ingredients and garnishments

S. Tsuruta
Explosion of very massive stars and the origin of intermediate mass black holes

Z. Kuncic
Ultraluminous X-ray sources: X-ray binaries in a high/hard state?

M. Cropper
On the nature of ultraluminous X-ray sources from optical/IR measurements

A. Wolter
Variability of ultraluminous X-ray sources in the Cartwheel Ring

Břevnov monastery (founded in 993 AD). Over the romanesque crypt, today's appearance includes St Margaret Church in baroque style.

Black Holes from Stars to Galaxies – Across the Range of Masses
Proceedings IAU Symposium No. 238, 2006
V. Karas & G. Matt, eds.
© 2007 International Astronomical Union
doi:10.1017/S1743921307005042

Recipes for ULX formation: necessary ingredients and garnishments

Roberto Soria[1,2]

[1]Harvard-Smithsonian CfA, 60 Garden st, Cambridge, MA 02138, USA

[2]MSSL, University College London, Holmbury St Mary, Dorking, RH5 6NT, UK

email: rsoria@cfa.harvard.edu

Abstract. We summarize the main observational features that seem to recur more frequently in the ULX population. We speculate that low metal abundance, and clustered star formation triggered by molecular cloud collisions are two fundamental physical requirements for ULX formation. In this scenario, most ULXs are formed from recent stellar processes, have black hole (BH) masses $< 100 M_\odot$, and do not require merger processes in super-star-clusters.

Keywords. Black hole physics – X-ray: binaries – galaxies: star clusters

1. Common features in the ULX population

The nature and formation mechanisms of ULXs remain unclear. Much of the uncertainty is due to the lack of direct mass estimates of the compact objects that power ULXs. It is difficult to unequivocally identify or take spectra (let alone phase-resolved spectra) of their optical counterparts, at distances $\gtrsim$ a few Mpc. This prevents the determination of ULX mass functions. Moreover, ULXs are an ill-defined class of systems, based simply on their apparent luminosity: they may include diverse physical objects. Nonetheless, it is possible and useful to summarize common features that appear associated with a majority of ULXs, to find out which of those phenomena are a clue to their physical nature. Here, we discuss some of them, with particular attention to the brightest ULXs (i.e., those with X-ray luminosities $\gtrsim 10^{40}$ erg s^{-1}, of which only ~ 20 are known).

Do ULXs require an accreting black hole?

In most cases, *Yes*. Some young supernova remnants (SNRs) can be misidentified as ULXs (e.g., Circinus Galaxy X-2). So can young high-energy pulsars, with magnetic fields $\sim 10^{14}$ Gauss and rotational periods ~ 10 ms (Stella & Perna 2004); they are also likely to be associated with young SNRs. But most ULXs show long-term fluctuations and flux variability inconsistent with the SNR or pulsar model, and lack X-ray emission lines, unlike typical SNR spectra.

Do ULXs require BHs more massive than typical Galactic BHs?

That is, with masses $> 20 M_\odot$? Probably *Yes*. Their apparent luminosity is up to 50 times higher, and at least a few of them are inconsistent with strong beaming. Models explaining such a high enhancement entirely with super-Eddington emission or beaming are not ruled out yet, but the simplest scenario consistent with the observations is to allow for higher BH masses. If so, it means that the main difference between ULXs and Galactic BHs is due to the compact object, rather than the companion star or the gas flow.

Do they require intermediate-mass BHs?

That is, with masses $\gtrsim 200 M_\odot$? This is a controversial issue. In the absence of kinematic masses, various indirect methods have been suggested: X-ray spectral modelling, timing analysis, breaks in the luminosity function, patterns of state transitions, X-ray–radio correlations. Timing features such as low-frequency QPOs and breaks in the power-density spectrum suggest masses either one or two orders of magnitude higher than stellar-mass BHs, depending on the assumed model. The break or cutoff in the ULX luminosity function at $\approx 3 \times 10^{40}$ erg s^{-1} (Swartz et al. 2004; Gilfanov et al. 2004) suggests an upper mass limit $\approx 200 M_\odot$ if the Eddington limit is adhered to, or less if super-Eddington emission is allowed.

The presence of a soft-excess in the X-ray spectra of the brightest ULXs, with a characteristic temperature ≈ 0.15 keV, was interpreted as evidence in favor of BH masses $\sim 1000 M_\odot$, if the emission comes from a standard accretion disk (Miller et al. 2004). However, we argued (Soria et al. 2006) that that argument is incorrect: when the dominant power-law component is also taken into account, the luminosity and temperature are consistent with masses $\sim 50 M_\odot$. It is also possible that the soft excess does not come from a disk, but from reprocessing in an ionized outflow (Gonçalves & Soria 2006).

In conclusion, the available observational evidence does *not* require intermediate-mass BHs with masses $\sim 1000 M_\odot$ (although they are not ruled out, either), and is still consistent with masses $\lesssim 100 M_\odot$. Stellar-evolution models predict that He cores with masses $64 M_\odot \lesssim M \lesssim 133 M_\odot$ are disrupted by the pair instability and do not collapse into a BH (Heger & Woosley 2002; Yungelson 2006). Therefore, there might be two subclasses of ULXs: one with BH masses $\lesssim 70 M_\odot$ (accounting for some mass increase due to accretion) and one with masses $\gtrsim 130 M_\odot$. There is no observational evidence of such dichotomy, and it is likely that most ULXs belong to the lower-mass group. However, a few (4 or 5) ULXs have been observed at least once with apparent luminosities ~ 5–12×10^{40} erg s^{-1} and have been labelled "hyperluminous X-ray sources" by some authors. One may speculate that they belong to the higher-mass group; alternatively, some of them could be nuclear BHs of disrupted satellite galaxies (King & Dehnen 2005).

Do they require young stellar environments?

Yes. Most ULXs brighter than a few 10^{39} erg s^{-1} are located in star-forming environments, rather than spiral bulges, halos and elliptical galaxies. This suggests that ULXs are scaled-up versions of high-mass X-ray binaries, with an OB donor star overflowing its Roche lobe. This enables a mass transfer rate $\gtrsim 10^{-6} M_\odot$ yr^{-1} for a few Myr (nuclear timescale), and suggests characteristic ULX ages $\sim 10^7$ yr. Low-mass donors could reach this level of mass transfer only during short-lived (thermal timescale) evolutionary phases, at a much later age (> 1 Gyr).

Do they require a donor star in a binary system?

Probably *Yes.* Models based on Bondi accretion from molecular clouds (Krolik 2005) cannot be ruled out in some cases, but are generally disfavoured by the low absorption seen in almost all ULX spectra (typically $< 10^{21}$ cm^{-2}) and low extinction in the surrounding stellar population. It is unlikely that all ULXs accreting from molecular clouds are at the very edge of them.

Do they require starburst environments?

No. Although starburst conditions are positively correlated with ULX formation, they do not seem to be a necessary condition. Some of the brightest ULXs are located in dwarf irregular galaxies with only localized star formation (such as those in Ho II and Ho IX, in the M 81 group). Others are located in normal star-forming (not starburst) galaxies, such as NGC 1313 and NGC 1365. A few are in relatively quiescent environments of starburst galaxies (for example, the two brightest ULXs in NGC 7714), many kpc away from the starburst region.

Does ULX formation require super-star-clusters (SSCs)?

No. Very few ULXs are found in SSCs, defined as young, compact clusters with stellar masses $\gtrsim$ a few $10^5 M_\odot$ and sizes $\lesssim$ a few pc. Among ULXs with $L_X > 10^{40}$ erg s^{-1}, the only examples are one in M 82 and one in NGC 7714; none are found in the Antennae; there may be some in the Cartwheel but it is too far for unequivocal identifications. In most cases, ULXs are near or inside OB associations or small open clusters, with no SSCs nearby. Characteristic stellar ages ($\sim 10^7$ yr) are too young to be consistent with the evaporation of a hypothetical parent SSC. Even if we assume that a parent cluster had time to disperse, the integrated mass of all the stars seen today within ~ 100 pc of a ULX does not generally add up to $10^5 M_\odot$; more typical values are $\lesssim 10^4 M_\odot$. Finally, when SSCs and ULXs are present in the same region, typical displacements are too large (a few hundred pc) to be consistent with cluster ejection.

SSCs were modelled as an ideal environment to form BHs as massive as ~ 500–$1000 M_\odot$ in the local universe, via runaway core-collapse and merger of O stars over a timescale $\lesssim 3$ Myr (Portegies Zwart & McMillan 2002). However, we have argued that there is no longer a compelling need to invoke intermediate-mass BHs in ULXs, and that the upper mass limit is likely to be somewhere between 50 and $200 M_\odot$. Correspondingly, if dynamical collapse and merger processes are still needed to form a very massive stellar progenitor ($> 100 M_\odot$), clusters as small as $\sim 10^4 M_\odot$ may do the job. We have also argued (Soria 2006) that collapse and merger processes can be more efficient at an earlier stage of cluster evolution, when its protostars are still surrounded by large, optically-thick envelopes, and are still accreting from neutral intra-cluster gas. Collapsing molecular clumps with masses $\sim 10^4 M_\odot$ are large enough to enable the formation of stars with masses $> 100 M_\odot$ via accretion and coalescence, and at the same time are small enough to disperse quickly after the most massive stars reach the main sequence (Kroupa & Boily 2003), leaving behind an open cluster or OB association, in agreement with the observations.

Does ULX formation require low metal abundance?

Almost certainly *Yes.* This is supported both by (still sketchy) empirical evidence and theoretical arguments. We leave a detailed discussion of the available metallicity data for ULX host galaxies to further work. It is of course more difficult to produce BHs at higher metallicities, because more mass is lost from the progenitor star via stellar winds. At solar metallicity, all O stars – including the Pistol star and η Carinae, despite their initial masses ≈ 150–$200 M_\odot$ – are predicted to leave behind only a neutron star. At $Z \sim 0.1$, they may produce BHs with masses $\sim 50 M_\odot$.

Does ULX formation require primordial abundances?

Probably *not.* Massive Pop-III stars were suggested as an alternative to local SSC scenarios for IMBH production. There may well be Pop-III BH remnants with masses up to $\sim 1000 M_\odot$ floating around in galactic halos, or slowly sinking towards the centres, but this does not explain the observed ULX correlation with young, star-forming environments. The Pop-III scenario requires that floating BH remnants capture an OB star while they cross a star-forming environment, perhaps after being thrown out of their halo orbits during tidal interactions and collisions. In the absence of independent evidence for the very existence of Pop-III remnants, it remains an unlikely (though interesting) conjecture, especially if IMBHs are not needed after all to explain the ULX luminosity.

Is ULX formation directly favoured by tidal interactions and collisions?

Apparently *Yes.* Many ULXs are found in tidal dwarfs, or colliding galaxies, or dwarf irregular galaxies located in tidally interacting groups. Spectacular examples include galaxies such as the Antennae, the Mice, the Cartwheel, NGC 7714/15, NGC 4485/90, and the M 81/M 82 group. Other bright ULXs are associated with local collisional events: NGC 4559 X-1 is in a ring of star formation (age ~ 20 Myr) probably caused by a small satellite galaxy splashing through the gas-rich disk; M 99 X-1 is apparently located where a large, fast HI cloud is impacting the outer disk; NGC 1313 may have undergone a recent collision with a satellite, near its ULX X-2.

Are these chance associations? One simple explanation could be that collisions enhance star formation, and a high star-formation rate (SFR) leads to more X-ray binaries and a larger probability to form ULXs – the normalization of the high-mass X-ray binary luminosity function being proportional to the SFR (Gilfanov *et al.* 2004). While this is probably part of the explanation, it cannot be the whole story. In various cases, the local SFR in a collisional or tidal feature is small, compared with the SFR in the rest of the galaxy or group; and yet ULXs seem to be directly associated to those environments (NGC 7714, M 99 and NGC 4559 are striking examples). I suggest that there can be a direct physical association between collisions and ULX formation, if collisions tend to produce a *qualitatively different kind of star formation, that is more likely to lead to the formation of relatively massive BH remnants*, and hence to some ULXs.

2. Outlining a plausible ULX scenario

Taking into account the previous arguments, I speculate that the following line of investigation appears the most promising. Most ULXs contain BHs with masses $\sim 50 M_\odot$ and in any case $\lesssim 100 M_\odot$, formed via direct core collapse from very massive stellar progenitors, and accrete from an OB star coeval with the BH progenitor. The luminosity enhancement with respect to Galactic BHs can be explained with a factor of ≈ 5–10 in mass, and ≈ 5 in super-Eddington emission, particularly outside the disk plane.

Progenitor stars with initial masses ~ 150–$200 M_\odot$ do exist (although they are very rare), and can be formed in clustered environments, via fast gas accretion and mergers of smaller protostars – this is also the way ordinary O stars are thought to form. The proto-cluster NGC 2264C in the Cone nebula is a textbook example of a molecular clump, with a gas mass $\approx 1700 M_\odot$ that is undergoing dynamical collapse (infall of $\gtrsim 10^{-3} M_\odot$ yr^{-1} over a free-fall timescale of $\sim 10^5$ yr) rather than turbulent fragmentation (Peretto *et al.* 2006). There is no need for such clusters to be more massive than $\sim 10^4 M_\odot$.

Such global collapses occur when star formation is triggered by an external pressure wave. Cloud-galaxy or galaxy-galaxy collisions provide ideal environment for triggered star formation and therefore also for massive stellar progenitors. Low metal abundance provides the second ingredient, ensuring that a massive BH remnant is formed.

The normalization of the high-mass X-ray binary luminosity function, and probably also the number of fainter ULXs with luminosities $\sim$ a few 10^{39} erg s^{-1}, is directly proportional to the SFR. However, the location of the upper-luminosity break, and hence the probability of forming ULXs with luminosities $> 10^{40}$ erg s^{-1}, depends more strongly on the two factors mentioned above: external triggers and low metal abundance. A key observational test would be to map the presence of very massive stars in nearby galaxies, although it may be difficult to distinguish them from unresolved stellar groups.

References

Gilfanov, M., Grimm, H.-J., Sunyaev, R. 2004, Nucl. Phys. Suppl., 132, 369

Gonçalves, A. C., Soria, R. 2006, MNRAS, 371, 673

Heger, A., Woosley, S. E. 2002, ApJ, 567, 532

King, A. R., Dehnen, W. 2005, MNRAS, 357, 275

Krolik, J. H. 2004, ApJ, 615, 383

Kroupa, P., Boily, C. M. 2002, MNRAS, 336, 1188

Miller, J. M., Fabian, A. C., Miller, M. C. 2004, ApJ, 614, L117

Peretto, N., André, P., Belloche, A. 2006, A&A, 445, 979

Perna, L., Stella, R. 2004, ApJ, 615, 222

Portegies, Zwart, S. F., McMillan, S. L. W. 2002, ApJ, 576, 899

Soria, R. 2006, in: E. J. A. Meurs & G. Fabbiano (eds.), Populations of High Energy Sources in Galaxies, IAU Symp. 230 (Cambridge University Press: Cambridge), p. 473 (astro-ph/0509573)

Soria, R., Gonçalves, A. C., Kuncic, Z. 2006, in: The Multicoloured Landscape of Compact Objects and Their Explosive Progenitors, Cefalú, Sicily (June 2006), AIP, in press

Swartz, D. A., Ghosh, K. K., Tennant, A. F., Wu, K. 2004, ApJSS, 154, 519

Yungelson, L. 2006, astro-ph/0610021

ANDREA MERLONI: How much does the estimated power fraction in the corona depends on the assumption used to fit the spectrum?

ROBERTO SORIA: If the "soft excess" is entirely due to cool disc emission, it represents a correction of only about 10–20% to the Comptonized component (see e.g. Stobbart *et al.* 2006). If it is due to reprocessing in an ionized outflow, then of course the direct component is even less than that.

CATHERINE CESARSKY: Conditions that you describe for forming ULXs are similar to what we see in large redshift gamma ray bursts. Perhaps there is some common physics or common evolution here?

ROBERTO SORIA: This is indeed a very good point, possibly related with very massive stars.

DANY VANBEVEREN: I think that it may be incorrect to state that the progenitor of a $100 M_\odot$ black hole is a $200 M_\odot$ main-sequence star in a low metallicity environment. Such a star in a low-Z environment produces electro-pair supernovae.

ROBERTO SORIA: I understand the exact mass threshold is still uncertain. Lev Yungelson's calculations suggest that massive metal-poor stars can grow a core up to 60–$70 M_\odot$ before collapsing to a black hole. Perhaps that will turn out to be a neat upper limit to "stellar" black-hole masses; it is high enough to explain ULXs.

ELAINE SADLER: The formation conditions you describe for ULXs (star-forming regions with low metallicity) sound very similar to those observed for high-redshift gamma-ray bursts. Do you think there could be links between the physics of ULXs and GRBs?

ROBERTO SORIA: Yes – especially if black holes in ULXs turn out to be $\sim 50 M_\odot$ objects, their birth process should be different form the more common types of core-collapse supernovae in the local universe. The key is to understand the difference between the black hole formation via supernovae explosions or by direct collapse.

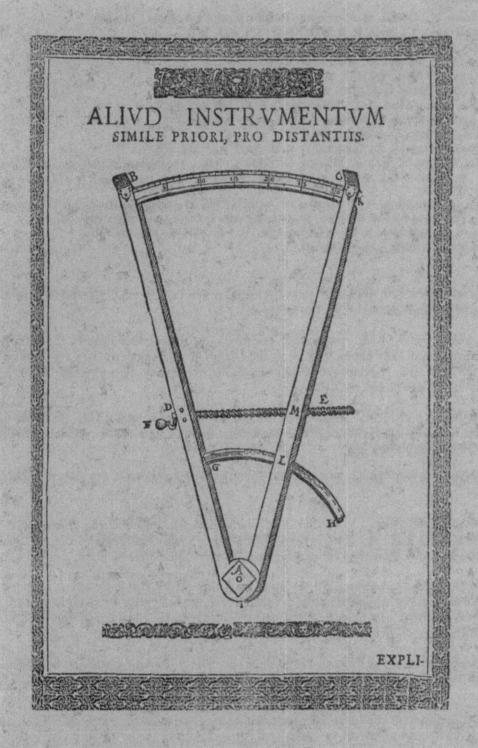

ALIVD INSTRVMENTVM
SIMILE PRIORI, PRO DISTANTIIS.

EXPLI-

Black Holes from Stars to Galaxies – Across the Range of Masses
Proceedings IAU Symposium No. 238, 2006
V. Karas & G. Matt, eds.

© 2007 International Astronomical Union
doi:10.1017/S1743921307005054

Explosion of very massive stars and the origin of intermediate mass black holes

Sachiko Tsuruta,[1] Takuya Ohkubo,[2] Hideyuki Umeda,[2] Keiichi Maeda,[3] Ken'ichi Nomoto,[2,4] Tomoharu Suzuki[2] and Martin J. Rees[5]

[1]Department of Physics, Montana State University, Bozeman, MT 59717, USA
email: uphst@gemini.msu.montana.edu

[2]Department of Astronomy, Univ. of Tokyo, Hongo, Bunkyo-ku, Tokyo, Japan

[3]Department of Earth Sci. and Astron., Univ. of Tokyo, Komaba, Meguro-ku, Tokyo, Japan

[4]Research Center for the Early Universe, Univ. of Tokyo, Hongo, Bunkyo-ku, Tokyo, Japan

[5]Institute of Astronomy, University of Cambridge, Madingley Road, Cambridge CB3 0HA, UK

email: ohkubo@astron.s.u-tokyo.ac.jp, umeda@astron.s.u-tokyo.ac.jp,
nomoto@astron.s.u-tokyo.ac.jp, suzuki@astron.s.u-tokyo.ac.jp, maeda@esa.c.u-tokyo.ac.jp,
mjr@ast.cam.ac.uk

Abstract. We calculate evolution, collapse, explosion, and nucleosynthesis of Population III very massive stars with $500 M_\odot$ and $1000 M_\odot$. It was found that both $500 M_\odot$ and $1000 M_\odot$ models enter the region of pair-instability but continue to undergo core collapse to black holes. For moderately aspherical explosions, the patterns of nucleosynthesis match the observational data of intergalactic and intercluster medium and hot gases in M82, better than models involving hypernovae and pair instability supernovae.

Our results suggest that explosions of Population III core-collapse very massive stars contribute significantly to the chemical evolution of gases in clusters of galaxies. The final black hole masses are about $500 M_\odot$ for our most massive $1000 M_\odot$ models. This result may support the view that Population III very massive stars are responsible for the origin of intermediate mass black holes which were recently reported to be discovered.

1. Introduction

One of the most interesting challenges in astronomy is to investigate the mass and properties of first generation 'Population III (Pop III)' stars, and how various elements have been synthesized in the early universe. Just after the Big Bang these elements were mostly only H, He and a small amount of light elements (Li, Be, B, etc). Heavier elements, such as C, O, Ne, Mg, Si and Fe, were synthesized during the evolution of later generation stars, and massive stars exploded as supernovae (SNe), releasing heavy elements into space.

The fate of zero-metalicity Pop III stars are: Stars lighter than $8 M_\odot$ form white dwarfs; those with $8 M_\odot$ – $130 M_\odot$ generally explode as the core-collapse supernovae leaving neutron stars or black holes as compact remnants; stars with $130 M_\odot$ – $300 M_\odot$ disrupt completely as the pair-instability supernovae (PISN) with no compact remnants left; stars with $300 M_\odot$ to $\sim 10^5 M_\odot$ (without rotation) collapse to black holes with no explosion; and those over $10^5 M_\odot$ collapse directly to black holes before reaching the main-sequence. (See, e.g., Fryer, Woosley & Heger 2001)

It has been suggested that the initial mass function (IMF) of Pop III first stars is probably more massive than, e.g., the Salpeter mass function (e.g., Abel, Bryan &

Norman 2000; Bromm, Coppi & Larson 2002; Qian, Sargent & Wasserburg 2002). Numerical simulations by, e.g. Bromm & Loeb (2004) indicate that the maximum mass of Pop III stars to be formed will be $\sim 300 M_\odot - 500 M_\odot$. Omukai & Palla (2003), however, point out that under certain conditions very massive stars much heavier than $300 M_\odot$ can be formed in the zero-metalicity environment. Another scenario for the formation of very massive stars for any metallicity has been presented by Ebisuzaki *et al.* (2001) and Portegies Zwart *et al.* (2004), where very massive stars are formed by merging of less massive stars in the environment of very dense star clusters.

In the present paper, we call the stars with $M = 130 M_\odot - 10^5 M_\odot$ 'Very-Massive Stars (VMSs)'. Among VMSs we define $M > 300 M_\odot$ stars as 'Core-Collapse Very-Massive Stars (CVMSs)', in order to clarify the distinction between the PISN mass range and the core-collapse range. Here we focus on CVMSs, and deal with $500 M_\odot$ and $1000 M_\odot$ models.

Stellar mass black holes ($\sim 10 M_\odot$) are formed as the central compact remnants of ordinary massive ($25 - 130 M_\odot$) stars at the end of their evolution. Supermassive black holes (SMBHs) ($\sim 10^5 - 10^9 M_\odot$) are now known to exist in the center of almost all galaxies (e.g., Kormendy & Richstone 1995), and there are several scenarios for their formation (e.g., Rees's contribution to this volume). As to intermediate mass black holes (IMBHs) with $\sim 5 \times 10^{(2-4)} M_\odot$, there is a strong possibility that some of these stars have been discovered as ULX (see, e.g., contribution to this volume by Makishima). The question of whether CVMSs ($\sim 300 M_\odot - 10^5 M_\odot$) actually existed is of great importance, for instance, to understand the origin of IMBHs.

Also, if CVMSs existed and if they were rotating, they might have released a large amount of heavy elements into space by mass loss and/or supernova explosions. If so, they might have significantly contributed to the early galactic chemical evolution because they could be the source of reionization of intergalactic H and He (e.g., Gnedin & Ostriker 1997; Venkatesan *et al.* 2003). The reionization of intergalactic He has traditionally been attributed to quasars. However, according to the results of the *the Wilkinson Microwave Anisotropy Probe* (*WMAP*) observation in 2003, reionization in the universe took place as early as 0.2-0.3 billion years after the Big Bang (redshift z $>\sim$ 20; Kogut *et al.* 2003). Then these Pop III CVMSs might provide a better alternative channel which could operate at redshifts higher than what is assumed for quasars (Bromm, Kudritzki & Loeb 2001).

2. Our Recent Research

2.1. *Our Models*

Motivated by the backgrounds as outlined in the previous section, we calculated evolution, core-collapse, explosion, and nucleosynthesis of Pop. III CVMSs (Ohkubo *et al.* 2006, hereafter O06). We start our evolutionary calculations by assuming that the stars have $500 M_\odot$ and $1000 M_\odot$ with zero-metalicity on the pre-main sequence. Without rotation these stars are expected to form black holes directly with no explosion at the end of evolution.

However, if the star is rotating the whole star probably will not become a black hole at once, but form an accretion disk around the central remnant. After forming an accretion disk, jet-like explosions may occur by extracting energy from the accretion disk and/or the black hole itself. Therefore, in our explosion and collapse calculations we adopt a two-dimensional approach including accretion along the equatorial direction and jets toward the polar direction.

To calculate pre-supernova evolution we adopt the stellar evolution code constructed by Umeda & Nomoto (2002) based on the Henyey method. We include 51 isotopes up to Si until He burning ends, and 240 up to Ge afterwards. Our evolutionary calculations are carried out from the pre-main sequence up to the iron-core collapse where the central density reaches as high as $2 \times 10^{10} \text{gcm}^{-3}$.

When the temperature reaches 5×10^9 K, where 'nuclear statistical equilibrium' is realized, the abundance of each isotope is determined for a given set of density, temperature, and the number of electrons per nucleon. Our CVMSs are assumed to explode in a form of bipolar jets, and we explored the required constraints. We adopted two types of models: Case (A) where almost all the energy of the jet is given as kinetic energy, and Case (B) where almost all the energy is given as thermal energy. Case (A) jets are better collimated, while Case (B) jets are only mildly aspherical. See O06 for further details.

2.2. Results and Discussion

Although our $500M_\odot$ and $1000M_\odot$ stars pass through the region of electron-positron pair-instability, both stars proceed to iron-core collapse, unlike a $300M_\odot$ star.

The region which experiences explosive silicon burning to produce iron-peak elements is more than 20% of the total mass, much larger than those of ordinary massive stars such as a $25M_\odot$ star. Note that for the metal-free $25M_\odot$ star model, this fraction is less than 10% (Umeda & Nomoto 2002). This is because for the $500M_\odot$ and $1000M_\odot$ models the density and temperature distributions are much flatter than those of $25M_\odot$ stars.

Typical explosion energy is of the order of 10^{54} erg for $1000M_\odot$ models for the parameter ranges in this study.

Masses of the remnant black holes we found are $\sim 500M_\odot$ for the $1000M_\odot$ star models. Note that such a black hole mass is very similar to those of IMBHs, e.g., a claimed $\sim 700M_\odot$ black hole in M82. It is quite possible that CVMSs could be the progenitors of IMBHs.

Nucleosynthesis yields of CVMSs have similar patterns of $[\alpha/\text{Fe}]$ to the observed abundance patterns of both intercluster medium (ICM) and hot gases in M82 if the contribution of the jet is small (Case B). Specifically, for Case B small ratios of $[\text{O}/\text{Fe}]$ and $[\text{Ne}/\text{Fe}]$ combined with large $[\text{Mg}/\text{Fe}]$, $[\text{Si}/\text{Fe}]$ and $[\text{S}/\text{Fe}]$ (i.e. large $[(\text{Mg, Si, S})/\text{O}]$) are generally *more consistent* with these observational data of ICM, than those of hypernovae and PISNe.

$[\text{C}/\text{Si}]$ of our CVMS models is compatible with that of intergalactic medium (IGM) at high redshift ($z = 5$), which is sufficiently higher and thus better than those of PISNe.

For Fe-peak elements, the main feature of the yields of our Case B CVMSs is that $[\text{Cr}/\text{Fe}]$ and $[\text{Mn}/\text{Fe}]$ are small while $[\text{Co}/\text{Fe}]$ and $[\text{Zn}/\text{Fe}]$ are large. This is consistent with the observed ratios in the extremely metal-poor (EMP) stars. The over-solar ratios of some α-elements, such as $[\text{Mg}/\text{Fe}]$ and $[\text{Si}/\text{Fe}]$, are also consistent with EMP stars.

We estimated the ionization efficiency of our CVMSs. It was found that the number of ionizing photons per baryon in the universe, generated in association with the IGM metalicity $Z_{\text{IGM}} \sim 10^{-4}$, is $N_{\text{Lyc}}/N_{\text{b}} \sim 150$, and so *CVMSs can contribute significantly to reionization of IGM in the early epoch.* Here we emphasize that our current result for CVMSs is contributed from the mass range with $\sim 300M_\odot - 1000M_\odot$ and thus the PISN (pair instability supernova) range is *not* included. On the other hand, Venkatesan & Truran (2003) give $N_{\text{Lyc}}/N_{\text{b}} \sim 10$ for $Z_{\text{IGM}} \sim 10^{-4}$ for the mass range $\sim 100M_\odot - 1000M_\odot$. This outcome therefore reflects the large contribution of PISNe to metal enrichment. Note that in their models CVMSs do not explode.

The relation between reionization and metal enrichment of IGM becomes clearer if we solve the equation for $N_{\text{Lyc}}/N_{\text{b}}$ (Eqn. 2 of Venkatesan & Truran 2003) for a given value of

Z_{IGM}. For a $1000 M_\odot$ star, $Z/Z_\odot \sim 10^{-3.4}$ and $10^{-4.4}$ for the required number of ionizing photons per baryon 10 and 1, respectively. This is about one order of magnitude smaller than the case for the mass range $100 M_\odot - 1000 M_\odot$ (mainly contribution from PISNe). The difference between CVMSs and PISNe is larger if we consider the enrichment of iron. The 260 $M_\odot$ PISN of Heger & Woosley (2002) gives $Z_{\text{Fe}}/Z_{\text{Fe},\odot} \sim 10^{-2} - 10^{-3}$, while our 1000 $M_\odot$ star gives $\sim 10^{-3.2} - 10^{-4.2}$.

A main critique against the existence of PISNe comes from the fact that we do not see the abundance patterns of PISNe in EMP stars (Tumlinson *et al.* 2004). The EMP abundances were suggested to be accounted for by hypernovae or faint supernovae from less massive stars of $<\sim 100 M_\odot$ (Umeda & Nomoto 2003). However, the apparent lack of evidence of VMSs by no means contradicts the existence of CVMSs at earlier epochs, if a majority of first stars in the earlier epoch has masses $>\sim 300 M_\odot$. First, in this case PISNe from stars of $<\sim 300 M_\odot$ will be just a minor fraction, which explains well the lack of the signature of PISNe. Second, $Z/Z_\odot$ expected from our CVMSs is smaller than PISNe. Namely, *the metal enrichment by CVMSs might be finished before ordinary core-collapse SNe become dominant.*

Here we have shown that the yields of our CVMSs can reproduce the abundance of IGM. In our scenario CVMSs are first formed in pre-galactic mini halos, and then subsequently ordinary core-collapse SNe took place in the galactic halo.

Acknowledgements

We would like to thank useful comments given by some colleagues in the audience, especially Dr. Portegies Zwart.

References

Abel, T., Bryan, G. L. & Norman, M. L. 2000, ApJ, 540, 39
Bromm, V., Coppi, P. S. & Larson, R. B. 2002, ApJ, 564, 23
Bromm, V., Kudritzki, R. P. & Loeb, A. 2001, ApJ, 552, 464
Bromm, V. & Loeb, A. 2004, NewA, 9, 353
Ebisuzaki, T., Makino, J., Tsuru, T. G., Funato, Y., Portegies, Zwart, S., Hut, P. *et al.* 2001, ApJ, 562, L19
Fryer, C. L., Woosley, S. E. & Heger, A. 2001, ApJ, 550, 372
Gnedin, N. Y. & Ostriker, J. P. 1997, ApJ, 486, 581
Heger, A. & Woosley, S. E. 2002, ApJ, 567, 532
Kogut, A., Spergel, D. N., Barnes, C., Bennett, C. L., Halpern, M. & Hinshaw, G. *et al.* 2003, ApJS, 148, 161
Kormendy, J. & Richstone, D. 1995, ARA&A, 33, 581
Ohkubo, T., Umeda, H., Maeda, K., Nomoto, K., Suzuki, T., Tsuruta, S. *et al.* 2006, ApJ, 645, 1352 (O06)
Omukai, K. & Palla, F. 2003, ApJ, 589, 677
Portegies, Zwart, S. F., Baumgardt, H., Hut, P., Makino, J. & McMillan, S. L. W. 2004, Nature, 428, 724
Qian, Y.-Z, Sargent, W. L. W. & Wasserburg, G. J. 2002, ApJ, 588, 1099
Tumlinson, J., Venkatesen, A. & Shull, J. M. 2004, ApJ, 612, 602
Umeda, H. & Nomoto, K. 2002, ApJ, 565, 385
Umeda, H. & Nomoto, K. 2003, Nature, 422, 871
Venkatesan, A., Tumlinson, J. & Shull J. M. 2003, ApJ, 584, 621
Venkatesan, A. & Truran, J. W. 2003, ApJ, 594, L1

ALEXANDER ZAKHAROV: 1. How did you choose initial conditions for 2-D explosion calculations using 1-D evolution calculations? 2. Have you solved the so-called "energetic problem" to explain supernova explosion?

SACHIKO TSURUTA: Answering these questions requires a complicated exploration; a referee of our ApJ paper asked exactly the same questions and our answers are given in detail there (Ohkubo *et al.*, ApJ, July 10, 2006). So, please, read this paper for the answers to your questions. The energy source is accretion and jets.

SIMON PORTEGIES ZWART: Comment: it is very interesting that you require rapid rotation in the pre-SN star. This is a natural consequence of the collision runaway scenario to make $\sim 1000 M_\odot$ stars in young and dense star clusters.

SACHIKO TSURUTA: In our next work we are hoping to include binary merging during the stellar evolution stages.

DANY VANBEVEREN: Very massive stars have very large stellar wind mass-loss rates; also low-Z very massive stars. What is the mass-loss rate prescription that you are using?

SACHIKO TSURUTA: At the next stage we will include mass loss, though we have not included it in our first approximation. Our guess is that large mass star (e.g. $\sim 5000 M_\odot$) are needed to produce a $500 M_\odot$ black hole. We discussed this problem in Ohkubo *et al.* (ApJ, July 10, 2006).

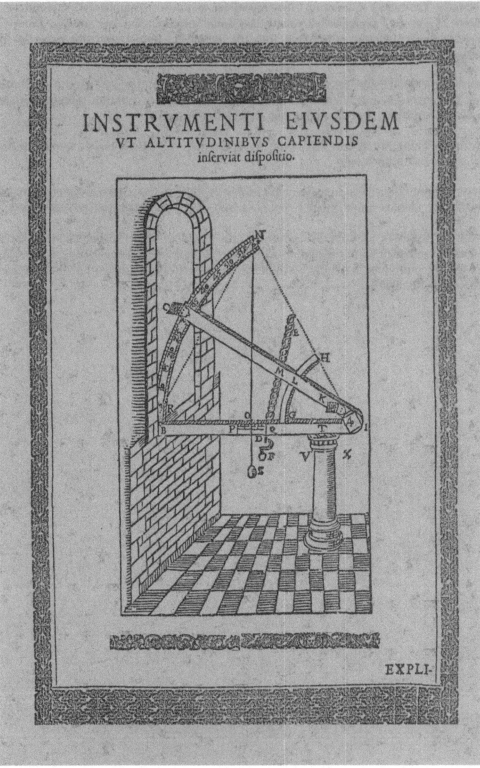

EXPLI-

Black Holes from Stars to Galaxies - Across the Range of Masses
Proceedings IAU Symposium No. 238, 2006
V. Karas & G. Matt, eds.
© 2007 International Astronomical Union
doi:10.1017/S1743921307005066

Ultra-luminous X-ray sources:
X-ray binaries in a high/hard state?

Z. Kuncic,[1] R. Soria,[2] C. K. Hung,[1] M. C. Freeland[1]
and G. V. Bicknell[3]

[1]School of Physics, University of Sydney, Sydney NSW 2006, Australia

[2]Mullard Space Science Laboratory, University College London, Holmbury St Mary,
Dorking, Surrey RH5 6NT, UK

[3]Research School of Astronomy and Astrophysics, Australian National University,
Mount Stromlo Observatory, Cotter Road, Canberra ACT 2611, Australia

email: z.kuncic@physics.usyd.edu.au, rs1@mssl.ucl.ac.uk, geoff@mso.anu.edu.au

Abstract. We examine the possibility that Ultraluminous X-ray sources (ULXs) represent the extreme end of the black hole X-ray binary (XRB) population. Based on their X-ray properties, we suggest that ULXs are persistently in a high/hard spectral state and we propose a new disk–jet model that can accomodate both a high accretion rate and a hard X-ray spectrum. Our model predicts that the modified disk emission can be substantially softer than that predicted by a standard disk as a result of jet cooling and this may explain the unusually soft components that are sometimes present in the spectra of bright ULXs. We also show that relativistic beaming of jet emission can indeed account for the high X-ray luminosities of ULXs, but strong beaming produces hard X-ray spectra that are inconsistent with observations. We predict the beamed synchrotron radio emission should have a flat spectrum with a flux density $\lesssim 0.01\,\mathrm{mJy}$.

Keywords. accretion disks – black hole physics – X-rays: binaries

1. Introduction

The exceptionally high X-ray luminosity of ULXs, $L_{0.5-8\,\mathrm{keV}} \approx (0.2-10) \times 10^{40} \mathrm{erg\,s}^{-1}$, emerges almost entirely as a hard power-law with photon indices $1.5 \lesssim \Gamma \lesssim 2.5$ (Irwin, Bregman & Athey 2004; Swartz *et al.* 2004; Liu & Bregman 2005; Stobbart, Roberts & Wilms 2006). ULXs are statistically consistent with representing a sub-population of XRBs (Swartz *et al.* 2004). However, known Galactic black hole XRBs have never been observed to enter a spectral state in which they simultaneously have an X-ray luminosity that is at least Eddington and a power-law spectrum that is harder than $\Gamma \approx 2.5$. The XRB spectral state that perhaps most closely resembles ULXs is the short-lived steep power-law (or very high) state, in which $L_{\mathrm{x}} \approx L_{\mathrm{Edd}}$, but $\Gamma \gtrsim 2.5$ (see McClintock & Remillard 2006). ULXs therefore appear to be in a persistently high/hard state.

It is unclear whether existing models for XRBs can explain a high/hard spectral state. ULXs appear to require a model with both a high $\dot{M}_{\mathrm{a}}$ and strong corona and/or jet. Furthermore, it is clear that most of the accretion power in ULXs is not being dissipated in a standard accretion disk. This implies that we cannot use the standard Multi-Colour Disk (MCD) model to fit any soft spectral component that might be present. We summarize the main results of the model proposed by Freeland *et al.* (2006) and we examine further the implications of a strong, accretion-powered jet for spectral fitting of accretion disk models to soft components evident in some ULX spectra.

2. Model Outline and Results

The details of our ULX model can be found in Freeland *et al.* (2006). It is based on the generalized theoretical framework of Kuncic & Bicknell (2004), which specifically addresses vertical angular momentum transport by a magnetic torque on the the accretion disk surface. This magnetic torque is identified as the mechanism responsible for jet/corona formation. It results in a modified radial structure of the accretion disk and hence, a modified disk spectrum, which we refer to as an Outflow Modified Multi-Colour Disk (OMMCD). Let us assume here that the total power extracted by the torque is injected into a relativistic jet and partitioned into magnetic and kinetic energies.

Modified disk spectrum. The radiative energy flux of an accretion disk modified by a magnetized jet is

$$\sigma T^4 = F(r) = \frac{3GM_\bullet \dot{M}_a}{8\pi r^3} \left[f_{ss}(r) - f_j(r) \right], \tag{2.1}$$

where $f_{ss} = 1 - (r/r_i)^{-1/2}$ is the small-r correction factor for a standard disk and where the jet correction factor is directly related to the magnetic torque acting on the disk surface:

$$f_j(r) = \frac{1}{\dot{M}_a r^2 \Omega} \int_{r_i}^r 4\pi r^2 \frac{B_\phi B_z}{4\pi} \, dr. \tag{2.2}$$

Thus, the jet drains energy from the disk and modifies the radial temperature profile. As a result, the total disk spectrum is modified in the soft X-ray bandpass for stellar-mass black holes (see also Soria, Kuncic & Gonçalves 2006, this volume). This means that unusually soft spectral components seen in some bright ULXs may be interpreted as an accretion disk spectrum modified by a jet that is responsible for the dominant power-law spectral component.

Figure 1 shows the *XMM-Newton*/EPIC spectrum of ULX X-7 in NGC 4559. The spectrum of this bright ULX cannot adequately be fitted with a single power-law model. A broken power-law plus MCD model provides an acceptable fit. The low inner disk temperature required by the MCD fit to the soft component implies a black hole mass $M_\bullet \sim 1.4 \times 10^3 M_\odot$. We also show in Fig. 1 an almost identical spectral fit to the data using a similar broken power-law plus an OMMCD model. The best fit model parameters are listed in Table 1. The OMMCD parameters correspond to a $M_\bullet \approx 155 M_\odot$ black hole accreting at $\dot{M}_a \approx 3\dot{M}_{Edd}$, but with only $L_d \approx 0.3 L_{Edd}$ being emitted by the disk. According to this model, the bulk of the accretion power is removed from the disk and injected into a jet, which is responsible for the hard power-law spectral component. However, only a small fraction of the jet power is dissipated in the form of radiation, since jets are highly inefficient emitters. Relativistic beaming is then responsible for boosting the X-ray emission to the high observed levels.

Jet emission. We calculated the jet spectral energy distribution for two different relativistic beaming scenarios (Körding, Falcke & Markoff 2002) : 1. the microblazar scenario, where the jet is pointing close to our line-of-sight and the observed emission is thus strongly beamed; and 2. the microquasar scenario, where the jet is directed at larger angles and hence, there is less contribution from relativistic beaming to the observed X-ray emission. We used a simple radiative transfer model to take into account the synchrotron optical depth. The details of the microblazar and microquasar spectral models can be found in Freeland *et al.* (2006).

The spectral modelling results shown in Figure 2 confirm that both the microblazar and microquasar scenarios can produce X-ray luminosites sufficiently high to be consistent with ULXs. The microblazar model, however, predicts strong deviations from a power-law in the $0.2-10$ keV bandpass resulting from strongly beamed nonthermal Comptonization.

Table 1. Best fit parameters for the spectral models used in Figure 1.

broken power-law+MCD model		broken power-law+OMMCD model	
$N_{\rm H}(\times 10^{21}\,{\rm cm}^{-2})$	2.3	$N_{\rm H}(\times 10^{21}\,{\rm cm}^{-2})$	2.3
$kT_{\rm in}$ (keV)	0.16	$M_{\bullet}(M_{\odot})$	155
normalization	162.2	$\dot{M}_{\rm a}(\times 10^{-5} M_{\odot}\,{\rm yr}^{-1})$	1.3
Γ_1	2.11	$L_{\rm d}/L_{\rm Edd}$	0.3
$E_{\rm break}$(keV)	4.66	normalization $(\times 10^{-6})$	1.0
Γ_2	3.11	Γ_1	2.04
normalization $(\times 10^{-4})$	2.2	$E_{\rm break}$(keV)	4.05
χ^2/dof	202.6/195	Γ_2	2.79
		normalization $(\times 10^{-4})$	1.99
		χ^2/dof	264.7/196

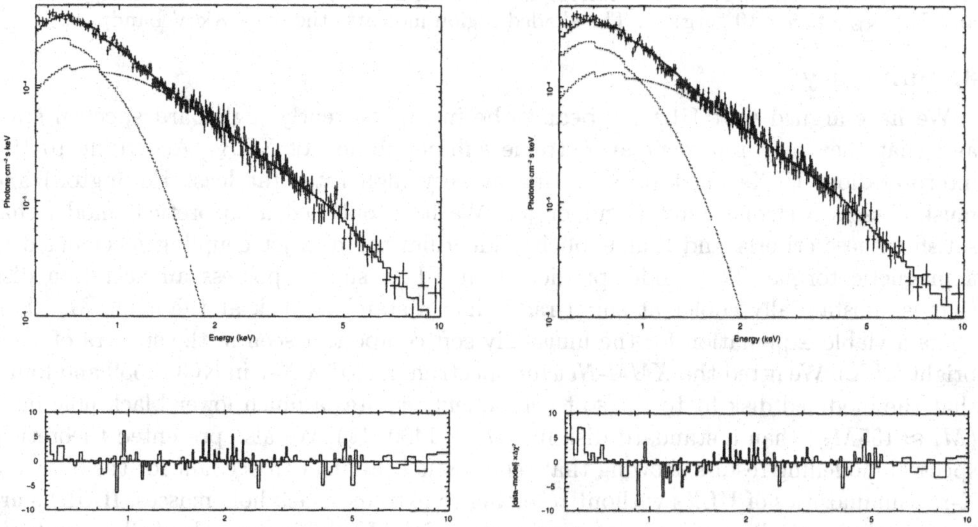

Figure 1. *XMM-Newton*/EPIC unfolded spectra of ULX NGC 4559 X-7. The solid line is the total model spectrum. The dotted lines are the broken power-law plus MCD models (left) and the broken power-law plus OMMCD models (right). χ^2 residuals are shown underneath. Best fit model parameters are shown in Table 1.

This is inconsistent with the observational data to date. The microquasar model, on the other hand, predicts an approximately power-law hard X-ray spectrum.

Both models predict similar radio properties. The synchrotron radio spectra are approximately flat at 5 GHz and the specific radio power is $L_{\nu} \approx 10^{22}\,{\rm erg\,s}^{-1}{\rm Hz}^{-1}$. This corresponds to a flux density $S_{\nu} \approx 10\,\mu{\rm Jy}$ at at distance of 1 Mpc. This is more than two orders of magnitude below the levels measured for the few cases where radio sources have been found associated with ULXs. Our theoretical results support other pieces of evidence ruling out beamed jet emission as the origin of ULX radio counterparts. Our results are also consistent with the overwhelming excess of non-detections over detections found in the deepest ULX radio counterpart search to date, down to detection limits $\approx 60\,\mu{\rm Jy}$ with the VLA (Körding, Colbert & Falcke 2005). Note that a typical Galactic XRB placed at a distance of 1 Mpc would have a flux density of only $\simeq 1\,\mu{\rm Jy}$.

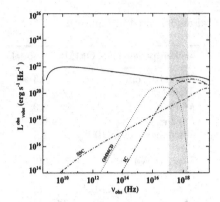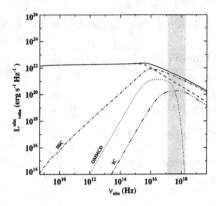

Figure 2. The predicted spectral energy distributions for the microblazar (left) and microquasar (right) scenarios for ULXs. The microblazar model has $M_\bullet = 5 M_\odot$, $\dot{M}_a = \dot{M}_{Edd}$, $\delta = 8.4$ and $L_{0.5-8\,keV} \approx 3 \times 10^{39}\,erg\,s^{-1}$. The microquasar model has $M_\bullet = 20 M_\odot$, $\dot{M}_a = \dot{M}_{Edd}$, $\delta = 1.6$ and $L_{0.5-8\,keV} \approx 5 \times 10^{39}\,erg\,s^{-1}$. The shaded region indicates the $0.5 - 8\,keV$ bandpass.

3. Summary

We have argued that ULXs appear to be in a persistently high/hard spectral state and that they may represent an extreme sub-population of XRBs. According to this interpretation, ULXs must be accreting at very high rates (at least Eddington) and must possess a strong corona and/or jet. We have outlined a theoretical model that satisfies these criteria and that explicitly identifies the disk-jet coupling mechanism as a magnetic torque. The model predicts that ULXs should possess an accretion disk that is substantially cooler at small radii than a standard disk at the same $\dot{M}_a$. This offers a viable explanation for the unusually soft component seen in the spectra of some bright ULXs. We fitted the *XMM-Newton* spectrum for ULX X-7 in NGC 4559 and found that the modified disk fit to the soft component requires a much lower black hole mass ($M_\bullet \approx 155 M_\odot$) than a standard disk fit ($M_\bullet \simeq 1400 M_\odot$). We also presented theoretical spectral modelling results showing that relativistic beaming can account for the observed X-ray luminosities of ULXs without resorting to extreme black hole masses. If ULXs are indeed relativistically beamed, we predict they should exhibit unresolved, flat-spectrum radio cores with fluxes $\lesssim 0.01\,mJy$. Such radio counterparts have not yet been detected.

Acknowledgements

ZK acknowledges a University of Sydney R&D Grant. RS acknowledges a NASA *Chandra* grant.

References

Freeland, M. C., Kuncic, Z., Soria, R. & Bicknell, G. V. 2006, MNRAS, 372, 630
Irwin, J. A., Bregman, J. N. & Athey, A. E. 2004, ApJ, 601, L143
Körding, E., Falcke, H. & Markoff, S. 2002, A&A, 382, L13
Körding, E., Colbert, E. & Falcke, H. 2005, A&A, 436, 427
Kuncic, Z. & Bicknell, G. V. 2004, ApJ, 616, 669
Liu, J.-F. & Bregman, J. N. 2005, ApJS, 157, 59
McClintock, J. E. & Remillard, R. A. 2006, in: W. H. G. Lewin & M. van der Klis (eds.), Compact Stellar X-ray Sources (Cambridge Univ. Press: Cambridge), p. 157
Stobbart, A.-M., Roberts, T. P. & Wilms, J. 2006, MNRAS, 368, 397
Swartz, D. A., Ghosh, K. K., Tennant, A. F. & Wu, K. 2004, ApJS, 154, 519

Black Holes from Stars to Galaxies – Across the Range of Masses
Proceedings IAU Symposium No. 238, 2006
V. Karas & G. Matt, eds.
© 2007 International Astronomical Union
doi:10.1017/S1743921307005078

On the nature of ultra-luminous X-ray sources from optical/IR measurements

Mark Cropper,[1] Chris Copperwheat,[1] Roberto Soria[2,1] and Kinwah Wu[1]

[1]Mullard Space Science Laboratory, University College London, Holmbury St Mary, Dorking, Surrey RH5 6NT, UK

[2]Centre for Astrophysics, Harvard Smithsonian Astrophysical Observatory, 60 Garden St, Cambridge, MA 02138, USA

Abstract. We present a model for the prediction of the optical/infra-red emission from ULXs. In the model, ULXs are binary systems with accretion taking place through Roche lobe overflow. We show that irradiation effects and presence of an accretion disk significantly modify the optical/infrared flux compared to single stars, and also that the system orientation is important. We include additional constraints from the mass transfer rate to constrain the parameters of the donor star, and to a lesser extent the mass of the BH. We apply the model to fit photometric data for several ULX counterparts. We find that most donor stars are of spectral type B and are older and less massive than reported elsewhere, but that no late-type donors are admissible. The degeneracy of the acceptable parameter space will be significantly reduced with observations over a wider spectral range, and if time-resolved data become available.

Keywords. Black hole physics – X-rays: galaxies – X-rays: stars – accretion, accretion discs

1. Introduction

Ultra-luminous X-ray sources (ULXs) are non-nuclear X-ray sources in nearby galaxies with inferred luminosity $>$few$\times 10^{39}$ ergs s^{-1}. This luminosity exceeds the Eddington luminosity of a $20 M_\odot$ black hole (BH) (an observational overview is available in Fabbiano, 2004). While these objects are generally agreed to be binary systems, the nature of their constituents is still controversial. Their emission could be as a result of sub-Eddington accretion rates onto intermediate mass black holes (IMBH) with masses $\sim 200 - 1000 M_\odot$, (Colbert & Mushotzky, 1999), super-Eddington accretion onto stellar-mass BH (Begelmann 2002, King 2001) or Eddington accretion onto BH with masses in the range $\sim 50 - 200\ M_\odot$ (Soria & Kuncic 2006).

Recently, reasonably secure optical counterparts for these systems have been identified, mostly using *HST* observations. This has opened a new channel of investigation into the nature of ULX. The optical/infrared emission is derived from the irradiated mass donor star and disk, so it is essential to model these appropriately in both the spectral and time domain if system parameters such as the mass and radius of the mass donor star and the mass of the BH are to be constrained. This paper describes such a model for the optical/infrared emission, and summarises some of the constraints that can be derived from its application to optical/infrared data. The objectives of this work are (a) to provide constraints on the possible optical counterparts of ULXs, eliminating those candidates which are inconsistent with the predicted colours/variability; (b) to determine the characteristics of the ULX constituent parts as accurately as possible; (c) to constrain the origin of ULXs and (d) to make predictions for future observations. More detailed expositions can be found in Copperwheat *et al.* (2005, 2006).

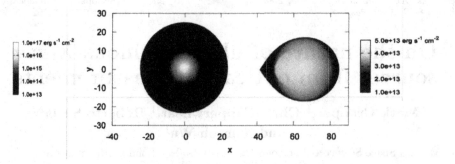

Figure 1. The variation in intensity $B(\tau)$ with $\tau = 2/3$ for (right) an irradiated O5V star and (left) a disk for a BH mass of $150M_\odot$. The plot shows the view onto the orbital plane with the labeled distances in units of $R_\odot$. Note that the intensity scales are logarithmic for the disk and linear for the star.

2. The model

The compact object in the model is a BH of mass in the range $10 - 1000M_\odot$. The mass donor star fills its Roche lobe, and accretion takes place through Roche lobe overflow into an accretion disk. We assume the disk to be a standard thin disk, tidally truncated at a radius 0.6 of that of the distance to the L_1 point. The mass donor star evolves according to the isolated star evolutionary tracks of Lejeune & Schaerer (2001). We ignore the effects of mass transfer on the star. We assume also that mass transfer is driven by the nuclear evolution of the mass donor. The model includes the Roche geometry, gravity and limb darkening, disk shadowing, radiation pressure according to the prescription of Phillips & Podsiadlowski (2002), the evolution of the companion and system orientation effects (inclination and binary phase). The X-ray irradiation is assumed to be isotropic.

The irradiation of the disk and star is handled according to a formulation by Wu *et al.* (2001) which is based originally on the grey stellar irradiation model of Milne (1926) and incorporates the different opacities of the irradiated surface to hard and soft X-rays (the X-ray hardness ratio is an input parameter). The effective temperature of the irradiated star or irradiated disk is a superposition of the irradiated and natural temperatures *i.e.* $T_{eff} = \left(\frac{\pi}{\sigma} B_x(2/3) + T_{unirr}^4 \right)^{1/4}$ where $B_x(\tau)$ is derived in Copperwheat *et al.* (2005). Figure 1 provides an example of the irradiated disk and star.

Using this model we can predict the different contributions of the constituents of the ULX as a function of (for example) BH mass and donor star spectral type, as shown in Figure 2. For high-mass BHs, the disk is large, and hence disk emission dominates. In addition, irradiation effects are much larger for late-type stars. We make predictions of the optical/infrared flux. We find that there is generally better system parameter discrimination at infrared wavelengths. We also predict the variability timescale. Because of the axial symmetry of our disks, these are dominated by the (modified) ellipsoidal variations from the Roche-lobe-filling donor stars. The variability timescale is typically days, dependent on BH mass.

The photometric predictions can be compared to observations of ULX counterparts in different wavebands, and acceptable model parameter regimes determined from χ^2 fitting to the observations. If only single-epoch photometric observations are available in 2 or 3 bands the model fits are under-constrained, leading to degenerate solutions. An example is shown in Figure 3(left). Nevertheless, even with limited photometric data much of the possible parameter space can be eliminated, and with additional wavelength coverage particularly in the infrared, degeneracies can be reduced further.

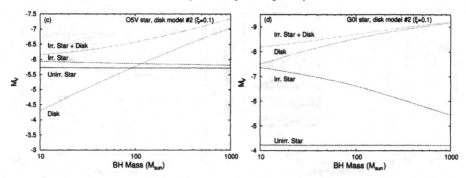

Figure 2. The V band absolute magnitudes for an un-irradiated and irradiated O5V star and accretion disk (left) and for a G0I supergiant with disk (right), shown as a function of BH mass. Here $L_x = 10^{40}$ ergs s^{-1}, $\cos i = 0.5$ and the star is at superior conjunction.

We add a further constraint for the χ^2 fitting using the mass transfer rate, which is determined by the evolution of the stellar radius (from the evolutionary tracks) compared with the evolution of the Roche lobe radius (Wu, 1997, Ritter, 1988). We measure the mass transfer rate from the X-ray luminosity assuming an accretion efficiency $\eta = 0.1$ appropriate for a BH, and select only those secondary stars which are evolving in radius on nuclear timescales in such a way as to provide the measured mass transfer rate.

3. Fits to data

We have gathered *HST* and VLT photometric data available up to mid-2006 on the most luminous ULX counterparts, where we can be reasonably certain that mass transfer is driven by Roche lobe overflow. The input data for M81 X-6, NGC 4559 X-7, M101 ULX-1, NGC 5408 ULX, Holmberg II ULX, NGC 1313 X-2 (C1) and NGC5204 ULX are given in table 1 of Copperwheat *et al.* (2006) and for M51 X5/9 in Copperwheat (2007). We fitted our model to these photometric data, using the additional mass transfer constraint from X-ray measurements.

Figure 3 (right) shows an example of the allowed parameter space in the donor star mass vs BH mass plane from the χ^2 fitting. We also produced similar plots for the donor star radius and donor star age, which we determined from the mass-radius relation from the evolutionary tracks. More details can be found in Copperwheat *et al.* (2006).

These fits provide the current spectral type of the mass donor star. By tracing to earlier times along the evolutionary tracks, the ZAMS spectral type can be predicted assuming the mass loss has not significantly altered the evolution.

4. Main outcomes

The optical/infrared emission from our binary ULX model is significantly different from models assuming un-irradiated companion stars and no disks, hence it is not adequate to assume standard colours from single stars to determine the mass donor characteristics in ULXs.

The model fits to the currently available data do not provide strong constraints on the BH mass, mainly because of the unknown orientation of the system. Depending on inclination, upper or lower limits can, however, be set. For example, the BH mass in NGC 5204 is < 240 $M_\odot$ for $\cos i = 0.5$, while that in NGC 1313 X-1 is < 100 $M_\odot$. The reason is that the signature of a disk can be almost entirely hidden for $\cos i = 0.0$, in which case little information can be obtained on the size of the primary Roche lobe.

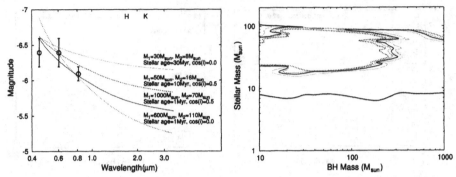

Figure 3. (left) Absolute magnitude as a function of wavelength for different combinations of BH (M_1) and donor star (M_2) and inclination i compared to photometric measurements for NGC5408 ULX, indicating the degeneracy of possible solutions when no mass transfer constraints are applied. (right) Acceptable parameter space for the mass donor star projected onto the donor mass/BH mass plane when the mass transfer constraint is included. Contours are at 68, 90 (solid line), 95 and 99% confidence levels.

The mass, radius and age of the donor star are, however, more tightly constrained. Typical ages range from $10^7 - 10^8$ years, with typical ZAMS masses $5 - 10$ $M_\odot$, with some up to 50 $M_\odot$. In general we find that the mass donor stars are less massive and older than generally quoted in the literature from less comprehensive modeling – this is to be expected given the effects of irradiation. We find that none of the systems are found to contain late-type mass donor stars, whatever disk component is admitted.

The preference for donor stars of spectral type B is interesting. The high mass transfer rates and modest donor star masses require that ULX lifetimes are short (this is true also for higher mass donors which in any case have short lifetimes). The duration that the donor star has been in contact with its Roche lobe is an important parameter for ULXs. If it has been in contact for some Myr, especially in the case of B-type ZAMS stars, then binary evolution models will be required.

The diagnostic capability of our model for the BH mass and donor star characteristics will improve significantly as more filter bands, and particularly time-resolved data become available.

Acknowledgements

R. Soria acknowledges support from a Marie Curie Fellowship from the EC.

References

Begelman, M. C. 2002, ApJ, 568, L97

Colbert, E. J. M. & Mushotzky, R. F. 1999, AJ, 519, 89

Copperwheat, C., Cropper, M., Soria, R. & Wu K. 2005, MNRAS, 362, 79

Copperwheat, C., Cropper, M., Soria, R. & Wu K. 2006, MNRAS, submitted

Copperwheat, C. 2007, PhD Thesis, University of London, submitted

Fabbiano, G. 2004, Rev. Mex. A. A. (Serie de Conferencias), 20, 46

King, A. R. 2002, MNRAS, 335, L13

Lejeune, T. & Schaerer D. 2001, A&A, 366, 538

Milne, E. A. 1926, MNRAS 87, 43

Phillips, S. N. & Podsiadlowski, P. 2002, MNRAS 337, 431

Ritter, H. 1988, A&A 202, 93

Soria, R. S. & Kuncic, Z. 2006, Adv. Space. Res. (Cospar special issue) submitted

Wu, K. 1997 Accretion Phenomena and Related Outflows; IAU Colloquium 163., ASP Conference Series Vol. 121; ed. D. T. Wickramasinghe; G. V. Bicknell & L. Ferrario (1997), p.283

Wu, K., Soria, R., Hunstead, R. W. & Johnston, H. M. 2001, MNRAS 320, 177

Black Holes from Stars to Galaxies – Across the Range of Masses
Proceedings IAU Symposium No. 238, 2006
V. Karas & G. Matt, eds.
© 2007 International Astronomical Union
doi:10.1017/S174392130700508X

Variability of ultraluminous X-ray sources in the Cartwheel Ring

Anna Wolter,[1] Ginevra Trinchieri[1] and Monica Colpi[2]

[1]INAF, Osservatorio Astronomico di Brera, via Brera 28, 20121 Milano, Italy

[2]Dipartimento di Fisica G. Occhialini, Università degli Studi di Milano Bicocca,
Piazza della Scienza 3, 20126 Milano, Italy

email: anna.wolter@brera.inaf.it, ginevra.trinchieri@brera.inaf.it, Monica.Colpi@mib.infn.it

Abstract. The Cartwheel is one of the most outstanding examples of a dynamically perturbed galaxy where star formation is occurring inside the ring–like structure. In previous studies with *Chandra*, we detected 16 Ultra Luminous X-ray sources lying along the southern portion of the ring. Their Luminosity Function is consistent with them being in the high luminosity tail of the High Mass X-ray Binaries distribution, but with one exception: source N.10. This source, detected with *Chandra* at $L_X = 1 \times 10^{41} \mathrm{erg\,s^{-1}}$, is among the brightest non–nuclear sources ever seen in external galaxies. Recently, we have observed the Cartwheel with *XMM-Newton* in two epochs, six months apart. After having been at its brightest for at least 4 years, the source has dimmed by at least a factor of two between the two observations. This fact implies that the source is compact in nature. Given its extreme isotropic luminosity, there is the possibility that the source hosts an accreting intermediate–mass black hole. Other sources in the ring vary in flux between the different datasets. We discuss our findings in the context of ULX models.

Keywords. Accretion, accretion disks – black hole physics – galaxies: individual (Cartwheel) – galaxies: stellar content – X-rays: binaries – X-rays: galaxies

1. Introduction

Very luminous off-nuclear X-ray sources in nearby galaxies are known since the *Einstein* satellite times. They have been named Ultra-Luminous X-ray sources (ULXs) because their isotropic X-ray luminosity is significantly higher than the Eddington limit for a solar mass black hole ($L_X \sim 1.4 \times 10^{38}$ erg s^{-1}). The name itself is only a phenomenological description of their L_X, since their nature is not clear yet. They do not appear to have an equivalent among Galactic sources, and this may be related to the low Star-Formation rate of the Milky Way since ULXs are mostly found associated with Star-Forming regions. Explanations of their nature involve beamed emission from an accreting stellar mass compact object, or super-Eddington emission, or isotropic accretion onto an intermediate–mass black hole.

An extraordinary example of ULX is the source N.10 detected in the narrow, gas-rich star–forming ring of the Cartwheel galaxy with isotropic luminosity of $L_{0.5-10\mathrm{keV}} \sim 1.3 \times 10^{41}$ erg s^{-1} (Wolter *et al.* 1999; Wolter & Trinchieri 2004 - hereafter WT04; Gao *et al.* 2003). This is the brightest of a number of very bright individual sources that also appear to reside in the ring, with isotropic luminosities in excess of $L_{0.5-10\mathrm{keV}} = 3 \times 10^{39}$ erg s^{-1} (WT04).

Source N.10 is one of the brightest ULXs known, with observed L_X comparable with the peak L_X of the best studied example of the class, source X-1 in M82 (Ptak & Griffiths 1999; Matsumoto *et al.* 2001; Kaaret *et al.* 2001; Strohmayer & Mushotzky 2003; Dewangan *et al.* 2006). The spectral and temporal variability of X-1 (typically

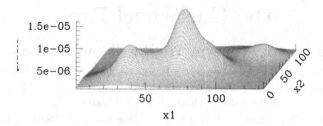

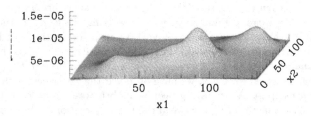

Figure 1. A 3-d representation of the two *XMM-Newton* pn observations [101] (top) and [201] (bottom) with a six months time separation. The smoothed images have been normalized to the respective exposure time. It is evident that the peak located at the position of source N.10 is very less prominent in the second epoch.

bursts of about a month duration) have led to the estimate of $200M_\odot$ for the accreting object responsible for the X-ray emission (Dewangan *et al.* 2006).

The lack of detailed information on variability for source N.10 in the Cartwheel has instead prevented up to now the exclusion of an extended nature. We present here new *XMM–Newton* observations which have now confirmed its compact nature.

2. *XMM–Newton* observations

We have observed the source with *XMM-Newton* in two epochs (December 2004 and May 2005). A detailed description of the results obtained for source N.10 is presented in Wolter, Trinchieri & Colpi (2006). A 3-d representation of the two smoothed datasets is plotted in Figure 1.

The detection of variability is important because it is our strongest proof that we are detecting a single source, namely an accreting binary, and not a collection of less luminous unresolved X-ray sources. It is also relevant because it can provide essential information about the details of the accretion process and, most important, about the masses involved in the process (donor and accretor), giving the possibility to infer the presence of an intermediate-mass black hole.

The spectrum in the two different *XMM-Newton* observation (the first, [101] of 24 ksec, and the second [201], of 42 ksec exposure) is shown in Figure 2. The *XMM-Newton* PSF is such that the extraction radius (even if we used a smaller than customary value of $10''$) contains part of the ring emission, which is described by the sum of a diffuse gas component (thermal spectrum) and the contribution from the unresolved point-like sources (X-ray binaries; power law spectrum. See WT04 for details). Even if the statistics is low, therefore, a multicomponent spectrum has been used to fit the emission: we have

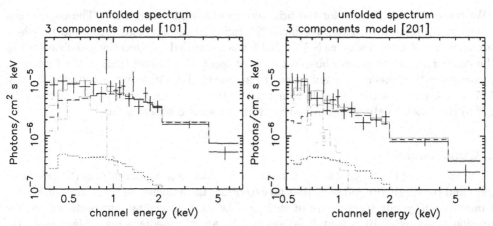

Figure 2. Unfolded spectrum, from pn data only for clarity, showing the three components (green dot-dashed line = gas; blue short dashed line = unresolved binaries; red long dashed line = ULX) at the best fit values. The difference between the left [101] and the right [201] datasets is only the normalization of the 'ULX' component [besides the binning scheme due to the different statistics at the two epochs].

added a power law describing the ULX power law emission (as seen in *Chandra*) to the two ring components. The spectra of the two observations 6 months apart clearly show that the thermal component (gas in the ring) is constant in time and that the factor of 2 variability of the ULX is quite evident.

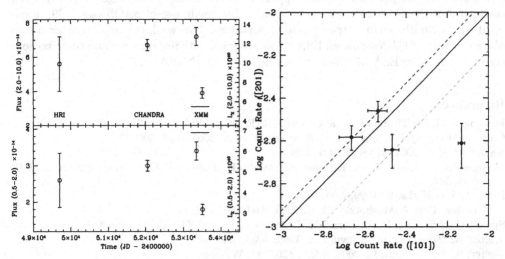

Figure 3. Left: long term light curve in the soft (lower panel) and hard (upper panel) energy band, over an interval of about 10 years. The two *XMM–Newton* points that define the variation are not subject to cross-calibration uncertainties. All fluxes are computed with the same spectral shape, i.e. power law with $\Gamma = 1.6$ and $N_H = 3.6 \times 10^{21}$ cm^{-2}; see fit to *Chandra* data in WT04. Right axis reports luminosities computed assuming the Cartwheel distance. From Wolter *et al.* 2006. Right: flux comparison of the brightest sources in the first and second epoch.

The long term behavior of N.10 shows a nearly constant luminosity for about 10 years (see Figure 3 – left), and then a rapid dimming of at least a factor of 2 in 6 months. This is quite different from the variability pattern observed in the best studied bright ULX, namely X–1 in the starburst galaxy M82 (Dewangan *et al.* 2006).

We compared count rates for the brightest source neighboring N.10. The comparison is shown in Figure 3 (Right). We have determined that the source next to it, which corresponds to *Chandra* sources N.13 & N.14, has not varied between the two observations (the count rate in the second observation is at most 15% higher than in the first one). If we assume that this is constant, then the source to the NW, corresponding to N.16 & N.17, is also constant (the same 15% increase in the count rate), but the SE source (N.7 & N.9) has faded to about half its strength in the second observation.

3. Conclusions

The whole ring of the Cartwheel consists of bubbles and condensations (Struck *et al.* 1996) and the neighborhood of N.10 is no exception; the association with an environment of massive and young stars is almost certain. The most appealing interpretation for the emission of source N.10 is that it is powered by an intermediate–mass black hole. We cannot rule out the presence of Super-Eddington accretion. Other sources have been observed to vary in the same time frame of N.10. Further studies on variability will help us in distinguishing between different accretion modes. It is evident that the Cartwheel ring is an excited (and exciting!) environment, in which star formation is violently at work. The low metallicity measured for the optical gas might be a key ingredient for the formation of sources capable of emitting very high X-ray luminosity, whatever the mechanism is.

Acknowledgements

We acknowledge partial financial support from the Italian Space Agency (ASI) under contract ASI-INAF I/023/05/0. This research has made use of SAOImage DS9, developed by the Smithsonian Astrophysical Observatory. This work is based on observations obtained with *XMM-Newton*, an ESA science mission with instruments and contributions directly funded by ESA Member States and the USA (NASA).

References

Dewangan, G. C., Titarchuk, L. & Griffiths, R. E. 2006, ApJ, 637, L21

Gao, Y., Wang, Q. D., Appleton, P. N. & Lucas, R. A. 2003, ApJ, 596, 171

Kaaret, P. *et al.* 2001, MNRAS, 321, L29

Matsumoto, H., Tsuru, T. G., Koyama, K., Awaki, H., Canizares, C. R., Kawai, N. *et al.* 2001, ApJ, 547, 25

Ptak, A. & Griffiths, R. 1999, ApJ, 517, L1

Strohmayer, T. E. & Mushotzky, R. F. 2003, ApJ, 586, L61

Struck, C., Appleton, P. N., Borne, K. D. & Lucas, R. A. 1996, AJ, 112, 1868

Wolter, A., Trinchieri, G. & Iovino, A. 1999, A&A, 342, 41

Wolter, A. & Trinchieri, G. 2004, A&A, 426, 787, WT04

Wolter, A., Trinchieri, G. & Colpi, M. 2006, MNRAS, 373, 1627

Supermassive Black Holes and Their Galaxies I

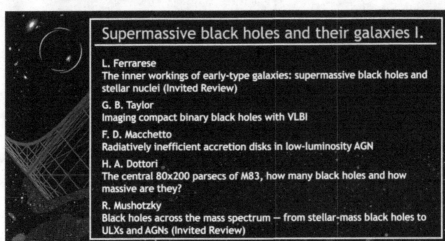

Supermassive black holes and their galaxies I.

L. Ferrarese
The inner workings of early-type galaxies: supermassive black holes and stellar nuclei (Invited Review)

G. B. Taylor
Imaging compact binary black holes with VLBI

F. D. Macchetto
Radiatively inefficient accretion disks in low-luminosity AGN

H. A. Dottori
The central 80x200 parsecs of M83, how many black holes and how massive are they?

R. Mushotzky
Black holes across the mass spectrum — from stellar-mass black holes to ULXs and AGNs (Invited Review)

Night view of the Old Town from the Clementinum Astronomical Tower, originally *Observatorium Astronomicum.*

Black Holes from Stars to Galaxies – Across the Range of Masses
Proceedings IAU Symposium No. 238, 2006
V. Karas & G. Matt, eds.

© 2007 International Astronomical Union
doi:10.1017/S1743921307005108

The inner workings of early-type galaxies: cores, nuclei and supermassive black holes

Laura Ferrarese and Patrick Côté

National Research Council of Canada, Herzberg Institute of Astrophysics, 5071 West Saanich Road, Victoria, BC, V9E 2E7, Canada

email: laura.ferrarese@nrc-cnrc.gc.ca, patrick.cote@nrc-cnrc.gc.ca

Abstract. Recent years have seen dramatic progress in the study of the core and nuclear properties of galaxies. The structure of the cores has been shown to vary methodically with global and nuclear properties, as cores respond to the mechanisms by which galaxies form/evolve. The dynamical centers of galaxies have been found capable of hosting two seemingly disparate objects: supermassive black holes (SBHs) and compact stellar nuclei. In a drastic departure from previous beliefs, it has been discovered that both structures are common: galaxies lacking SBHs and/or stellar nuclei are the exception, rather than the norm. This review explores the connection between cores, SBHs and stellar nuclei in early-type galaxies, as revealed by the ACS Virgo Cluster Survey.

Keywords. Galaxies: elliptical and lenticular, cD – galaxies: dwarf – galaxies: fundamental parameters – galaxies: photometry – galaxies: structure – galaxies: nuclei – galaxies: bulges

1. Introduction

Cores — a term which we will use loosely to describe the central few hundred parsecs of galaxies — act as recording devices of the process of galaxy evolution. Because dynamical timescales are shorter than elsewhere in the galaxy, the morphology, dynamics, and stellar populations of the core regions contain fossil records of the gas, dust and stellar systems (satellite galaxies, star clusters, etc) that were drawn to the bottom of the gravitational potential well over cosmic timescales. Moreover, it has now become clear that the core and global properties of galaxies are linked through a number of intriguing scaling relations. In particular, those involving supermassive black holes (SBHs) – which, in high-luminosity galaxies at least, appear to be a generic subcomponent of the cores – underscore the importance of nuclear feedback in galaxy evolution (Ferrarese & Merritt 2000; Gebhardt *et al.* 2000; Graham *et al.* 2001; Ferrarese 2002; Haring & Rix 2004).

The study of galactic cores received a tremendous push forward with the deployment of the Hubble Space Telescope. HST images brought into focus a plethora of structural features, including nuclear stellar disks, bars, "evacuated" regions (possibly scoured out by the evolution of SBH binaries), and an entire spectrum of dust features, from small irregular patches to large, highly organized disks. In early-type galaxies, cores were found to fall into two distinct classes: those with a shallow surface brightness profile, and those whose surface brightness kept increasing, in roughly a power-law fashion, to the inner-most radius accessible given the resolution limit of the instrument (Ferrarese *et al.* 1994; Lauer *et al.* 1995,2005; Ravindranath *et al.* 1996; Rest *et al.* 2001). Galaxies falling into the first class have come to be known (somewhat unfortunately) as "core" galaxies, and those falling into the second class as "power-law". The division between the two classes was found to correlate neatly with galaxy luminosity, with core galaxies being exclusively bright giant ellipticals, while fainter galaxies are (with few exceptions) classified as power-laws. The stark separation between the two classes has been attributed to differing formation/evolutionary histories. Power-law galaxies have been claimed to be the result of dissipation during galaxy formation, with some authors further claiming that all power-law galaxies host stellar disks; while core-galaxies are believed to be the result of the merging of fainter (power-law) galaxies, and of their central SBHs.

SBHs have now been detected dynamically in roughly three dozen galaxies (see Ferrarese & Ford 2005 for a review). Indeed, balancing the SBH mass function from the QSO epoch to the present day requires virtually all local galaxies brighter than a few $0.1L^*$ to host a SBH (e.g. Shankar et al. 2004; Marconi et al. 2004). Recent observations, however, have made it clear that SBHs are not the only objects to enjoy a privileged position at a galaxy's dynamical center. Stellar nuclei, or nuclear star clusters, have recently been detected in a large fraction (70% to 90%) of galaxies of all Hubble types and luminosities (Böker et al. 2002; Lotz et al. 2004; Grant et al. 2005). Follow-up spectroscopy of stellar nuclei in spiral galaxies (Walcher et al. 2005, 2006; Rossa et al. 2006) has shown them to be massive, dense objects akin to compact star clusters. Luminosity-weighted ages range from 10 Myr to 10 Gyr, younger than the age of the galactic disk, and with the younger clusters found preferentially in the later type spirals.

This review explores the connection between cores, nuclei and supermassive black holes in light of recent results from the ACS Virgo Cluster Survey (ACSVCS), the largest HST imaging survey to date of early-type galaxies in the local universe. This survey was designed specifically to provide an unbiased characterization of the core structure of early-type galaxies using well understood selection criteria.

2. The ACS Virgo Cluster Survey

The ACSVCS (Côté et al. 2004) consists of HST imaging for 100 members of the Virgo Cluster, supplemented by imaging and spectroscopy from WFPC2, Chandra, Spitzer, Keck and KPNO. The program galaxies span a range of ≈ 460 in B-band luminosity and have early-type morphologies: E, S0, dE, dE,N or dS0. All images were taken with the Advanced Camera for Surveys (ACS; Ford et al. 1998) using a filter combination roughly equivalent to the g and z bands in the SDSS photometric system. The images cover a $\approx 200'' \times 200''$ field with $\approx 0''.1$ resolution ($\approx$8pc at the distance of Virgo, $D = 16.5$ Mpc).

This review summarizes results from the subset of ACSVCS papers which deal with the morphology, isophotal parameters and surface brightness profiles for early-type galaxies (Ferrarese et al. 2006a), their central nuclei (Côté et al. 2006) and scaling relations for nuclei and SBHs (Ferrarese et al. 2006b). Other ACSVCS papers have discussed the data reduction pipeline (Jordán et al. 2004a), the connection between low-mass X-ray binaries and globular clusters (Jordán et al. 2004b), the measurement and calibration of surface brightness fluctuation distances (Mei et al. 2005ab), the connection between globular clusters and ultra-compact dwarf galaxies (Haşegan et al. 2005), the luminosity function, color distributions and half-light radii of globular clusters (Jordán et al. 2005, 2006; Peng et al. 2006a), and diffuse star clusters (Peng et al. 2006b).

3. The core structure of early-type galaxies

Over the three-decade radial range between a few tens of parsecs and several kiloparsecs (i.e. to the largest radii covered by the ACSVCS images), the surface brightness profiles of the ACSVCS early-type galaxies are well described by a simple Sérsic model (Sérsic 1968) with index n increasing steadily with galaxy luminosity. By contrast, a "Nuker" model (Lauer et al. 1995; 2005) fails to reproduce the curved surface brightness profiles that real galaxies exhibit and often grossly overestimates the luminosity at large and intermediate radii. Notable, and systematic, deviations from a Sérsic model are however registered in the innermost regions. For eight of the 10 brightest galaxies ($M_B \lesssim -20.3$) the measured inner profiles (typically within $0''.5$ to $2''.5$, corresponding to 40 to 200pc) are shallower than expected based on an inward extrapolation of the Sérsic model constrained by the region beyond. For these galaxies, the surface brightness profile is best fitted by joining the outer Sérsic profile to an inner, shallower, power-law component (such composite models are referred to as "core-Sérsic", Graham *et al.* 2003; Trujillo *et al.* 2004),

The opposite is seen in fainter galaxies, where $\approx 80\%$ of the sample galaxies show an upturn, or inflection, in the surface brightness profile within (typically) the innermost few tens of parsec region (see Figure 1 of P. Côté, these proceedings). The upturn signals the presence of a stellar nucleus that is most likely structurally distinct from the main body of the underlying galaxy. When a nucleus is present, the inner surface brightness is, by definition, larger than the inward extrapolation of the outer Sérsic model.

The picture that has emerged from the ACSVCS is therefore one in which, in moving down the luminosity function from giant to dwarf early-type galaxies, the innermost 100-parsec region undergoes a systematic and smooth transition from light (mass) "deficit" (relative to the overall best fitting Sérsic model) to light "excess". Although the subset of ACSVCS "core-Sérsic" galaxies coincides with the galaxies that were classified as "cores" in previous investigations, there are critical differences between our study and the ones that preceded it. Compared to previous work, the ACSVCS has emphasized the role of stellar nuclei; the fact that the frequency, luminosities and sizes of the ACSVCS nuclei are in remarkable agreement with those measured (using different techniques and assumptions) by recent independent surveys in both early and late type galaxies, supports the robustness of the ACSVCS analysis. Recognizing the nuclei as separate components has allowed us to revisit the issue of the division of early-type galaxies into "core" and "power-law" types. Such division was based on the fact that the distribution of the logarithmic slopes, $\gamma = -d\log I/d\log r$, of the inner surface brightness profile had been found to show various degrees of bimodality. Such bimodality is absent in the γ distribution of the ACSVCS galaxies.

In agreement with previous studies, in galaxies brighter than $M_B \approx -20.3$, γ is indeed found to decrease with galaxy luminosity, while the opposite trend is seen for fainter galaxies, however, the transition is smooth, rather than abrupt. In a further departure from previous studies, we find that the low-γ end of the distribution (corresponding to the galaxies with the shallowest surface brightness profiles) is occupied mostly by the faintest dwarf systems, rather than by the brightest giant ellipticals. We note here that the absence of a bimodal behaviour in γ does not automatically invalidate a picture in which brighter galaxies evolve mainly through merging while fainter systems are largely left untouched. Indeed, such picture does not necessarily explain the perceived stark separation of galaxies in "core" and "power-law" types for which it was formulated. The extent to which structural parameters are compromised by merging of galaxies (and their supermassive black holes) depends on the the the masses of the progenitors (e.g., Bournaud *et al.* 2005; Milosavljevic & Merritt 2001); given a continuous distribution for the latter, combined with a galaxy luminosity function heavily biased towards low-mass systems, allows for the possibility of a smooth transition between progenitors and merger products.

4. Compact stellar nuclei in the ACSVCS

At the outset of the ACSVCS, it was known that at least $\approx 25\%$ of the program galaxies contained nuclei, based on ground-based, photographic classifications from the

VCC (Binggeli *et al.* 1985). Stellar nuclei in the ACSVCS images were identified using a variety of indicators, including direct inspection of the ACS frames, color changes in the $g - z$ color images, and sudden upturns in the surface brightness profiles. Based on these criteria, 60 to 80% of ACSVCS galaxies host stellar nuclei (with the precise fraction depending on galaxy magnitude), in line with the fraction reported in both spiral and elliptical galaxies based on recent high-resolution surveys (Carollo, Stiavelli & Mack 1998; Matthews *et al.* 1999; Böker *et al.* 2002, 2004; Balcells *et al.* 2003; Lotz *et al.* 2004; Graham & Guzman 2003; Grant *et al.* 2005), but a factor ~ 3 higher than expected based on the VCC.

Our analysis shows that surface brightness selection biases in the VCC data are largely responsible for the difference. In galaxies with central g-band surface brightnesses lower than ≈ 20.5 mag arcsec^{-2}, the agreement between the ACSVCS and VCC is nearly perfect, while above 19.5 mag arcsec^{-2}, virtually all nuclei were missed by the ground-based survey. Selection effects might, of course, still be at work in the ACSVCS sample, implying that our estimate for the frequency of nucleation, $f_n \approx 60 - 80\%$, is almost certainly a lower limit on the true frequency. As will be discussed shortly, the luminosity and half-light radii of stellar nuclei correlate strongly with the magnitude of the host galaxy; it is therefore possible, for each galaxy classified as non-nucleated, to determine whether a nucleus, if present, could have gone undetected.

Based on these tests, with very few exceptions, the only galaxies for which the existence of a nucleus can be confidently excluded are those brighter than $M_B \approx -20.3$ mag. These are the same galaxies with central light "deficits" for which the surface brightness profiles are well represented by "core-Sérsic" rather than Sérsic models (§3).

Scaling relations for stellar nuclei. For 51 galaxies in the ACSVCS the sharp upturn in the surface brightness within $\approx 1''$ is conspicuous enough that a measurement of the nucleus' photometric and structural parameters is possible. These parameters are recovered by adding a central King model (King 1966) to the underlying Sérsic component when fitting the brightness profile.

The luminosity function of nuclei follows a Gaussian distribution with dispersion in the range $1.5 - 1.8$ mag and peak absolute $g-$band magnitude ≈ -10.7 mag, a factor $\approx 25\times$ brighter than the peak of the globular cluster luminosity function. With a half-dozen exceptions, nuclei in the ACSVCS galaxies are spatially resolved (thereby ruling out a non-thermal AGN origin), with individual sizes ranging from 62 pc down to the resolution limit of 2 pc, and a median half-light radius of $\langle r_h \rangle = 4.2$ pc. Unlike globular clusters, for which size is independent of magnitude, nuclear sizes are found to scale with luminosity according to the relation $r_h \propto \mathcal{L}^{0.50\pm0.03}$ (Figure 1, left panel).

One of the most popular models for the formation of nuclei posits that globular clusters are drawn to the bottom of the potential well by dynamical friction, where they then coalesce (e.g. Tremaine *et al.* 1975). While the size–magnitude relation observed for the ACSVCS nuclei is consistent with the prediction of such model (Bekki *et al.* 2004), a more complex picture is suggested by the observations that nuclei, again unlike globular clusters, display a color–magnitude relation (Figure 1, right panel). Monte Carlo simulations show that mergers of globular clusters through dynamical friction are unable to explain the observed color–magnitude relation; indeed the existence of this relation suggests that the chemical enrichment of nuclei is governed by local or internal factors, along the lines of the various "gas accretion" models (e.g. Mihos & Hernquist 1996). Note that the nuclei's color–magnitude relation is better defined for galaxies fainter than $M_B \approx -17.6$ mag, while the nuclei belonging to brighter galaxies frequently show very red colors, $(g - z) \sim 1.5$. If confirmed (measurements are more uncertain for these nuclei,

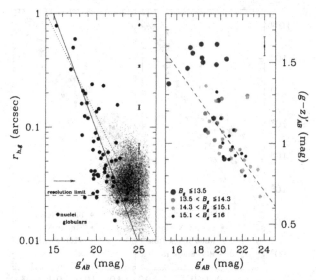

Figure 1. Left panel: the size–magnitude relation, in the g–band, for the 51 ACSVCS nuclei for which structural parameters could be measured (solid circles) and the sample of globular clusters from Jordán *et al.* (2005) (points). Typical error-bars for the nuclear sizes are shown in the right hand side of the panel. The arrow shows the "universal" half-light radius of $0\overset{''}{.}033$ (≈ 2.7 pc) for globular clusters in Virgo (Jordán *et al.* 2005), while the dashed line shows a conservative estimate for the resolution limit of the ACS images. The solid diagonal line shows the best fitting relation for the nuclei ($r_h \propto L^{0.5}$), while the dotted line shows the prediction of the globular cluster merging model of Bekki *et al.* (2004). Right panel: color–magnitude diagram for the ACSVCS nuclei. The size of the symbol for the nuclei is proportional to the magnitude of the host galaxy, as shown in the legend. The dashed line shows the best fit relation for the nuclei of galaxies fainter than $B_T = 13.5$ mag.

due to the high underlying galaxy surface brightness), this observation might suggest that these nuclei may constitute a separate class following a different formation route.

A third model, namely nuclear formation through two-body relaxation around a black hole, is inconsistent with the observation that nuclei are spatially resolved in most of the ACSVCS galaxies. Nuclei formed through this mechanism are predicted to extend to $\approx$ 1/5th of the SBH sphere of influence (e.g. Merritt & Szell 2005), and would therefore be spatially unresolved by the ACS in all of the ACSVCS galaxies, in clear contradiction with our observations. Finally, we note that the luminosity function and size distribution of the ACSVCS nuclei shows remarkable agreement with those of the "nuclear star clusters" detected in spiral galaxies (Böker *et al.* 2002, 2004). This points to a formation mechanism for the nuclei that is largely independent on both intrinsic and extrinsic factors, such as host magnitude and Hubble type, and immediate environment.

5. Stellar Nuclei and Supermassive Black Holes

The ubiquity of SBHs and stellar nuclei, and their unique location at the dynamical centers of galaxies, are reasons to suspect that a connection between the two might exist. The ACSVCS data strongly support this view (Ferrarese *et al.* 2006b). The left panel of Figure 2 shows a recent characterization of the relation between the masses of SBHs (black circles) and the stellar velocity dispersion of the host bulge, originally discovered by Ferrarese & Merritt (2000) and Gebhardt *et al.* (2000).

The $\mathcal{M} - \sigma$ relation is one of the tightest, and therefore most fundamental, of the scaling relations for SBHs, and has been used extensively to constrain the joint evolution of SBHs and galaxies (e.g., Haehnelt 2004 and references therein). The ACSVCS stellar

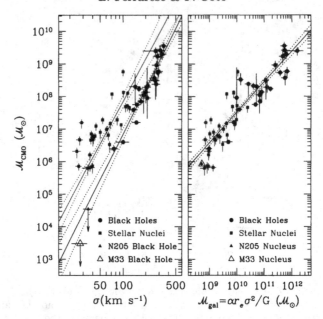

Figure 2. Left panel: mass of the CMO (stellar nuclei as red squares and SBHs as black circles) plotted against the velocity dispersion of the host galaxy. The solid red and black lines show the best fits to the nuclei and SBH samples, respectively, with 1σ confidence levels shown by the dotted lines. Right panel: CMO mass plotted against galaxy virial mass. The solid line is the fit obtained for the combined nuclei and SBH sample.

nuclei (shown as red squares) obey an $\mathcal{M} - \sigma$ relation with the same slope, although different normalization, as the one defined by SBHs. This is a notable and unexpected finding, suggesting close similarities in the formation and evolutionary history of these two radically different structures (McLaughlin *et al.* 2006). Furthermore, when σ is combined with the effective radius r_e to produce a measure of the galaxy's virial mass, $\mathcal{M}_{gal} \propto \sigma^2 r_e/G$, SBHs and stellar nuclei are found to obey an identical $\mathcal{M}-\mathcal{M}_{gal}$ relation (right panel of Figure 2; see also Côté *et al.* 2006). Remarkably, the same relation is also found to hold in spiral galaxies (Rossa et al. 2006) and to extend to dEs as faint as $M_B \approx -11.7$ mag (Wehner & Harris 2006).

These findings can be summarized as follows: a constant fraction, $\mathcal{M}_{CMO}/\mathcal{M}_{gal} \approx$ 0.3%, of a galaxy total mass is used in the formation of a nuclear structure, or "central massive object" (CMO). This holds true for galaxies spanning a factor 10^4 $M_\odot$ in mass, all Hubble types, luminosities and environments. In spite of their obviously different nature, SBHs and stellar nuclei may be nothing but complementary incarnations of CMOs – they likely share a common formation mechanism and follow a similar evolutionary path throughout their host galaxy's history. From the perspective of a theoretical framework of galaxy evolution, the commonalities between SBHs and stellar nuclei imply that both are equally important constraints on galaxy formation models: as the characterization of SBHs has been instrumental in furthering our understanding of galaxy evolution (via AGN feedback), so too promises to be the characterization of stellar nuclei (via superwinds and stellar feedback).

Several questions remain unanswered at this stage. Perhaps the most intriguing is whether the formation of SBHs and stellar nuclei are mutually exclusive. Nuclei are not present in the brightest ACSVCS galaxies, the prototypical objects in which SBHs are expected to reside, and for which a "mass deficit" has been attributed to the evolution of supermassive black hole binaries (Milosavljevic & Merritt 2001). At the other extreme of the luminosity range spanned by the ACSVCS galaxies, NGC205 and M33, for which there is no evidence of a SBH (Merritt *et al.* 2001; Gebhardt *et al.* 2001; Valluri *et al.* 2003), host stellar nuclei that follow the same scaling relations as the nuclei detected in the ACSVCS galaxies (Figure 2, right panel).

It is possible that nuclei form in every galaxy, but are subsequently destroyed in the brightest system as a consequence of the evolution of SBH binaries. Alternatively, it is possible that collapse to a SBH takes place preferentially in the brightest galaxies, while in fainter systems, the formation of a stellar nucleus is the most common outcome. In the latter case, nuclei could represent "failed black holes" – low-mass counterparts of the SBHs detected in the brightest galaxies.

The ACSVCS collaboration is currently pursuing several programs aimed at studying the dynamics and stellar population of the ACSVCS galaxies and nuclei; a similar investigation is underway for a sample of 43 early-type galaxies in the Fornax Cluster (Jordán *et al.* 2006b, in preparation). These projects promise to shed further light on the core structure of early-type galaxies, their nuclei and their relationships to SBHs.

Appendix

In a recent submission to astro-ph (astro-ph/0609762), Lauer *et al.* have reiterated their claim for a bimodal distribution of the inner profile slopes γ, and have questioned the reliability of the ACSVCS results. Their assertions are based on a heterogeneous sample of early-type galaxies — at widely differing distances and observed with a variety of instruments/filters/spatial resolutions — whose surface brightness profiles have been parametrized as "Nuker" laws (which has been shown to produce biased results and to be inadequate in describing the full extent of the surface brightness profiles of early-type galaxies; e.g., Graham *et al.* 2003).

In their posting, Lauer *et al.* further claim that the parametric form used by Ferrarese *et al.* (2006a) to fit the profiles of the ACSVCS galaxies forces the introduction of *ad hoc* stellar nuclei. These claims are difficult to understand in light of the fact that Lauer *et al.* themselves define nuclei as an inner excess against an adopted (Nuker) model, despite the fact that such models fail to match the data at small *and* large scales. More details on the gross misrepresentation of the ACSVCS data and analysis by Lauer *et al.* can be found on the ACSVCS website: http://www.cadc.hia.nrc.gc.ca/community/ACSVCS/

Acknowledgements

More rewarding than the results themselves has been to work with a great team. The friendship, hard work and dedication of each member of the ACSVCS team is very gratefully acknowledged.

References

Bekki, K., *et al.* 2004, ApJ, 610, L13
Best, P.N. *et al.* 2006, MNRAS, 368, 67
Binggeli, B., Sandage A. & Tammann G. A. 1985, AJ, 90, 1681
Böker, T. *et al.* 2002, AJ, 123, 1389
Böker, T. *et al.* 2004, AJ, 127, 105
Bournaud, F., Jog, C. J. & Combes, F. 2005, A&A, 437, 69
Côté, P. *et al.* 2004, ApJS, 153, 223
Côté, P. *et al.* 2006, ApJS, 165, 57
Ferrarese, L. & Merritt, D. 2000, ApJL, 539, 9
Ferrarese, L. 2002, ApJ, 578, 90
Ferrarese, L., Ford, H. C. 2005, Sp.Sc.Rev.,116, 523

Ferrarese, L. *et al.* 1994, AJ, 108, 1598
Ferrarese, L. *et al.* 2006a, ApJS, 164, 334
Ferrarese, L. *et al.* 2006b, ApJL, 644, L21
Gebhardt, K. *et al.* 2000, ApJ, 539, L13
Graham, A. *et al.* 2001, ApJ, 563, L11
Graham, A. W., Erwin, P., Trujillo, I. & Asensio, R. A. 2003, AJ, 125, 2951
Grant, N. I. *et al.* 2005, MNRAS, 363, 1019
Haehnelt, M., 2004, in oevolution of Black Holes and Galaxies 405
Haring, N., Rix, H.-W. 2004, ApJ, 604, 89
Hasegan, M. *et al.* 2005, ApJ, 627, 203
Jordán, A. *et al.* 2004a, ApJS, 154, 509
Jordán, A. *et al.* 2004b, ApJ, 613, 279
Jordán, A. *et al.* 2005, ApJ, 634,1002
Jordán, A. *et al.* 2006, ApJL, in press
Lauer, T. *et al.* 1995, AJ, 110, 2622
Lauer, T. *et al.* 2005, AJ, 129, 2138
Lotz, J., Miller, B. W & Ferguson, H. C. 2004, ApJ, 613, 262
Marconi, A. *et al.* 2004, MNRAS, 351, 169
McLaughlin, D. E., King, A. R. & Nayakshin, S. 2006, ApJL, in press (astro-ph/0608521)
Mei, S. *et al.* 2005a, ApJS, 156, 113
Mei, S. *et al.* 2005b, ApJ, 625, 121
Merritt, D., Ferrarese, L. & Joseph, C. 2001, Science, 293, 1116
Merritt, D. & Szell, A. 2005, ApJ, 648, 890
Milosavljevic, M. & Merritt, D. 2003, ApJ, 596, 860
Mihos, C. & Hernquist, L. 1996, ApJ, 464, 461
Peng, E. W. *et al.* 2006a, ApJ, 639, 95.
Peng, E.W. *et al.* 2006b, ApJ, 639, 838.
Ravindranath, S. *et al.* 2001, AJ, 122, 653
Rest, A. *et al.* 2001, AJ, 121, 2431
Rossa, J. *et al.* 2006, AJ, 132, 1074
Shankar, F. *et al.* 2004, MNRAS, 354, 1020
Tremaine, S. D., Ostriker, J. P. & Spitzer, L., Jr. 1975, ApJ, 196, 407
Valluri, M., Ferrarese, L., Merritt, D. & Joseph, C. 2005, ApJ, 628, 137
Walcher, C. J. *et al.* 2005, ApJ, 618, 327
Walcher, C. J. *et al.* 2006, ApJ, in press (astro-ph/0604138)
Wehner, E. & Harris, W. E. 2006, ApJL, 644, L17

DEBORAH DULTZIN-HACYAN: Could you explain what kind of (long-slit) spectrum one has in the nucleus?

LAURA FERRARESE: Indeed we use long-slit spectra. Even with HST these nuclei are barely resolved. Therefore, separating nuclear spectra from the galaxy is rather demanding. For the time being the nucleus definition is based on morphology.

MITCHELL BEGELMAN: The SLOAN survey of type 2 AGNs seems to find a transitions galaxy mass below which there is no activity. How does your transition mass between nuclei and black holes compare to that mass?

LAURA FERRARESE: It is of about the same value, a few times $10^{10} M_\odot$.

Black Holes from Stars to Galaxies – Across the Range of Masses
Proceedings IAU Symposium No. 238, 2006
V. Karas & G. Matt, eds.
© 2007 International Astronomical Union
doi:10.1017/S174392130700511X

Imaging compact supermassive binary black holes with Very Long Baseline Interferometry

G. B. Taylor,[1] C. Rodriguez,[1] R. T. Zavala,[2] A. B. Peck,[3] L. K. Pollack[4] and R. W. Romani[5]

[1]Department of Physics and Astronomy, Univ. of New Mexico, Albuquerque, NM 87131, USA

[2]United States Naval Observatory, Flagstaff Station, AZ 86001, USA

[3]Harvard-Smithsonian CfA, SMA Project, 645 N. A'ohoku Pl, Hilo, HI 96721, USA

[4]Department of Astronomy & Astrophysics, University of California at Santa Cruz, Santa Cruz, CA 95064, USA

[5]Department of Physics, Stanford University, Stanford, CA 94305 USA

email: gbtaylor@unm.edu

Abstract. We report on the discovery of a supermassive binary black-hole (SBBH) system in the radio galaxy 0402+379, with a projected separation between the two black holes of just 7.3 pc. This is the most compact SBBH pair yet imaged by more than two orders of magnitude. These results are based upon multi-frequency imaging using the Very Long Baseline Array (VLBA) which reveal two compact, variable, flat-spectrum, active nuclei within the elliptical host galaxy of 0402+379. Multi-epoch observations from the VLBA also provide constraints on the total mass and dynamics of the system. The two nuclei appear stationary while the jets emanating from the weaker of the two nuclei appear to move out and terminate in bright hot spots. The discovery of this system has implications for the number of compact binary black holes that might be sources of gravitational radiation. The VLBI Imaging and Polarimetry Survey (VIPS) currently underway should discover several more SBBHs.

Keywords. Black hole physics – gravitational waves – galaxies: active – radio continuum: galaxies

---

1. Introduction

Given that most galaxies harbor supermassive black holes at their centers, and that galaxy mergers are common, binary black holes should likewise be common. Yet very few systems have been found, perhaps because they proceed rapidly to parsec-scale separations which cannot be resolved by current X-ray or optical telescopes. Fortunately, in the case where both black holes are radio loud, they can be imaged using Very Long Baseline Interferometry (VLBI). An understanding of the evolution and formation of these systems is important for an understanding of the evolution and formation of galaxies in general.

Our ability to resolve both supermassive black holes in any given binary system depends on the separation between them, on their distance from Earth, and on the resolving power of the telescope used. It is believed that the longest timescales in the evolution of a supermassive binary black hole system leading up to coalescence is the stage in which the system is closely bound ($\sim 0.1 - 10$ pc), meaning that in most of these systems the black hole pair can only be resolved by VLBI observations, which provides resolutions of milliarcseconds and finer.

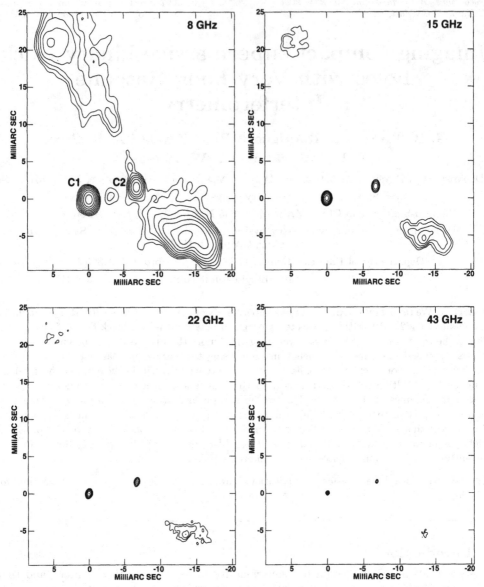

Figure 1. Naturally weighted 2005 VLBA images of 0402+379 at 8, 15, 22 and 43 GHz. Contours are drawn beginning at 3σ and increase by factors of 2 thereafter. The labels shown in the 8 GHz map indicate the positions of the two strong, compact, central components. See Rodriguez *et al.* (2006) for further details.

Some source properties like X-shaped radio galaxies and double-double radio galaxies, helical radio-jets, double-horned emission line profiles, and semi-periodic variations in lightcurves have been taken as indirect evidence for compact binary black holes though other explanations are possible (see review by Komossa 2003 detailing observational evidence for supermassive black holes binaries). Some wider systems have, however, been found more directly. The ultra luminous galaxy NGC 6240, discovered by the Chandra X-ray observatory, was found to have a pair of active supermassive black holes at its center (Komossa *et al.* 2003), separated by a distance of 1.4 kpc. Another system that has been known for some time is the double AGN (7 kpc separation) constituting the radio source 3C 75, which was discovered by the VLA to have two pairs of radio jets (Owen *et al.* 1985).

Recently, the radio galaxy 0402+379 was found to contain two central, compact, flat spectrum, variable components (designated C1 and C2 - see Fig. 1), a feature which has not been observed in any other compact source. Multi-frequency VLBA observations are described in greater detail in a recent paper by Rodriguez *et al.* (2006).

At the redshift of 0402+379 of 0.055, for $H_0=75$ km s^{-1} Mpc^{-1}, and $q_0 = 0.5$, gives a scale of 1 mas = 1.06 pc.

2. Constraints on the Orbital Parameters

A total mass of a few 10^8 $M_\odot$ was estimated for the SBBH system in 0402+379 using VLBA HI observations (Maness *et al.* 2004) and HET spectroscopy of Hα (Rodriguez *et al.* 2006). The HET spectrum shows a red shoulder with a velocity of 300 km s^{-1}. At the observed separation of 7.3 pc the implied mass for the system is $1.5 \times 10^8 (v/300$ km s$^{-1})^2$ $(r/7.3$ pc) $M_\odot$. Using the HI velocity separation of 1000 km s^{-1} on scales of 20 pc gives a larger mass of 5×10^9 $M_\odot$, but the two velocity systems might be probing line-of-sight velocities in the products of a recent merger, and not the orbital motions of the two black holes. Higher resolution observations using a Global VLBI array are planned.

Using a system mass estimate equal to 1.5×10^8 $M_\odot$ and the projected radial separation between them derived from the VLBA images (7.3 pc), we find from Kepler's Laws that the period of rotation for such a binary supermassive black hole system should be $\sim 1.5 \times 10^5$ y. This period corresponds to a relative projected velocity between components C1 and C2 of ~ 0.001 c. The upper limit found for the relative motion between components C1 and C2, < 0.088c, is consistent with the expected relative velocity assuming a stable, Keplerian orbit. To actually constrain the masses of the black holes would require observations over a longer time baseline (~ 100 y).

The natural gravitational wave frequency (Hughes 2003) for this system is approximately 2×10^{-13} Hz. This is well below the expected minimum frequency of LISA. Although 0402+379 may be a long way from generating a detectable gravitational wave signal it may represent a source of noise important for future observations of cosmologically produced gravitational radiation. Ultra low frequency gravitational radiation generated during inflation (Hughes 2003) has an upper limit of 10^{-13} Hz. Thus, a population of black hole binaries like 0402+379 may generate substantial noise which could interfere with the detection of the physics of inflation. The current merger time assuming only orbital decay due to gravitational radiation is $\sim 10^{18}$ y. Some other loss of angular momentum (e.g., dynamical friction) will be necessary if this system is to merge in less than a Hubble time.

3. Statistics of SBBH Systems

We have identified one SBBH system out of the entire Caltech-Jodrell Bank Flat-spectrum survey (CJF) of 293 AGN. Our current incidence is thus 1/293 or 0.34% with very large uncertainty. Assuming that 10% of luminous elliptical galaxies host an AGN, and that every galaxy undergoes a merger, then the duration of the SBBH phase should be $\sim 3\%$ of the lifetime of the elliptical host galaxy or 3×10^8 years. This seems rather long though it is possible that the merger has stalled. Merritt (2006) finds typical stalling radii for Virgo galaxies with comparable black hole masses to those in 0402+379 of ~ 7pc assuming black hole mass ratios of 2:1.

It is not clear whether, at present, dynamical friction losses or gas dissipative effects (see Komossa 2003, Merritt & Milosavljević 2005) are strong enough to bring the binary to the gravitational radiation loss regime within a Hubble time. This relates to the issue of the probability of catching the binary at its present (modest) separation. Are we seeing a recent merger in the act of core coalescence or has the binary stalled and must now await loss-cone replenishment and/or re-supply of nuclear gas? The fact that both nuclei are active (radio bright) suggests on-going accretion and implies dissipation today. Only a larger sample of imaged active nuclei can address the fraction of the population in this state. The VLBA Imaging and Polarization Survey (VIPS, Taylor *et al.* 2005) will image 1127 sources. So far we have identified twenty candidate SBBH systems, and observing time on the VLBA has been approved for all of them.

4. Conclusions

Based on the compactness, motion, variability, and spectra of the two central components in 0402+379 we conclude that they are both active nuclei of a single galaxy. This pair of AGN forms the closest binary black hole system yet discovered with a projected separation of 7.3 pc. The total mass of the system is estimated to be 1.5×10^8 $M_\odot$, and the gravitational radiation frequency to be 2×10^{-13} Hz. Energy losses due to gravitational radiation are not yet significant, so that other mechanisms must be invoked if the orbit is to decay. 0402+379 may be the tip of an iceberg for a population of supermassive black hole binaries with parsec scale separations. Such a population may collectively produce significant gravitational wave radiation which may need to be considered for the detection of gravitational radiation in the ultra to very low frequency bands.

Acknowledgements

The VLBA is operated by the the National Radio Astronomy Observatory, a facility of the National Science Foundation operated under a cooperative agreement by Associated Universities, Inc.

References

Hughes, S. A. 2003, Annals of Physics, 303, 142

Komossa, S. 2003, AIP Conf. Proc. 686: The Astrophysics of Gravitational Wave Sources, 686, 161

Komossa, S., Burwitz, V., Hasinger, G., Predehl, P., Kaastra, J. S. & Ikebe, Y. 2003, ApJL, 582, L15

Maness, H. L., Taylor, G. B., Zavala, R. T., Peck, A. B. & Pollack, L. K. 2004, ApJ, 602, 123

Merritt, D. 2006, submitted, astro-ph/0603439

Merritt, D., Milosavljević, M. 2005, Living Reviews in Relativity, 8, 8

Owen, F. N., Odea, C. P., Inoue, M. & Eilek, J. A. 1985, ApJL, 294, L85

Rodriguez, C., Taylor, G. B., Zavala, R. T., Peck, A. B., Pollack, L. K. & Romani, R. W. 2006, ApJ, 646, 49

Taylor, G. B. *et al.* 2005, ApJS, 159, 27

XIAN CHEN: Why are the binary systems discovered by VLBI all compact systems?

GREGORY TAYLOR: Actually, some of them are extended. Future low-frequency observations may reveal more extended structures.

Black Holes from Stars to Galaxies – Across the Range of Masses
Proceedings IAU Symposium No. 238, 2006
V. Karas & G. Matt, eds.
© 2007 International Astronomical Union
doi:10.1017/S1743921307005121

Radiatively inefficient accretion disks in low-luminosity AGN†

Ferdinando D. Macchetto and Marco Chiaberge

Space Telescope Science Institute, 3700 San Martin Dr., Baltimore, MD 21218
email: macchetto@stsci.edu

Abstract. We study a complete and distance-limited sample of 25 LINERs, 21 of which have been imaged with the Hubble Space Telescope to study their physical properties and to compare their radio and optical properties with those of other samples of local AGNs, namely Seyfert galaxies and low-luminosity radio galaxies (LLRG). Our results show that the LINERs population is not homogeneous, as there are two subclasses: i) the first class is similar to LLRG, as it extends the population of radio-loud nuclei to lower luminosities; ii) the second is similar to Seyferts, and extends the properties of radio-quiet nuclei towards the lowest luminosities. The different nature of the various classes of local AGN are best understood when the fraction of the Eddington luminosity they irradiate, $L_o/L_{\rm Edd}$, is plotted against the nuclear radio-loudness parameter: Seyferts are associated with relatively *high* radiative efficiencies $L_o/L_{\rm Edd} \gtrsim 10^{-4}$ (and high accretion rates onto *low* mass black holes); LLRG are associated with *low* radiative efficiencies (and low accretion rates onto *high* black hole masses); all LINERs have low radiative efficiency (and accretion rates), and can be radio-loud or radio quiet depending on their black hole mass.

Keywords. galaxies: active — accretion, accretion disks — galaxies: individual (NGC 4565)

---

1. Introduction

Low luminosity active galactic nuclei (LLAGN) are believed to be powered by accretion of matter onto the central supermassive black hole, similarly to powerful AGN. In a large fraction of LLAGN, the central black hole is as massive as in powerful distant quasars ($M_{\rm BH} \sim 10^8$–$10^9 M_\odot$), thus their very low nuclear luminosity implies that accretion occurs with very low radiative efficiency (or at very low rates); Ho (2004), Chiaberge *et al.* (2005). If so, the physics of the accretion process may be different from the "standard" optically thick, geometrically thin accretion disks. Because of the very low radiation they emit at all wavelengths, these objects are very difficult to observe. While the AGN nature of optical nuclear components seen in HST images of a sample of LLAGN have been unambiguously established it is still unclear whether the radiation is from a jet or from the accretion flow. LLAGN have also been found to lie on the so-called "fundamental plane of black hole activity" (Merloni *et al.* 2003; Falcke *et al.* 2004), which attempts to unify the emission from all sources around black holes, over a large range of masses and luminosities, from Galactic sources to powerful quasars. But the origin of such a "fundamental plane" and its relationship with the origin of the radiation is still a matter of debate.

† Based on observations obtained at the Space Telescope Science Institute, which is operated by the Association of Universities for Research in Astronomy, Incorporated, under NASA contract NAS 5-26555.

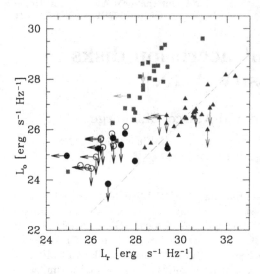

Figure 1. Optical nuclear luminosity vs. radio core luminosity for LINERs (circles), Seyferts (squares) and FR I radio galaxies (triangles). The dashed line is the correlation between the two quantities found for 3CR FR I sample. Open circles are LINERs in late-type hosts, filled circles are LINERs in early-type hosts.

2. LINERs in the framework of the local AGN population

Different accretion disk models are expected to show the largest difference in spectral shape in the IR-to-UV region. RIAFs should lack both the "big blue-bump" and the IR (reprocessed) bump, which instead characterize optically thick, geometrically thin accretion disk emission and the surrounding heated dust. For example, in low luminosity radio galaxies non-thermal emission from the jet dominates the optical nuclear radiation (Chiaberge et al. 1999), while the Galactic center is not visible in the optical because it is hidden by a large amount of dust.

We have studied a complete and distance-limited sample of 25 LINERs, 21 of which have been imaged with the Hubble Space Telescope. In nine objects we detect an unresolved nucleus. In order to study their physical properties, we compare the radio and optical properties of the nuclei of LINERs with those of other samples of local AGNs, namely Seyfert galaxies and low-luminosity FR I radio galaxies (LLRG, Fig. 1). The radio-optical correlation found for FR I, which is best explained as the result of a single emission process in the two bands (i.e. non-thermal synchrotron emission from the base of the relativistic jet), provides us with a powerful tool to investigate the origin of the nuclei. We have shown that in the radio-optical plane of the nuclei there is a clear separation between Seyferts and radio galaxies. For similar radio core luminosity, Seyfert 1 are significantly brighter in the optical than FR I. Therefore, although most Seyferts have $R = L_{5GHz}/L_B > 10$, radio-quiet and radio-loud AGN appear to be still well differentiated. This implies that the nuclear physical properties of the two classes are significantly different. Our results show that the LINERs population is not homogeneous, as there are two subclasses: i) the first class is similar to LLRG, as it extends the population of radio-loud nuclei to lower luminosities; ii) the second is similar to Seyferts, and extends the properties of radio-quiet nuclei towards the lowest luminosities. Furthermore, all radio-loud LINERs have $M_{BH}/M_\odot \gtrsim 10^8$, while Seyferts and radio-quiet LINERs have $M_{BH}/M_\odot \lesssim 10^8$.

We have derived the radiative efficiency of the accretion process around the central black holes in our samples of local AGN. All of them emit only a small fraction of the Eddington luminosity. Although the determination of the bolometric luminosity is uncertain because of the lack of detailed spectral information, the accretion process in LINERs appears to take place on a highly sub-Eddington regime ($L_o/L_{Edd} < 10^{-5}$, and can be as low as $\sim 10^{-8}$). Such low values are clearly not compatible with the expectations

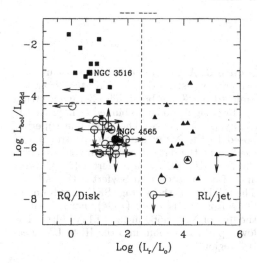

Figure 2. Optical to Eddington luminosity ratio plotted against the radio to optical ratio (the "nuclear radio-loudness") for the sample of nearby LLAGN; cf. Chiaberge *et al.* (2005). Seyfert 1s are plotted as squares, low luminosity radio galaxies are triangles, LINERs are empty circles. NGC 4565 (filled circle) and the Seyfert 1 galaxy NGC 3516 (large square) are also marked in the figure. The dashed lines are only used to guide the eye and divide objects of high and low Eddington ratio (top and bottom of the figure) and radio-quiet (disk-dominated nuclei, left) or radio loud (jet dominated nuclei, right).

from a standard optically thick and geometrically thin (quasar-like) accretion disk. Thus, low accretion rates and/or low efficiency processes appears to be required in all LINERs. Our results are qualitatively in agreement with Ho (2004), who made use of the $H\alpha$ emission line as an indicator of the AGN power for a large sample of Seyferts and LINERs.

The different nature of the various classes of local AGN are best understood when the fraction of the Eddington luminosity they irradiate is plotted against the nuclear radio-loudness (Fig. 2). Our objects populate three different quadrants, according to their physical properties. We identify Seyferts and radio-quiet LINERs as the high and low accretion rate counterparts, respectively. For low accretion regimes, the nuclei appears to be "radio-loud" only when a more massive black hole ($M_{BH} > 10^8 M_\odot$) is present. We speculate that the fourth quadrant, which appears to be "empty" in the local universe, would contain radio-loud nuclei with high L_o/L_{Edd}, readily identified with radio loud quasars.

We further tested this picture by studying the nuclear spectral energy distribution of a galaxy, NGC 4565, that seems to be a perfect candidate for hosting a RIAF around the central supermassive black hole. The object is part of the "Palomar sample" of LLAGN (Ho *et al.* 1997a), and it is included in both the Merloni *et al.* (2003) and Falcke *et al.* (2004) samples that were used to define the "fundamental plane of black hole activity". It is worth mentioning that NGC 4565 does not show any significant peculiarity in that plane. NGC 4565 is a nearby (d=9.7 Mpc) LLAGN classified as a Seyfert 1.9 because of the possible presence of a faint, relatively broad (FWHM = 1750 km s^{-1}) $H\alpha$ line. Although it is a Type 2 Seyfert, this object is only moderately absorbed, and the nuclear radiation is visible in the optical spectral region. NGC 4565 may thus represent the first clear example of low-luminosity accretion onto a supermassive black hole in the optical band.

The SED (Fig. 3) is peculiar, as it is almost flat in a $\log \nu - log(\nu F\nu)$ representation, with no sign of both a UV bump and thermally reprocessed IR emission. The very low luminosity of the source associated with a relatively high central black hole mass imply an extremely small value of the Eddington ratio ($L_o/L_{Edd} \sim 10^{-6}$). This, together with the position occupied by this object on diagnostic planes for low luminosity AGN, represents clear evidence for a low radiative efficiency accretion process at work in its innermost regions.

The fact that the [OIII] emission line flux is substantial in this object implies that an extended narrow line region, similar to other Seyfert galaxies, is still present in NGC 4565.

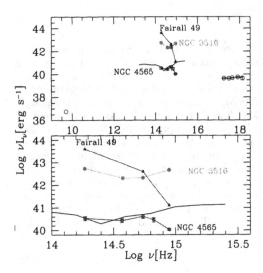

Figure 3. Absorption corrected nuclear spectral energy distribution of NGC 4565 from the radio to the X-ray band. The X-ray spectrum has been significantly re-binned to improve the clarity of the figure. The solid line superimposed to the X-ray data is a spectral model used to fit the data. For comparison, we show the IR-to-UV SED of a Seyfert 1 (NGC 3516) and of a Compton-thin Seyfert 2 (Fairall 49). The solid line is the average SED of radio-quiet QSO from Elvis *et al.* (1994), normalized to the flux of NGC 4565 in the F814W filter. The lower panel is a zoom into the IR-to-UV spectral region.

A possible intriguing scenario is that the active nucleus has recently "turned-off", switching from a high efficiency, standard, accretion disk, to a radiative inefficient accretion process. However, since the EW of the [OIII]5007 emission line is rather small, with the present data we cannot rule out that the amount of ionizing photons from the RIAF is sufficient to produce the observed [OIII] flux.

3. Conclusions

The scenario we propose needs further investigation, since optical detections of the nuclei are available for a minority of the LINER's sample. Thus, deep imaging with high spatial resolution (achievable only with the Hubble Space Telescope) are crucial. In particular, when suitable observations of a large number of LINERs will be available, it will be possible to address whether the dichotomy persists or some of the low black hole mass objects with optical upper limits mix with the population of "radio-loud" LINERs. Clearly this would falsify our scenario for the role of the black hole mass in determining the radio-loudness of the nuclei. Deep imaging of a larger complete sample would also address the issue of whether there is continuous transition between the two classes. This is indeed a subject of great interest in the study of high luminosity quasars and for a more complete understanding of the overall subject should be also extended to lower end of the AGN luminosity function.

Acknowledgements

We acknowledge R. Gilli, A. Capetti and W. Sparks for a productive collaboration.

References

Chiaberge, M., Capetti, A. & Celotti, A. 1999, A&A, 349, 77
Chiaberge, M., Capetti, A. & Macchetto, F. D. 2005, ApJ, 625, 716
Elvis, M. *et al.* 1994, ApJ Suppl., 95, 1
Falcke, H., Körding, E. & Markoff, S. 2004, A&A, 414, 895
Ho, L. C., Filippenko, A. V. & Sargent, W. L. W. 1997a, ApJ Suppl., 112, 315
Ho, L. C. 2004, in Multiwavelength AGN Surveys, 153
Merloni, A., Heinz, S. & di Matteo, T. 2003, MNRAS, 345, 1057

Black Holes from Stars to Galaxies – Across the Range of Masses
Proceedings IAU Symposium No. 238, 2006
V. Karas & G. Matt, eds.

The central 80×200 pc of M83: how many black holes and how massive are they?

H. Dottori,[1] R. Díaz,[2] I. Rodrigues,[1] M. P. Agüero[3] and D. Mast[3]

[1]Instituto de Física, UFRGS, cp: 15051, cep: 91501-970, Porto Alegre, Brazil

[2]Gemini Observatory, Southern Operations Center, Chile

[3]Observatório Astronómico de Córdoba, Laprida 854, Córdoba, Argentina

email: dottori@if.ufrgs.br, rdiaz@gemini.edu

Abstract. The central $\approx 80 \times 200$ pc of the barred spiral galaxy M83 (NGC 5236) has been observed with the GEMINI-S+CIRPASS configuration which produced 490 spectra with a spectral resolving power of 3200, centered at 1.3 microns, oriented NW-SE. We determine the kinematics of this region with 0.36″ sampling and sub-arcsec resolution. Disk-like motions have been detected in Paβ at parsec scales around: a) the optical nucleus, b) the center of the external K-band isophotes coincident with that of the CO velocity map both also tracing the center of the bulge, and c) a hidden condensation located at (R,θ)=(158pc, 301°). The present resolution allows to detect other minor whirls, not discussed here. The disk around (a) has a radius of ~ 8 pc and the two around (b) and (c) can be traced approximately up to 50–60 pc from their kinematical centers.

The rotation curves have been fitted by Satoh like spheroids+disk indicating masses of $\approx 4 \times 10^6~M_\odot$, 60–$70 \times 10^6 M_\odot$ and 30–$40 \times 10^6 M_\odot$ respectively. Limit to the masses of central BHs can be set by supposing that the central unresolved line broadening inside each condensations is dominated by the BH as far as allowed by the central error bars. The BH+Satoh models were smoothed with a 9 pc Gaussian. The upper mass limit derived for the BH is for (a) $\sim 10^6 / \sin(i) M_\odot$, and for (b) and (c) 0.2–$1.0 \times 10^6 / \sin(i) M_\odot$. Many questions arise from this interesting nucleus: are we witnessing a unique phenomenon or simply a barred galaxy with ongoing strong SF in our backyard? Does each one of the condensations host a BH indeed? or is there only one at the bulge center? Which is the fate of this complex scenario?

Keywords. Galaxies: nuclei – black hole physics

---

1. Introduction

It is generally accepted that galaxies host Super Massive Black Holes at their center, as one of the sessions of this symposium testify. The cannibalism of small neighbors by large galaxies is a phenomenon that also deserve increasing attention nowadays. Questions that naturally arise in these cases are if accreted satellites carry their own BHs, how massive they are and how much do they contribute to the large host galaxy central BH mass.

M 83 is a unique case for this type of studies, because of its proximity and the richness of phenomena occurring in their central 300 pc. Indeed, the condensation traditionally recognized as its nucleus (hereinafter, ON, for optical nucleus) is observable at optical wavelengths and is off-centered by 4″ to the NE with respect to the center of the external K band isophotes of the bulge which can be fitted by a de Vaucouleurs' law in an annular region between $\approx 10″$ and $40″$ (Jensen *et al.* 1981, Gallais *et al.* 1991). Thatte, Tecza & Genzel (2000) demonstrated that the bulge center host an obscured condensation (kinematical center, KC). These authors claims that if KC and ON are virialized systems, they should have comparable masses $M_{KC} = M_{ON} \approx 13.2 \times 10^6~M_\odot$ inside 5.4 pc in a cluster or a super massive core. CO interferometry (Sakamoto *et al.* 2004) shows

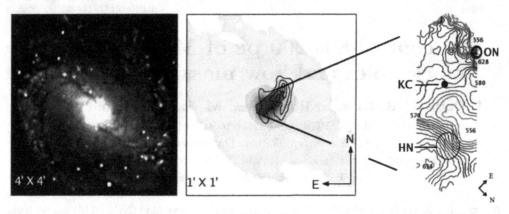

Figure 1. Image of M83 with zoomed central $1' \times 1'$ showing the position of CIRPASS array (ached area) superposed to the 10 cm continuum isophotes. The iso-velocities map is presented in the last panel. The positions of ON, KC and HN are detached as well as some iso-velocity levels. In Díaz *et al.* (2006) ON is also referred as visible nucleus, KC as bulge geometrical center, and HN as intruder and dark rotation center.

steep velocity gradients across the ON and a mass of $3 \times 10^8 M_\odot$ within a radius of 40 pc, nevertheless, the spatial resolution in CO observations is of the order of $3''$ and inhibits conclusions about the precise position of the rotation center (see Díaz *et al.* 2006). Elmegreen *et al.* (1997) determined for ON a photometric mass of $M_{ON} \approx 4 \times 10^6 M_\odot$ (Hot Spot 1 in their table 4). More recently, Mast, Díaz & Agüero (2006) demonstrated the existence of a third condensation, also hidden at optical wavelengths, which coincides with the maximum of continuum emission at 6 cm and 10 cm of a giant cloud (Telesco 1988) $7''$ to the WSW of KC. Díaz *et al.* (2006) propose that HN is a small galaxy been cannibalized. We discuss here high resolution 3-D spectroscopy of the central $4.7'' \times 13''$ enclosing the three condensations and the implications in connection with the putative Black Holes at the center of M83.

2. Observations

We used the Cambridge IR Panoramic Survey Spectrograph (Parry *et al.* 2000, 2004) installed on the GEMINI South telescope in March 2003. These observations, performed in queue mode, were taken with an integral field unit (IFU) sampling of $0.36''$ (7.8 pc) in an elliptical array with size $4.7'' \times 13''$. The IFU was oriented at PA 120° and centered slightly to the NW of the kinematical center, according to Mast *et al.* (2002). It was exposed during 45 minutes and covers an spectral range between 1.2–1.4μm, including Paβ 1.3μm and the [FeII] 1.26μm. The spectral resolution is approximately 3200. During the observations the peripheral wavefront sensor of GEMINI active optics was used to achieve a FWHM of approximately $0.5''$, therefore the focal plane was slightly sub-sampled by the configuration constrain.

The data were reduced using IRAF, ADHOC and SAO. Details about the reduction are presented by Díaz *et al.* (2006) and the general technique is thoroughly discussed by Díaz *et al.* (1999) and Mast *et al.* (2002).

3. 2-D kinematics of M 83 central region

In Figure 1 we show the Paβ iso-velocity contours on the 2-D radial velocity map. Three spider diagrams can be well differentiated on the complex 2-D radial velocity map.

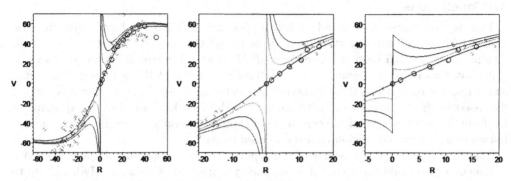

Figure 2. HN rotation curve. The squares are more than 50 velocity points obtained from the spider diagram. Circles are velocities mean values from receding and approaching side. Full line represent Satoh's best fit to the circles. The curves resulting from adding point like masses of $0.2\times$, $1\times$ and $2\times10^6\,M_\odot$ to the Satoh's spheroid are also shown. The Satoh+Kepler velocity diagrams were filtered with a 9 pc Gaussian, to mimic the central velocity dispersion (bar length at r=0) which can then be compared to the observed one (table 1).

Table 1. Condensations and putative BHs properties

object	radius [pc]	v [km/s]	σ [km/s]	M_k [$10^6 M_\odot$]	M_{BH} [$10^6 M_\odot$]
ON	8 ± 1	46 ± 9	110 ± 10	3.7 ± 2	$\leqslant 1.0$
KC	65 ± 5	68 ± 8	82 ± 10	66 ± 1.2	$0.2 - 1.0$
HN	45 ± 8	58 ± 11	96 ± 10	36 ± 1.2	$0.2 - 1.0$

One coincides with the optical nucleus (ON), a second condensation is coincident with the bulge center (KC), and a strongly absorbed third one (HN) has the largest local radial velocity gradient. A detailed description of the M83 central region astrometry can be found in Díaz *et al.* (2006).

The three condensations show well defined spider diagrams indicating that the P_β emitting gas is disk-like ordered (figure 1). KC shows an inclination of $68° \pm 10°$ is The HN inclination is $50° \pm 10°$, Díaz *et al.* (2006). In figure 2 we present the rotation curve for this body which was fitted with a Satoh's spheroid to which three Keplerian masses were added, representing a putative BH. A 9 pc mask was applied to the cases of Satoh+Kepler rotation curve in order to reproduce the system spatial resolution. To decide which is the highest acceptable BH mass we use two criteria, namely, the velocities error bar and the observed central velocity dispersion. Observed kinematical data and deduced masses are presented in table 1. Reasonable BH upper mass limits are between 0.2 and $1.0 \times 10^6 M_\odot$ for KC and HN.

Based on the kinematical data we have run numerical simulations using GADGET2 Springel (2005), in order to study the dynamical evolution in the central region of M83. We considered the three condensations: KC, ON and HN, and followed the evolution of their stellar and gaseous components. KC, used as reference, was placed at the center of a large disk representing the inner part of M83 bulge ($2 \times 10^8\,M_\odot$). A total of 10^5 particles have been used and different orbits have been tested. The main result is that ON, KC and HN will merge, forming a single massive core in less than 50 Myr.

4. Conclusions

The spatial resolution achieved with this observations allows to better constrain the masses of possible central very dense objects in the ON and KC (Thatte *et al.* 2000). The presence of a BH more massive than $10^6 M_\odot$ within the condensations seems to be ruled-out. Our numerical simulations show that ON, KC and HN will merge forming a single massive core in less than 50 Myr. The existence of two distinct resonance circles, discussed in the literature (e.g. Elmegreen *et al.* 1997; Sakamoto *et al.* 2004) seems to be, from the point of view of the central region stability, totally irrelevant, since they are not able to preserve the region identity, as seen today.

The short dynamical timescales of the reported merger make of this phenomenon a unique one (probability less than 1% per galaxy). Since the Centaurus-Hydra group to which M83 belongs together with NGC 5128, NGC 5236 and others minor galaxies is so rich in activity, it is natural to rise the question on why should not have occurred a previous capture of this type by M83. It is important to detach from this point of view that object 28 in Maddox *et al.* (2006) list is a radiogalaxy projected on M83 disk which presents radio-lobes (objects 27 and 29 in the list) aligned with the galactic bulge at a projected distance between 1 and 2 kpc. Although the object has been characterized as a background galaxy with radio-lobes of type FII (Soria & Wu 2003), the curious alignment leads to think on the possibility of a kicked-off BH (Micic, Sirgudson & Abel 2005; also Madau's and Rees' talks at this Symposium) produced in a previous merger, similar to that occurring presently with the HN.

References

Díaz, R. J., Dottori, H., Mediavilla, E., Agüero, M. & Mast, D. 2006, New Astron. Rev., 49, 547
Díaz, R., Carranza, G., Dottori, H. & Goldes, G. 1999, ApJ, 512, 623
Elmegreen, D. M., Chromey, F. R. & Warren, A. R. 1998, AJ, 116, 2834
Galais, P., Rouan, P., Lacombe, D., Tiphene, D. & Vauglin, L. 1991, A&A, 243, 309
Harris, J., Calzetti, D., Gallagher, J., Conselice, C. & Smith, D. 2001, AJ, 122, 3046
Heap, S., Holbrook, J., Malumuth, E., Shore, S. & Waller, W. 1993, BAAS, 182, 3104
Jensen, E. B., Talbot, R. J. Jr. & Dufour, R. J 1981, ApJ, 243, 716
Maddox, L., Cowan, J., Kilgard, R., *et al.* 2006, AJ, 132, 310
Mast, D., Díaz, R. & Agüero, M. P. 2006, AJ, 131, 1394
Mast, D., Díaz, R., Agüero, M., Weidmann, W., Carranza, G. & Gimeno, G. 2002, BAAA, 45, 74
Micic, M., Sigurdsson, S. & Abel, T. 2005, BAAS, 37, 1355
Telesco, C. M. 1988, ARAA, 26, 343
Thatte, N., Tecza, M. & Genzel, R. 2000, A&A, 364, 47
Parry, I., Mackay, C. D., Johnson, R. A., *et al.* 2000, SPIE, 4008, 1193
Parry, I., Mackay, C. D., Johnson, R. A., *et al.* 2004, SPIE, 5492, 1135
Pastoriza, M. G. 1975, APSS, 33, 173
Sakamoto, K., Matsushita, S., Peck, A. B., Wiedner, M. & Iono, D. 2004, ApJ, 616, L59
Soria, R., Wu, K. 2003, A&A, 410,535
Springel, V. 2005, MNRAS, 364, 1105

Supermassive Black Holes and Their Galaxies II

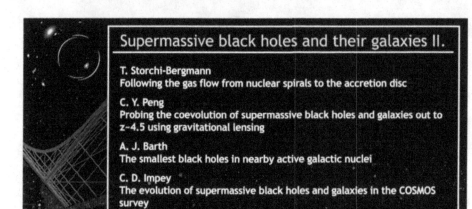

Supermassive black holes and their galaxies II.

T. Storchi-Bergmann
Following the gas flow from nuclear spirals to the accretion disc

C. Y. Peng
Probing the coevolution of supermassive black holes and galaxies out to z~4.5 using gravitational lensing

A. J. Barth
The smallest black holes in nearby active galactic nuclei

C. D. Impey
The evolution of supermassive black holes and galaxies in the COSMOS survey

J. Woo
Cosmic evolution of black holes and galaxies to z=0.4

Prague Castle panorama and Lesser Quarter in the dusk. A view from Old Town Bridge Tower guarding the entrance to the Charles Bridge.

Black Holes from Stars to Galaxies – Across the Range of Masses
Proceedings IAU Symposium No. 238, 2006   © 2007 International Astronomical Union
V. Karas & G. Matt, eds.   doi:10.1017/S1743921307005157

Following the gas flows from nuclear spirals to the accretion disk

Thaisa Storchi-Bergmann

Instituto de Física, Universidade Federal do Rio Grande do Sul, Porto Alegre, RS, Brazil
email: thaisa@ufrgs.br

Abstract. A recent analysis of HST optical images of 34 nearby early-type active galaxies and of a matched sample of 34 inactive galaxies – both drawn from the Palomar survey – shows a clear excess of nuclear dusty structures (filaments, spirals and disks) in the active galaxies. This result supports the association of the dusty structures with the material which feeds the supermassive black hole (hereafter SMBH). Among the inactive galaxies there is instead an excess of nuclear stellar disks. As the active and inactive galaxies can be considered two phases of the "same" galaxy, the above findings and dust morphologies suggest an evolutionary scenario in which external material (gas and dust) is captured to the nuclear region where it settles and ends up feeding the active nucleus and replenishing the stellar disk – which is hidden by the dust in the active galaxies – with new stars. This evolutionary scenario is supported by recent gas kinematics of the inner few hundred parsecs of NGC 1097, which shows streaming motions (with velocities $\sim$50 km s^{-1}) towards the nucleus along spiral arms. The implied large scale mass accretion rate is much larger than the one derived in previous studies for the nuclear accretion disk, but is just enough to accumulate one million solar masses over a few million years in the nuclear region, thus consistent with the recent finding of a young circumnuclear starburst of one million solar masses within 9 parsecs from the nucleus in this galaxy.

Keywords. Galaxies: active – galaxies: nuclei – galaxies: kinematics and dynamics – galaxies: individual (NGC 1097) – galaxies: ISM

---

1. Introduction

The relation between the morphologies of host galaxies and the presence of nuclear activity has been investigated for many years. One of the first studies arguing for a difference between active and inactive galaxy hosts is the one of Simkin, Su & Schwarz (1980). These authors have claimed that Seyfert galaxies hosts are more distorted than inactive galaxies, showing an excess of bars, rings and tails. More recent studies did not confirm the excess of bars in Seyferts (e.g. Mulchaey & Regan 1997) while others have found a small excess (e.g. Knapen *et al.* 2000).

A number of studies using Hubble Space Telescope images of nearby galaxies have revealed a trend for active galaxies always showing a lot of dust structure in the nuclear region. Van Dokkum & Franx (1995) have found that radio-loud early-type galaxies have more dust than radio-quiet ones. Pogge & Martini (2002) and later Martini *et al.* (2003) found that Seyfert galaxies almost always present dusty filaments and spirals in the nuclear region, while Xilouris & Papadakis (2002) found that, among early Hubble types, active galaxies present more dust structure than inactive galaxies. Recently, Lauer *et al.* (2005) have also argued that dust in early-type galaxies is correlated with nuclear activity.

The goal of the present paper is to discuss the results of a recent study (Lopes *et al.* 2007) which provides a robust analysis of the relation between nuclear dust structures and activity in galaxies on the basis of optical HST images. The particular case of NGC 1097 is then discussed, for which we have obtained, in addition, kinematics of the gas associated with the nuclear dust structures (Fathi *et al.* 2006) which reveal streaming motions towards the nucleus. We then use results of previous studies (Storchi-Bergmann *et al.* 2003, Nemmen *et al.* 2006) to relate the mass accretion rate reaching the accretion disk to the larger scale mass flow rate derived from the observed streaming motions.

2. Correlation between circumnuclear dust and nuclear activity

We (Lopes *et al.* 2007) have recently selected a sample comprising all active galaxies from the Palomar survey (Ho *et al.* 1995) which have optical images in the HST archive, as well as a pair-matched sample of inactive galaxies, and constructed "structure maps", a technique proposed by Martini & Pogge (1999) to enhance both absorption and emission structures in the images. The total sample comprises 34 matched pairs of early-type galaxies (T⩽0) and 31 pairs of late-type galaxies (T>0).

The results for the 34 early-type pairs of our sample are illustrated in Fig. 1 for a subsample of 10 pairs: while all (100%) active galaxies present some dust structure, only 26% (9 of 34) of the inactive galaxies possess some dust. Another difference is the finding that at least 13 of the 34 (38%) early-type inactive galaxies present bright *stellar disks*, while only one of the active galaxies show such disks. Most dust structures and stellar disks are seen within a few hundred parsecs from the nucleus. A different result was obtained for the late-type pairs: all galaxies show circumnuclear dust regardless the presence or absence of nuclear activity, and stellar disks were found only in a couple of galaxies.

The above findings imply a strong correlation between circumnuclear dust and nuclear activity, indicating that the dust is connected to the material which is currently feeding the active nucleus, and is probably tracing this material on its way to the nucleus. The morphologies of the dust structures range from chaotic filaments to regular dusty spirals and disks, suggestive of a "settling scenario" for the dust, as proposed by Lauer *et al.* (2005). Our finding of nuclear stellar disks in inactive galaxies suggests that there is one further step in this evolutionary scenario: the stellar disk, which shows up in the inactive phase of the galaxy. The evolutionary scenario can be described as follows: externally acquired matter is traced by the chaotic filamentary structure which gradually settles into more regular nuclear spirals and disks. Stars then form in the dusty disks, and when the gas and dust are fed to the black hole, nuclear activity ceases and and the stellar disks are unveiled (Ferrarese *et al.* 2006). As stellar disks should be longer lived than the gaseous dusty disks, the stellar disks are probably present also in the active phase, but are obscured by dust. In the evolutionary scenario proposed above, the episodic accretion of matter then replenishes the disk, which grows together with the mass of the nuclear black hole.

On the theoretical side, support for the evolutionary scenario includes the work of Maciejewski (2004) who demonstrated that, if a central SMBH is present, nuclear disks of gas and dust can develop spiral shocks and generate gas inflow compatible to the accretion rates observed in local active galaxies. On the observational side, kinematic evidence for inflow along nuclear dusty spiral arms has been found so far in one case: NGC 1097 (Fathi *et al.* 2006).

3. Streaming motions along nuclear spirals in NGC 1097

The observations of NGC 1097 were obtained with the Integral-Field Unit of the Gemini Multi-Object Spectrograph and allowed the mapping of the gas kinematics in the inner few hundred parsecs (Fathi *et al.* 2006). After subtracting a circular velocity model, streaming motions along spiral arms with inward velocities of up to $50 \, \mathrm{km \, s^{-1}}$ were

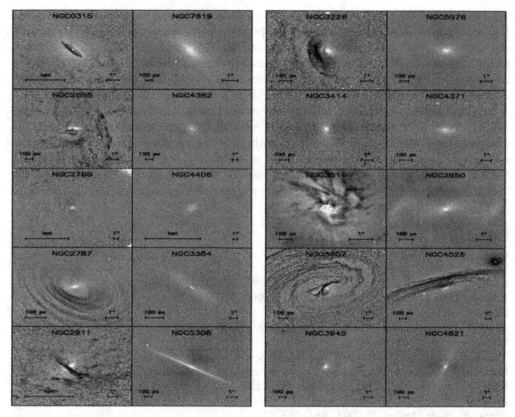

Figure 1. Structure maps for 10 matched pairs of early-type galaxies. Each image covers 5% of the galaxy diameter D25. North is up and East to the left. In each pair the active galaxy is shown to the left and the inactive matched pair to the right. From Lopes *et al.* 2007.

found. Another relevant finding on this galaxy is the young obscured starburst recently discovered (Storchi-Bergmann *et al.* 2005) very close to the nucleus (within $\sim$9 pc), in agreement with the suggestion that inflowing gas and dust gives birth to stars in the nuclear dusty spiral or disk (Ferrarese *et al.* 2006).

The velocity observed for the streaming motions along the nuclear spiral allows an estimate of a few Myr for the gas to flow from a few hundred parsecs to the nucleus. For an estimated gas density of $\sim$500 protons cm^{-3}, and estimated circular cross-section of 3 spiral arms at 100 pc from the nucleus (opening angle 20°), we conclude that the mass flow rate along the nuclear spiral arms in NGC 1097 is $\sim$35 times the one at the accretion disk of $dM/dt = 1.1 \times 10^{24}$ g s^{-1} (Nemmen *et al.* 2006), and allows the accumulation in the nuclear region of $\sim 10^6 M_\odot$ in a few Myr, which can provide the necessary matter to give birth to the nuclear starburst (Storchi-Bergmann *et al.* 2005).

4. Concluding remarks

We have found a strong correlation between circumnuclear dust and nuclear activity, and between nuclear stellar disks and absence of nuclear activity in early-type galaxies. As both the active and inactive galaxies can be thought of as the "same galaxy" observed in different phases, our findings suggest an evolutionary scenario in which the nuclear activity is triggered by capture of dusty gas to the nuclear region. The origin of the gas and dust is still not clear, but the absence of dust in the inactive phase suggests that in cannot originate from continuous mass loss from stars and is most probably external. In our proposed evolutionary scenario, once the gas and dust are captured to the nuclear region they gradually settle into a nuclear spiral or disk, where new stars are born while the excess gas and dust is accreted by the nuclear SMBH. Replenishment of a nuclear stellar disk (observed in the inactive galaxies) is the final product after activity ceases. As a result, both the stellar component of the galaxy and its SMBH at the nucleus grow after each activity cycle, as implied by the M–σ relation (e.g. Tremaine *et al.* 2002).

The evolutionary scenario is supported by recent kinematic observations of the nuclear spiral in NGC 1097, which reveal streaming motions along the spiral arms, providing an accretion flow rate which is enough to both form the starburst recent observed surrounding the nucleus (Storchi-Bergmann *et al.* 2005) in a few Myr and feed the nuclear SMBH. As similar dusty structures are observed in most active galaxies (Lopes *et al.* 2007), streaming motions along nuclear spirals may be the main mechanism for black hole feeding in active galaxies.

References

Fathi, K. *et al.* 2006, ApJ, 641, L25
Ferrarese, L. *et al.* 2006, ApJ, 164, 334
Ho, L. C., Filippenko, A. V. & Sargent, W. L. W. 1995, ApJ, 98, 477
Knapen, J. H., Shlosman, I. & Peletier, R. F. 2000, ApJ, 529, 93
Lauer, T. R. *et al.* 2005, AJ, 129, 2138
Lopes, R. S., Storchi-Bergmann, T. & Saraiva, M. F. O. 2007, ApJ, in press [astro-ph/0610380]
Maciejewski, W. 2004, MNRAS, 354, 883
Martini, P. & Pogge, R. W. 1999, AJ, 118, 2646
Martini, P., Regan, M. W., Mulchaey, J. S. & Pogge, R. W. 2003, ApJ, 589, 774
Mulchaey, J. S. & Regan, M. 1997, ApJ, 482, L135
Nemmen, R. S. *et al.* 2006, ApJ, 643, 652
Pogge, R. W. & Martini, P. 2002, ApJ, 569, 624
Simkin, S. M., Su, H. J. & Schwarz, M. P. 1980, ApJ, 237, 404
Storchi-Bergmann, T. *et al.* 2003, ApJ, 598, 956
Storchi-Bergmann, T. *et al.* 2005, ApJ, 624, 13
Tremaine, S. *et al.* 2002, ApJ, 574, 740
Van Dokkum, P. G. & Franx, M. 1995, AJ, 110, 2027
Xilouris, E. M. & Papadakis, I. E. 2002, A&A, 387, 441

MITCHELL BEGELMAN: Are you able to distinguish flow along a bar from flow along a spiral? If so, do you see evidence of bars?

THAISA STORCHI-BERGMANN: We would be able to distinguish a bar from a spiral if there was a bar. We do not see gaseous/dusty bars. Bars have been seen in continuum images, but we do not see them in our structure maps.

YIPING WANG: Have you done any statistics of your sample to see a relationship between the star-formation rate and the accretion rate for the inflow? It seems that star-formation in the nuclear region and the accretion are linked. We propose a model aimed to explain the black hole–bulge mass relation in which this relation is important.

THAISA STORCHI-BERGMANN: The accretion rate is $\sim 0.6 M_\odot$ per year in NGC 1097 (the only case so far in which we were able to calculate the large scale inflow rate), so we can accumulate enough mass in one million years to form the young star cluster observed at the nucleus and produce necessary star-formation and feeding. We plan more work on our sample – this is very important, but we do not have enough data right now to perform such analysis.

Black Holes from Stars to Galaxies - Across the Range of Masses
Proceedings IAU Symposium No. 238, 2006
V. Karas & G. Matt, eds.
© 2007 International Astronomical Union
doi:10.1017/S1743921307005169

A survey of AGN and supermassive black holes in the COSMOS Survey

Chris D. Impey,[1]† Jon R. Trump,[1] Pat J. McCarthy,[2] Martin Elvis,[3]
John P. Huchra,[3] Nick Z. Scoville,[4] Simon J. Lilly,[5] Marcella Brusa,[6]
Günther Hasinger,[6] Eva Schinnerer,[7] Peter Capak[4] and Jared Gabor[1]

[1]Steward Observatory, University of Arizona, Tucson, AZ 85721, USA

[2]Carnegie Observatories, 813 Santa Barbara Street, Pasadena, CA 91101, USA

[3]Harvard-Smithsonian Center for Astrophysics, 60 Garden Str., Cambridge, MA 02138, USA

[4]California Institute of Technology, MC 105-24, Pasadena, CA 91101, USA

[5]Department of Physics, Swiss Federal Institute of Technology, CH-8093 Zurich, Switzerland

[6]Max Planck Institut für Extraterrestrische Physik, D-85478, Garching, Germany

[7]Max Planck institut für Astronomie, Königstuhl 17, Heidelberg, D-69117, Germany

Abstract. The Cosmic Evolution Survey (COSMOS) is an HST/ACS imaging survey of 2 square degrees centered on RA = 10:00:28.6, Dec = +02:12:21 (J2000). While the primary goal of the survey is to study evolution of galaxy morphology and large scale structure, an extensive multi-wavelength data set allows for a sensitive survey of AGN. Spectroscopy of optical counterparts to faint X-ray and radio sources is being carried out with the Magallen (Baade) Telescope and the ESO VLT. By achieving ~80% redshift completeness down to $I_{AB} = 23$, the eventual yield of AGN will be ~1100 over the whole field.

Early results on supermassive black holes are described. The goals of the survey include a bolometric census of AGN down to moderate luminosities, the cosmic evolution and fueling history of the central engines, and a study of AGN environments on scales ranging from the host galaxy to clusters and superclusters.

Keywords. Active galactic nuclei – supermassive black holes – surveys

---

1. Introduction

The Cosmic Evolution Survey (COSMOS) is the first survey with the ideal combination of depth and area to study the coupled evolution of large scale structure, star formation, and nuclear activity in galaxies, see Scoville et al. (2007a). The core data set is a 2 deg^2 HST/ACS mosaic made up of single orbit exposures in the F814W (I) band that reach a 5σ depth of $I_{AB} = 28.6$ for point sources. The field is aligned N-S, E-W, and is centered at RA = 10^h 00^m 28.6^s, Dec = $+02°$ $12'$ $21''$. Astrometry across the field has a relative precision of 0.05 arcseconds (absolute to ~0.2 arcseconds) and relative fluxes are accurate to 1% (absolute to ~10%). Parallels with WFPC2 and NICMOS cover 55% and 5% of the field, respectively. Details of the HST observations are in Scoville et al. (2007b).

The COSMOS field is accessible to all major observing facilities and is the subject of an extensive campaign of multi-wavelength observations. Ground-based imaging from Subaru, CFHT, UKIRT, and NOAO in 7 bands from U to K_s has yielded $\sim 8 \times 10^5$

† Based on observations with the NASA/ESA *Hubble Space Telescope*, obtained at the Space Telescope Science Institute, which is operated by AURA Inc, under NASA contract NAS 5-26555; also based on data collected at the Magellan Telescope, which is operated by the Carnegie Observatories, and the VLT, which is operated by the European Southern Observatory.

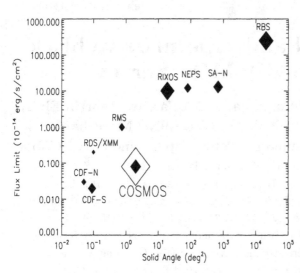

Figure 1. The power of the COSMOS survey for AGN studies is shown when the depth and solid angle of various X-ray surveys are compared. Symbol size is proportional to the number of spectroscopically-confirmed AGN; the solid diamond for COSMOS is the yield for this paper and the companion paper of Trump *et al.* (2006) and the open diamond shows the expected yield after the spectroscopic survey is complete. From left to right, the surveys plotted are the twin Chandra Deep Fields, the ROSAT/XMM deep survey of the Lockman Hole, the ROSAT Medium Deep Survey, ROSAT International X-ray Optical Survey, the ROSAT North Ecliptic Pole and Selected Areas North Surveys, and the ROSAT Bright Survey.

photometric redshifts from among the $\sim 2 \times 10^6$ galaxies in the ACS catalog, and VLT observations over the next 2-3 years will deliver $\sim 38,000$ galaxy redshifts. The VLT and Magellan telescopes are being used to measure spectroscopic AGN redshifts, as described here. Beyond the optical and near infrared parts of the spectrum, full coverage of the field has been obtained at radio (VLA), infrared (Spitzer/IRAC), ultraviolet (GALEX), and X-ray (XMM) wavelengths, and partial sub-millimeter coverage will come from the CSO and IRAM facilities. Deep Spitzer/MIPS and Chandra observations are scheduled. A more complete description of ancillary data is presented by Scoville *et al.* (2007a).

The significance of the COSMOS survey for AGN is conveyed in Figure 1, which shows the X-ray flux limit and areal coverage of a number of recent surveys, with a symbol size proportional to the number of spectroscopically-confirmed Type 1 AGN. The statistics for other surveys are taken from a compilation of Hasinger, Miyaji & Schmidt (2005). The filled diamond for COSMOS reflects the outcome of the 2005 season of Magellan observing, as described by Trump *et al.* (2007), with 106 new AGN adding to 40 existing SDSS AGN. The open diamond projects the eventual yield from our single pass survey, 350 Type 1 AGN, easily exceeding any other contiguous deep field.

2. Optical spectroscopy

The majority of the AGN candidates are being observed with the Magellan (Baade) telescope at Las Campanas, using the IMACS imager and spectrograph. We use IMACS with the 200 line prism and a 565–920 (or OG 570) filter, giving wavelength coverage of 5400–9200 Å (or 5500–9100 Å). The dispersion was 2 Å pixel^{-1} and the FWHM of an unresolved line was 5 pixels, giving effective resolution of 10 Å, or ~ 1000 km s^{-1}, sufficient to distinguish broad emission line (Type 1) from narrow emission line (Type 2) AGN. IMACS is most effective when used in nod–and–shuffle mode, see Abraham *et al.* (2004). We generally adopted a slit 11 arcseconds long and 1 arcsecond wide, with a 1.8 arcsecond nod between the two positions. Additional details of slit placement efficiency and redshift yield are given in the companion paper that presents spectra from the first season of observing, see Trump *et al.* (2007).

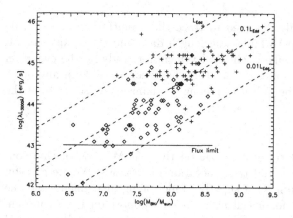

Figure 2. Optical luminosity versus black hole mass for Type 1 AGN in the COSMOS field from the IMACS (diamonds) and SDSS (crosses) surveys. The black hole masses are calculated from scaling relations applied to broad line velocity widths: CIV, MgII and Hβ depending on the redshift, using the techniques of McLure & Jarvis (2002) and Vestergaard & Peterson (2006). Tracks of luminosity relative to the Eddington value are shown as dashed lines. The solid horizontal line corresponds to the flux limit of the spectroscopic survey.

3. Black hole science

After one season of spectroscopic follow-up, the survey has produced 86 new Type 1 AGN and 130 new Type 2 AGN with high reliability classification and redshifts, as detailed by Trump *et al.* (2007). An additional 18 Type 1 AGN and 11 Type 2 AGN had lower reliability classification. The projection is ~850 AGN at the end of the one-pass Magellan survey in 2007, counting only the high reliability redshifts. The total yield of AGN with all qualities of classification and redshift will be ~1100, and the VLT multi-pass survey will add to this number by going deeper and picking up the brighter objects that IMACS will not target. This total includes ~90 AGN from the SDSS spectroscopic coverage of the COSMOS field.

A first look at the black hole demographics revealed by the COSMOS survey is shown in Figure 2. Monochromatic optical luminosity is plotted against black hole mass for 63 Type 1 AGN with IMACS observations and 59 Type 1 AGN from SDSS observations of the COSMOS field. We will move to working with the (more physically meaningful) bolometric luminosities now that recent Spitzer data are in hand. The luminosity is rest-frame, calculated using a power law fit to the continuum after subtraction of Fe blends, the same continuum fit is used for broad line fitting. The same procedure was used for SDSS and IMACS spectra. The black hole masses are based on broad line width scaling relations of CIV, MgII and Hβ, as described by McLure & Jarvis (2002) and Vestergaard & Peterson (2006). The dashed lines show tracks for 100%, 10%, and 1% of the Eddington luminosity, based on an assumption that the monochromatic optical luminosity at 3000Å is 20% of the bolometric luminosity.

This one region of sky yields SMBH spanning three orders of magnitude in mass and the fainter COSMOS sample extends to less efficient accretors than the SDSS. The data in Figure 2 occupy the same parameter-space as the three dozen local reverberation mapping calibrators, from work by Peterson *et al.* (2004). The spectroscopic flux limit of $I_{AB} = 23$ prevents the detection of the extremely low accretion efficiencies; it is represented approximately in Figure 2 as a horizontal line at a luminosity corresponding to $z = 0.6$, the 90% envelope of the AGN sample. Very low mass black holes are censored from the lower left part of the plot in a more complex way since the mass scales as $(FWHM)^2$ and $(\lambda L)^{0.7}$ and emission lines become undetectable or unmeasurable due to a combination of strength and breadth. Selection effects will be modeled in future work.

4. Conclusions

The COSMOS survey was primarily designed to trace the evolution and large scale structure of normal galaxies, but superb multi-wavelength data and the unique combination of areal coverage and depth make it an ideal vehicle for studying the evolution of supermassive black holes down to lower masses than has been possible in any other AGN survey. AGN activity will also be linked to fueling mechanisms on scales from the host galaxy to surrounding clusters.

Acknowledgements

We acknowledge the dedicated efforts of everyone on the COSMOS team who contributed to the reduction of ACS data and the generation of catalogs. We are grateful to Alan Dressler and the IMACS team for a superbly functioning instrument, and the staff at Las Campanas Observatory for excellent technical support on the mountain. We are very grateful to Marianne Vestergaard for giving us access to results ahead of publication. The HST COSMOS Treasury program was supported through NASA grant HST-GO-09822.

References

Abraham, R. G. *et al.* 2004, AJ, 127, 2455
Hasinger, G., Mijayi, T. & Schmidt, M. 2005, A&A, 441, 417
McLure, R. J. & Jarvis, M. J. 2002, MNRAS, 337, 109
Peterson, B. M. *et al.* 2004, ApJ, 613, 682
Scoville, N. Z. *et al.* 2007a, ApJS, in press
Scoville, N. Z. *et al.* 2007b, ApJS, in press
Trump, J. R. *et al.* 2007, ApJS, in press
Vestergaard, M. & Peterson, B. M. 2006, ApJ, 641, 689

DEBORAH DULTZIN-HACYAN: Comment: With respect to companions to AGN, there is an important difference between type 1 and type 2 objects. Recently, we (Koulouridis *et al.* 2006) confirmed, by applying a three-dimensional analysis, that Sy2s do have excess companions with respect to a control sample of non-active galaxies of the same redshift distribution, the same morphology apart from the nucleus, etc. On the other hand, Sy1s don't show such excess. Can you comment any preliminary results in this context?

CHRIS IMPEY: In our analysis we do not have enough statistics to perform such analysis of structures.

MARGARITA SAFONOVA: Can you comment on AGN without any host galaxies in your sample

CHRIS IMPEY: There have been several celebrated and refuted claims of naked AGN at low redshift – this could happen e.g. if one has a binary supermassive black hole of which one member gets ejected and is floating in the space – but with sufficient sensitivity people always found a host galaxy in the end.

SUZY COLLIN: Comment: Regarding the reverberation mapping studies, there can be a selection effect for small broad-line region which point toward small masses.

Black Holes from Stars to Galaxies - Across the Range of Masses
Proceedings IAU Symposium No. 238, 2006
V. Karas & G. Matt, eds.
© 2007 International Astronomical Union
doi:10.1017/S1743921307005170

Cosmic evolution of black holes and galaxies to $z = 0.4$

J.-H. Woo,[1] T. Treu,[1] M. A. Malkan[2] and R. D. Blanford[3]

[1]Department of Physics, University of California, Santa Barbara, CA 93106-9530

[2]Department of Physics, University of California, Los Angeles, CA 90095-1547 [3]Kavli Institute for Particle Astrophysics and Cosmology, Stanford, CA 94305

email: woo@physics.ucsb.edu, tt@physics.ucsb.edu, malkan@astro.ucla.edu, email: rdb@slac.stanford.edu

Abstract. We test the evolution of the correlation between black hole mass and bulge properties, using a carefully selected sample of 20 Seyfert 1 galaxies at $z = 0.36 \pm 0.01$. We estimate black hole mass from the $H\beta$ line width and the optical luminosity at 5100Å, based on the empirically calibrated photo-ionization method. Velocity dispersion are measured from stellar absorption lines around Mgb (5175Å) and Fe (5270Å) using high S/N Keck spectra, and bulge properties (luminosity and effective radius) are measured from HST images by fitting surface brightness. We find a significant offset from the local relations, in the sense that bulge sizes were smaller for given black hole masses at $z = 0.36$ than locally. The measured offset is $\Delta \log M_\bullet = 0.62 \pm 0.10, 0.45 \pm 0.13, 0.59 \pm 0.19$, respectively for $M_\bullet - \sigma$, $M_\bullet - L_{\text{bulge}}$, and $M_\bullet - M_{\text{bulge}}$ relations. At face value, this result implies a substantial growth of bulges in the last 4 Gyr, assuming that the local $M_\bullet$–bulge property relation is the universal evolutionary end-point. This result is consistent with the growth of black holes predating the final growth of bulges at these mass scales ($\langle \sigma \rangle = 170$ km s^{-1}).

Keywords. Black hole physics – galaxy evolution

1. Introduction

The correlation of the mass of the central black Hole ($M_\bullet$) with the spheroid velocity dispersion σ (Ferrarese & Merritt 2000; Gebhardt *et al.* 2000) links phenomena at widely different scales (from the pcs of the BH sphere of influence to the kpcs of the bulge). This connection between galaxy formation and Active Galactic Nuclei (AGNs) has inspired several unified formation scenarios (e.g. Kauffmann & Haenhelt 2000; Volonteri *et al.* 2003; Haiman, Ciotti & Ostriker 2004). One of the most powerful observational tests of the proposed explanations is to measure the evolution of empirical relations with redshift. Different scenarios – all reproducing the local $M_\bullet$–σ relation – predict different cosmic evolution. For example – for a fixed $M_\bullet$ – Robertson *et al.* (2005) predict an increase of σ with redshift, Croton (2006) predict a decrease, while the models of Granato *et al.* (2004) expect no evolution. Solid observational input is clearly needed to make progress. In this study, we test the evolution of the correlations between $M_\bullet$ and bulge properties, i.e. σ, luminosity, and mass, using a carefully selected sample of 20 Seyfert 1 galaxies at $z = 0.36 \pm 0.01$.

2. Sample selection

We selected relatively low luminosity AGNs where the fraction of stellar light in the integrated spectrum is non-negligible so that σ can be reliably measured. At the same time, virial $M_\bullet$ can be obtained from the integrated properties of the broad line region using the same spectrum.

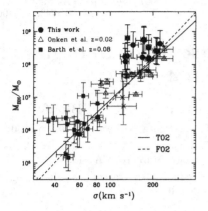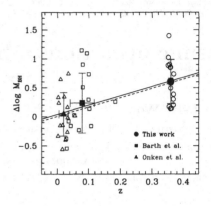

Figure 1. Left: The $M_\bullet$–σ relation of active galaxies. The symbols represent 14 Seyferts at $z = 0.36$ from this work (circles), 15 dwarf Seyfert galaxies at $z \sim 0.08$ from Barth et al. (2005; squares), 14 local AGNs with BH masses measured via reverberation mapping from Onken *et al.* (2004; triangles; two additional objects, excluded by Onken *et al.* and for consistency in our work, are shown as crosses). The local relationships of quiescent galaxies are shown for comparison as a solid (Tremaine *et al.* 2002) and dashed (Ferrarese 2002) line. Right: Offset from the local $M_\bullet$–σ relation. Large solid points with error bars represent the average and rms scatter for the three samples. The best linear fit to the data are shown as a solid line. The average offset of the $z = 0.36$ points is 0.62 ± 0.10 dex in $M_\bullet$. Note that $M_\bullet$ is estimated consistently with the same shape factor and therefore the relative position of the three samples along the y-axis is independent of the shape factor.

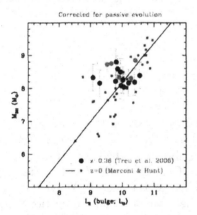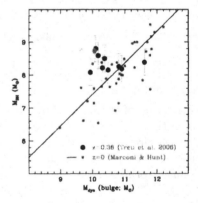

Figure 2. The $M_\bullet$–L_{bulge} (left) and the $M_\bullet$–M_{bulge} (right) relations at $z = 0.36$ are shown as blue circles (cyan circles are upper limits). The local relationship as measured by Marconi & Hunt (2003) is shown as solid line. The individual local points are shown as green squares. The bulges are found to be fainter or less massive at $z = 0.36$ for a fixed $M_\bullet$. The average offset from the local relation corresponds to 0.45 ± 0.13 dex and 0.59 ± 0.19 in M_{BH}, respectively, consistent with that found from the $M_\bullet$–σ relation, 0.62 ± 0.10 dex.

In order to minimize the systematic uncertainties related to sky subtraction and atmospheric absorption corrections, it is convenient to select specific redshift windows where the relevant emission and absorption lines (Hβ, Mgb, and Fe) fall in clean regions of the Earth's atmosphere. Accordingly, we selected the "clean window" $z = 0.36 \pm 0.01$, which corresponds to a look-back time of ~ 4 Gyrs. Based on the redshift range and the width of Hβ, we selected 20 AGNs from the SDSS DR4 for this study (see Woo *et al.* 2006).

3. Measuring the $M_\bullet$–bulge property relations at $z = 0.36$

Estimating the black hole mass. To estimate $M_\bullet$, we adopt the latest calibrations (Onken *et al.* 2004; Kaspi *et al.* 2005) of the local reverberation mass shape factor of the R_{BLR}–L_{5100} relationship:

$$M_\bullet = 2.15 \times 10^8 M_\odot \times \left(\frac{\sigma_{H\beta}}{3000 \text{kms}^{-1}} \right)^2 \left(\frac{\lambda L_{5100}}{10^{44} \text{ergs}^{-1}} \right)^{0.69}, \tag{3.1}$$

where $\sigma_{H\beta}$ is the second moment of the Hβ line profile and L_{5100} is the monochromatic luminosity at 5100Å, in the rest frame. Based on comparisons of reverberation data and single-epoch data, it is estimated that the intrinsic uncertainty associated with this method is approximately a factor of 2.5, i.e. 0.4 dex, on the $M_\bullet$ (Vestergaard 2002). This uncertainty dominates the errors on our input quantities, $\sigma_{H\beta}$ and L_{5100}, and we adopt it as our total uncertainty on the $M_\bullet$. We emphasize that when measuring the *evolution* of the $M_\bullet - \sigma$ relation, we will use the *same* shape factor for the local and distant sample, so that the specific choice of the shape factor is irrelevant.

The $M_\bullet$–σ relation. For 14 objects, reliable stellar velocity dispersions were measured from high S/N ratio spectra, obtained with the LRIS spectrograph at the Keck-I telescope. We used spectral regions including Mgb 5175Å and Fe 5270Å for comparing with stellar template. Figure 1 shows shows our $M_\bullet$–σ relation in comparison with the local relationship as measured by Tremaine *et al.* (2002) and by Ferrarese (2002). All objects in our sample are above the local relationship, that is smaller velocity dispersion for a fixed $M_\bullet$. In order to improve the measurement of the offset from the local relationship, we compare our sample with two samples of AGNs at lower redshift: the 14 reverberation mapped AGNs with mean redshift of 0.02 from Onken *et al.* (2004), and the 15 dwarf Seyfert galaxies with mean redshift of 0.08 from Barth *et al.* (2005). $M_\bullet$ of our sample and Barth *et al.* are consistently estimated using Eq. 3.1, which is calibrated on the reverberation masses (Onken *et al.* 2004). In other words, a change in the shape factor will move the three samples vertically by the same amount.

By design, the Onken *et al.* points straddle the local relationships. The Barth *et al.* points tend to lie preferentially above the local relationships with an average offset, of which the exact amount depends on the local slope. The $z = 0.36$ points are definitely above the local relationship. The offset is clearly detected and appears to increase with redshift. The average offset of our sample is 0.62 ± 0.10 dex in $M_\bullet$, corresponding to 0.15 ± 0.03 in $\Delta \log \sigma$.

$M_\bullet$–L_{bulge} and $M_\bullet - M_{\text{bulge}}$ relations. High quality HST images were obtained for our sample (GO-10216; PI Treu) and used to measure host galaxy properties. Using the GALFIT program (Peng *et al.* 2002), we fit 3-4 components, point source, bulge, disk, and bar, at the same time and determine the best fit parameters (Treu *et al.* in preparation). Bulge luminosity is measured for 17 objects (including upper limits for 5 objects) while for 3 objects we could not determine reliable bulge properties due to the presence of dust lanes. We correct the passive evolution of luminosity due to the aging of stellar populations in order to compare our sample with the local sample. Combining the measured σ and the effective radius, we derive dynamical masses of bulges for 10 objects. Figure 2 shows $M_\bullet$ relation with bulge luminosity (left panel) and bulge mass (right panel), comparing with the local relations of quiescent galaxies from Marconi & Hunt (2003). For given $M_\bullet$, bulges at $z = 0.36$ are fainter than the local sample, with average offset 0.45 ± 0.13 dex in $M_\bullet$. Considering upper limits of luminosity for 5 objects, this offset could be a lower limit. A consistent result is found in the $M_\bullet$–M_{bulge} relation, which shows an average offset of 0.59 ± 0.19 dex in $M_\bullet$.

4. Discussion and conclusions

All three relations between $M_\bullet$ and bulge properties show a significant offset from the local relationship. Three possible explanations are 1) systematic errors; 2) selection effects; 3) cosmic evolution. Systematic errors are unlikely to account for the offset, which is significantly larger than the overall systematic uncertainty (~ 0.2–0.3 dex). Selection effects could be present both in our sample (selected against low luminosity AGNs and thus small $M_\bullet$), and in the local sample (favoring more evolved systems), possibly resulting in the observed evolution of the $M_\bullet$–bulge property relation. Larger samples of AGNs with determined $M_\bullet$, σ, and host galaxy properties are needed both locally and at high-redshift to improve the understanding of selection effects. If cosmic evolution is the correct explanation, the observed offset would support earlier growths of supermassive BHs in galaxies with mass scales of $\langle\sigma\rangle=170$ km s^{-1}. This could be evidence for 'downsizing' in the BH-galaxy coevolution i.e. more massive galaxies arrive at the local relationship early in time. This scenario can be further investigated with a sample of AGN host galaxies with a range in mass at fixed redshifts.

References

Barth, A. J., Greene, J. E. & Ho, L. C. 2005, ApJL, 619, L151
Croton, D. J. 2005, MNRAS, 369, 1808 (astro-ph/0512375)
Ferrarese, L. & Merritt, D. 2000, ApJL, 539, L9
Ferrarese, L. 2002, in Proc. of the 2nd KIAS Astrophysics Workshop, eds. Chang-Hwan Lee & Heon-Young Chang (Singapore: World Scientific), p. 3
Gebhardt, K. *et al.* 2000, ApJL, 539, L13
Granato, *et al.* 2000, ApJ, 600, 580
Kaspi, S. *et al.* 2005, ApJ, 629, 61
Kauffmann, G. & Haehnelt, M. 2000, MNRAS, 311, 576
Marconi, A. & Hunt, L. K. 2003, ApJ, 589, L1
Onken, C. A. *et al.* ApJ, 615, 645
Peng, C. *et al.* 2002, AJ, 124, 266
Robertson, B. *et al.* 2006, ApJ, 641, 90
Tremaine, S. *et al.* 2002, ApJ, 574, 740
Volonteri, M., Haardt, F. & Madau, P. 2003, ApJ, 582, 559
Woo, J.-H., Treu, T., Malkan, M. A. & Blandford, R. D. 2006, ApJ, 645, 900

DEBORAH DULTZIN-HACYAN: Some authors use the width of narrow lines such as [OIII] and [NII] lines as tracers of velocity dispersion. You have the data to check if this is correct.

JONG-HAK WOO: We find that depending on how [OIII] line is fitted (Gaussian fits, double-Gaussian, etc) σ_* derived from [OIII] show a systematic offset. Also, there is a large scatter between measured velocity dispersions and [OIII]-derived velocity dispersions as also other authors find.

SUZY COLLIN: Isn't there a contradiction with what we heard in a talk by Dr. Peng, namely, that there is evolution at $z \sim 4$ while you find no evolution below $z \sim 2$?

JONG-HAK WOO: I think the main reason for the difference between samples is the mass dependency. Peng's sample has more massive galaxies whereas our objects are about order of magnitude less massive. Perhaps by selecting more massive objects at low redshift would give similar results as the Peng's sample.

Black Holes Across the Mass Spectrum

Black holes across the mass spectrum

J. M. Hameury
The disc instability model for dwarf novae in the AGN context

K. Ohsuga
Radiation hydrodynamic simulations of super-Eddington accretion flows

M. Tuerler
Synchrotron outbursts in galactic and extra-galactic jets, any difference?

F. Mirabel
Closing: Present status and future developments of black holes across the range of masses

Clementinum – the former site of Jesuit College and the site of National Library at present.

Astronomical Clock, among the oldest horologes of its kind (1410; moving sculptures added later).

Black Holes from Stars to Galaxies – Across the Range of Masses
Proceedings IAU Symposium No. 238, 2006
V. Karas & G. Matt, eds.

© 2007 International Astronomical Union
doi:10.1017/S1743921307005194

The thermal–viscous disk instability model in the AGN context

Jean-Marie Hameury,[1] Jean-Pierre Lasota[2] and Maxime Viallet[1]

[1]Observatoire de Strasbourg, CNRS/Université Louis Pasteur,
11 rue de l'Université, 67000 Strasbourg, France

[2]Institut d'Astrophysique de Paris, CNRS/Université Pierre et Maris Curie,
98bis Bd. Arago, 75015 Paris, France

email: hameury@astro.u-strasbg.fr; viallet@astro.u-strasbg.fr

Abstract. Accretion disks in AGN should be subject to the same disk instability responsible for dwarf novae outbursts and soft X-ray transients in cataclysmic variables (CVs) and LMXBs. It has been suggested that this thermal/viscous instability can account for long term variability of AGNs. We analyze here the application of the DIM to the AGN case, using our adaptive grid numerical code developed in the context of CVs, enabling us to fully resolve the disk radial structure. We show that in AGNs, the width of heating and cooling fronts is so small that they cannot be resolved by standard codes, and that they propagate on time scales much shorter than the viscous time. As a result, transition fronts propagate back and forth in the disk, leading only to small luminosity variations. Truncation of the inner part of the disk by e.g. an ADAF will not alter this result, but enables the presence of quiescent states.

Keywords. accretion, accretion disks – instabilities – galaxies: active – stars: dwarf novae

1. Introduction

Accretion disks are found in a large variety of astronomical objects, from young stars to active galactic nuclei. Among these, close binaries deserve special attention, because they are nearby, and vary on short timescales that enable time-dependent studies of their light curve. In particular, a number of these systems show large outbursts, as for example dwarf novae (DN), which are a subclass of cataclysmic variables. These outbursts are believed to be due to a thermal–viscous accretion disk instability which arises when the disk temperature becomes of order of 10^4K, enough for hydrogen to become partially ionized and opacities to depend strongly on temperature (Lasota 2001). Similarly, soft X-ray transients, a subclass of low-mass X-ray binaries, also show outbursts, their amplitude being, in general, larger and the time scales longer than for dwarf novae. The ionization instability of the accretion disk is also thought to be the cause of the outbursts, the difference with DNe being due to larger masses of the compact objects (and thus to deeper gravitational potential wells) and to the effect of illumination of the disk, much more important in the case of X-ray binaries (Dubus *et al.* 2001).

It was realized long ago (Lin & Shields 1986) that the same instability could be present in accretion disks in AGNs; it was found that, at radii $\sim 10^{15-16}$ cm, where the effective temperature is indeed of a few thousand degrees, the disk should be unstable. For the parameters of AGNs, the implied timescales are of order of $10^4 - 10^7$ yr, making impossible the direct observation of the instability in a particular object, but implying that in many systems the disk is not in viscous equilibrium and that many AGNs are in a quiescent state. It was also immediately realized that, as in DNe, the character of putative AGN outbursts strongly depends on assumptions one makes about the disk viscosity (Mineshige

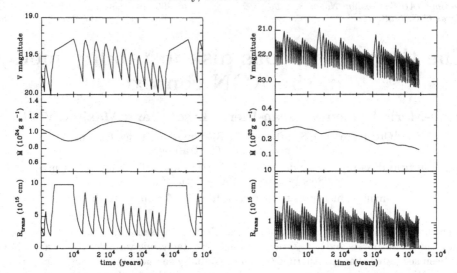

Figure 1. Left: Time evolution of an accretion disk with the following parameters: black hole mass: 10^8 $M_\odot$, inner and outer radius: 10^{14} and 10^{16} cm respectively, mean mass transfer rate: 10^{24} g s^{-1}. Top panel: visual magnitude, intermediate panel: accretion rate onto the black hole, lower panel: radius at which the transition between the hot and cold regimes takes place. Right: same parameters except that $\dot{M} = 2\ 10^{22}$ g s^{-1}.

& Shields 1990). However, while in the case of DNe one is guided in fixing the viscosity prescription by the *observed* outburst properties, in the case of AGN it is not even clear that outbursts are present as the variability of these objects could be due just to mass-supply variations. (In many cases it is not even clear that a disc is present.) This state of affairs gave rise to various, more or less arbitrary, prescriptions of how viscosity varies (or not) with the state of the accretion flow (Mineshige & Shields 1990; Menou & Quataert 2001; Janiuk *et al.* 2004). In addition, results of numerical calculations of AGN outbursts were marred by the insufficient resolution of grids used. As showed by Hameury *et al.* (1998) low grid resolution often leads to unreliable, unphysical, results.

We revisit here the application of the disk instability model in the context of AGNs, using the code we have developed in the context of dwarf novae and soft X-ray transients.

2. Disk evolution

We solve the standard equations for mass and angular momentum conservation in a geometrically thin accretion disk. The heat conservation involves the heating and cooling rates of the unsteady disk; in other words, one must know the effective temperature $T_{\rm eff}$ as a function of radius r, surface density Σ and central temperature $T_{\rm c}$. We use the standard approximation that $T_{\rm eff}$ does not depend on any other parameter than r, Σ and $T_{\rm c}$, and is thus found by solving for the vertical structure of the disk in steady state, the value of the Shakura-Sunyaev parameter α being not constrained.

We calculate a grid of vertical structures in the same way as in Hameury *et al.* (1998). We do not include self gravity, which can be important at large distances from the black hole, typically $1 - 2\ 10^{16}$ cm for a 10^8 $M_\odot$ black hole (Cannizzo 1992; Cannizzo & Reiff 1992). When self gravity dominates, the disk becomes gravitationally unstable, which may lead to fragmentation if the cooling time is short enough, or significantly change the angular momentum transport by introducing non local terms (see e.g. Lin & Pringle 1987; Balbus & Papaloizou 1999). In any case, the thermal-viscous instability model can

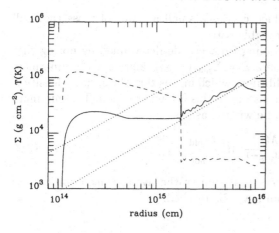

Figure 2. Radial structure of a disk. The blue solid line represents the surface density, the red dashed one the central temperature. The dotted lines are the critical $\Sigma_{\min}$ and $\Sigma_{\max}$.

no longer apply, and we always make sure that, in our calculations, self gravity is never important.

In the standard DN model, it is assumed that the α-parameter changes rapidly when the disk temperature reaches the ionization instability; this is required for the amplitude of the modelled instability to be comparable to the observed one. It has been suggested that α might remain constant in AGN disks (Gammie & Menou 1998; Menou & Quataert 2001); however, this proposal is based on numerical simulations which are far from providing us with a complete treatment of the turbulent dynamics of disks (Balbus, private communication), and we use here the standard DN model for viscosity.

Fig. 1 shows the evolution of a disk around a 10^8 M$_\odot$ black hole, for two different values of the mass transfer rate: 1% and .01% of the Eddington limit, assuming a 10% efficiency. In both cases, $\alpha_{\rm hot} = 5\alpha_{\rm cold} = 0.2$, and the inner and outer disk radius are 10^{14} and 10^{16} cm respectively. The disk is indeed unstable; however, this instability does not result in large outbursts, but rather in rapid oscillations reminiscent of what is found in dwarf novae when α is assumed to be constant.

The basic reason for this is that transition fronts propagates at approximately α times the sound speed (Menou *et al.* 1999), i.e. on a time scale

$$t_{\rm front} = \frac{r}{\alpha c_{\rm s}} = \frac{r}{h}t_{\rm th} \tag{2.1}$$

where $t_{\rm th}$ is the thermal time scale. $t_{\rm front}$ is shorter than the viscous time $t_{\rm visc} = (r/h)^2 t_{\rm th}$ by a factor r/h, i.e. by several orders of magnitude. The cooling front therefore propagates so rapidly that the surface density at smaller radii does not change; to a first approximation, it cannot propagate in regions where $\Sigma > \Sigma_{\max}(\alpha_{\rm cold})$, the maximum surface density for a cold stable disk. In the CV case, $t_{\rm front}$ is shorter than $t_{\rm visc}$, but not by such a large amount and strong gradients in the disk make the effective viscous time comparable to the the front propagation time. When $\Sigma_{\max}$ is reached, the cooling front is reflected into a heating front, which propagates outwards until the outer disk edge is reached, or Σ is less than $\Sigma_{\min}(\alpha_{\rm hot})$ (the minimum surface density of a hot stable disk). The radial profile of the disk is shown in Fig. 2, in the case where $\dot{M}$ is low. There are two stable inner and outer regions which are clearly seen, where the disk is always hot (inner zone) or cold (outermost regions). The intermediate, unstable zone is divided in two regions: an inner unstable part, and an outer marginally stable one, where $\Sigma = \Sigma_{\min}(r)$, resulting from the successive passage of heating fronts that die at radii decreasing with time; a leftover of the death of these fronts is a little wiggle in Σ that get smoothed with time as a result of diffusion, or when a heat front is able to cross a large distance. Note also the spike in the unstable region, which carries a small amount of mass that will cause the small wiggles in the marginally stable region.

The minimum radius reached by the cooling front can de determined by noting that the front propagates down to a point where $\Sigma = \Sigma_{max}(\alpha_{cold})$, and that the innermost parts of the disk are in quasi viscous equilibrium, which means that the surface density is determined by the accretion rate, almost constant in this hot inner region. The minimum radius reached by cooling fronts can then be written as:

$$r = 1.7 \times 10^{15} \Big(\frac{\dot{M}}{10^{23}\text{g s}^{-1}} M_8 \frac{\alpha_{cold}}{\alpha_{hot}} f \Big)^{0.4} \text{cm} \qquad (2.2)$$

where $f = 1 - (r_{in}/r)^{1/2} \sim 1$; this shows that, unless the average accretion rate is extremely small, a cold quiescent state can never be reached.

3. Discussion

We have shown that the straightforward application of the disk instability model to AGN does lot lead to large amplitude outbursts, but to some sort of flickering. This is because in AGN, the Mach number is so large that the propagation time of heating and cooling fronts is much less than the viscous time. Evaporation of the innermost parts due e.g. to the formation of an ADAF will change this conclusion, as low states will occur, as shown by Eq. 2.2, since if the disk is truncated at a few Schwarzschild radii, cooling fronts can reach the inner disk edge even for relatively high $\dot{M}$; the flickering behaviour in high states will remain though. The amplitude of the outbursts, as well as the maximum luminosity reached remain to be determined. Irradiation of the disk by X-rays might also affect the outcome of the instability; Burderi *et al.* (1998) showed that irradiation does not prevent the instability, and, because it does not affect the propagation time of the front nor the viscous time, it in unlikely that it can have a major effect.

References

Balbus, S. A. & Papaloizou, J. C. B. 1999, ApJ, 376, 214
Burderi, L., King, A. R. & Szuszkiewicz, E. 1998, ApJ, 509, 85
Cannizzo, J. K. 1992, ApJ, 385, 94
Cannizzo, J. K. & Reiff, C. M. 1992, ApJ, 385, 87
Dubus, G., Hameury, J.-M. & Lasota, J.-P. 2001, A&A, 373, 251
Gammie, C. F. & Menou, K. 1998, ApJ, 492, L75
Hameury, J.-M., Menou, K., Dubus, G., Lasota, J.-P. & Huré, J. M. 1998, MNRAS, 298, 1048
Janiuk, A., Czerny, B., Siemiginowska, A. & Szczerba, R. 2004, ApJ, 602, 595
Lasota, J.-P. 2001, New Astron. Revs, 45, 449
Lin, D. N. C. & Pringle, J. E. 1986, MNRAS, 225, 607
Lin, D. N. C. & Shields, G. 1986, ApJ, 305, 28
Menou, K. & Quataert, E. 2001, ApJ, 552, 204
Menou, K., Hameury, J.-M. & Stehle, R. 1999, MNRAS, 305, 79
Mineshige, S. & Shields, G. A. 1990, ApJ, 351, 47

SUZY COLLIN: How do you avoid to enter in the ADAF/ADIOS/... regime at mass–rate that is typically below 1/100 of the Eddington limit?

JEAN-MARIE HAMEURY: It is not known at present what is the minimum accretion rate for the standard Shakura–Sunayev solution to apply; the 1/100 limit you quote refers to a maximum value for ADAF to be possible. In any case, the formation of an ADAF/ADIOS etc. will result in truncation of the inner disc, and this is included in the calculations I have shown.

Black Holes from Stars to Galaxies - Across the Range of Masses
Proceedings IAU Symposium No. 238, 2006
V. Karas & G. Matt, eds.
© 2007 International Astronomical Union
doi:10.1017/S1743921307005200

Radiation hydrodynamic simulations of super-Eddington accretion flows

Ken Ohsuga

Department of Physics, Rikkyo University, Toshimaku, Tokyo 171-8501, Japan

email: k_ohsuga@rikkyo.ac.jp

Abstract. We perform the two-dimensional radiation-hydrodynamic simulations to study the radiation pressure-dominated accretion flows around a black hole (BH). Our simulations show that the highly supercritical accretion flow (mass accretion rate is much larger than the critical value) is composed of the disk region and the outflow region above the disk.

The radiation force supports the thick disk and drives the outflow. The photon trapping plays an important role within the disk, reducing the disk luminosity. On the other hand, in the case that mass accretion rate moderately exceeds the critical value, we find that the disk is unstable and exhibits the limit-cycle oscillations. The disk oscillations in our simulations nicely fit to the variation amplitude and duration of quasi-periodic luminosity variations observed in the GRS 1915+105 microquasar.

Keywords. Accretion, accretion disks – black hole physics

---

1. Introduction

Recent observations have discovered very bright objects which may undergo supercritical (or super-Eddington) accretion flows. Ultraluminous X-ray sources (ULXs; Fabbiano 1989; Ebisawa *et al.* 2003) and narrow-line Seyfert 1 galaxies (NLS1s; see Boller 2004 for a review) are good examples. Although the ULXs might be powered by the sub-critical accretion onto the intermediate mass BHs (Makishima *et al.* 2000), a piece of evidence of the supercritical flow is reported (Vierdayanti *et al.* 2006). On the other hand, NLS1s have in general large Eddington ratios (luminosity over Eddington luminosity) and some of them seem to fall in the supercritical accretion regimes (Mineshige *et al.* 2000).

The supercritical accretion onto the central BH is thought to play important roles for the evolution of their host galaxies. King (2003), as well as Silk & Rees (1998), suggested that the strong outflow from the supercritical accretion flow regulates the evolution of the supermassive BH and its host galaxy, leading to the correlation between the velocity dispersion of the bulge stars ($\sigma_\star$) and the BH mass ($M - \sigma_\star$ relation; Gebhardt *et al.* 2000; Ferrarese & Merritt 2000). Despite growing evidence indicating the existence of supercritical accretion flows in the universe, theoretical understanding is far from being complete.

What makes the supercritical accretion flows distinct from the standard disk type flow is the presence of photon trapping (Begelman 1978; Ohsuga *et al.* 2002). By the photon trapping, photons generated via the viscous process are advected inward with accreting gas without being radiated away. In addition, the multi-dimensional gas motion, such as convective or large-scale circulation, would occur in the supercritical disk accretion flows. Strong outflow might also be generated due to the strong radiation force. We thus need at least two-dimensional treatment to investigate the supercritical disk accretion flows.

When the advective cooling is predominant over the radiative cooling, the disks are stabilised. But, the supercritical disks are not always stable. If the mass accretion rate is comparable to or moderately exceeds the critical value, thermal and secular instability is though to arise in the radiation-pressure dominant region (Lightman & Eardley 1974; Shibazaki & Hōshi 1975).

In this paper, we investigate the steady structure of the stable supercritical accretion flows and time evolution of the unstable supercritical accretion disks by performing the two-dimensional radiation hydrodynamic (RHD) simulations.

2. Highly supercritical flows; stable disks

We solve the full set of RHD equations including the viscosity term with using an explicit-implicit finite difference scheme on the Eulerian grids. Our viscosity model is basically the same as the α prescription of the viscosity proposed by Shakura & Sunyaev (1973) The matter is added continuously from the outer disk boundary to the initially (nearly) empty space at the rate of $10^3 L_E/c^2$. The RHD equations and our methods are shown in Ohsuga et al. (2005) and Ohsuga (2006) in detail.

By performing long-term numerical calculations, we for the first time succeed in reproducing the quasi-steady state of the supercritical accretion flows. Although the research history of such simulations stems back to the late 1980's, when Eggum et al. (1987) for the first time performed numerical simulations, their calculations were restricted to the first few sec (see also Kley 1989; Okuda et al. 1997).

Figure 1 displays the cross-sectional view of the density distributions (colors), overlaid with the velocity vectors (arrows) in the quasi-steady state. We find that the flow structure is divided into two regions: the disk region around the equatorial plane (orange) and the outflow region above the inflow region (blue). The disk is geometrically and optically thick, and it is supported by the radiation force. We find a number of cavities

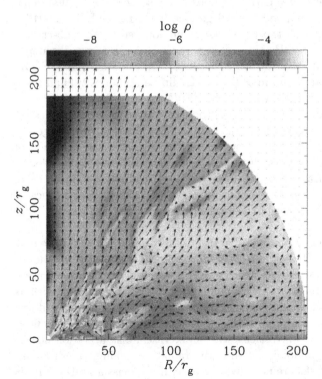

Figure 1. The density distribution in the quasi-steady state overlaid with the velocity vectors. Here, we set the mass input rate to be $10^3 L_E/c^2$ and the mass of the BH to be $10 M_\odot$.

and the circular motion in the disk region. The radiatively-driven outflow appears along the rotation axis. The Kelvin-Helmholtz instability is caused around the disk surface.

In the quasi-steady state, the mass accretion rate onto the BH is $\sim 10^2 L_{\mathrm{E}}/c^2$. In other words, only 10% of the inflowing material is finally swallowed by the BH; i.e., 90% of the mass input rate gets stuck in the dense, disk-like structure around the equatorial plane, or transform into the known collimated high-velocity outflows perpendicular to the equatorial plane or into the low-velocity outflows with wider opening angles, respectively.

The resulting luminosity is about three times larger than the Eddington luminosity. Thus, the energy conversion efficiency, ~ 0.03, is much smaller than the prediction of the standard disk theory, ~ 0.1. This is because the large amount of photons generated inside the disk is swallowed by the BH without being radiated away. This is so-called photon trapping effects. Since the slim-disk model (Abramowicz *et al.* 1988) does not fully consider the photon trapping effects, this model overestimates the disk luminosity.

Since the supercritical disk is geometrically and optically thick, its emission is mildly collimated. Although the disk luminosity is $\sim 3L_{\mathrm{E}}$, the apparent luminosity exceeds $10L_{\mathrm{E}}$ in the face-on view. It implies that the large luminosities of the ULXs are explained by the supercritical accretion onto the stellar-mass BHs.

3. Moderately supercritical flows; unstable disks

Here, we set the mass input rate to be $10^2 L_{\mathrm{E}}/c^2$, although we assume it to be $10^3 L_{\mathrm{E}}/c^2$ in the previous simulation. In this case, the disk is unstable and exhibits quasi-periodic state transitions. Figure 2 represents the time evolution of the mass accretion rate onto the BH, $\dot{M}_{\mathrm{acc}}$ (blue), the outflow rate, $\dot{M}_{\mathrm{out}}$ (magenta), the disk luminosity, L (red), and the trapped luminosity, L_{trap} (green). As shown in this figure, the mass accretion rate onto the BH ($\dot{M}_{\mathrm{acc}}$) drastically varies. It suddenly rises, retains high value for about 40 s, and then decays. Such time variation of the mass accretion rate causes the quasi-periodic luminosity changes. Whereas the luminosity (L) is not more than $0.3L_{\mathrm{E}}$ in the low-luminosity state, it reaches around $2L_{\mathrm{E}}$ in the high-luminosity state. The burst duration is also about 40 s. The physical mechanism of such limit-cycle oscillations is the disk instability in the radiation-pressure dominant region.

Resulting light curve in our simulations gives a nice fit to the time variation of the luminosity of GRS 1915+105. Based on the analysis of the data of GRS 1915+105 taken by

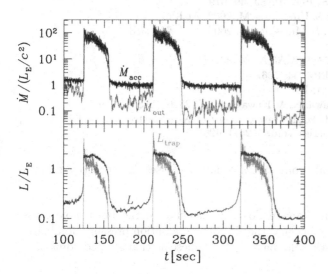

Figure 2. The time evolution of the mass accretion rate (blue), outflow rate (magenta), the luminosity (red), and trapped luminosity (green). Here, we set the mass input rate to be $10^2 L_{\mathrm{E}}/c^2$.

RXTE and ASCA, Yamaoka *et al.* (2001) have produced the light curve. The luminosity is several times higher in the high-luminosity state than in the low-luminosity state. The duration of the high-luminosity state is about $30 - 50$ s, and there is sharp edge between the high and low states. Our numerical results succeed in reproducing these observed features.

In Figure 2, the outflow rate ($\dot{M}_{out}$) suddenly rises and decays with the luminosity. It implies that the radiately-driven outflow forms intermittently. The typical velocity of the outflow is 30% of the light speed. The photon-trapping effects appear during the high-luminosity state. It is found that the trapped luminosity (L_{trap}), which means the radiation energy swallowed by the BH per unit time, is comparable to the luminosity in the high-luminosity state. The energy conversion efficiency is around 0.03 in the high-luminosity state, although it is ~ 0.1 in the low-luminosity state. We stress that the multi-dimensional effects, e.g., outflow and photon trapping, are significant in the high-luminosity state.

Acknowledgements

The calculations were carried out by a parallel computer at Rikkyo University and Institute of Natural Science, Senshu University. We acknowledge Research Grant from Japan Society for the Promotion of Science (17740111).

References

Abramowicz, M. A., Czerny, B., Lasota, J. P. & Szuszkiewicz, E. 1988, ApJ, 332, 646
Begelman, M. C. 1978, MNRAS, 184, 53
Boller, T. 2004, PThPS, 155, 217
Ebisawa, K., Zycki, P., Kubota, A., Mizuno, T. & Watarai, K. 2003, ApJ, 597, 780
Eggum, G. E., Coroniti, F. V. & Katz, J. I. 1987, ApJ, 323, 634
Fabbiano, G. 1989, ARA&A, 27, 87
Ferrarese, L., Merritt, D. 2000, ApJ, 539, L9
Gebhardt, K. *et al.* 2000, ApJ, 539, L13
King, A. 2003, ApJ, 596, L27
Kley, W. 1989, A&A, 222, 141
Lightman, A. P. & Eardley, D. M. 1974, ApJ, 187, L1
Makishima, K. *et al.* 2000, ApJ, 535, 632
Mineshige, S., Kawaguchi, T., Takeuchi, M., & Hayashida, K. 2000, PASJ, 52, 499
Okuda, T., Fujita, M. & Sakashita, S. 1997, PASJ, 49, 679
Ohsuga, K., Mineshige, S., Mori, M. & Umemura, M. 2002, ApJ, 574, 315
Ohsuga, K., Mori, M., Nakamoto, T. & Mineshige, S. 2005, ApJ, 628, 368
Ohsuga, K. 2006, ApJ, 640, 923
Shakura, N. I. & Sunyaev, R. A. 1973, R.A.A&A, 24, 337
Shibazaki, N. & Hōshi, R. 1975, PThPh, 54, 706
Silk, J. & Rees, M. J. 1998, A&A, 331, L1
Vierdayanti, K., Mineshige, S., Ebisawa, K. & Kawaguchi, T. 2006, PASJ, 58, 915
Yamaoka, K., Ueda, Y. & Inoue, H. 2001, New Century of X-ray Astronomy, ed H. Inoue., H. Kunieda. (ASP Conference Series Volume 251), 426

SUZY COLLIN: Is it possible to scale your model to larger masses, i.e. to active galactic nuclei?

KEN OHSUGA: The instability mechanism does not depend on the black hole mass. However, there is some dependency because the size of the radiation pressure dominated region depends slightly on the black hole mass, so the timescale changes accordingly.

Black Holes from Stars to Galaxies – Across the Range of Masses
Proceedings IAU Symposium No. 238, 2006
V. Karas & G. Matt, eds.

© 2007 International Astronomical Union
doi:10.1017/S1743921307005212

Synchrotron outbursts in Galactic and extragalactic jets, any difference?

Marc Türler[1,2] and Elina J. Lindfors[3,4]

[1]INTEGRAL Science Data Centre, ch. d'Ecogia 16, 1290 Versoix, Switzerland

[2]Geneva Observatory, University of Geneva, ch. des Maillettes 51, 1290 Sauverny, Switzerland

[3]Tuorla Observatory, Väisälä Institute of Space Physics and Astronomy, University of Turku, 21500 Piikkiö, Finland

[4]Metsähovi Radio Observatory, Helsinki University of Technology, 02540 Kylmälä, Finland

email: marc.turler@obs.unige.ch, elilin@utu.fi

Abstract. We discuss differences and similarities between jets powered by super-massive black holes in quasars and by stellar-mass black holes in microquasars. The comparison is based on multi-wavelength radio-to-infrared observations of the two active galactic nuclei 3C 273 and 3C 279, as well as the two galactic binaries GRS 1915+105 and Cyg X-3. The physical properties of the jet are derived by fitting the parameters of a shock-in-jet model simultaneously to all available observations. We show that the variable jet emission of galactic sources is, at least during some epochs, very similar to that of extragalactic jets. As for quasars, their observed variability pattern can be well reproduced by the emission of a series of self-similar shock waves propagating down the jet and producing synchrotron outbursts. This suggests that the physical properties of relativistic jets is independent of the mass of the black hole.

Keywords. Radiation mechanisms: nonthermal – shock waves – stars: individual (GRS 1915+105, Cyg X-3) – galaxies: active – galaxies: jets – quasars: individual (3C 273, 3C 279)

1. Introduction

Quasars and microquasars – their galactic analogs – are relativistic jet sources thought to be powered by super-massive and stellar-mass black holes, respectively (see Mirabel & Rodríguez 1998 for a review). Although the radio emission of microquasars tends to show a greater variety of behaviours, during some activity periods their radio variability pattern resembles that of quasars, but on much shorter timescales (hours or days instead of years). It is now rather well established for quasars that outbursts identified in their radio lightcurves are related to moving structures in the jet imaged with radio interferometric techniques. There is also growing evidence that these structures emitting synchrotron radiation are propagating shock waves along the jet flow rather than discrete ejected plasma clouds (Kaiser *et al.* 2000; Türler *et al.* 2004).

Based on these facts, a natural step forward is to confront theoretical predictions of shock wave models with the observed evolution of the outbursts. The approach we developed since several years is to try to fit a series of self-similar model outbursts to the multi-wavelength monitoring observations of some of the best studied objects. Until now, we studied data sets of four objects: the quasar 3C 273 (Türler *et al.* 1999; 2000), the blazar 3C 279 (Lindfors *et al.* 2006) and the microquasars GRS 1915+105 (Türler *et al.* 2004) and Cyg X-3 (Lindfors *et al.* in prep.). In this short contribution, we try to compare and discuss the jet properties derived for these four sources as one of the first attempts to sketch out a more global picture of black hole jets across the range of mass.

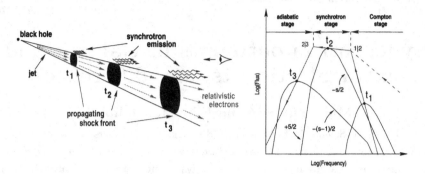

Figure 1. Schematic view of a propagating shock wave in a relativistic jet and the three-stage evolution of the associated synchrotron outburst according to the model of Marscher & Gear (1985) and with the modification proposed by Björnsson & Aslaksen (2000) (dashed line).

2. Data and Method

A description of the data sets and the method can be found in our previous publications listed above. We just recall here that for 3C 279 and 3C 273 we use data spanning one and two decades, respectively, whereas for GRS 1915+105 we use the data of 15 May 1997 in the plateau/state C/class χ state (Mirabel & Rodríguez 1998) and for Cyg X-3 the outbursts of Feb.–Mar. 1994 (Fender *et al.* 1997). The main change with respect to previous studies is that we use here consistently for all sources the modification proposed by Björnsson & Aslaksen (2000) for the initial Compton stage of the Marscher & Gear (1985) shock model in the case of only first-order Compton cooling. This flattens the initial rise of the spectral turnover with decreasing frequency, as shown schematically in Fig. 1. In the case of 3C 273 the jet parameters are such that this modification made the Compton stage completely flat or even inverted (decreasing turnover flux with time), which would make the model incompatible with infrared data. This problem could however be solved by assuming that the shocked material is viewed sideways (Marscher *et al.* 1992). Such an assumption is not unreasonable in sources with highly superluminal jets, as the jet angle θ to the line of sight for this to happen is defined by $\cos(\theta) = \beta \Leftrightarrow \sin(\theta) = 1/\gamma$, which is the same condition that maximizes the apparent transverse velocity β_{app} for a given bulk velocity $\beta = v/c$. We therefore use here a sideways orientation of the jet for both extragalactic sources and a face-on orientation for both microquasars.

3. Results and Discussion

Due to lack of space we cannot present here the derived evolution of the average outburst in the four sources. The main differences we obtain with the changes to the model described above, is that in general we do not obtain an almost flat synchrotron stage, but one with a much steeper increase of flux with turnover frequency, which becomes more difficult to distinguish from the now shallower Compton stage. There is therefore some uncertainty whether the synchrotron stage exists at all or whether there is a direct transition from the Compton to the adiabatic stage. Another interesting point is that for the initial phases of the outburst evolution we suspect the spectral turnover ν_m not to be due to synchrotron self-absorption as assumed, but to a low energy cut-off of the electron energy distribution. Evidence for this is the shallow spectral slope we find below the turnover frequency and which we describe in our modelling by an inhomogeneous synchrotron source. However, one would rather expect the source to become more inhomogeneous with time as the source increases in size, whereas we observe the

Table 1. Physical jet properties of the four objects derived for the average model outburst at the time when it reaches a maximum flux (i.e. around the transition to the final decay stage).

Object	M_{BH} [$M_\odot$]	Δt_{obs} [s]	Δl [AU]	θ_{src} [mas]	B [G]	K [erg^{s-1}/cm^3]	s	u_e [erg/cm^3]	u_B/u_e
3C 273 [1]	$6.6\,10^9$	$3.1\,10^7$	$3.1\,10^6$	0.39	0.13	$6.7\,10^{-7}$	2.52	$1.3\,10^{-4}$	5.6
3C 279 [2]	$3\,10^8$	$3.1\,10^7$	$4.7\,10^6$	0.25	0.26	$8.5\,10^{-7}$	2.09	$1.4\,10^{-5}$	195
Cyg X-3 [3]	<3.6	$8.6\,10^4$	$6.6\,10^2$	4.6	0.83	$1.9\,10^{-4}$	1.68	$7.9\,10^{-5}$	344
GRS 1915 [4]	14	$1.8\,10^3$	3.1	0.02	0.15	$6.7\,10^2$	1.79	$1.4\,10^3$	$6.3\,10^{-7}$
GRS 1915 [5]	14	$1.8\,10^3$	6.7	0.07	7.5	$3.4\,10^{-1}$	1.79	$4.7\,10^{-1}$	4.8

[1] M_{BH} from Paltani & Türler (2005) and with $\beta_{app} = 10$ & $\theta = 10°$ (Savolainen *et al.* 2006).
[2] M_{BH} from Wang *et al.* (2004) and with $\beta_{app} = 10$ (Jorstad *et al.* 2004) & $\theta = 5°$ (see text).
[3] M_{BH} from Stark & Saia (2003) and $\beta = 0.81$, $\theta = 14°$ & $d = 10\,$kpc (Mioduszewski *et al.* 2001).
[4] M_{BH} from Greiner *et al.* (2001) and $\beta = 0.6$, $\theta = 61°$ & $d = 9\,$kpc (Türler *et al.* 2004).
[5] alternative with $\beta = 0.9$, $\theta = 55°$ & $d = 7\,$kpc (uncertainty on d, see Fender *et al.* 2003).

opposite behaviour. An alternative explanation is that this flatter spectral index mimics the characteristic $\nu^{1/3}$ slope expected when self-absorption becomes important at a lower frequency than the synchrotron frequency associated to the electrons with minimal energy (see e.g. the spectral shape in the upper panel of Fig. 1 of Granot & Sari 2002). This interpretation could also explain the too rapid evolution of the turnover frequency with time that we observe in both galactic sources. Currently, this is described in our modelling by a non-conical jet with an opening angle increasing with time, but the physical justification for such a trumpet-like shape is not clear.

More quantitatively, we made a first attempt to derive the physical properties of the jets, which we summarize in Table 1. These values have been derived for the particular point when the synchrotron spectrum of the average outburst reaches its maximum flux, which happens here for all sources at the transition to the final adiabatic cooling phase. A critical parameter is the angular size θ_{src} of the source of synchrotron emission, which we assume to be equal to the width of the jet. By assuming a same jet opening half-angle of 2° for all sources, θ_{src} can thus be calculated from the length along the jet Δl traveled by the shock during the observed time interval Δt_{obs} needed for the outburst to reach a maximal flux. This length itself depends on good estimates of the object's distance d, the jet angle θ to the line of sight, and the apparent $\beta_{app} = v_{app}/c$ or real $\beta = v/c$ jet speed. We can then simply use Eqs. (3) to (5) of Marscher (1987) to calculate the magnetic field strength B, the normalization K of the electron energy distribution $N(E) = K\,E^{-s}$ and the energy density u_e of relativistic electrons. For the latter we use a ratio ν_2/ν_1 of 10^4 between the highest ν_2 and the lowest ν_1 frequency considered for integration. Although these formula apply strictly to a homogeneous and spherical synchrotron source, we do not expect the emitting material in the jet to depart a lot from this ideal case at the start of the final decay stage. Finally, we calculate the energy density of the magnetic field $u_B = B^2/(8\pi)$ and the ratio u_B/u_e, which is extremely sensitive to the source size as: $u_B/u_e \propto \theta_{src}^{17}$. It is thus quite surprising to get values close to equipartition (i.e. $u_B \approx u_e$) for all objects when assuming a same jet opening half-angle of 2° and realistic values for the jet orientation and speed as given in the footnotes of Table 1. For GRS 1915+105 there is an important uncertainty on the distance d and jet velocity β (Fender *et al.* 2003) and actually we find that higher jet speeds and smaller distances favor equipartition. For 3C 279 we note that taking a viewing angle θ of 1° (Jorstad *et al.* 2004) instead of 5° results in extreme values of $B = 4.6\,10^2\,$G and $u_e = 3.8\,10^{-14}\,$erg cm^{-3} leading to a ratio u_B/u_e of $2\,10^{17}$, which seems very unrealistic.

4. Conclusion

The observed properties of synchrotron outbursts in microquasars appears to be quite similar, at least during some period of activity, with the behaviour of quasars. We find that timescales and physical sizes of the jet do scale with the black hole mass, but not strictly linearly as expected (e.g. Mirabel & Rodríguez 1998). For the observed timescales we find very roughly a square root dependence $\Delta t_{obs} \propto M_{BH}^{1/2}$, becoming more linear for the intrinsic length-scale Δl of the jet, which is corrected for orientation and relativistic effects. Apart from this scaling the physical properties of the jets, like the magnetic field and the electron energy density, are found to be very similar in all sources except GRS 1915+105, which has significantly higher values. The only clear difference we find between galactic and extragalactic jets is a harder electron energy distribution in microquasars, with an index s being clearly below 2 in Cyg X-3 and GRS 1915+105.

Our method of fitting multi-wavelength lightcurves with model outbursts is getting to the point where we can test different models and constrain the physics of relativistic jets. There are indications that the standard shock model of Marscher & Gear (1985) has to be modified to describe the observed evolution of synchrotron outbursts. Apart from the modifications proposed by Björnsson & Aslaksen (2000), we find evidence that the spectral turnover might rather be due to a low energy cut-off of the electron energy distribution than to synchrotron self-absorption during the initial phases of the outburst. More information, figures and animations at: http://isdc.unige.ch/~turler/jets/

References

Björnsson, C.-I. & Aslaksen, T. 2000, ApJ, 533, 787

Fender, R. P. 2003, MNRAS, 340, 1353

Fender, R. P., Bell Burnell, S. J., Waltman, E. B., Pooley, G. G., Ghigo, F. D. & Foster, R. S. 1997, MNRAS, 288, 849

Granot, J. & Sari, R. 2002, ApJ, 568, 820

Greiner, J., Cuby, J. G. & McCaughrean, M. J. 2001, Nature, 414, 522

Jorstad, S. G., Marscher, A. P., Lister, M. L., Stirling, A. M., Cawthorne, T. V., Gómez, J.-L. & Gear, W. K. 2004, AJ, 127, 3115

Kaiser, C. R., Sunyaev, R. & Spruit, H. C. 2000, A&A, 356, 975

Lindfors, E. J., Türler, M., Valtaoja, E., Aller, H., Aller, M., Mazin, D. et al. 2006, A&A, 456, 895

Marscher, A. P. 1987, in: Zensus, J. A. Pearson, T. J. (eds.), Superluminal Radio Sources (Cambridge Uversity Press: Cambridge), p. 280

Marscher, A. P. & Gear, W. K. 1985, ApJ, 298, 114

Marscher, A. P., Gear, W. K. & Travis, J. P. 1992, in: Valtaoja E., Valtonen M. (eds.), Variability of Blazars (Cambridge University Press: Cambridge), p. 85

Mioduszewski, A. J., Rupen, M. P., Hjellming, R. M., Pooley, G. G. & Waltman, E. B. 2001, ApJ, 553, 766

Mirabel, I. F., Dhawan, V., Chaty, S., Rodríguez, L. F., Martí, J. et al. 1998, ApJ 330, L9

Mirabel, I. F. & Rodríguez, L. F. 1998, Nature, 392, 673

Paltani, S. & Türler, M. 2005, A&A, 435, 811

Savolainen, T., Wiik, K., Valtaoja, E. & Tornikoski, M. 2006, A&A, 446, 71

Stark, M. J. & Saia, M. 2003, ApJ, 587, L101

Türler, M., Courvoisier, T. J.-L., Chaty, S. & Fuchs, Y. 2004, A&A, 415, L35

Türler, M., Courvoisier, T. J.-L. & Paltani, S. 1999, A&A, 349, 45

Türler, M., Courvoisier, T. J.-L. & Paltani, S. 2000, A&A, 361, 850

Wang, J.-M., Luo, B. & Ho, L. C. 2004, ApJ, 615, L9

FELIX MIRABEL: Comment: The square of the mass dependency you find for the timescales is very interesting. It appears to be consistent with a similar relation that has been claimed for 3C 120.

Black Holes from Stars to Galaxies – Across the Range of Masses
Proceedings IAU Symposium No. 238, 2006
V. Karas & G. Matt
© 2007 International Astronomical Union
doi:10.1017/S1743921307005224

Black holes: from stars to galaxies

I. Félix Mirabel †

European Southern Observatory, Alonso de Cordova 3107, Santiago, Chile
email: fmirabel@eso.org

Abstract. While until recently they were often considered as exotic objects of dubious existence, in the last decades there have been overwhelming observational evidences for the presence of stellar mass black holes in binary systems, supermassive black holes at the centers of galaxies, and possibly, intermediate-mass black holes observed as ultraluminous X-ray sources in nearby galaxies. Black holes are now widely accepted as real physical entities that play an important role in several areas of modern astrophysics.

Here I review the concluding remarks of the IAU Symposium No 238 on Black Holes, with particular emphasis on the topical questions in this area of research.

Keywords. Black hole physics – galaxies: nuclei – galaxies: jets – stars: general

---

1. The aim of IAU Symposium No 238

The interaction of black holes with their surroundings produces analogous phenomena in AGN and stellar black hole binaries. The scales of length and time of the phenomena are proportional to the mass of the black hole, and the whole phenomenological diversity that takes place around black holes can be described by the same physical concepts, however, the observed phenomena exhibit enormous complexity (see figure 1). Quasars and microquasars can eject matter several times, whereas collapsars form jets only once. When the jets point to the Earth these objects appear as microblazars, blazars and gamma-ray bursts, respectively. Synergy between research on stellar mass and supermassive black holes has become essential for our understanding of the underlying physics.

With this in mind, the aim of IAU Symposium No 238 has been to discuss the relations between different types of astrophysical black holes in a broader evolutionary context, bringing together astronomers working on AGN with those working on compact stellar binaries. Several groups that actively search for the elusive intermediate-mass black holes have had an active participation.

2. Stellar black holes

Current physics suggests that compact objects in stellar binaries with mass functions larger than 4 solar masses must be black holes. There are about 20 known objects with such mass functions. They are believed to be the tip of an iceberg, since it is estimated that in the Milky Way alone there should be at least 1000 dormant black hole X-ray transients, while the total number of stellar-mass black holes (isolated and in binaries) could be as large as 100 million. The number statistics of known stellar black holes is still very small and at present remain open the following questions:

Is the gap in the black hole mass function between 2.2 and 4.0 solar masses real? If so, which is the physical reason? Why in the Milky Way it has not been found a stellar

† On leave from CEA, France.

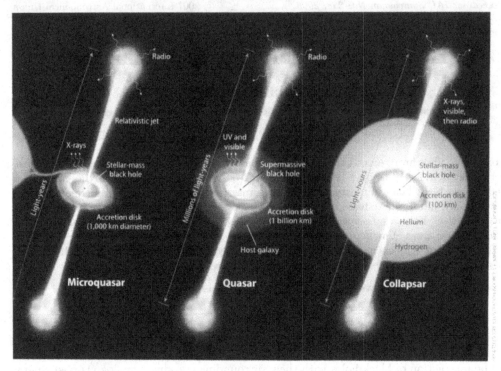

Figure 1. The same physical mechanism can be responsible for the three different types of objects: microquasars, quasars/AGN and massive stars that collapse ("collapsars") to form a black hole producing Gamma-Ray-Bursts. Each one of these objects contains a black hole, an accretion disk and relativistic particles jets. Credit: Mirabel & Rodríguez (2002).

black hole with mass larger than 14 solar masses? Is this due to poor statistics or to large mass losses in stellar winds by the metal rich progenitors in the Milky Way?

Kinematic studies of black hole binaries suggest that some stellar black holes form associated to very energetic supernova explosions (see figure 2). In some cases, this kinematic evidence is reinforced by the chemical composition of the donor star, when it contains elements produced in supernova explosions. However, kinematic studies suggest that black holes may also form by direct collapse, namely, in the dark (figure 3). Therefore, it is an open question when stellar black holes form in energetic supernovae and when by direct core collapse; more specifically, whether the presence of an energetic explosion depends or not on the mass of the collapsing stellar core. In fact, the kinematics of the microquasars Cygnus X-1 and GRS 1915+105, which contain black holes of ~ 10 and ~ 14 solar masses, respectively, did not receive strong kicks from natal energetic supernova explosions.

This question on the explosive or implosive black hole formation can be approached by observations of nearby gamma-ray bursts of long duration, which are believed to take place when stellar black holes are form by core collapse of massive stars. Therefore, the question on whether all collapsar GRBs are associated with supernovae of type Ia/b, or some are not, is of topical interest for the understanding of the very last phases of massive stellar evolution and black hole formation.

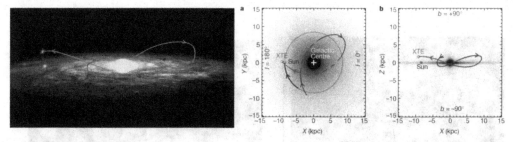

Figure 2. Galactic trajectory of a black hole wandering in the Galactic halo. This black hole may have been shoot out from the Galactic plane by an energetic, natal supernova explosion. Credit: Mirabel *et al.* (2001).

Figure 3. Kinematic evidence for the direct formation of the black hole in Cygnus X-1. Optical image of the sky around the black hole binary Cyg X-1 and the parent association of massive stars Cyg OB3. The red arrow shows the motion in the sky of the radio-counterpart of Cyg X-1 for the past 0.5 million years. The yellow arrow shows the average Hipparcos motion of the massive stars of Cyg OB3 (circled in yellow) for the past 0.5 million years. Cyg X-1 moves along with the parent association of massive stars indicating that the compact object did not received an energetic kick from a natal supernova. Credit: Mirabel & Rodrígues (2003).

3. The supermassive black hole at the Galactic Center

Dynamics is the most direct method to determine the mass of astrophysical compact objects, and therefore, the best evidence for the existence of a black hole. The first unambiguous dynamic evidence for a supermassive black hole was found following the motion of water masers around the center of the galaxy NGC 4258.

More robust evidence has been obtained by the motion of stars around the dormant black hole of 3-4 million solar masses at the center of our Galaxy. These stars seem to be distributed in two randomly inclined disks of 0.04 pc and 0.5 pc radii. The unexpected discovery of a compact cluster of massive stars in the central parsec of our Galaxy (see figure 4) poses new questions and may open new horizons for our understanding of massive black hole formation and its relation with massive star formation. This central cluster exhibits a flat mass function and it is only 6 million years old. At present, it is not clear how it got there. Perhaps it was formed in situ. How this could take place under the strong tidal forces from the central black hole remains a mystery.

4. Supermassive black holes in external galaxies

Supermassive black holes are ubiquitous at the centers of galaxies. Their mass is correlated with the mass of the host galaxy, and in particular with that of the stellar bulge. This indicates that massive black hole and host galaxy formation are tided up.

The most massive black holes are found at $z \geqslant 6.0$ implying that they assembled very early, in less than one billon years after the Big Bang. On the other hand, supermassive black holes of lower mass have formed more slowly by merging at $z \leqslant 3$. The peak

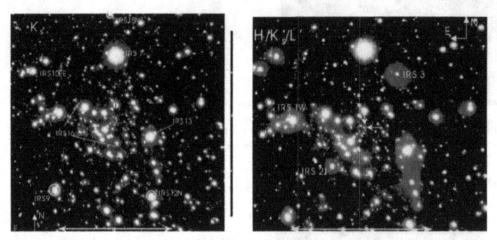

Figure 4. Left: The K-band near-infrared adaptive optics image of the central 20 arcsec of the Milky Way. Right: HKL colour composite image. Credit: Genzel *et al.* (2003).

accretion rate into black holes occurred rather late at z ∼ 0.7. Interestingly, this redshift is similar to the redshift of the peak density of luminous infrared galaxies at z ∼ 0.8, which suggests the association of maximum rate of accretion into black holes with dust enshrouded starbursts in luminous infrared galaxies, which are known to be mergers of gas-rich galaxies.

Although several models have been proposed, there is no general concensus on how supermassive black holes were form. More specifically, it is not known how could the most massive black holes in the universe form so rapidly with no feedback. Did supermassive black holes form before, after or coevally with the bulges?

The kick velocity due to gravitational recoil in merging supermassive black holes may displace the merged black hole from the dynamic center of the host galaxies. For dwarf galaxies the estimated kick velocity is larger than the typical escape velocities of 10-20 km/s, which may result in the ejection to the intergalactic medium of naked, massive black holes. How could we identify these runway black holes?

5. Black holes of intermediate mass

The existence of black holes of intermediate mass remains an open question. From the spectral properties of ultraluminous X-ray sources in nearby galaxies it has been suggested that some of these sources contain black holes of hundreds and perhaps thousands of solar masses. However, no dynamic evidence from the motion of satellite objects around such objects has been found. Furthermore, the luminosity function of black hole binaries as well as the properties of the most luminous x-ray black hole binaries in our Galaxy indicate that most of the super Eddington sources observed in nearby galaxies are a natural extension of the luminosity function of X-ray binaries, most of them being black holes with masses smaller than 100 solar. It has been argued that the hyper-accreting Galactic sources SS 433 and GRS 1915+105 seen from a different angle would be classified as ultraluminous X-ray sources.

If some ultraluminous x-ray sources are black holes of intermediate mass, one may ask why none has been identified in our Galaxy and/or the Magellanic Clouds, where the mass functions of compact objects can be determined. Black holes of intermediate mass may exist but are difficult to find.

6. Correlations among black holes of all masses

a) The fundamental plane: At low accretion rates, correlations between radio and x-ray luminosities are found for black holes across the whole range of masses. At low rates of accretion most the radiation power is dominated by synchrotron, compact, flat spectrum jets, from the x-rays to radio waves. For black holes in the jet dominated state, universal correlations between the x-ray luminosity, radio luminosity, and black hole mass have been found. Using this black hole fundamental plane it is expected that the masses of black holes could be determined from the x-ray and radio luminosities.

b) Accretion–jet coupling: Because of their proximity and rapid variability, micro-quasars have become the most adequate objects to study the connection between instabilities in the accretion disks and the genesis of relativistic jets. Figure 5 shows this connection as observed in an interval of time of a few tens of minutes. This sequence has also been observed in quasars, but on scales of a few years instead of tens of minutes.

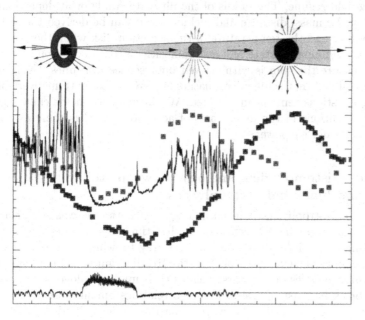

Figure 5. Real time observations in a microquasar of the connection between instabilities in the accretion disk and the genesis of jets in time scales of tens of minutes. Analogous disk–jet coupling has been observed in quasars, but on time scales of years. Credit: Mirabel *et al.* (1998).

c) Massive outflows: Across all mass scales, black holes in high luminous states exhibit X-ray absorption lines that reveal sub-relativistic winds with mass outflows as large as 0.3 times the accretion mass. Super-Eddington accretion makes massive outflows inevitable because all the accreting energy cannot be radiated.

In the case of black holes binaries the joint action of jets and massive outflows heat and blow away the interstellar medium, as does the black hole binary SS433 in the host nebula W50. X-ray observations with Chandra and XMM-Newton show that AGN produce analogous impact on the intergalactic medium, which has solved the long standing cooling flow paradox.

In this conference it was proposed that massive outflows may result from the inter-action of relativistic jets with warped accretion disks, in both stellar and supermassive accreting black holes. Furthermore, it was proposed that bended disks may explain several intriguing observational results, such as the large numbers of obscured AGN, as well as the re-direction of precessing jets, as in SS 433. The following questions remain open:

Are the relativistic jets in AGN and microquasars purely leptonic or hadronic? Are the hadrons in the jets of SS433 a result of entrainment by the relativistic jets in the massive winds? Is the nebula W50 that hosts SS433 the remnant of the natal supernova of the compact object, or a blown away cavity by the super-winds, or both?

d) Quasi-periodic oscillations: Sgr A* accretes at low mass rates, and quasi-periodic, polarized flares on scales of tens of minutes have now been observed at X-rays, infrared, sub-millimeter and radio waves. At longer waves the flares are polarized up to 10%. From the time lag at longer wavelengths it has been proposed that these flares could be synchrotron self Compton emission from adiabatically expanding clouds. It is unclear whether these expanding plasma clouds in Sgr A* are rotating blobs in the accretion disk, or whether they are collimated expanding jets as observed in microquasars.

e) Spin: Black holes are the simplest objects in the universe. They are defined by only three parameters: the mass, the spin and the electric charge. Because much of the radiation emerges within 6 gravitational radii they provide a unique opportunity to probe gravity in the strong-field regime. The radius of the ultimate stable orbit depends from the spin, and knowing the mass, distance and inclination, it can be derived the spin by measuring the X-ray flux and the temperature of the accretion disk when the accretion disk is in the thermally dominated state and at luminosities $\leqslant 0.3$ Eddington.

Another way to measure the spin is with skewed fluorescence iron lines. Using both methods it has been claimed that some microquasars and AGN host extreme Kerr black holes. The following questions remain unanswered: Would magnetic fields change completely the physical conditions in the inner parts of accretion disks? Is there a general correlation between spin and jet power?

7. Historical and epistemological analogy between stellar astrophysics and black hole astrophysics

At present, black hole astrophysics is in an analogous situation as was stellar astrophysics in the first decades of the XX century. At that time, well before reaching the physical understanding of the interior of stars and the way by which they produce and radiate their energy, empirical correlations such as the HR diagram were found and used to derive fundamental properties of the stars, such as their mass. Analogous approaches are taking place in black hole astrophysics. Using correlations among observables such as the radiated fluxes in x-rays and radio waves, quasi-periodic oscillations, flickering frequencies, etc, fundamental parameters that describe astrophysical black holes such as the mass and spin of the black holes are being derived.

References

Genzel, R., Schödel R., Ott, T. *et al.* 2003, ApJ, 594, 812

Mirabel, I. F., Dhawan, V., Chaty, S. *et al.* 1998, A&A, 330, L9

Mirabel, I. F., Dhawan, V., Mignani, R. P., Rodrigues, I. & Guglielmetti, F. 2001, Nature, 413, 139

Mirabel, I. F. & Rodríguez, L. F. 2002, Sky & Telescope, 103 (May), 32

Mirabel, I. F. & Rodríguez, L. F. 2003, Science, 300. 1119

View of Prague

Historical dome of the Astronomical Institute in Ondřejov

Poster Papers

Poster session

Detail of the Astronomical Clock dial at the Old Town Hall.
Bust of Albert Einstein on a house in Lesnická Str, where he lived with Mileva and their sons, 1911–1912.

Internal mechanism of the Astronomical Clock.
Memorial tablet on the house facade of the Fantas' salon (Old Town Square).

Black Holes from Stars to Galaxies – Across the Range of Masses
Proceedings IAU Symposium No. 238, 2006
V. Karas & G. Matt, eds.
© 2007 International Astronomical Union
doi:10.1017/S1743921307005248

X-ray variability of accreting black hole systems: propagating-fluctuation scenario

Patricia Arévalo, Phil Uttley and Ian McHardy

University of Southampton, Southampton, SO17 1DH, UK

email: patricia@astro.soton.ac.uk

Abstract. Propagating fluctuation models can reproduce fundamental properties of the variability observed in the X-ray light curves of accreting black hole systems. We explore this type of model and show how extended emitting regions introduce at the same time energy dependent power spectral densities (PSD) and time lags between different energy bands.

Keywords. Galaxies: active – X-rays: binaries, galaxies – accretion disks

---

X-ray light curves from AGN and XRBs show fluctuations on a very broad range of timescales, which is hard to reconcile with the compact nature expected for the X-ray emitting region. Lyubarskii (1997) proposed a solution where fluctuations in the accretion rate produced at different radii in the accretion flow propagate inward, to finally modulate the emission at the central, X-ray emitting region.

Propagating fluctuations produce a broad range of variability timescales and higher variability amplitudes at higher flux levels. Following Lyubarskii (1997), we assumed that independent fluctuations are introduced at geometrically spaced radii, with a radially-dependent characteristic timescale and that they propagate inward with viscous velocity (Arévalo & Uttley 2006). Light curves produced numerically following this procedure (Fig. 1a) have a PSD $\propto 1/f$ (Fig. 1b) with a high frequency cutoff, as observed in many AGN and XRBs in the high/soft state. The method also reproduces the linear relation between rms variability and flux (Fig. 1c) seen in both types of objects (Uttley & McHardy 2001).

Different emissivity profiles for different energy bands produce energy-dependent power spectra and time lags between the bands. Observationally, softer X-ray light curves show less variability on short timescales and tend to lead the variability of harder bands. This behaviour can be achieved by the model if the energy spectrum of the emitting region hardens towards the centre, i.e. if the softer X-rays are emitted by a more extended region. Figure 2a shows the PSDs of two simulated light curves, which differ only by the extent of their emitting regions, the solid line corresponds to a radial emissivity profile $\epsilon(r) \propto r^{-3.5}$ and the dots correspond to $\epsilon(r) \propto r^{-3}$. The flatter the emissivity profile, the more high frequency variability power is suppressed. Also, as the fluctuations propagate towards the central black hole, they are seen first in the more extended emitting region (soft) and later in the more concentrated emitting region (hard), which produces hard time-lags, that increase both with the timescale of the fluctuations and with the difference in emissivity indices. The lag spectrum calculated between the simulated light curves that produce the PSDs in Fig. 2a is shown in Fig. 2b.

PSDs showing peaks can be produced by the propagating fluctuation model if only a few radii contribute to the variability. The effect of the extended emitting region is to create a stepped lag spectrum (Fig. 2b), from the lag value associated with one of these radii to the lag of the next. This might be the case in the Seyfert 1 galaxy Ark 564 (Fig. 2c) where the PSD appears to contain two components (tentatively shown by dotted and

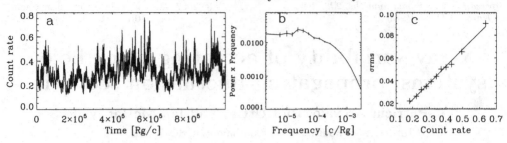

Figure 1. Simulated light curve (a), its Power spectrum (b) in terms of Power×f, shows a flat top (corresponding to Power $\propto 1/f$), that cuts off at high frequencies due to the filtering effect of an extended emitting region. Simulated light curves also reproduce the rms–flux relation (c).

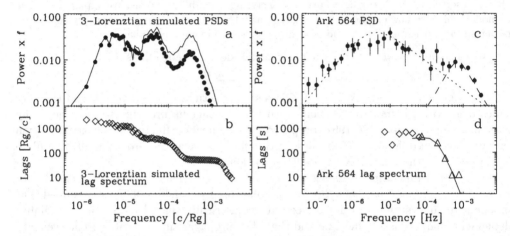

Figure 2. PSD (a) and lag spectrum (b) of two simulated light curves produced by assuming only three annuli in the accretion flow contribute to the variability. The light curves are calculated with different emissivity profiles (see text) which introduce the difference in the high frequency power and the time lags. PSD (c, from McHardy *et al.* in prep) and lag spectrum (d, from Arévalo *et al.* 2006) of the Seyfert 1 galaxy Ark 564. The PSD shape suggests two main variability components, which is supported by the marked drop in the lag spectrum at the frequency where the components overlap.

dashed lines), and the lag spectrum shows a marked drop at the frequency where the components overlap (Fig. 2d).

The general propagating fluctuation model can reproduce several observed variability properties, PSD shape and energy dependence, rms-flux relation and time lags. This approach connects the energy dependence of the PSD with the lag spectrum, so both measurements can be used jointly to disentangle different components of the variability.

References

Arévalo, P. & Uttley, P. 2006, MNRAS 367, 801
Arévalo, P., Papadakis, I., Uttley, P., McHardy, I. & Brinkmann, W. 2006, MNRAS 372, 401
Lyubarskii, Y. 1997, MNRAS 292, 679
Uttley, P. & McHardy, I. 2001, MNRAS 323, 26

Black Holes from Stars to Galaxies – Across the Range of Masses
Proceedings IAU Symposium No. 238, 2006
V. Karas & G. Matt, eds.
© 2007 International Astronomical Union
doi:10.1017/S174392130700525X

Monte Carlo simulations of dusty gas discs around supermassive black holes

Maarten Baes[1] and Sławomir Piasecki[1,2]

[1]Universiteit Gent, Belgium, email: maarten.baes@ugent.be

[2]Adam Mickiewicz University, Poznań, Poland

Abstract. Using detailed Monte Carlo radiative transfer simulations in realistic models for galactic nuclei, we critically investigate the influence of interstellar dust in ionised gas discs on the rotation curves and the resulting black hole mass measurements. We find that absorption and scattering by interstellar dust leaves the shape of the rotation curves basically unaltered, but slightly decreases the central slope of the rotation curves. As a result, the "observed" black hole masses are systematically underestimated by some 10 to 20% for realistic optical depths.

Keywords. Galaxies: nuclei – black hole physics

1. Introduction

Measuring the kinematics of ionised gas disks has become one the most important methods to determine the masses of supermassive black holes (SMBHs) in the nuclei of nearby galaxies. The state-of-the-art modeling techniques are refined to a high degree and take into account the major relevant astrophysical processes as well as the main instrumental effects. Applying such modeling techniques to high-quality HST data, a formal SMBH mass measurement uncertainty of 25% or better has been claimed with well-behaved disks (e.g. Barth *et al.* 2001; Marconi *et al.* 2003; Atkinson *et al.* 2005; Häring-Neumayer *et al.* 2006). Many ionised gas disks contain substantial amounts of interstellar dust. As optical radiation is easily absorbed and scattered by interstellar dust grains, these effects might affect the observed rotation curve and hence the SMBH mass estimate. The goal of our investigation is to determine the importance of this effect through detailed Monte Carlo simulations.

2. The modelling approach

We have set up a detailed radiative transfer model to investigate the effects of neglecting dust absorption and scattering on ionised gas disc rotation curves and the corresponding SMBH masses in a typical galactic nucleus. Our model consists of a thin (but not infinitesimally thin) axisymmetrical double-exponential disc rotating in a galactic nucleus. The intrinsic rotation velocity of the disc is determined by the combined gravitational potential of a stellar distribution and a central SMBH. We consider various SMBH masses. An additional isotropic bulk velocity dispersion is taken into account. We assume that a fraction of the disc consists of interstellar dust with typical Milky Way dust optical properties. The dust is assumed to be coupled to the ionised gas and has exactly the same spatial and velocity distribution. The SMBH mass $M_\bullet$, the thickness z_0 of the disc and the V-band optical depth τ_V, a quantity equivalent to the dust mass, are free parameters in our models.

For each model, we calculate the observed line-of-sight velocity field (data cubes) at several inclination angles. We use the SKIRT code, a 3D radiative transfer code that has

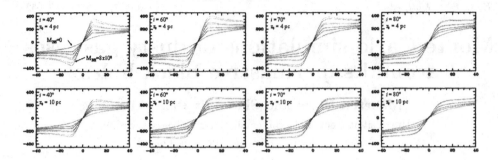

Figure 1. Nuclear rotation curves for dusty ionised gas discs. Each panel corresponds to a particular choice for the inclination i and the disc scale height z_0. In each panel we plot the rotation curves for different black hole masses (ranging between 0 and 8×10^8 $M_\odot$) and for different optical depths (ranging between 0 and 2). For any i, z_0 and $M_\bullet$, the slope of the rotation curve slightly decreases with increasing optical depth.

been developed explicitly to model the kinematics of dusty galaxies (Baes & Dejonghe 2002; Baes et al. 2003). The effects of absorption and scattering are properly taken into account. From the data cubes, we calculate rotation velocity maps and velocity dispersion maps. We focus in particular on the major axis rotation curves.

3. Results and conclusions

In Figure 1 we plot a set of nuclear rotation curves for our dusty ionised gas discs. The different panels correspond to different values for z_0 and the inclination angle. Within each individual panel, rotation curves are shown for different values of $M_\bullet$ and τ_V. For a fixed spatial resolution and inclination, the shape of the rotation curve of dust-free models depends on the SMBH mass and disc thickness. In particular, increasing disc thickness has a significant influence due to the larger integration along the line of sight. Modeling ionised gas disks as infinitesimally thin disks might introduce a bias.

To investigate the effect of interstellar dust absorption and scattering, we must compare models with different values of the optical depth. The curves corresponding to various τ_V basically coincide: apparently absorption and scattering by interstellar dust leaves the shape of the rotation curves qualitatively unaltered. However, close inspection shows that an increasing amount of interstellar dust slightly decreases the central slope of the rotation curves. We estimate the bias one makes by neglecting the effect of interstellar dust on SMBH mass estimates by translating the observed rotation curve central slopes into SMBH masses. Neglecting dust systematically leads to an underestimate of the black hole mass. For $\tau_V \sim 0.2$ the SMBH is underestimated by about 10%, for $\tau_V > 1$ the black hole mass is underestimated by 10 to 20%. This effect is more or less independent of the SMBH mass and the inclination of the disc. Our study demonstrates that the systematic effect of dust attenuation should be taken into account in SMBH demographics studies.

References

Atkinson, J. W., Collett, J. L., Marconi, A. et al. 2005, MNRAS, 359, 504
Baes, M. & Dejonghe, H. 2002, MNRAS, 335, 441
Baes, M., Davies, J. I., Dejonghe, H. et al. 2003, MNRAS, 343, 1081
Barth, A. J., Sarzi, M., Rix, H.-W. et al. 2001, ApJ, 555, 685
Häring-Neumayer, N., Cappellari, M., Rix, H.-W. et al. 2006, ApJ, 643, 226
Marconi, A., Axon, D. J., Capetti, A. et al. 2003, ApJ, 586, 868

Black Holes from Stars to Galaxies – Across the Range of Masses
Proceedings IAU Symposium No. 238, 2006
V. Karas & G. Matt, eds.

Dipole–vortex structure of the obscuring tori in active galactic nuclei

Elena Yu. Bannikova and Victor M. Kontorovich

Karazin Kharkov National University, 4 Svobody Sqr., 61077 Kharkov, Ukraine

and

Institute of Radio Astronomy of the National Academy of Science of Ukraine,
4 Chervonopraporna Str., 61002 Kharkov, Ukraine

email: bannikova@astron.kharkov.ua, vkont@ira.kharov.ua

Abstract. A "matrjoshka" scheme for obscuring vortex torus structure is proposed, which could help to explain the AGN radiation flares, variability and evolution.

Keywords. Galaxies: active – instabilities

1. Introduction

Starting with the Antonucci and Miller's outstanding work, tori have been considered as a necessary element of the AGN-structures forming the basis of the AGN unified model. A brilliant achievement was the first direct observation of the obscuring tori described by Jaffe, Meisenheimer, Röttgering *et al.* (2004). Existence of tori was confirmed by observation with VLT optical interferometer equipped with MIDI IR-camera.

We propose to consider a torus as a dynamic object with its proper vortex motion. It is well known that two independent rotation types are possible in a torus: "orbital" – along the torus big circle (with radius R), and "vortical" – along its small circle (with radius r). This latter will be of our major interest. Vortical motion in a self-gravitating torus

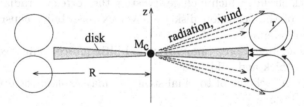

Figure 1. Proposed scheme of the dipol–toroidal vortex in the center of AGN: section by symmetry plane ortogonal to disk; z-axis of symmetry. Arrows show the possible matter movement.

(see discussion in Bliokh & Kontorovich 2003) principally differs from pure orbital one. For the near Eddington luminosity $L \approx L_{\rm Edd}$, when the gravitation is largely balanced by light pressure, this motion in AGN is not so noticeable and can be quite neglected at first. Originating from or being sustained by the central source radiation or wind "twist", the vortical torus motion is capable of "replenishing" the accretion disk mass, thereby adjusting the process of accretion and introducing feedback (Fig. 1; see Bannikova & Kontorovich 2006). Here the dipole structure of a toroidal vortex, defined by the symmetry of radial-outflowing wind and radiation, is important. Note that the structure of the lines of flow across such dipole vortex resembles the structure and topology of streamlines in the well-studied hydrodynamic models, such as Hill and Lamb's vortices.

At the same time, each component of a toroidal dipole, taken separately, resembles the Maxwell vortex, but with the opposite direction of rotation.

2. Discussion

The aforesaid proves that positive feedback and related instability with linear increment $\alpha = \xi\xi_1\zeta c/(2R)$ are possible in the considered system. Here c is light velocity, R is the torus big radius, $\xi \sim 0.1$ means the efficiency of conversion of accretion energy into radiation, ξ_1 is a portion of small torus radius characterizing the matter inflow from torus to disk, factor ζ takes into account the effect of torus shape, absorption and angular dependence of radiant flux. Note that the time delay τ in a feedback circuit of accretion-wind instability reduces the increment as $\alpha \rightarrow \alpha f(\alpha\tau)$, where universal function $f < 1$. The accretion-wind instability increment determines the characteristic time scale of its development. For the 3.5-year time scale of the observable burst duration in quasars (see Babadzhanyants & Belokon 1985) and $\xi_1 \sim \zeta \sim 0.1$ we should take the space scale $R \sim 10^{16}$ cm.

Since the preliminary observational data Jaffe, Meisenheimer, Röttgering et al. (2004) point at significantly larger torus sizes, it should be natural to suggest the "matrjoshka" scheme: there are tori of smaller radii within the outer big torus. In the case of Eddington luminosity, the mass of torus that replenishes the accretion disk is proportional to its big radius. In particular, for $R \sim 10^{16}$ cm, the mass of central black hole $10^9 M_\odot$ and accepted above parameters – the mass of inner torus $M_{\text{torus}} \sim 10^3 M_\odot$. The inner toroidal vortex may be responsible for the variability of AGN, the development of instability, etc. On the contrary, in the shadow of the minor-radius torus nearest to the center, a preference exists for the center-falling matter, which stretches into a torus.

3. Conclusions

1) A dipole–toroidal vortex can be an essential element of AGN-structure, which replenishes the accretion disk.

2) In the feedback circuit, which includes twisting the vortex by radiation (or by wind) and replenishment ζ of the accretion disk by a vortex, instability causing the bursts in active nuclei may develop.

3) Presence of centrifugal force in a vortex and lifting force connected with wind flow allow existence of a thick and cold torus.

4) The "matrjoshka" scheme of toroidal structure may explain the evolution and variability effects in AGNs.

Acknowledgements

This work is partly supported by the Ukraine President's grant GP/F8/0051.

References

Jaffe, W., Meisenheimer, K., Röttgering, H. J. A., Leinert, Ch., Richichi, A., Chesneau, O. et al. 2004, Nature 429, 47

Bliokh, K. Yu. & Kontorovich, V. M. 2006, JETP 123, No 6, 1123 (astro-ph/0407320)

Bannikova, E. Yu. & Kontorovich, V. M. 2006, Radio Physics & Radio Astronomy 11, No 1, 42 (in Russian); Problems of Atomic Sci & Techn. No 5, 146

Babadzhanyants, M. K. & Belokon, E. T. 1985, Astrofizika 23, No 3, 459

Black Holes from Stars to Galaxies – Across the Range of Masses
Proceedings IAU Symposium No. 238, 2006
V. Karas & G. Matt, eds.
© 2007 International Astronomical Union
doi:10.1017/S1743921307005273

Black hole masses in type II AGNs from the Sloan Digital Sky Survey

W. Bian,[1,2] Q. Gu[3] and Y. Zhao[3]

[1]Dept. of Physics and Inst. of Theor. Phys., Nanjing Normal Univ., Nanjing 210097, China

[2]Key Laboratory for Particle Astrophysics, Institute of High Energy Physics, Chinese Academy of Sciences, Beijing 100039, China

[3]Department of Astronomy, Nanjing University, Nanjing 210093, China

email: bianwh@ihep.ac.cn

Abstract. The stellar population synthesis method is used to model the stellar contribution for a sample of 209 type II AGNs at $0.3 < z < 0.83$ from the Sloan Digital Sky Survey. The reliable stellar velocity dispersions are obtained for 33 type II AGNs with significant stellar absorption features. We use the formula of Greene & Ho to obtain the corrected stellar velocity dispersions (σ_*^c). 20 of which can be classified as type II quasars. The SMBHs masses and the Eddington ratios are calculated. We measure the gas velocity dispersion (σ_g) from NLRs, and find that the relation between σ_g and σ_*^c becomes much weaker at higher redshifts than at smaller redshifts. We find that the deviation of σ_g from σ_*^c is correlated with the Eddington ratio.

Keywords. Galaxies: active – galaxies: nuclei – quasars: emission lines

1. Stellar bulge velocity dispersion in type II AGNs

According to the unification model, AGNs are classified into two classes depending on whether the central engine and broad-line regions (BLRs) are viewed directly (type I AGNs) or are obscured by medium (type II AGNs). Type II quasars are the luminous analogs of low-luminosity type II AGNs. Some methods have been used to discover type II quasars, but only a handful have been found. The sparsity of type II quasars is possibly due to the torus evolution with the luminosity. Here we used the sample of type II Quasars at redshifts $0.3 < z < 0.83$ from the Sloan Digital Sky Survey (SDSS) (Zakamsa *et al.* 2003). The fibers in the SDSS survey have a diameter of 3", and contains the host galaxy light. It is the key point to accurately measure the stellar velocity dispersion (σ_*).

We first modelled the stellar contribution in the SDSS spectra through the stellar population synthesis code, STARLIGHT (version 2.0, Cid Fernandes *et al.* 2001). The code does a search for the linear combination of Simple Stellar Populations (SSP) to match a given observed spectrum O_λ. We focus on the strongest stellar absorption features of Ca II K and the G-band. We find that the fitting goodness (chi-square value) depends not only on the S/N (> 5), but also on the absorption lines equivalent widths (EW of Ca II K line $> 1.5\text{Å}$). At last we select 33 type II AGNs, which had shown significant stellar absorption features and are well fitted to derive reliable measurements of stellar velocity dispersion. For more details, please refer to Bian *et al.* (2006).

2. The extinction correction of the [O III]λ5007 luminosity

We used the [O III]λ5007 line as a tracer of AGNs activity. However, the [O III]λ5007 luminosity is subject to extinction by interstellar dust in the host galaxy and in our Galaxy. The extinction correction is usually corrected by using the Balmer decrement, which is regarded as the best approximation. For 18 out of these 33 objects with available

Hα measurements in the stellar-light subtracted spectra, we used the relation $L^c_{[OIII]} = L_{[OIII]}[\frac{(H_\alpha/H_\beta)_{obs}}{(H_\alpha/H_\beta)_0}]^{2.94}$ to obtain the extinction correction of the [O III]λ5007 luminosity. If we used the $L_{[OIII]}$ criterion of 3×10^8 L$_\odot$, which corresponds to the intrinsic absolute magnitude $M_B < -23$ (Zakamsa $et~al.$ 2003), 14 out of these 18 objects can then be classified as type II quasars. Using the SDSS AGN catalogue at MPA/JHU (Kauffmann et al 2003), we obtained the extinction factor at 5007Å, $\log(\frac{L^c_{[OIII]}}{L_{[OIII]}})$. We tried to obtain a relation between $\frac{L^c_{[OIII]}}{L_{[OIII]}}$ and other observational parameters, such as σ_*, $L_{[OIII]}$, $L_{H\beta}$. However, no correlation is found. More careful work is required to deal with this problem in the future.

3. The SMBHs masses and Eddington ratios in type II AGNs

There are some methods to calculate the SMBHs masses in AGNs. With the broad emission lines from BLRs (e.g. Hβ, Mg II, CIV; Hα), the reverberation mapping method and the empirical size-luminosity relation can be used to derive the virial SMBHs masses in AGNs. The obscuration of BLRs makes above method infeasible to derive SMBHs masses. However, we can use the well-known $M_{BH} - \sigma_*$ relation to derive SMBHs masses if we can accurately measure the stellar bulge velocity dispersion (σ_*).

Recently, Greene & Ho (2006) used the direct-fitting method to study the the systematic biases of σ_* from different regions around Ca II triplet, Mg Ib triplet, and CaII H+K stellar absorption features. Here we used the following formula to obtain the corrected velocity dispersion σ^c_* (Greene & Ho 2006), $\sigma^c_* = (1.40 \pm 0.04)\sigma_* - (71 \pm 5)$. From this corrected velocity dispersion, we calculated the SMBHs masses. We used the extinction-corrected [O III] luminosity to calculate the bolometric luminosity L_{bol}, $L_{bol} = 3500 L^c_{[OIII]}$. At last we calculated the Eddington ratio, L_{bol}/L_{Edd}, where $L_{Edd} = 1.26 \times 10^{38} M_{BH}/M_\odot$ ergs s^{-1}.

4. The σ_g–σ_* relation

The good correlation between stellar velocity dispersion (σ_*) and ionized gas velocity dispersion (σ_g) suggests that the gaseous kinematics of NLRs in Seyfert galaxies be primarily dominated by the bulge gravitational potential.

Following Greene & Ho (2005), we study the $\sigma_g - \sigma^c_*$ relation for 33 Type II AGNs at redshifts $0.3 < z < 0.83$ after measuring the gas velocity dispersion (σ_g) from the narrow emission lines from NLRs, and the stellar velocity dispersion (σ^c_*) from the CaII H+K, G-band absorption feature. In Fig. 6, we showed the relation between σ_g and σ^c_*. The relation between σ_g and σ^c_* becomes much weaker at higher redshifts.

This work has been supported by the NSFC (Nos. 10403005, 10473005, 10325313, 10233030 and 10521001) and the Science-Technology Key Foundation from Education Department of China (No. 206053).

References

Bian, W., Gu, Q., Zhao, Y., Chao, L. & Cui, Q. 2006, MNRAS, 372, 876
Cid Fernandes, R., Sodre, L., Schmitt, H. R. & Leao, J. R. S. 2001, MNRAS 325, 60
Greene, J. E., Ho, L. & C. 2005, ApJ 627, 721
Greene, J. E. & Ho, L. C. 2006, ApJ 641, 117
Kauffmann, G. $et~al.$ 2003, MNRAS 346, 1055
Zakamsa, N. L. $et~al.$ 2003, AJ 126, 2125

Black Holes from Stars to Galaxies – Across the Range of Masses
Proceedings IAU Symposium No. 238, 2006
V. Karas & G. Matt, eds.
© 2007 International Astronomical Union
doi:10.1017/S1743921307005285

Variations of geometrical and physical characteristics of innermost regions of active galactic nuclei on time-scale of years

Nikolai G. Bochkarev[1] and Alla I. Shapovalova[2]

[1]Sternberg Astronomical Institute of Lomonosov Moscow State University
Universitetskij prosp., 13, Moscow, 119992, Russia

[2]Special Astrophysical Observatory of Russian Academy of Science,
Nizhnij Arkhys, Karachaevo-Cherkesia, 369167 Russia

email: boch@sai.msu.ru, ashap@sao.ru

Abstract. Several examples of observational evidence of possible variations of geometrical and physical characteristics of the Broad Line Regions (BLR) of AGN on the time-scale of years are listed. Cases of accretion disk size variations and variable input of jets in BLR formation are presented.

Keywords. Galaxies: active – galaxies: individual (NGC 4151, NGC 5548, 3C390.3, Mrk 79, PG 1211+143, Fairall 9) – jets – nuclei – emission lines – accretion disks

---

1. Accretion disk size variations

Analysis of long-term (1968–2003) variations of NGC 4151 continuum in UBV(RI) bands and K-band ($\approx 2\mu$) shows that the slow changes in the optical continuum are connected with accretion disk (AD) formation and dissipation (Lyuty *et al.* 1998; Oknyanskij *et al.* 1999; Lyuty 2005). During 1984–1989 deep minimum the former AD is assumed to totally disappear; a new AD begins to form in 1988–1989, its brightness reaching maximum in 1996 and considerably decreasing by 2003. At the same time, the fast variations ($30 - 100^d$) of the optical continuum are due to flares in AD (Lyuty 2005). Properties of the new disk are different from the previous one: it is twice as bright and shows much more intensive variations (Lyuty, Taranova & Shenavrin 1998; Lyuty 2005).

The best evidence of BLR size changes was found on the base of "AGN Watch" program data. The 13 year long intensive multi-wavelength spectral and photometric monitoring of NGC 5548 clearly shows significant variations of size (from 6 to 26 light days) of BLR emitting in Hβ spectral line on a time scale $1 - 3$ years (Peterson *et al.* 2002; Peterson *et al.* 2004).

2. Possible variable near-jet component of BLR in radio galaxy 3C390.3 and other AGN

Two maxima (flat-top peaks near $\approx 30^d$ and $\approx 100^d$) on the cross-correlation function (CCF) for Hβ broad component variation delay in respect to the optical continuum based on 3C390.3 monitoring during 1996-2001 have been found by Shapovalova *et al.* (2005). Sometimes only one of the two maxima is present, and sometimes both. The time-scale of the maxima variability is 1-2 years. Similar conclusions were drawn by Sergeev *et al.* (2002).

On the other hand, Arshakian *et al.* (2005) have found for 3C390.3 an observational evidence of connection between variations in UV and optical continua and radio-jets inner structure.

According to their scenario during the periods of the nucleus maximal brightness most of the continuum variable radiation is emitted from near-jet region at about 0.4 pc from the nucleus. The jet ionizes the surrounding gas and creates near-jet broad line emission region located rather far from the accretion disk and the "classical" BLR – another BLR appears (BLR-2). During the brightness minima the jet contribution to the ionizing continuum is decreases and the main part of broad line emission comes from the "classical" BLR (BLR-1: AD and the surrounding gas).

So, the broad-line emission is likely to be generated both near the disk (Peterson *et al.* 2002) (BLR-1, ionized by the emission from the accretion disk or/and its hot corona) and in a rotating sub-relativistic outflow surrounding the jet (BLR-2, ionized by the emission from the relativistic plasma in the jet).

Such a scenario explains the two variable maxima ($\approx 30^d$ and $\approx 100^d$) found in the CCF describing the time-lag of broad line variations relatively to continuum on the base of the results of 3C390.3 optical monitoring (Shapovalova *et al.* 2001; Sergeev *et al.* 2002). The first maximum ($\approx 30^d$) is probably connected with BLR near the accretion disk (BLR-1). The second one is assumed to indicate location of BLR-2 produced by jet activity. The time-scale of the maxima variability (1–2 years) agrees with time intervals between appearance of new knots in radio jet innermost parts.

On the base of a complete and consistent reanalysis of broad emission-line reverberation-mapping data for 35 AGN Peterson *et al.* (2004) found some similar cases of the continuum-line CCF has two peaks (at least for some time intervals and some spectral lines). One case is Hβ line in Mrk 79 during the period JD 2,449,996-2,450,220 (the highest peaks correspond to time-lags of about 6 and about 40 days). Another case is Hγ line in PG 1211+143 (about 50 and 250 days correspondingly).

Having analyzed of variations of the ratio line intensities along line profiles and the ratio variability based on UV and optical spectral line observations of Fairall 9 for the period May 1994-Jan. 1995 Nazarova & Bochkarev (2006) also arrived at the idea that it is necessary to take into account the inputs of different, spatially separated BLRs (AD and jet).

Acknowledgements

This work was supported by grant RFBR 06-02-16843.

References

Arshakian, T. G., Lobanov, A. P., Chavushyan, V. H. *et al.* 2005, astro-ph/0512393

Lyuty, V. M. 2005, Astron. Lett. 31, 645 (PAZ, 31, 723)

Lyuty, V. M., Taranova, O. G. & Shenavrin, V.I. 1998, Astron. Lett. 24, 199 (PAZ, 24, 243)

Nazarova, L. S. & Bochkarev, N. G. 2006, Proc. The Central Engine of Active Galactic Nuclei (Xi'an, China), in preparation

Oknyanskii, V. L., Lyuty, V. M., Taranova, O. G. & Shenavrin, V. I. 1999, Astron. Lett. 24, 483 (PAZ, 25, 563)

Peterson, B. M., Berlind, P., Bertram, R. *et al.* 2002, ApJ, 581, 197

Peterson, B. M., Ferrarese, L., Gilbert, K. M., *et al.* 2004, ApJ, 613, 682

Sergeev, S. G., Pronik, V. I., Peterson, B. M., *et al.* 2002, ApJ, 576, 660

Shapovalova, A. I., Burenkov, A. N., Carrasco, L., *et al.* 2001, A&A, 376, 775

Black Holes from Stars to Galaxies – Across the Range of Masses
Proceedings IAU Symposium No. 238, 2006
V. Karas & G. Matt, eds.
© 2007 International Astronomical Union
doi:10.1017/S1743921307005297

Accretion in the broad line region of active galactic nuclei

E. Bon,[1] L. Č. Popović[1] and D. Ilić[2]

[1] Astronomical Observatory, 11160 Belgrade, Serbia

[2] Department of Astronomy, University of Belgrade, 11000 Belgrade, Serbia

email: ebon@aob.bg.ac.yu, lpopovic@aob.bg.ac.yu, dilic@matf.bg.ac.yu

Abstract. We modeled the single-peaked Broad Emission Lines (BELs) with two-component model (accretion disk, with surrounding spherical region), comparing it with observational line profiles for a number of Active Galactic Nuclei (AGN). We find that the accretion in the Broad Line Region (BLR) can be present even if the profiles of BELs are single-peaked.

Keywords. Galaxies: nuclei – galaxies: Seyfert – accretion, accretion disks – quasars: emission lines

1. Introduction

The BLR is a very attractive subject for studding since the most widely accepted model for AGN includes a super-massive black hole and an accretion disk. Some AGN emit double-peaked lines, indicating the existence of the disk emission, but their number is small and statistically insignificant (around 5%, see Eracleous & Halpern 2003). Majority of AGN with one-peaked BELs, but it does not necessary indicate that the contribution of the disk emission can be neglected. Even if the emission in a spectral line comes only from a disk, the parameters of the disk (e.g. for a small inclination angle) can produce a single-peaked line (Chen & Halpern 1989, Dumont & Collin-Souffrin 1990, Kollatschny & Bischoff 2002, Kollatschny 2003).

Recently (Popović et al. 2001, Popović et al. 2002, Popović et al. 2003, Popović et al. 2004, Ilić, D. et al. 2006, Bon et al. 2006) showed that broad emission lines of several AGN, can be well fitted with a model which has two components.

In this work we study the possibility to have the disk emission in the BLR of an AGN with the single-peaked BELs.

2. Two-component model for the BLR

The two component model consists of an accretion disk (Keplerian relativistic model Chen & Halpern 1989), and emission of surrounding spherical region. Two component model is described in Popović et al. 2004. Here we did a simulation of the line profiles with different parameters of the model.

We simulated and measured parameter $k = W_{1/10}/W_{1/2}$ ($W_{1/2}$ = full width at a half maximum, $W_{1/10}$ = full width at a tenth of maximum) for different flux contributions of the disk and spherical region. We concluded that the parameter k can be very useful as an indicator of the disk emission. Even if we have a BEL with more prominent single-peak, there is still a possibility that the disk emission is present. Moreover, we find that the disk emission can be detected if the ratio of the disk emission flux (F_{sph}) and the flux of the spherical region (F_{disk}) is $F_{disk}/F_{sph} > 0.5$.

E. Bon, L. Č. Popović & D. Ilić

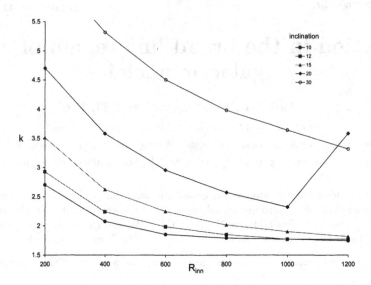

Figure 1. Response of the inner radius of the fixed ring in accretion disk ($R_{inn} - R_{out} = 800R_G$), for different inclinations, to the measured parameter $k = W_{1/10}/W_{1/2}$ of simulated spectra.

Using this technique we found better accuracy than by fitting of the line shapes. Large number of parameters in the fit causes that we could get only rough estimation of the model parameters (as already shown in Popović *et al.* 2004, Bon *et al.* 2006).

We note that it is really hard to estimate the contribution of the disk emission in the one-peaked line profiles. With this approach we showed that one should take into account disk emission, and also we showed domains where we could expect this influence

As a future work we plane to simulate even denser greed of parameters, for finer determination of parameter domains. We also tend to apply this model on a larger number of AGN spectra.

Acknowledgements

The work was supported by the Ministry of Science, Technologies and Development of Serbia through the project 146002.

References

Bon, E., Popović, L. Č., Ilić, D. & Mediavilla, E. G. 2006, New Ast. Rev., 50, 716
Chen, K. & Halpern, J. P. 1989, ApJ, 344, 115
Dumont, A. M. & Collin-Souffrin, S. 1990, A&AS 83, 71
Ilić, D., Popović, L. Č., Bon, E., Mediavilla, E. G. & Chavushyan, V. H. 2006, MNRAS, 371, 1610
Kollatschny, W. & Bischoff, K. 2002, A&A, 386, L19
Kollatschny, W. 2003, A&A, 407, 461
Eracleous, M. & Halpern, J. P. 2003, ApJ, 599, 886
Murray, N. & Chiang, J. 1997, ApJ, 474, 91
Popović, L. Č., Stanić, N., Kubičela, A. & Bon, E. 2001, A&A, 367, 780
Popović, L. Č., Mediavilla, E.G., Kubičela, A. & Jovanović, P. 2002, A&A, 390, 473
Popović, L. Č., Mediavilla, E. G., Bon, E., Stanić, N. & Kubičela, A. 2003, ApJ, 599, 185
Popović, L. Č., Mediavilla, E.G., Bon, E. & Ilić, D. 2004, ApJ, 423, 909

Black Holes from Stars to Galaxies – Across the Range of Masses
Proceedings IAU Symposium No. 238, 2006
V. Karas & G. Matt, eds.
© 2007 International Astronomical Union
doi:10.1017/S1743921307005303

Triaxial orbit-based model of NGC 4365

Remco C. E. van den Bosch,[1] Glenn van de Ven,[2]
Michele Cappellari[1] and P. Tim de Zeeuw[1]

[1]Sterrewacht Leiden, Universiteit Leiden, Postbus 9513, 2300 RA Leiden, The Netherlands
[2]Institute for Advanced Study, Einstein Drive, Princeton, NJ 08540, USA
email: Bosch@strw.leidenuniv.nl

Abstract. We have developed an orbit-based method for constructing triaxial models of elliptical galaxies, which fit their observed surface brightness and kinematics (van den Bosch *et al.*). We have tested this extended Schwarzschild method (1979) against analytical models with general distribution functions (DF) and find that we can recover the DF (van de Ven *et al.*). Here, we present a model of NGC 4365.

Keywords. galaxies: elliptical and lenticular, cD – galaxies: kinematics and dynamics

1. NGC 4365

NGC 4365 is an E3 galaxy with a kinematically de-coupled core (KDC). Its rotation axis is misaligned with respect to that of the main body of the galaxy (Surma & Bender 1994; Davies *et al.* 2001; Statler *et al.* 2004). The bulk of the galaxy appears to rotate around the major axis, but the stars in the central $6''$ rotate around the minor axis. These properties can be explained by an intrinsically triaxial shape.

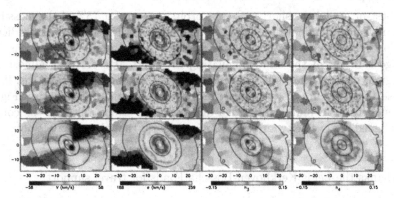

Figure 1. Stellar kinematics of NGC 4365 on the top row. From left to right: the mean velocity, velocity dispersion and the higher order moments h_3 and h_4 (measuring deviations from Gaussian). The contours are representative isophotes. The best-fitting model is shown on the bottom row and it is a good fit, reproducing the observed features in the symmetrized kinematics (middle row) with great accuracy (χ^2/degrees-of-freedom ~ 1).

We measured the stellar kinematics of NGC 4365 (Fig. 1) with the SAURON integral-field spectrograph (Bacon *et al.* 2001). We constructed triaxial models using both the integral-field stellar kinematics and HST and ground-based imaging and including a central black hole. Our best-fit model (Fig. 1) is close to oblate with axis ratios of 1:0.97:0.73 (see also Statler *et al.* 2004) and surprisingly consists primarily of orbits that rotate around the short axis.

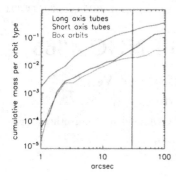

Figure 2. The cumulative mass per orbit-type as a function of radius. The modest fraction of stars on long axis tube orbits all have the same sense of rotation, and are responsible for the observed mean motion.

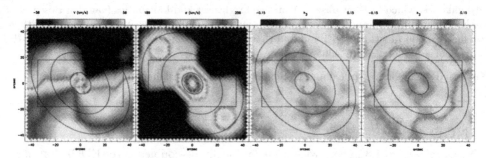

Figure 3. Kinematics of the model up to a radius of $60''$. Only the part inside the rectangular box is constrained by the kinematic observations, the rest is extrapolated. The kinematics outside the box are thus not very reliable, but they do allow one to see the extended structure.

2. Orbit types

In our best-fit model the short axis tube orbits account for 75% of the mass inside $30''$ (Fig. 2) and hence dominate even outside the KDC. This seems in contradiction with the rotation seen around the major axis at large radius in the observations. However, the average rotation of the stars populating the short axis orbits is almost negligible in this region, causing the high dispersion measured along the major axis. To be able to visualise the structure of the model we show the extended kinematics (Fig. 3). The extended structure, including the sigma dip along the minor axis and the continuation of the unaligned zero velocity curve show more clearly that the galaxy is oblate/triaxial.

Acknowledgements

RvdB thanks the Leids Kerkhoven-Bosscha Fonds for travel support.

References

Bacon R. *et al.* 2001, MNRAS, 326, 23
Davies R. L., *et al.* 2001, ApJ, 548, L33
Schwarzschild M. 1979, ApJ, 232, 2
Surma P. & Bender R. 1995, A&A, 298, 405
Statler T. S., Emsellem E., Peletier R. F. & Bacon R. 2004, MNRAS, 353, 1
van den Bosch R. C. E., van de Ven G., Verolme E. K., Cappellari M. & de Zeeuw P. T. 2006, MNRAS, submitted
van de Ven G., de Zeeuw P. T. & van den Bosch R. C. E. 2006, MNRAS, submitted

Black Holes from Stars to Galaxies – Across the Range of Masses
Proceedings IAU Symposium No. 238, 2006
V. Karas & G. Matt, eds.
© 2007 International Astronomical Union
doi:10.1017/S1743921307005315

Modulation of high-frequency quasi-periodic oscillations by relativistic effects

Michal Bursa

Astronomical Institute, Academy of Sciences, Boční II, CZ-14131 Prague, Czech Republic
email: bursa@astro.cas.cz

Abstract. Two distinct frequencies are often present in frequency power spectra of low-mass X-ray binary lightcurves. These frequencies are closely related to each other and there is a strong evidence that they have origin in the orbital motion in an accretion disc. A mechanism has to be in act that provides a modulation of emergent flux at observed frequencies – not necessarily the actual frequencies at which the fluid oscillates. Using numerical ray-tracing we show that strong gravity has an effect on light modulation and may itself be responsible for significant part of variations in observed lightcurves.

Keywords. Accretion disks – gravitational lensing – X-rays: binaries

1. Introduction

Observations show that the solely presence of a thin accretion disc is not sufficient to produce the high-frequency quasi-periodic oscillations (HFQPO), because they are exclusively connected to the spectral state, where the energy spectrum is dominated by a steep power law with some weak thermal disc component. A model is more appropriate, where an outer cool disc is continuously transitioned into or sandwiched by a hot, thick, but optically thin flow (Esin *et al.* 1998). An optically thin advection-dominated accretion flow is mostly transparent for photons, and therefore general relativistic light bending and lensing effects may gain a particular importance. Significant temporal variations in the observed flux can then be accomplished by global oscillations of such geometrically thick flow, fluid tori.

In order to explore, whether it is possible to obtain some flux modulation just by effects of strong gravity, we set up a model of a possible accretion configuration, largely simplified to a hot and optically thin luminous torus, optionally surrounded by a cool opaque disc. The torus is considered in a slender approximation, i.e. with its cross-section being much smaller than its distance from the gravity centre (Bursa 2006).

2. Slender torus model

We impose on the torus rigid and axisymmetric sinusoidal oscillations in the vertical direction, i.e. parallel to its axis, as well as in the perpendicular radial direction. Such assumption will serve us to model the possible basic global modes found by (Abramowicz *et al.* 2006). The torus is rigidly displaced from its equilibrium, so that the position of the central circle of maximal pressure varies in time periodically, $r(t) = r_0 + \delta r \sin(\omega_r t)$, $z(t) = \delta z \sin(\omega_z t)$, with the frequencies of epicyclic motion at a given radius.

Two modes are assumed in numerical calculations described hereby: an incompressible mode and a compressible mode. In the incompressible mode, the equipotential structure and all thermodynamical quantities describing the torus do not vary in time as the torus moves. The compressible mode allows for the redistribution of gas in the torus in a response to changes in the radial distance of the torus centre.

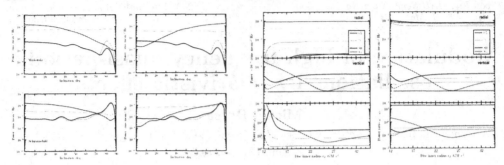

Figure 1. Left pair of panels: the effect of changing observer inclination on the distribution of powers in the radial (red) and vertical (blue) oscillations considering Minkowski and Schwarzschild geometry and the effect of g-factor (solid/dashed lines). The left-most panel shows the compressible mode, the right panel shows the incompressible mode. Right pair of panels: the case of an opaque disc as a function of the torus distance from the gravity centre. Different viewing angles are plotted. The disc lies in the the equatorial plane and goes from infinity down to some terminal radius r_d, which is a parameter of the model. The disc can extend as close as to the torus, but does not penetrate into it. Left panel shows the compressible mode, the right-most panel shows the incompressible mode. For details, see Bursa (2006).

3. Inclination and an opaque-disc effect

Figure 1 compares the dependence of the radial and vertical oscillation powers on changing inclination if the torus is incompressible (left) or compressible (right). We can see that the power in the vertical oscillation stays unchanged, while the radial power is largely affected, particularly if the inclination is changed. There is a clear difference between the red curve progression in the left and right panel. It is caused by the inversion of the luminosity dependence on the torus displacement, which in combination with the effect of g-factor results in a reverse trend of the ω_r power.

There can likely be an outer cool disc surrounding the torus, from which the torus is formed, and which can have a substantial effect on light modulation. The Shakura–Sunyaev disc is optically thick and blocks photons crossing the equatorial plane beyond its terminal radius. Most of the stopped photons have been strongly bent and they carry information predominantly about the vertical mode; thus, the presence or lack of an opaque disc may be important for the power distribution in QPO modes, namely the vertical one. Figure demonstrates how the power of the oscillation modes is changed if an opaque disc is present. The presence of a thin disc is important if it does not end far from the torus, but rather within a distance of ~ 5 gravitational radii from it, and when the viewing angle is moderate to high. Under these conditions the effect of the torus obscuration by an optically thick medium is capable to substantially change powers in the oscillations, and in particular in the vertical mode.

Acknowledgements

The author is affiliated with Centre for Theoretical Astrophysics in Prague and his work is supported by the Czech Science Foundation (ref. 205/07/0052) and the Academy of Sciences (ref. KJB300030703).

References

Abramowicz, M. A., Blaes, O. M., Horák J., Kluźniak, W. & Rebusco, P. 2006, Classical and Quantum Gravity, 23, 1689

Bursa, M. 2006, High-frequency quasi-periodic oscillations and their modulation by relativistic effects, PhD Thesis (Charles University, Prague)

Esin, A. A., Narayan, R., Cui, W., Grove, J. E. & Zhang, S.-N. 1998, ApJ, 505, 854

Black Holes from Stars to Galaxies – Across the Range of Masses
Proceedings IAU Symposium No. 238, 2006
V. Karas & G. Matt, eds.
© 2007 International Astronomical Union
doi:10.1017/S1743921307005327

XMM-Newton RGS spectra of four Seyfert 1 galaxies

M. V. Cardaci,[1,2] M. Santos-Lleó[2] and A. I. Díaz[1]

[1]Universidad Autónoma de Madrid, Cantoblanco, 28049-Madrid, Spain

[2]XMM-Newton Science Operation Centre, ESAC, ESA, PO Box 50727, 28080 Madrid, Spain

email: Monica.Cardaci@esa.int

Abstract. Using the spectral fitting technique we have studied the components contributing to the soft X-ray spectra of four AGN. The selected objects are the Seyfert 1 galaxies: HE 1143-1810, Mkn 110, CTS A08.12 and ESO 359-G19. The high-resolution X-ray spectra analysed were taken with the Reflection Grating Spectrometers on board the XMM-Newton satellite. In contrast to the results for other well-studied Seyfert 1 galaxies, we have found that the spectra of the four galaxies lack of significant absorption features. Hence, there are no signs of high-column-density partially-ionised absorbing material in the vicinity of the active nucleus in these galaxies.

Keywords. galaxies: Seyfert – galaxies: individual (HE 1143-1810, CTS A08.12, ESO 359-G19, Mkn 110) – X-rays: galaxies

---

We have performed a simultaneous fitting of the RGS1 and RGS2 spectra of each of the sources in Table 1. To fit the continuum we used an absorbed power law with column density fixed to the Galactic values. For three of the sources, the addition of an intrinsic neutral absorption component significantly improved the fit. Finally, we added some Gaussian emission lines. Table 2 lists the values of the best fitting parameters. Figure 1 shows the spectra and the best model for Mkn 110.

As it can be seen in Table 2 the power-law indexes found are in the range of 2.2 to 3.7. These values are similar to those found in other AGN showing a soft-X-ray excess. Therefore, the indexes in Table 2 are probably indicating the presence of soft excess emission. For Mkn 110, CTS A08.12 and ESO 359-G19 we have found signatures of an intrinsic cool absorbing material with column densities between 0.1 and 3.6 $\times 10^{21}$ cm^{-2}. We have found oxygen emission lines in three of the four objects. The large errors on the OVII triplet lines in HE 1143-1810 and Mkn 110 do not allow charaterisation of electronic temperatures and densities of the emitting medium as explained in Gabriel & Jordan (1969) and Porquet & Dubau (2000). There are no indications of the presence of warm absorbers in the line-of-sight towards any of these four active nuclei.

Table 1. Observation details. In the last column we have included the galactic column densities estimated for the positions of the objects.

Name	Obs. ID	Obs. date	exp. time (s)	10^{-2} counts/s RGS1	10^{-2} counts/s RGS2	Gal n_H (cm^{-2})
HE 1142-1810	0201130201	8 Jun 2004	31000	90.4 ±0.6	98.7 ±0.6	3.4 × 10^{20}
Mkn 110	0201130501	15 Nov 2004	47000	81.3 ±0.4	88.4 ±1.8	1.42 × 10^{20}
CTS A08.12	0201130301	30 Oct 2004	46000	8.1 ±0.3	9.6 ±0.3	4.07 × 10^{20}
ESO 359-G19	0201130101	9 Mar 2004	24000	2.8 ±0.3	3.4 ±0.3	1.02 × 10^{20}

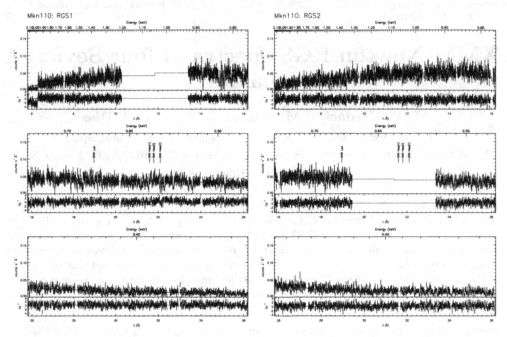

Figure 1. RGS spectra, plotted in rest frame, and the model fit (in red) for Mkn 110.

Table 2. Best fit parameters obtained for the spectra of the four active nuclei. The line fluxes are given in 10^{-14} erg cm^{-2} s^{-1}; power-law normalizations in 10^{-4} photons keV^{-1} cm^{-2} s^{-1}.

Component	Parameter	HE 1143-1810	Mkn 110	CTS A08.12	ESO 359-G19
Intrins. mat.	n_H $\left(\frac{10^{21}}{\text{cm}^{-2}}\right)$	—	0.11±0.03	1.25±0.09	3.6±0.9
Power law	Γ	2.42±0.02	2.24±0.04	2.39±0.06	3.7±0.6
	norm	120±1	104±1	22.3±0.5	15^{+4}_{-3}
Ovii(r) line	E (keV)	0.5717±0.0005	0.5734±0.0003	—	—
	flux	6±3	3^{+3}_{-2}	—	—
Ovii(i) line	E (keV)	0.5663	0.5680	—	—
	flux	3^{+4}_{-3}	6±2	—	—
Ovii(f) line	E (keV)	0.5587	0.5604	—	—
	flux	9^{+5}_{-6}	8^{+2}_{-3}	—	—
Oviii Lyα line	E (keV)	—	$0.6530^{+0.0136}_{-0.0008}$	$0.6531^{+0.0072}_{-0.0001}$	—
	flux	—	4±2	6±2	—
Fit statistic	χ^2/dof	5074.6/5219	5667.6/5353	5010.4/5190	5142.1/5217
Flux 0.5–2.0 keV	$\left(\frac{\text{erg}}{10^{12}\text{cm}^2\text{s}}\right)$	23.91 ± 0.08	21.36 ± 0.05	3.0 ± 0.2	$1.1^{+0.2}_{-0.5}$

References

Gabriel, A. H. & Jordan, C. 1969, MNRAS, 145, 241

Porquet, D. & Dubau, J. 2000, A&A, 143, 495

Black Holes from Stars to Galaxies – Across the Range of Masses
Proceedings IAU Symposium No. 238, 2006
V. Karas & G. Matt, eds.
© 2007 International Astronomical Union
doi:10.1017/S1743921307005339

Analytical binary black hole model of Sgr A* and its implications

Tapan K. Chatterjee

University of the Americas, Department of Physics, A. P. 1316, Puebla, Mexico

e-mail: chtapan@yahoo.com

Abstract. Surrounding the Galactic center is a molecular cloud with a central black hole. This raises the question as to how the hole was formed. While clouds containing HCN are circulating around the Galactic center, they do not all move at the same speed and so collide mutually and should make a more uniform motion and distribution after a period of some 100 kyrs (e.g., Ekers, van Gorkom, Schwartz *et al.* 1983). Hence some very energetic phenomena must have occurred within that period.

Keywords. Black hole physics – galaxies: nuclei

---

1. Introduction and Theory

The radio source at the Galactic center, Sgr A*, is an excellent candidate for a black hole, as the flat sub-mm spectrum with synchrotron emission requires high compactness (e.g., Zylka, Mezger, Warg-Thompson *et al.* 1995; Duschl & Lesch 1994). However, its radiation intensity is orders of magnitude weaker. Several observations indicate non-homogeneous source structure: different core sizes at different mm wavelengths; different variability at sub-mm and radio wavelengths; varying simultaneous spectral indices (e.g., Lacy, Townes & Hoolenbach 1982). The radial velocities of the molecular hydrogen emission regions, straddling the Galactic center, exhibit an asymmetry between the two peaks, corresponding to a radial velocity component of 40 km/s, with respect to the Sun (Clube 1986). No such effect is noticeable from the distribution of radial velocities of the stars around the Galactic center. This indicates rotational motion with practically no offset in radial velocity (e.g., Kent 1992). An interpretation of the radial asymmetry is that it is due to the orbital motion around the black hole.

2. Results and Conclusions

In this context we model analytically a binary black hole (Magalinsky & Chatterjee 2002). Using the variational principle (adopted from the semi-classical approach) a criterion is formulated (which tends towards energy minimization and gravity softening) to study the orbital motion of the pair. An analysis of the expected distribution of orbital parameters, on the basis of statistical mechanics, indicates that the system is characterized by a state preferring high eccentricities.

The system is found to slowly circularize its orbit in the adiabatic regimen, but in a large timescale. This has been confirmed by numerical simulations (Chatterjee & Magalinsky 2002). Thus the separation of the black holes decreases gradually, followed by coalescence after which new pairs of black holes are likely to be established. The system exhibits poor accretion efficiency, as the accretion rate varies radially (Chatterjee & Magalinsky 2006). This is because most of the mass, from the stellar wind appears to fall in the center of mass of the system; and is not accreted by either component of the binary.

Preliminary results indicate that the model explains the low intensity of radiation of Sag A The binarity of a black hole system is very hard to confirm observationally. (only VLBI might detect radio radiation from both components). A recourse has to be taken to indirect methods. A regular pattern of variability of optical emission lines (Halpern & Fillippenko 1988), the precessional behavior of radio jets, (Rees 1984) and the changing precessional period due to orbital motion of the binary (Roos 1988; Roos, Kaastra & Hummel 1993) are expected.

References

Chatterjee, T. K. & Magalinsky, V. B. 2002, Astron. Astrophys. Trans., 21, 79

Chatterjee, T. K. & Magalinsky, V. B. 2006, Astron. Astrophys. Trans., submitted

Clube, S. V. M. 1986, The Observatory, 106, 166

Duschl, W. J. & Lesch, H. 1994, A&A, 286, 431

Ekers, R. D., van Gorkom, J. H., Schwartz, U. J. & Goss, W. M. 1983, A&A, 122, 143

Halpern, J. P. & Fillippenko, A. V. 1988, Nature, 331, 46

Kent, S. M. 1992, ApJ, 387, 181

Lacy, J. H., Townes, C. H. & Hoolenbach, D. J. 1982, ApJ, 262, 120

Magalinsky, V. B. & Chatterjee, T. K. 2002, Astron. Astrophys. Trans., 18, 807

Rees, M. J. 1984, ARA&A, 22, 123

Roos, N. 1988, ApJ, 334, 95

Roos, N., Kaastra, J. S. & Hummel, C. A. 1993, ApJ, 409, 130

Zylka, R., Mezger, P. G., Warg-Thompson, D., Duschl, W. & Lesch, H. 1995, A&A, 297, 83

Black Holes from Stars to Galaxies – Across the Range of Masses
Proceedings IAU Symposium No. 238, 2006
V. Karas & G. Matt, eds.
© 2007 International Astronomical Union
doi:10.1017/S1743921307005340

Evolution of black-hole intermediate-mass X-ray binaries

Wen-Cong Chen and Xiang-Dong Li

Department of Astronomy, Nanjing University, Nanjing 210093, China
email: chenwc@nju.edu.cn, lixd@nju.edu.cn

Abstract. We propose a plausible mechanism for orbital angular momentum loss in black-hole intermediate-mass X-ray binaries, assuming that a small fraction of the transferred mass form a circumbinary disc. The disc can effectively drain orbital angular momentum from the binary, leading to the formation of compact black-hole low-mass X-ray binaries. This scenario also suggests the possible existence of luminous, persistent black hole low-mass X-ray binaries.

Keywords. Black holes – X-ray binaries – circumstellar matter

1. Introduction

There are currently nine compact black hole low-mass X-ray binaries (BHLMXBs). Their short orbital periods ($\leqslant$ 0.5 d) imply that they must have undergone secular orbital angular momentum loss. If their progenitor systems contain a low-mass secondary initially, it is not clear whether the secondary star has enough energy to eject the envelope of the black hole progenitor during the common envelope evolution phase. Justham, Rappaport & Podsiadlowski (2006) suggested that part of black hole intermediate-mass X-ray binaries (BHIMXBs) may evolve to compact BHLMXBs via magnetic braking. Here we explore an alternative possibility. We suggest that a fraction δ of the transferred matter may form a circumbinary (CB) disc, which can drive BHIMXBs to short period BHLMXBs.

2. Results

We have calculated the evolution of BHIMXBs consisting of an intermediate-mass donor star and a black hole (of mass $10M_\odot$). Neglecting the magnetic braking mechanism, we take into account three types of angular momentum loss from the binary system, i.e., gravitational wave radiation, isotropic winds from the donor star, and the tidal torque by the CB disc. Detail description of the model can be found in Chen & Li (2006).

Figure 1 shows the examples of the evolutionary sequence with an initial orbital period of 1 d. The solid and dotted lines correspond to two different values of δ. As can be seen in the figure, mass transfer is initially driven by nuclear evolution of the secondary, leading to expansion of the orbits. But this tendency is held up when angular momentum loss via the CB disc becomes sufficiently strong. The mass transfer drops into a "plateau" phase at a rate $\sim 10^{-9} M_\odot \text{yr}^{-1}$ for a few 10^8 yr. These features are different from those in Podsiadlowski, Rappaport & Han (2003), in which the orbits always increase secularly. After that the mass transfer rates increase sharply as the secondary ascends the giant branch, but the final orbital evolution depends on the adopted value of δ. With the higher value of δ, a compact BHLMXB will be finally produced after a few 10^8 yr mass transfer.

In Fig. 1 the thick and thin curves correspond to stable and unstable mass transfer during the evolution of BHXBs. It is interesting to see that the model BHLMXBs are

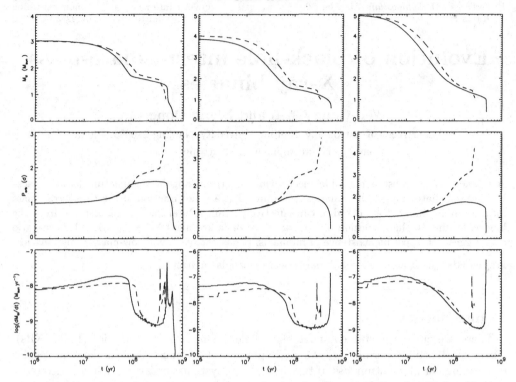

Figure 1. Evolution of the donor masses M_d (top), orbital periods P_{orb} (middle), and mass transfer rates $\dot{M}_d$ (bottom) for BHIMXBs with the initial orbital period $P_{orb} = 1$ d. The left, middle, and right panels correspond to $M_d = 3M_\odot$ (solid curve $\delta = 0.0055$, dashed curve $\delta = 0.004$), $4M_\odot$ (solid curve $\delta = 0.007$, dashed curve $\delta = 0.005$), and $5M_\odot$ (solid curve $\delta = 0.007$, dashed curve $\delta = 0.005$), respectively. Stable and unstable mass transfer processes are plotted with thick and thin curves, respectively.

likely to be persistent X-ray sources. This is not compatible with the observations of Galactic BHLMXBs, suggesting that δ does not need to be constant but changes during the evolution. However, *Chandra* observations of the 12.6 hr ultraluminous X-ray source in the elliptical galaxy NGC3379 suggest that the current on phase has lasted ~ 10 yr (Fabbiano *et al.* 2006). This source could be a persistent X-ray source as suggested by our calculations.

Similar as in Justham, Rappaport & Podsiadlowski (2006), our results encounter the difficulty that the calculated effective temperatures are not consistent with those of the observed donor stars in BHLMXBs. If this effective temperature problem can be solved, the CB disc mechanism may provide a plausible solution to the BHLMXB formation problem, without requiring anomalous magnetic fields in the donor stars.

Acknowledgements

This work was supported by the National Science Foundation of China under grant 10573010.

References

Chen, W.-C. & Li, X.-D. 2006, MNRAS, 373, 305
Fabbiano, G., Kim, D.-W., Fragos, T. *et al.* 2006, ApJ, 650, 879
Justham, S., Rappaport, S. & Podsiadlowski, Ph. 2006, MNRAS, 366, 1415
Podsiadlowski, Ph., Rappaport, S. & Han, Z. 2003, MNRAS, 341, 385

Black Holes from Stars to Galaxies – Across the Range of Masses
Proceedings IAU Symposium No. 238, 2006
V. Karas & G. Matt, eds.
© 2007 International Astronomical Union
doi:10.1017/S1743921307005352

Relationship between X-shaped radio sources and double-double radio galaxies

Xian Chen and Fukun Liu

Department of Astronomy, Peking University, 100871, Beijing, China
email: fkliu@bac.pku.edu.cn

Abstract. Both the X-shaped radio galaxies and double-double radio galaxies (DDRGs) are suggested in the literature to be due to the binary-accretion disk interaction or to the coalescence of SMBBHs. These models suggest some relationship between the two types of radio sources. In this paper, we collected data from literatures for two samples of X-shaped and double-double radio galaxies together with a control sample of FRII radio galaxies and statistically investigate their properties.

We find that the wings of X-shaped radio galaxies and the outer and inner lobes of DDRGs tend to be perpendicular to the major axis of the host galaxy (or dust structures), while the active lobes orient randomly. Both X-shaped and double-double radio galaxies are low luminous FRII or FRI/FRII transitional radio sources with the similar dimensionless accretion rate $\dot{m} \sim 0.01$, which is about the transitional accretion rate given in the literature.

All the statistic results can be reconciled if there is an evolutionary relationship between X-shaped and double-double radio galaxies, in the sense that X-shaped radio galaxies may be due to the interaction of active SMBBHs and accretion disk and DDRGs due to the removal of inner disk region and the coalescence of SMBBHs.

Keywords. Accretion, accretion disks – black hole physics – galaxies: active – galaxies: interactions – galaxies: jets

1. Introduction

Supermassive binary black holes (SMBBHs) at galactic centers are expected by the hierarchical galaxy formation model in cold dark matter (CMD) cosmology (Haehnelt & Kauffmann 2000; Kauffmann & Haehnelt 2000). Both the X-shaped radio galaxies and double-double radio galaxies (DDRGs) are suggested in the literature to be due to the binary-accretion disk interaction or to the coalescence of SMBBHs (Merritt & Ekers 2002; Liu *et al.* 2003; Liu 2004). These models suggest some relationship between the two types of radio sources. In this paper, we statistically investigate the observations of the two types of radio sources and discuss the implications to different models.

2. Method

We collected the observational data of radio power, linear size and projected position angles of radio feature and host galaxy for two samples of X-shaped and double-double radio galaxies and a control sample of FRII radio galaxies. We also estimated the central black hole mass and bolometric luminosity, and calculated the dimensionless accretion rate. We investigated the distributions of the three samples sources by employing the K–S test.

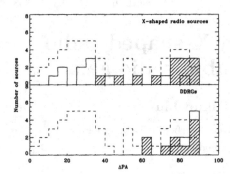 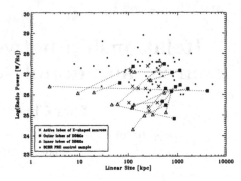

Figure 1. Left: Distributions of relative position angles (ΔPAs). The thin dashed line is for 3CRR control sample. The thick solid line and the shaded area are, respectively, for active lobes and wings of X-shaped radio sources (upper panel) and for the inner and outer lobes of DDRGs (lower panel). Right: Linear sizes of radio structures vs. radio power of the samples. Radio power is in W Hz^{-1} at 178 MHz. Cross is the active lobes of X-shaped radio sources. Squares and triangles are, respectively, the outer and inner lobes of DDRGs and, if belong to the same object, are connected with a dotted line. Dots are 3CRR control sample sources.

3. Results

The relative position angles between different radio structure and the major axis of host galaxy are presented in the left panel of Fig. 1. It is clear that the wings of X-shaped radio galaxies and the outer and inner lobes of DDRGs tend to be perpendicular to the major axis of the host galaxy (or dust structures), while the active lobes orient randomly.

The linear size and radio power of the sample sources are presented in the right panel of Fig. 1. It shows that both X-shaped and double-double radio galaxies are low luminous FRII or FRI/FRII transitional radio sources. The inner lobes of DDRGs and the active lobes of X-shaped radio sources have similar radio extension, but the outer lobes of DDRGs are significantly larger than wings of X-shaped radio sources. We also find that both types of sources have the similar dimensionless accretion rate $\dot{m} \sim 0.01$, which is about the transitional accretion rate given in the literature.

4. Conclusion

Our results are consistent with the scenarios for X-shaped and double-double radio galaxies that the X-shaped feature in winged sources is due to swift jet reorientation and the double-lobed radio structure in DDRGs is due to the interruption of jet formation. All the statistic results can be reconciled if there is an evolutionary relationship between X-shaped and double-double radio galaxies, in the sense that X-shaped radio galaxies may be due to the interaction of active SMBBHs and accretion disk and DDRGs due to the removal of inner disk region and the coalescence of SMBBHs.

Acknowledgements

This work is supported by the National Natural Science Foundation of China (NSFC 10573001).

References

Haehnelt, M. G. & Kauffmann, G. 2000, MNRAS, 318, L35
Kauffmann, G. & Haehnelt, M. G. 2000, MNRAS, 311, 576
Liu, F. K. 2004, MNRAS, 347, 1357
Liu, F. K., Wu, X. B. & Cao, S. L. 2003, MNRAS, 340, 411
Merritt, D. & Ekers, R. D. 2002, Science, 297, 1310

Black Holes from Stars to Galaxies – Across the Range of Masses
Proceedings IAU Symposium No. 238, 2006
V. Karas & G. Matt, eds.

General relativistic results for a galactic disc in a multidimensional space-time

Carlos H. Coimbra-Araújo[1] and Patricio S. Letelier[2]

[1]Instituto de Física Gleb Wataghin, Universidade Estadual de Campinas, CP 6165,
13083-970 Campinas, SP, Brazil

[2]Departamento de Matemática Aplicada, Instituto de Matemática, Estatística e Computação
Científica, Universidade Estadual de Campinas, 13083-970 Campinas, SP, Brazil

Abstract. We construct an exact and simple general relativistic model to describe a galactic disc based on a Schwarzschild disc immersed in a six dimensional space-time. The stability of this configuration is studied and we present results for the calculation of circular geodesic orbits.

Keywords. Gravitation – relativity – dark matter

1. Introduction

In general relativity, besides the solutions that currently offer important support to understand black holes and cosmology, actually the interest is extended for the case where the universe has more than four dimensions. Recently was proposed that a plausible reason for the gravitational force appear to be so weak can be its dilution in possibly existing extra-dimensions related to a bulk (the multidimensional space-time), where branes (the 4D space-time) are embedded (Dienes 1997; Lykken & Randall 2000). In amendment, solutions of Einstein field equations for axially symmetric configurations have great astrophysical interest, because they can be used to model galaxies. In this sense, a long range of discs solutions was derived (follow for instance references in Vogt & Letelier 2005). In the present work, as a first approach, we study the properties of a multidimensional axially symmetric matter configuration whose total density is located at the plane $z = 0$ and is living in a 6D universe.

2. Einstein equations and rotation curves for a static axially symmetric configuration with extra-dimensions

The metric for an axial symmetric 6D space-time can be written in quasi-cylindrical coordinates as

$$ds^2 = -e^{-\phi}dt^2 + f(dr^2 + dz^2) + h^2 e^{\phi}d\varphi^2 + \psi e^{\nu}dx^2 + e^{-\nu}dy^2, \qquad (2.1)$$

where $\phi = \phi(r,z)$, $f = f(r,z)$, $h = h(r,z)$, $\psi = \psi(r,z)$ and x and y are the extra-dimensional coordinates. The vacuum Einstein equations $R_{AB} = 0$, $(A, B = 0, 1, \ldots, 5)$ are obtained. We do $G = 1$ and $c = 1$.

To obtain a solution for Einstein equations which represents a thin disc located on $z = 0$, we can introduce discontinuities in first derivatives of f and ϕ by doing the replacement $z \to |z| + a$, where a is a constant. In first approximation one can consider that the particles in the disc move along geodesics. By solving the integral of motion for the metric (2.1), we find the equation for rotation curves V_C in function of specific extra-dimensional parameters. The Newtonian potential associated to a rod of constant density is used. In Fig. 1 we show the rotation curves for a disc in a extra-dimensional space-time, the stability of the model and in Fig. 2 the comparison to observational data and other models. The density profiles are the same as in the 4D case and we find that the extra-dimensions do not affect the density and the azimuthal and radial pressures.

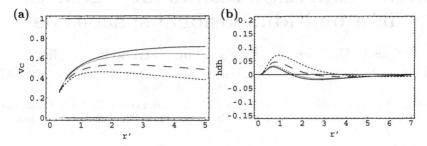

Figure 1. (a) Disc rotation curves with extra-dimensional parameters $C_x = 0$, $C_y = 0$, i.e., usual Newtonian profile (dotted line); $C_x = 0.1$, $C_y = 0.7$ (dashed line); $C_x = 0.1$, $C_y = 0.9$ (gray line); $C_x = 0.1$, $C_y = 0.95$ (full line). (b) Graphic showing the stability of the model. We take $r' = r/m$ and the value used for a is $a = 1.5$ (the stability is reached when $a > 1$ and when $hdh > 0$).

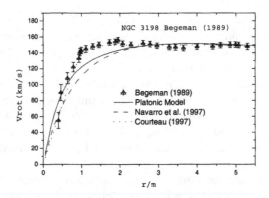

Figure 2. Different models – our model (*Platonic*), Navarro *et al.* (1997) and Courteau (1997) – for the rotation curves of the NGC 3198 spiral galaxy. Observational data from Begeman (1989).

In this work we show with a simple example the construction of a disc embedded in a multidimensional space-time. The circular geodesic orbits of our model are stable in the region of interest and have the necessary accuracy to fit the rotation curves of many spiral galaxies.

Acknowledgements

C.H. C.-A. thanks CAPES and IAU for financial support and P.S.L thanks CNPq and FAPESP for partial financial support.

References

Begeman, K. G. 1989, A&A, 223, 47
Courteau, S. 1997, Aj, 114, 2402
Dienes, K. R. 1997, Phys. Rep., 287, 447
Lykken, J. & Randall, L. 2000, J. High En. Phys., 06, 014
Navarro, J. F., Frenk, C. S. & White, S. D. M. 1997, Apj, 490, 493
Vogt, D. & Letelier, P. S. 2005, mnras, 363, 268

Black Holes from Stars to Galaxies - Across the Range of Masses
Proceedings IAU Symposium No. 238, 2006
V. Karas & G. Matt, eds.

Formation history of supermassive black holes in a viable cosmological window

Carlos H. Coimbra-Araújo[1] and Amâncio C. S. Friaça[2]

[1]Instituto de Física Gleb Wataghin, Universidade Estadual de Campinas, CP 6165,
13083-970 Campinas, SP, Brazil

[2]Departamento de Astronomia, Instituto de Astronomia, Geofísica e Ciências Atmosféricas,
Universidade de São Paulo, 05508-900 São Paulo, SP, Brazil

Abstract. We show, performing a viable cosmological window, that only the magneto hydro-dynamic (MHD) disk model is capable to explain how an intermediate mass black hole (IMBH) (with masses $\sim 10^3 M_\odot$) grows unto a supermassive black hole (SMBH) (with masses $\sim 10^7 M_\odot$). We still calculate the supermassive stars sequence of stability. Those stars, with synthetized helium or oxygen cores, collapse to form IMBHs. In our calculation we show that the primordial stars must have rapid rotation if they are in the stable part of the sequence.

Keywords. Black holes – quasars – first stars

1. First stars: sequence of stability

We consider that to form an intermediate mass black hole (IMBH) it is necessary a progenitor as a supermassive star. The existence of a such object can be tested by studying its stability. We perform this study by modelling a star with rotation, composed by a hydrogen envelop plus a helium or oxygen core. Geometrically the core is a Maclaurin ellipsoid with range of mass around 150–5000$M_\odot$ with an oblate parameter $e_\lambda = (a_3/a_1)^{2/3}$ – where a_3 and a_1 are respectively the equatorial and the polar radius of the ellipsoid – and where we apply post- and post-post-Newtonian criteria in the energy functional of the star. In Fig. 1 we show the sequence of stability for our star and the behavior of the rotation energy $T/|W|$ with the oblate parameter. The star necessarily has rapid rotation if it is in the stable part of the sequence (the left region of the Fig. 1b).

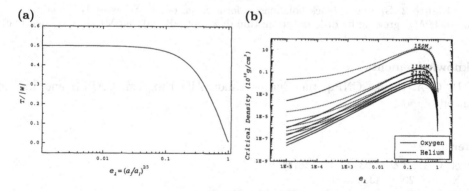

Figure 1. (a) The rotation energy $T/|W|$ versus the defined e_λ eccentricity. (b) Graphic showing the stability of the star. The behavior is different for both helium and oxygen cores.

2. Accretion onto black holes

In what follows we calculate how much time is necessary for a $10^3 M_\odot$ IMBH formed from our supermassive star to grow unto a $10^8 M_\odot$ supermassive black hole (SMBH). Here we adopt a ΛCDM model and use the WMAP cosmological parameters (Bennet *et al.* 2003; Spergel *et al.* 2003). The oldest QSO has redshift $z = 6.43$ (QSO $1148 + 525$, Fan *et al.* 2003) what means $t_{QSO} = 0.87$Gyrs. We consider that the accretion cosmological window begins at the creation of first stars ($z \sim 40$ or $t_{stars} \sim 0.067$Gyrs) and finish at the QSO $1148 + 525$ formation epoch. According to our calculation, the maxim life time of a first supermassive star is around $t \approx 0.01$Gyrs. An important parameter to unravel how supermassive black holes grow comes from the ratio R between the observed AGNs luminous density and the local SMBHs mass density. This parameter has a profound correlation with the efficiency η_M to convert rest mass in luminous energy in accretion disks. Recent measurements suggest that $0.1 < R < 0.2$ (Yu & Tremaine 2002). It is possible that the efficiency η_M is not a fixed parameter, but presents a variation related to the black hole spin. Following the model of McKinney & Gammie (2004) for MHD discs, we find that for a high speed rotating black hole, $\eta_M \to 0.19$. The deduction for the black hole mass function gives $M(t) = M_0 \exp\left[\frac{\eta_L (1-\eta_M)}{\eta_M} \frac{t-t_0}{\tau}\right]$, where η_L is the ratio between the object luminosity and Eddington luminosity (here we consider $\eta_L \to 1$), t_0 is the initial time (here we suppose t_0 is constrained by the first stars at $z \sim 40$) and τ is a temporal measure related to the chemical environment. In Fig. 2 we show that only MHD disks can explain how IMBHs become SMBHs in our cosmological window.

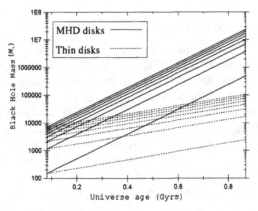

Figure 2. Growth of black hole masses for a initial range of masses M_0 equal to $150 - 6000 M_\odot$, growing by disk accretion: thin disk (dotted line) and MHD disk (solid line).

Acknowledgements

C.H. C.-A. thanks CNPq, the Europe Union Alfa Program, CAPES and IAU for financial support.

References

Bennet, C. L. *et al.* 2003, ApJS, 148, 97
Fan, X. *et al.* 2003, AJ, 125, 1649
McKinney, J. C. & Gammie, C. F. 2004, ApJ, 611, 977
Spergel, D. N. *et al.* 2003, ApJS, 148, 175
Yu, Q. & Tremaine, S. 2002, MNRAS, 335, 965

Black Holes from Stars to Galaxies – Across the Range of Masses
Proceedings IAU Symposium No. 238, 2006
V. Karas & G. Matt, eds.
© 2007 International Astronomical Union
doi:10.1017/S1743921307005388

The central cluster and X-ray emission from Sgr A*

Robert F. Coker[1] and Julian M. Pittard[2]

[1]Los Alamos National Laboratory, Los Alamos, NM 87545, USA

[2]Department of Physics and Astronomy, University of Leeds, Leeds, LS2 9JT, UK

email: email: robc@lanl.gov, jmp@ast.leeds.ac.uk

Abstract. At the centre of the Milky Way is Sgr A*, a putative 3 million solar mass black hole with an observed luminosity that is orders of magnitude smaller than that expected from simple accretion theories. The number density of early-type stars is quite high near Sgr A*, so the ensemble of their stellar winds has a significant on the black hole's environment.

We present results of 3D hydrodynamic simulations of the accretion of stellar winds onto Sgr A*. Using the LANL/SAIC code, RAGE, we model the central arc-second of the Galaxy, including the central cluster stars (the S-stars) with orbits and wind parameters that match observations. A significant fraction of the winds from the S stars becomes gravitationally bound to the black hole and thus could provide enough hot gas to produce the X-ray emission seen by Chandra. We perform radiative transfer calculations on the 3D hydrodynamic data cubes and present the resulting synthetic X-ray spectrum.

Keywords. Galaxy: center – stars: winds, outflows – X-rays: ISM

---

1. Introduction

At the dynamical center of the Galaxy lurks Sgr A*, a strong, compact radio source, whose emission is thought to be due to gas accreting onto a 3.6×10^6 $M_\odot$ black hole (see Melia & Falcke 2001 for a review). Chandra X-ray images show a point source, presumably due to thermal bremsstrahlung and up-scattered magnetic bremsstrahlung radio emission, within the ~ 1" astrometric error box of Sgr A*. This point source flares and is surrounded by diffuse (~ 0.6") emission in the 2–10 keV Chandra band.

There are dozens of OB stars with $K < 17$ mag located within ~ 1" of Sgr A* but only 6 have measured orbital parameters (Eisenhauer et al. 2005). The winds of the stars in this 'Sgr A* IR cluster' contribute to the observed X-ray flux of Sgr A*. So many stars in a small volume results in strong wind collisions and dissipation of kinetic energy. We want to determine how much of the Chandra X-ray flux is from the black hole and how much is directly from colliding stellar winds.

Using the 3D Eulerian code RAGE (Baltrusaitus et al. 1996), we treat the 6 individual OB stars as moving mass ($1.0 \times 10^{-6} M_\odot$ yr^{-1}) and energy sources (81.4 $L_\odot$ yr^{-1}; Aerts & Lamers 2003). Based on the orbits determined from IR observations, we calculate the stars' positions and velocities as a function of time and inject the appropriate mass, momentum, and energy. The volume of solution is taken to be 1.2" (at the GC, 1" $\sim$ 0.04 pc), the size of the extended Sgr A* emission seen by Baganoff et al. (2003).

We calculate emission and absorption coefficients for each cell and perform a radiative transfer calculation along the line-of-sight to obtain synthetic X-ray images, spectra, and luminosities. We include the gravitational potential from the black hole as well as the underlying stellar cluster. We use boundary conditions such that high velocity gas can escape and external gas can fall in. The calculation discussed here used a resolution

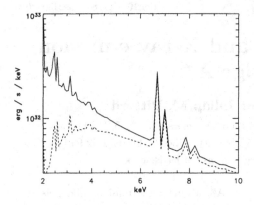

Figure 1. Simulated intrinsic (solid) and absorbed (dashed) 2–10 keV spectrum of the central 1.2" of the Galaxy. The 2–10 keV absorbed luminosity, attenuated by a 10^{23} cm^{-2} column, is 4.7×10^{32} erg s^{-1}. Note the prominent 7 keV Fe line which is also seen in the data.

of 10^{15} cm with a total of 320^3 cells. The stellar mass, energy, and momentum are injected uniformly in 2.5×10^{15} cm radius spheres at each star. The ISM is taken to be a $\gamma = 5/3$ gas with $n = 10^3$ cm^{-3} and $T = 10^4$ K in order to represent the hot, dense GC environment. The central black hole region is 'frozen' (replaced with initial conditions after every time-step) within $\sim 500 R_g$ to remove inflowing mass and energy. Optically thin radiative cooling, based on MEKAL cooling rates and assuming solar abundances, is included. We assume an equilibrium temperature of 10^4 K so that cold cells slowly heat up due to the hot emission from the young, massive stars in the region.

2. Results and Discussion

The colliding wind regions in the simulation reach $\sim 10^7$ K while the central region attains more than 10^8 K. The capture radius of these winds is ~ 0.03 pc so almost all of the gas is bound to the black hole. Wind speeds of ~ 2000 km s^{-1} (e.g. due to O stars) would be mostly unbound. Since the 6 stars are spectrally identified as B stars, their 'slow' winds should be accumulating in the central 1", as they do in the simulation (although they do reach saturation). With a column density of $\sim 10^{23}$ cm^{-2}, all of the intrinsic X-ray luminosity of Sgr A* (2.4×10^{33} erg s^{-1}) can be produced by *just* these 6 OB stars; their simulated intrinsic luminosity is 3.3×10^{33} erg s^{-1}. The predicted intrinsic 2–10 keV photon index (~ 3.0) also matches the observations (~ 2.7). Although the predicted emission is more compact than observed, including the other two dozen OB stars in the region should extend it (while permitting a lower mass loss rate for the stars). Baganoff *et al.* (2003) rejected the OB stars as the source of the Sgr A* emission based on Pittard & Stevens (1997), which required binaries or stellar separations of less than $\sim 10^{14}$ cm. However, the presence of a potential well enhances the density and temperature such that the OB stars alone are sufficient to explain the observed quiescent X-ray luminosity of Sgr A*. Thus, accretion models which predict constant significant X-ray emission from within ~ 500 R_g of the black hole Sgr A* may need to be revised.

References

Aerts, C. & Lamers, H. J. G. L. M. 2003, A&A, 403, 625
Baganoff, F. K. *et al.* 2003, ApJ, 591, 891
Baltrusaitis, R. M. *et al.* 1996, Physics of Fluids, 8, 2471
Eisenhauer, F. *et al.* 2005, ApJ, 628, 246
Melia, F. & Falcke, H. 2001, ARAA, 39, 309
Pittard, J. M. & Stevens, I. R. 1997, MNRAS, 292, 298

Black Holes from Stars to Galaxies – Across the Range of Masses
Proceedings IAU Symposium No. 238, 2006
V. Karas & G. Matt, eds.
© 2007 International Astronomical Union
doi:10.1017/S174392130700539X

Upper limits on the mass of supermassive black holes from HST/STIS archival data

E. M. Corsini,[1] A. Beifiori,[1] E. Dalla Bontà,[1] A. Pizzella,[1]
L. Coccato,[2] M. Sarzi[3] and F. Bertola[1]

[1]Dipartimento di Astronomia, Università di Padova, Padova, Italy
[2]Kapteyn Astronomical Institute, University of Groningen, Groningen, The Netherlands
[3]Centre for Astrophysics Research, University of Hertfordshire, Hatfield, UK

Abstract. The growth of supermassive black holes (SBHs) appears to be closely linked with the formation of spheroids. There is a pressing need to acquire better statistics on SBH masses, since the existing samples are preferentially weighted toward early-type galaxies with very massive SBHs. With this motivation we started a project aimed at measuring upper limits on the mass of the SBHs in the center of all the nearby galaxies ($D < 100$ Mpc) for which STIS/G750M spectra are available in the HST archive. These upper limits will be derived by modeling the central emission-line widths observed in the Hα region over an aperture of $\sim 0.1''$. Here we present our results for a subsample of 22 S0-Sb galaxies within 20 Mpc.

Keywords. black hole physics – galaxies: kinematics and dynamics – galaxies: structure

1. Introduction

The census of supermassive black holes (SBHs) is large enough to probe the links between mass of SBHs and the global properties of the host galaxies (see Ferrarese & Ford 2005). However, accurate measurements of SBH masses ($M_\bullet$) are available for a few tens of galaxies and the addition of new determinations is highly desirable.

To this purpose we started a project aimed at measuring upper limits on $M_\bullet$ in the center of all the nearby galaxies ($D < 100$ Mpc) for which STIS/G750M spectra are available in the HST archive. We retrieved data for 213 galaxies spanning over all the morphological types. This will extend previous works by Sarzi *et al.* (2002) and Verdoes Kleijn *et al.* (2006). Here we analyze a subsample of 22 galaxies. They have been selected to be S0-Sb within 20 Mpc and to have a measured stellar velocity dispersion ($\sigma_\star$). All the galaxies were observed as part of the HST/SUNNS project (PI: H.-W. Rix, GO-7361), except for NGC 4435 (Coccato *et al.* 2006).

2. Data reduction and analysis

The STIS/G750M spectra were obtained with the $0.2'' \times 52''$ slit crossing either the galaxy nucleus along a random position angle (SUNNS) or along the galaxy major axis (NGC 4435). The observed spectral region includes the [N II]$\lambda\lambda6548, 6583$Å, Hα, and [S II]$\lambda\lambda6716, 6730$Å emission lines. Data reduction was performed with the STIS pipeline, which we implemented to clean cosmic rays and hot pixels. For each galaxy we obtained the nuclear spectrum by extracting a $0.25''$-wide (< 25 pc) aperture centered on the continuum peak. The ionized-gas velocity dispersion was measured by the fitting Gaussians with the same width and velocity to the narrow component of the [N II] and [S II] emission lines. The Hα line and broad components were fitted with additional Gaussians.

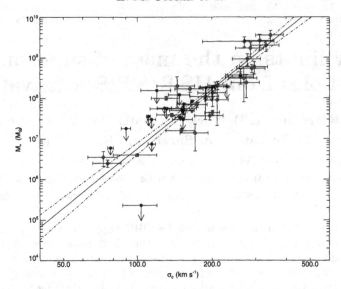

Figure 1. Comparison between our $M_\bullet$ upper limits (filled circles) and the $M_\bullet - \sigma_\star$ relation. The edges of the arrows correspond to upper limits obtained with $i = 33°$ and $i = 81°$, respectively. The open circles show the galaxies with accurate measurements of $M_\bullet$ (Ferrarese & Ford 2005).

We assumed that the ionized gas resides in a thin disk and moves onto circular orbits. The local circular velocity is dictated by the gravitational influence of the putative SBH. To derive the upper limits on $M_\bullet$ we first built the gaseous velocity field and projected it onto the sky plane according to the disk orientation. Then we observed it by simulating the actual setup of STIS and effects related to the STIS PSF, slit width and the charge bleeding as done by Coccato *et al.* (2006). Including the mass of the stellar component would lead to tighter upper limits on $M_\bullet$. We have no information on the orientation of the gaseous disk within the central aperture. For this pilot project we assumed that the disk is either nearly face on ($i = 33°$) or edge on ($i = 81°$) and the slit is placed along its major axis. We derived the intrinsic flux radial profile of the gaseous disk by fitting a Gaussian function to the narrow emission-line fluxes taking into account of disk inclination and STIS PSF.

3. Results

The resulting upper limits are consistent with those derived by Sarzi *et al.* (2002) for the SUNNS galaxy sample and by Coccato *et al.* (2006) for NGC 4435. Therefore, we are confident to obtain reliable estimates of the upper limit on $M_\bullet$ for all the sample galaxies with emission lines in their spectra. This will allow to increase the statistical significance of the relationships between $M_\bullet$ and galaxy properties and identify peculiar objects worthy of further investigations. This is the case of the $M_\bullet$ of NGC 3351, NGC 4314, and NGC 4143, which lie below the $M_\bullet - \sigma_\star$ relation (Fig. 1).

References

Coccato, L. *et al.* 2006, MNRAS, 366, 1050
Ferrarese, L. & Ford, H. 2005, Sp. Sci. Rev., 116, 523
Sarzi, M. *et al.* 2002, ApJ, 567, 237
Verdoes Kleijn, G. A., van der Marel, R. P. & Noel-Storr, J. 2006, AJ, 131, 1961

Black Holes from Stars to Galaxies – Across the Range of Masses
Proceedings IAU Symposium No. 238, 2006
V. Karas & G. Matt, eds.

The nuclear properties of early-type galaxies in the Virgo and Fornax clusters

Patrick Côté

Herzberg Institute of Astrophysics, National Research Council of Canada, Victoria, Canada
Email: Patrick.Cote@nrc-cnrc.gc.ca

Abstract. We summarize findings from imaging surveys of 143 early-type galaxies in the Virgo and Fornax Clusters. Using deep HST images in the F475W (g) and F850LP (z) band-passes obtained with the Advanced Camera for Surveys (ACS), we have examined the central structure of the program galaxies, finding clear evidence for a compact stellar nucleus in $\sim$ 75% of the sample. The formation of early-type galaxies of low- and intermediate-mass is often – but not always – accompanied by the formation of a compact stellar nucleus.

Keywords. galaxies: clusters – galaxies: elliptical and lenticular – galaxies: nuclei

1. Introduction

The ACS Virgo and Fornax Cluster Surveys [5,11] consist of HST imaging for 143 early-type (E, S0, dE, dE,N or dS0) members of the Virgo and Fornax Clusters, supplemented by imaging and spectroscopy from WFPC2, Chandra, Spitzer, Keck, VLT, Magellan, KPNO, and CTIO. All HST images were taken with the ACS [9] using a filter combination equivalent to the g and z band-passes in the SDSS photometric system. The images cover a $200'' \times 200''$ field with $\approx 0.1''$ resolution (corresponding to a physical resolution of 8 pc).

2. Results from the ACS Virgo and Fornax Surveys

For each galaxy, azimuthally averaged surface brightness profiles were derived as described in [6,7]. Nuclei were identified through a direct inspection of the ACS images and by fitting the surface brightness profiles, where nuclei appear as an "excess" over the inward extrapolation of the best-fitting galaxy model [10,15] (see Figure 1). Our main findings can be summarized as follows:

☐ The frequency of nucleation in early-type galaxies brighter than $M_B \approx -15$ falls in the range $66 \lesssim f_n \lesssim 82\%$, roughly three times higher than previous estimates [1]. Earlier ground-based surveys missed significant numbers of nuclei due to surface brightness selection effects, limited sensitivity and poor spatial resolution.

☐ A search for nuclei offset from the photo-centers of their host galaxies reveals only a handful of candidates (i.e., $\lesssim 7\%$ of the sample) with displacements of $0.5''$ (40 pc) or larger, all in dwarf galaxies. In each case, though, the evidence suggests that these "nuclei" are, in fact, globular clusters projected close to the galaxy photo-center.

☐ The nuclei have a median half-light radii of $r_h = 4.2$ pc, with the sizes of individual nuclei ranging from 62 pc down to $\leqslant 2$ pc (i.e., unresolved in our images). Excluding the half dozen unresolved objects, the nuclei sizes are found to depend on nuclear luminosity according to the relation $r_h \propto \mathcal{L}^{0.50\pm0.03}$.

☐ The large majority of nuclei are resolved, so we can rule out low-level, non-thermal AGN as an explanation for the central luminosity excess in almost all cases.

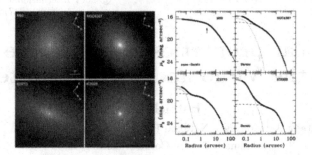

Figure 1. Left Panels: F475W images of the central regions of four galaxies from the ACS Virgo Cluster Survey. Right Panels: Azimuthally-averaged g-band surface brightness profiles for these same four galaxies. M60 is an example of a non-nucleated "core-Sérsic" galaxy; the best core-Sérsic model is shown as a solid curve. The vertical arrow shows the radius, r_b, at which the outer Sérsic profile "breaks" to an inner power-law. For the remaining galaxies, we show the best-fit model which consists of a central King model for the nucleus (dotted curve) and a Sérsic model for the underlying galaxy (dashed curve). The solid curve shows the composite model.

☐ The colors of the nuclei in galaxies fainter than $M_B \approx -17.6$ are tightly correlated with their luminosities, and less so with the luminosities of their host galaxies, suggesting that their chemical enrichment histories were governed by local or internal factors.

☐ Comparing the nuclei to the "nuclear clusters" found in late-type spiral galaxies [2-4,12-14,16] reveals a close match in terms of size, luminosity and overall frequency. A formation mechanism that is rather insensitive to the detailed properties of the host galaxy is required to explain this ubiquity and homogeneity.

☐ The mean of the frequency function for the nucleus-to-galaxy luminosity ratio in our nucleated galaxies is indistinguishable from that of the black hole-to-bulge mass ratio calculated in 23 early-type galaxies with detected supermassive black holes: i.e., $\approx 0.2\%$ by mass; [6,8,17]. The compact stellar nuclei found in our program galaxies may thus be low-mass counterparts to the supermassive black holes detected in the bright galaxies.

Acknowledgements

It is a pleasure to thank my collaborators on the ACS Virgo and Fornax Cluster Surveys for their hard work and enthusiasm during the past few years.

References

[1] Binggeli, B., Sandage, A. & Tammann, G. A. 1985, AJ, 90, 1681 (BST85)
[2] Böker, T. et al. 2002, AJ, 123, 1389
[3] Böker, T. et al. 2004, AJ, 127, 105
[4] Carollo, C. M., Stiavelli, M. & Mack, J. 1998, AJ, 116, 68
[5] Côté, P. et al. 2004, ApJS, 153, 223
[6] Côté, P. et al. 2006, ApJS, 165, 57
[7] Ferrarese, L. et al. 2006a, ApJ, 164, 334
[8] Ferrarese, L. et al. 2006b, ApJ, 644, L21
[9] Ford, H. C. et al. 1998, Proc. SPIE, 3356, 234
[10] Graham, A. W. et al. 2003, AJ, 125, 2951
[11] Jordán, A. et al. 2006, in preparation
[12] Matthews, L. D. et al. 1999, AJ, 118, 208
[13] Phillips, A. C. et al. 1996, AJ, 111, 1566
[14] Seth, A. et al. 2006, AJ, in press
[15] Sérsic, J.-L. 1968, Atlas de Galaxias Australes (Córdoba: Obs. Astron., Univ. Nac. Córdova)
[16] Walcher, C. J. et al. 2005, ApJ, 618, 237
[17] Wehner, E. H. & Harris W. E. 2006, ApJ, 644, L17

Black Holes from Stars to Galaxies – Across the Range of Masses
Proceedings IAU Symposium No. 238, 2006
V. Karas & G. Matt, eds.
© 2007 International Astronomical Union
doi:10.1017/S1743921307005418

Accreting corona model of the X-ray variability in soft state GBH and AGN

Bożena Czerny[1] and Agnieszka Janiuk[1,2]

[1]Copernicus Astronomical Center, Bartycka 18,00-716 Warsaw, Poland

[2]Department of Physics, University of Nevada, Las Vegas, NV89154, USA

email: bcz@camk.edu.pl, ajaniuk@physics.unlv.edu

Abstract. We develop a two-flow model of accretion onto a black hole which incorporates the effect of the local magneto-rotational instability. The flow consists of an accretion disk and an accreting corona, and the local dynamo affects the disk/corona mass exchange. The model is aimed to explain the power spectra of the sources in their soft, disk-dominated states. The local perturbations of the magnetic field in the disk are described as in King *et al.* (2004) and Mayer & Pringle (2006), but the time-dependent local magnetic field is assumed to affect the local supply of the material to the corona. The accreting corona model can reproduce the broad power spectra of Soft State X-ray binaries and AGN. The model, however, predicts that (i) sources undergoing radiation pressure instability (GRS 1915+105) should have systematically steeper power spectra than other sources, (ii) AGN should have systematically steeper power spectra than GBH. More measurements of power spectra of Soft State sources are clearly needed.

Keywords. Accretion, accretion disks – galaxies: active – X-rays: binaries – X-rays: galaxies

1. Introduction

The X-ray emission of the Galactic Black Holes (GBH) and active galactic nuclei (AGN) is strongly variable. Aperiodic character of this variability makes the interpretation of the physical nature of the phenomenon rather difficult. The key question is : why long timescale variability, plausibly related to the processes in the outer parts of the disk contributes so significantly to the overall variability although the energy dissipated in the outer parts of the disk is very low?

Here we propose to accommodate the basic picture of Lyubarskii (1997) within the frame of the accreting corona model. In our model the variations due to the local dynamo modify the accretion rate within the corona. The disk perturbations propagate in disk viscous timescale and therefore are smeared out, but the viscous timescale in the corona is much shorter than in the disk. This allows the coronal perturbations to be observed.

2. Model

We consider the case of an accretion disk extending down to the marginally stable orbit, surrounded by the hot optically thin corona. We postulate that the corona itself can also transport the angular momentum and can exchange mass and angular momentum with the underlying optically thick disk. The model was originally proposed in the context of disk evaporation in cataclysmic variables by Meyer-Hofmeister & Meyer (1994).

Since the exact mechanism is under discussion, we adopt three viable parameterizations of the evaporation process: $\dot{m}_z$ equal to $\frac{B^2}{8\pi} v_A \frac{\mu H}{kT_{vir}}$ (case A), $\frac{B^2}{8\pi} c \frac{\mu H}{kT_{vir}}$ (case B), or $\frac{B^2}{8\pi} v_A \frac{\mu H}{kT_{vir}} \frac{1}{f(r)}$ (case C). The variations of the local magnetic field, B, are described with

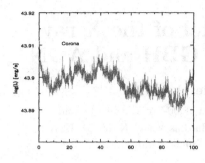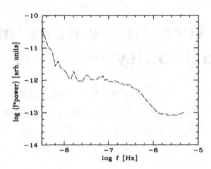

Figure 1. Model: $M = 10^8 M_\odot$, $\dot{m} = 0.22$, type Cb, with αP_{gas} viscosity law.

the use of the Markoff chain, as in King *et al.* (2004) and Mayer & Pringle (2006). We consider three choices of the size of a magnetic cell: Δr equal to $\sqrt{r}$ case (a), $h(r)$ case (b), or r case (c). The disk structure is described assuming either αP_{tot} or αP_{gas} torque, so some models undergo radiation pressure instability and show a limit cycle behaviour.

The time evolution of the disk-corona system is followed using the modified code of Janiuk & Czerny (2005). The disk and corona lightcurves are obtained, and the power spectra (PSD) are calculated for each model.

3. Results and Discussion

Fig. 1 (left) shows an example of a lightcurve (left) and PSD (right) for a corona above a stable disk (αP_{gas} viscosity law), so all the variability is due to the dynamo action.

We calculated many more examples of the disk and corona lightcurves. The evaporation law (B) overproduced the strength of the corona and can be rejected. Some of the models (A) and (C) are acceptable both from the point of view of the disk/corona luminosity ratio and the PSD slope. Observed slopes in GBH and AGN in their soft states are typically within the range 0.8–1.3. However, for a given evaporation law, viscosity prescription and the magnetic cell size, *the results are never universal*, i.e. the PSD slope depends on the black hole mass and accretion rate since these parameters affect the disk structure, and consequently, the corona variability.

The radiation pressure instability in the underlying accretion disk leads to steeper power density spectrum of its fluctuations, as well as to the larger amplitudes.

The PSD spectra corresponding to the accreting supermassive black holes are systematically steeper than for the galactic black holes. Further verification of the models requires more detailed observations and analysis of the long-term variability in AGN.

Part of this work was supported by grants 1P03D00829 and NNG06GA80G.

References

Janiuk, A. & Czerny B. 2005, MNRAS, 356, 205
King, A. R. *et al.* 2004, MNRAS, 348, 111
Lyubarskii, Yu. E. 1997, MNRAS, 292, 679
Mayer, M. & Pringle J. E. 2006, MNRAS, 368, 379
Meyer-Hofmeister, E. & Meyer F. 1994, A&A, 380, 739

Black Holes from Stars to Galaxies – Across the Range of Masses
Proceedings IAU Symposium No. 238, 2006
V. Karas & G. Matt, eds.

© 2007 International Astronomical Union
doi:10.1017/S174392130700542X

Supermassive black holes in BCGs

E. Dalla Bontà,[1,2] L. Ferrarese,[2] J. Miralda-Escudé,[3] L. Coccato,[4] E. M. Corsini[1] and A. Pizzella[1]

[1]Dipartimento di Astronomia, Università di Padova, Padova, Italy

[2]Herzberg Institute of Astrophysics, Victoria, Canada

[3]Institut de Ciencies de l'Espai (CSIC-IEEC)/ICREA, Barcelona, Spain

[4]Kapteyn Astronomical Institute, University of Groningen, Groningen, The Netherlands

Abstract. We observed a sample of three Brightest Cluster Galaxies (BCGs), Abell 1836-BCG, Abell 2052-BCG, and Abell 3565-BCG, with the Advanced Camera for Surveys (ACS) and the Imaging Spectrograph (STIS) on board the Space Telescope. For each target galaxy we obtained high-resolution spectroscopy of the Hα and [N II] $\lambda6583$ emission lines at three slit positions, to measure the central ionized-gas kinematics. ACS images in three different filters (F435W, F625W, and FR656N) have been used to determine the optical depth of the dust, stellar mass distribution near the nucleus, and intensity map. We present supermassive black hole (SBH) mass estimates for two galaxies which show regular rotation curves and strong central velocity gradients, and an upper limit on the SBH mass of the third one. For the SBHs of Abell 1836-BCG and Abell 3565-BCG, we derived $M_\bullet = 4.8^{+0.8}_{-0.7} \times 10^9$ M$_\odot$ and $M_\bullet = 1.3^{+0.3}_{-0.4} \times 10^9$ M$_\odot$ at 1σ confidence level, respectively. For the SBH of Abell 2052-BCG, we found $M_\bullet \leqslant 7.3 \times 10^9$ M$_\odot$.

Keywords. Black hole physics – galaxies: kinematics and dynamics – galaxies: structure

1. Introduction

The link between the evolution of SBHs and the hierarchical build-up of galaxies is imprinted in a number of scaling relations connecting SBH masses to the global properties of the host galaxies. The high SBH mass end of these relations has yet to be fully explored: the massive galaxies expected to host the most massive SBHs are generally at large distances, making a dynamical detection of SBHs observationally challenging. This is unfortunate, since black holes with mass in excess of 10^9 M$_\odot$ occupy an integral part in our understanding of the co-evolution of SBHs and galaxies: these are the systems that have undergone the most extensive and protracted history of merging; moreover, they represent the local relics of the high redshift quasars detected in optical surveys. To investigate the high-mass end of the SBH mass function, we selected three BCGs from the sample of Laine *et al.* (2003). Their large masses, luminosities and stellar velocity dispersions, as well as their having a merging history which is unmatched by galaxies in less crowded environments, make these galaxies the most promising hosts of the most massive SBHs in the local Universe.

2. Dynamical Modeling and Results

Following Coccato *et al.* (2006), a model of the gas velocity field for each galaxy is generated by assuming that the ionized-gas component is moving onto circular orbits in an infinitesimally thin disc centered at the galactic nucleus. The model is projected onto the plane of the sky for a grid of assumed inclination angles of the gaseous disk.

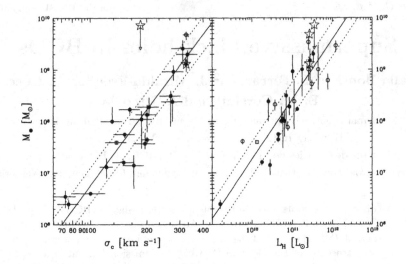

Figure 1. Location of the SBHs masses of our BCG sample galaxies (star symbols) with respect to the $M_\bullet - \sigma_c$ relation of Ferrarese & Ford (2005, left panel, σ_c for the BCGs are from Smith *et al.* 2000) and near-infrared $M_\bullet - L_{bulge}$ relation of Marconi & Hunt (2003, right panel). In the left panel, following Ferrarese & Ford (2005) we plot the SBH masses based on resolved dynamical studies of ionized gas (open circles), water masers (open squares), and stars (filled circles). In the right panel we plot the SBH masses for which $H-$band luminosity of the host spheroid is available from Marconi & Hunt (2003). In both panels, dotted lines represent the 1σ scatter in $M_\bullet$. No σ_c is yet available for Abell 1836-BCG.

Finally, the model is brought to the observational plane by accounting for the width and location (namely position angle and offset with respect to the disk center) of each slit, for the point spread function of the STIS instrument, and for the effects of charge bleeding between adjacent CCD pixels. The mass of the SBH is determined by finding the model parameters (SBH mass, inclination of the gas disk, and mass-to-light ratio of the stellar component) that produce the best match to the observed velocity curves.

For Abell 1836-BCG, we derive $M_\bullet = 4.8^{+0.8}_{-0.7} \times 10^9$ $M_\odot$, the largest SBH mass to have been dynamically measured to-date. The best fit inclination angle is $i = 76° \pm 1°$, while only an upper limit to the stellar mass-to-light ratio is found ($M/L_I \leqslant 4.0$ $M/L_{I\odot}$ at 1σ confidence level). For Abell 3565-BCG we determine $M_\bullet = 1.3^{+0.3}_{-0.4} \times 10^9$ $M_\odot$, with $i = 50° \pm 1°$ and $M/L_I = 9.0 \pm 0.8$ $M/L_{I\odot}$ (1σ confidence level). In the case of Abell 2052-BCG we find $M_\bullet \leqslant 7.3 \times 10^9$ $M_\odot$, following the method by Sarzi *et al.* (2002).

In Fig. 1 we show the location of our SBH masses determinations in the near-infrared $M_\bullet - L_{bulge}$ relation of Marconi & Hunt (2003) and the $M_\bullet - \sigma_c$ relation, as given in Ferrarese & Ford (2005). Implications of these observations will be discussed in a forthcoming paper (Dalla Bontà *et al.*, in preparation).

References

Coccato, L. *et al.* 2006, MNRAS, 366, 1050
Ferrarese, L. & Ford, H. 2005, Sp. Sci. Rev., 116, 523
Laine, S. *et al.* 2003, ApJ, 125, 478
Marconi, A. & Hunt, L. K. 2003, ApJ, 589, L21
Sarzi, M. *et al.* 2002, ApJ, 567, 237
Smith, R. J. *et al.* 2000, MNRAS, 313, 469

Black Holes from Stars to Galaxies – Across the Range of Masses
Proceedings IAU Symposium No. 238, 2006
V. Karas & G. Matt, eds.
© 2007 International Astronomical Union
doi:10.1017/S1743921307005431

The VSOP Survey: final aggregate results

R. Dodson,[1]† S. Horiuchi,[2,3,4] W. Scott,[5] E. Fomalont,[6] Z. Paragi,[7]
S. Frey,[8] K. Wiik,[1,9] H. Hirabayashi,[1] P. Edwards,[1,10] Y. Murata,[1]
G. Moellenbrock,[11] L. Gurvits,[7] Z. Shen[12] and J. Lovell[10]

[1] The Institute of Space and Astronautical Science, JAXA, 3-1-1 Yoshinodai, 229-8510, Japan
[2] Centre for Astrophysics and Supercomputing, Swinburne University, Vic. 3122, Australia
[3] National Astronomical Observatory, 2-21-1 Osawa, Mitaka, Tokyo 181-8588, Japan
[4] Jet Propulsion Laboratory, 4800 Oak Grove Drive, Pasadena, CA 91109, USA
[5] Physics and Astronomy Department, University of Calgary, Calgary, Canada
[6] National Radio Astronomy Obs., 520 Edgemont Road, Charlottesville, VA 22903, USA
[7] Joint Institute for VLBI in Europe, P.O. Box 2, 7990 AA, Dwingeloo, Netherlands
[8] FÖMI Satellite Geodetic Observatory, P.O. Box 546, H-1373, Budapest, Hungary
[9] Tuorla Observatory, Väisäläntie 20, FIN-21500 Piikkiö, Finland
[10] Australia Telescope National Facility, CSIRO, P. O. Box 76, Epping NSW 2122, Australia
[11] National Radio Astronomy Observatory, P.O. Box 0, Socorro, NM 87801, USA
[12] Shanghai Astronomical Observatory, Chinese Academy of Sciences, 80 Nandan Lu, China

Abstract. In February 1997 the Japanese radio astronomy satellite HALCA was launched to provide the space-borne element for the VSOP mission. HALCA provided linear baselines three-times greater than that of ground arrays, thus providing higher resolution and higher AGN brightness temperature measurements and limits. Twenty-five percent of the scientific time of the mission was devoted to the "VSOP survey" of bright, compact, extra-galactic radio sources at 5 GHz. A complete list of 294 survey targets were selected from pre-launch surveys, 91% of which were observed during the satellite's lifetime.

The major goals of the VSOP Survey are statistical in nature: to determine the brightness temperature and approximate structure, to provide a source list for use with future space VLBI missions, and to compare radio properties with other data throughout the electro-magnetic spectrum. All the data collected have now been analysed and is being prepared for the final image Survey paper. In this paper we present details of the mission, and some statistics of the images and brightness temperatures.

Keywords. Surveys – galaxies: active

---

1. Introduction

The VLBI Space Observing Program (VSOP) satellite HALCA provided the space baseline for the observation of a complete sample of bright compact Active Galactic Nuclei (AGN) at 5 GHz; the VSOP survey (Hirabayashi *et al.* 2000). Of this set of 294 AGNs: 102 were presented in Scott *et al.* (2004, hereafter P1), 140 will be presented in Dodson *et al.* (2007, hereafter P2), and 29 were not observed. The remaining 23 did not produce space fringes, where we expected to find them. Some of these may be correct, however we have erred on the side of caution and consider them to be failures.

The HALCA downlink had a bandwidth of 128 Mbps, or 32 MHz at Nyquist, two bit, sampling. A typical VSOP Survey experiment would consist of three Ground Radio Telescopes (GRTs) and a single tracking pass of HALCA; normally several hours. This leads to a nomimal fringe sensitivity of 100 mJy, and a image noise level of 10 mJy, for the space baselines. Figure 1a) shows the number of antennae in each experiment.

† Present address: Marie Curie Fellow, OAN, Alcalá de Heneres, Spain

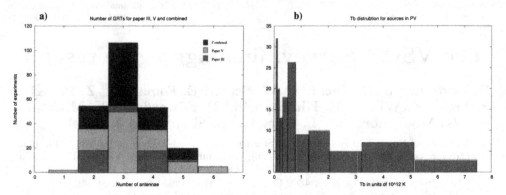

Figure 1. a) A histogram of the number of antennae for experiments in P1,2. b) A histogram of the upper limits to the source rest frame brightness temperatures from the imaged data in P2, in logarithmic bins.

One of the core goals of the VSOP Survey imaging program was to measure lower limits to the brightness temperatures (T_b) directly from the data (c.f. Horiuchi *et al.* 2004). The determined T_b (from VLBI) depends only on the physical baseline length, independent of frequency, so space baselines will always provide the highest possible limits. Figure 1b) plots the lower limits to T_b for the sources in P2, binned logarithmically. These are the source frame T_b values, unless the redshift is unknown. In which case the observer frame values are used, as a lower limit. Lower limits were estimated from the lowest brightness temperature model compatible with the data.

2. Conclusions

We have completed the VSOP survey imaging data reduction. The paper covering 140 sources is in preparation, and when it is published the imaging portion of the VSOP Survey Project will be completed. We have directly measured the source brightness temperatures, and produce a distribution of the lower limits to directly measured T_b.

Acknowledgements

RD, KJW and JEJL acknowledge support from the Japan Society for the Promotion of Science. WKS wishes to acknowledge support from the Canadian Space Agency. SH acknowledges support through an NRC/NASA-JPL Research Associateship. SF acknowledges the Bolyai Scholarship received from the Hungarian Academy of Sciences. RD wishes to thank S. Lorenzo. The NRAO is a facility of the National Science Foundation, operated under cooperative agreement by Associated Universities, Inc. The Australia Telescope Compact Array is part of the Australia Telescope which is funded by the Commonwealth of Australia for operation as a National Facility managed by CSIRO.

References

Dodson, R. *et al.* 2007, in preparation (P2)
Hirabayashi, H. *et al.* 2000, PASJ, 52, 997
Horiuchi, S. *et al.* 2004, ApJ, 616, 110
Scott, W. K. *et al.* 2004, ApJS, 155, 33 (P1)

Black Holes from Stars to Galaxies – Across the Range of Masses
Proceedings IAU Symposium No. 238, 2006
V. Karas & G. Matt, eds.

Polarization from an orbiting spot

Michal Dovčiak,[1] Vladimír Karas[1] and Giorgio Matt[2]

[1]Astronomical Institute, Academy of Sciences, Prague, Czech Republic

[2]Dipartimento di Fisica, Università degli Studi Roma Tre, Rome, Italy

email: dovciak@astro.cas.cz, vladimir.karas@cuni.cz, matt@fis.uniroma3.it

Abstract. The polarization from a spot orbiting around Schwarzschild and extreme Kerr black holes is studied. The time dependence of the degree and angle of polarization during the spot revolution is examined as a function of the observer's inclination angle and black hole angular momentum. The gravitational and Doppler shifts, lensing effect as well as time delays are taken into account.

Keywords. Black hole physics – accretion, accretion discs – polarization

1. Model

We assume Keplerian geometrically thin and optically thick disc around the Schwarz-schild or extreme Kerr black holes. The spot is two-dimensional and it rotates together with the disc. We assume the spot does not change its shape during its orbit. Only the zero order photons have been taken into account here. Only the emission from the spot is considered; the decrease of polarization degree due to the disc and corona emission is not accounted for.

We apply three different models of local polarization. In the first two models the local emission is totally polarized either in the direction normal to the disc or perpendicular to the toroidal magnetic field. In the case of partial local polarization the observed one will decrease proportionally.

The last model of local polarization describe a more realistic situation: the spot is considered a part of the disc illuminated from a flare (considered to be a point source) above it and moving with Keplerian velocity. The photons from the primary source (the flare) are scattered in the disc. Some of them are eventually reflected in the direction to the observer. The scattering is handled in Compton multiple scattering approximation.

For further details on the model and approximations used we refer the reader to Dovčiak, Karas & Matt (2004, 2006). For an exciting application to Sgr A*, see Eckart *et al.* (2006).

2. Results

The results of our computations are presented on figures 1 and 2. The polarization degree (figure 1) decreases in all models mainly in that part of the orbit where the spot moves close to the region where the photons that reach the observer are emitted perpendicularly to the disc. The polarization angle (figure 2) changes rapidly in this part of the orbit. The decrease in observed polarization degree for the local polarization perpendicular to the toroidal magnetic field happens also in those parts of the orbit where the magnetic field points approximately in the direction of photon's motion. The polarization has more complicated behaviour for the more realistic model. One of the most interesting features is the peak in polarization degree for the extreme Kerr black hole

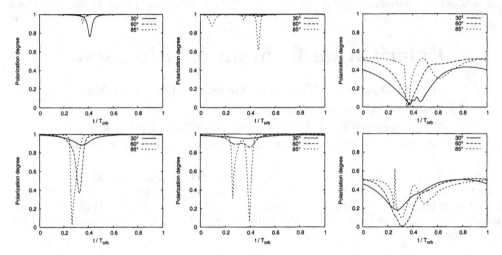

Figure 1. The time dependence of the polarization degree from a polarized orbiting spot in three different configurations of the local polarization vector — perpendicular to the disc, perpendicular to the toroidal magnetic field and Compton scattering approximation (from left to right) for Schwarzschild (top) and extreme Kerr (bottom) black holes and for three observer's inclination angles ($30°, 60°$ and $85°$). The spot is located at $r_s = 7GM_\bullet/c^2$ for Schwarzschild and $r_s = 3GM_\bullet/c^2$ for extreme Kerr black hole. The radius of the spot is approximately $\delta r = 0.87GM_\bullet/c^2$.

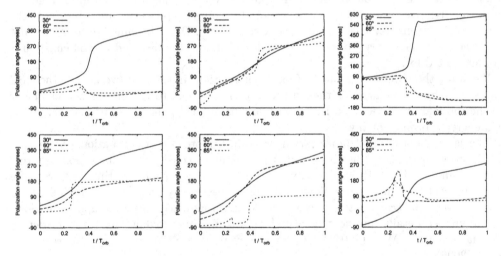

Figure 2. The same as in Fig.1 but for the polarization angle.

for large inclinations. It is due to the lensing effect when the emission with a particular polarization angle is enhanced. This is not visible for the Schwarzschild case because the spot is orbiting farther away from the centre.

M.D. has been supported by the Czech Science Foundation under grant 205/05/P525. We acknowledge the PECS project 98040.

References

Dovčiak, M., Karas, V., Matt, G. 2004, MNRAS, 355, 1005
Dovčiak, M., Karas, V., Matt, G. 2006, Astron. Nachrichten, 327, 993
Eckart, A., Schödel, R., Meyer, L. *et al.* 2006, ESO Messenger, 125, 2

Black Holes from Stars to Galaxies – Across the Range of Masses
Proceedings IAU Symposium No. 238, 2006
V. Karas & G. Matt, eds.
© 2007 International Astronomical Union
doi:10.1017/S1743921307005455

Optical, ultraviolet and X-ray analysis of the black hole candidate BG Geminorum

P. Elebert,[1] **P. J. Callanan,**[1] **L. Russell**[1] **and S. E. Shaw**[2]

[1]Department of Physics, University College Cork, Ireland
[2]School of Physics and Astronomy, University of Southampton, Southampton, SO17 1BJ, UK
email: p.elebert@ucc.ie

Abstract. We present the first high resolution optical spectrum of the black hole candidate BG Geminorum, as well as UV spectroscopy from the Hubble Space Telescope, and X-ray data from INTEGRAL. The UV spectra suggest the presence of material in BG Gem with velocities possibly as high as $\sim$1000 km s^{-1}, suggesting an origin at a distance of $\sim$0.7 R$_\odot$ from a 3.5 M$_\odot$ object; if real, this would be strong evidence that the primary in BG Gem is indeed a black hole. In contrast, the upper limit provided by the INTEGRAL data gives a maximum X-ray luminosity of only $\sim$0.1% of the Eddington luminosity.

Keywords. X-rays: binaries, accretion discs, stars: individual (BG Geminorum).

1. Introduction

In 1992, Benson *et al.* (2000) discovered an eclipsing and ellipsoidal variation in BG Gem with a period of 91.645 days. Combined with a sinusoidal secondary radial velocity curve, this suggested a binary system, with a K0 secondary feeding material into an accretion disc around the unseen primary. Kenyon *et al.* (2002) calculated the mass function to be (3.5 ± 0.5) M$_\odot$, greater than the theoretical minimum mass for a black hole, ~ 3M$_\odot$ (Rhoades & Ruffini 1974). However, because the primary cannot be observed directly (the inclination of the system is close to 90°) it could also be a B star (Kenyon *et al.* 2002). If confirmed to be a black hole, BG Gem would be the the longest period black hole binary system, by a factor of $\sim$3, as well as being the only known eclipsing black hole binary system in the Galaxy.

One method to investigate the nature of the primary is to acquire higher quality optical spectra, to determine the maximum rotational velocity of the accretion disc surrounding the primary. Large velocities would favour a black hole primary, as $\sim$700 km s^{-1} is the maximum velocity for a disc surrounding a B type star (Kenyon *et al.* 2002). If the primary is an un-obscured B star, then UV spectra should show a strong continuum and absorption lines from ionised Si. Detection of broad high ionisation emission lines would favour a black hole. Here, we present the first high resolution optical spectrum of this system, as well as the first UV spectra, and X-ray data from INTEGRAL.

2. Data

We acquired an optical spectrum with the ALFOSC instrument on the Nordic Optical Telescope (NOT) in La Palma on 2006 April 11. Kenyon (2001) acquired UV spectra with the Space Telescope Imaging Spectrograph on the Hubble Space Telescope (HST), on 4 nights between 2001 November and 2002 December. We have also examined archival X-ray data from the International Gamma-Ray Astrophysics Laboratory (INTEGRAL),

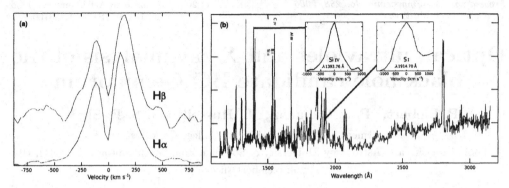

Figure 1. (a) Hα and Hβ emission lines normalised to the local continuum, and offset for clarity. (b) Combined HST UV spectrum from 2001 December 30.

consisting of 20–40 keV mosaics from the INTEGRAL Soft Gamma Ray Imager (ISGRI). The maximum exposure is ∼250 ksec, obtained between 2003 February and 2004 October.

3. Results and Discussion

Figure 1(a) shows double-peaked Hα and Hβ emission lines. The Balmer line emission in the wings is from material with velocities of $\lesssim 500$ km s^{-1}, low enough that the central object could be a main sequence B star. The UV spectra, such as that in Figure 1(b), show strong emission lines from Si, Al and Mg ions, with a very weak continuum – indicative of a black hole primary rather than a B star. The maximum velocity in the wings of S I λ1914.70 Å is possibly as much as 1000 km s^{-1} – if confirmed with higher resolution spectra this would be strong evidence that the primary is a black hole. However, the INTEGRAL observations yield an upper limit of $\lesssim 10^{33}$ erg s^{-1} in the 20–40 keV band. If BG Gem were intrinsically luminous in the X-rays, but obscured by its accretion disc because of the high orbital inclination, we might still expect some fraction (∼1%) of its intrinsic flux scattered into our line-of-sight by an accretion disc corona (White *et al.* 1981). In this scenario our upper limit would then correspond to a limit on the intrinsic X-ray flux of $\lesssim 10^{35}$ erg s^{-1}. This suggests that, if the primary is indeed a black hole, the system must be in quiescence.

Acknowledgements

The optical data presented here have been taken using ALFOSC, which is owned by the Instituto de Astrofisica de Andalucia. UV analysis is based on observations made with the NASA/ESA HST. S. E. Shaw is currently with the INTEGRAL Science Data Centre in Switzerland. We thank Tony Bird, of the IBIS survey team, for help with the INTEGRAL data. PE and PJC acknowledge support from Science Foundation Ireland.

References

Benson, P., Dullighan, A., Bonanos, A., McLeod, K. K. & Kenyon, S. J. 2000, AJ, 119, 890
Kenyon, S. J., Groot, P. J. & Benson, P. 2002, AJ, 124, 1054
Kenyon, S. J. 2001, HST Proposal, 5501K
Rhoades, C. E. & Ruffini, R. 1974, Phys. Rev. Lett., 32, 324
White, N. E., Becker, R. H., Boldt, E. A., Holt, S. S., Serlemitsos, P. J. & Swank, J. H. 1981, ApJ, 247, 994

Black Holes from Stars to Galaxies – Across the Range of Masses
Proceedings IAU Symposium No. 238, 2006
V. Karas & G. Matt, eds.
© 2007 International Astronomical Union
doi:10.1017/S1743921307005467

Dynamical evolution of rotating globular clusters with embedded black holes

José Fiestas and Rainer Spurzem

Zentrum für Astronomie, Astronomisches Rechen-Institut, Heidelberg, Germany
email: fiestas, spurzem@ari.uni-heidelberg.de

Abstract. Evolution of rotating globular clusters with embedded black holes is presented. The interplay between velocity diffusion due to relaxation and black hole star accretion is followed together with cluster rotation, using 2-dimensional, in energy and z-component of angular momentum, Fokker Planck numerical methods. Gravogyro and gravothermal instabilities drive the system to a faster evolution leading to shorter collapse times and a faster cluster dissolution in the tidal field of a parent galaxy.

Angular momentum transport and star accretion support the development of central rotation in relaxation time scales. Two-dimensional distribution (in the meridional plane) of kinematical and structural parameters (density, dispersions, rotation) are reproduced, with the aim to enable the use of set of models for comparison with observational data.

Keywords. Gravitation – stellar dynamics – globular clusters: general – black hole physics

---

The stellar system is assumed to be axisymmetric and in dynamical equilibrium. We use initially rotating King models of the form:

$$f_{\rm rk}(E, J_z) = {\rm const}[\exp(-\beta E) - 1] \exp(-\beta \Omega_0 J_z), \qquad (0.1)$$

$\omega_0 = \sqrt{9/4\pi G n_{\rm c}} \cdot \Omega_0$ is the dimensionless angular velocity, and a modified evolving potential $\phi(t) = \phi_{\rm cl}(t) - GM_{bh}(t)/r$, where $G = M_{\rm i} = r_{\rm c_i} = 1$. Using an initial $M_{\rm bh} = 5 \times 10^{-5} M_{\rm cl}$, we followed the evolution up to core collapse, where $M_{\rm bh} \sim 0.1 M_{\rm cl}$ was found. We confirmed the acceleration of core collapse in single-mass rotating systems with BH with respect to the non-rotating case. At collapse time, the BH-potential dominates the stellar distribution inside the influence radius (r_a). The density and velocity dispersion distributions evolve to a power-law of $n \propto r^\lambda$, $\lambda = -1.75$ and $\sigma \propto r^\gamma$, $\gamma = -0.5$ (Bahcall & Wolf (1976), Marchant & Shapiro (1980)). Post-collapse is driven through energy input from the central object. The rotational velocity and velocity dispersion distributions in the meridional plane reproduce the observed morphological structure of rotating clusters (see for example non-parametric fitting of kinematical data of ωCen in Merritt *et al.*, (1997), and compared to our data in Fiestas, Spurzem & Kim (2006)). Rotational velocity grows inside the BH influence radius strong influenced by the interplay between diffusion of angular momentum (gravogyro instability) and the redistribution of high energy, low J_z orbits close to the BH (loss-cone effect).

The models presented, are able to reproduce 2D distributions of kinematical and structural parameters, at any time of evolution and deep in the stellar cusp surrounding the central BH. Observational studies of globular clusters can be compared to evolutionary models, to elucidate theoretical predictions and understand the common evolution of star clusters and galaxies. With this aim, a detailed set of model data, covering a wide range of rotation rates and initial concentrations of rotating GCs, with and without BH, has been

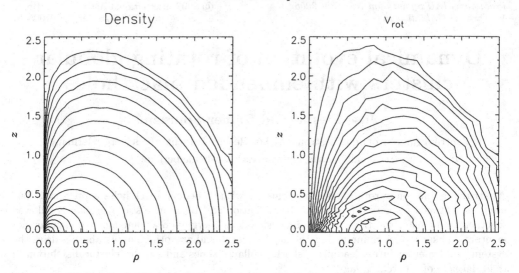

Figure 1. 2D density (left) and rotational velocity distribution (right) in the meridional plane for an initial King model ($W_0 = 6.0$, $\omega_0 = 0.9$), at time $t/t_{\mathrm{rhi}} = 3.45$. The BH ($M_{\mathrm{bh}} \sim 0.01 M_{\mathrm{cl}}(0)$) is located at the center of coordinates. Note the influence of rotation in the density distribution.

created, to enable observers to use them for comparisons with their data. The cluster database can be found on the web, at http://www.ari.uni-heidelberg.de/clusterdata/.

The influence of rotation in the evolution and redistribution of orbits in the cluster is illustrated in Fig. 1, for a high rotating model ($\omega_0 = 0.9$) at a late stage of evolution ($t/t_{\mathrm{rhi}} = 3.45$). Observe the increase of central rotation in torus-like contours of density and v_{rot}.

Although our models agree with published theoretical studies of dense stellar systems in the spherical case and with the observed BH-mass in globular clusters, our models are still idealized and simplified. Physical scalings, such as time scales, may change if more effects such of a mass spectrum and binaries are included (items partly in work).

Acknowledgements

We thank the Local Organizing Committee and ZAH Heidelberg for the financial support.

References

Bahcall, J. N. & Wolf, R. A. 1976, ApJ, 209, 214
Fiestas, J., Spurzem, R. & Kim, E. 2006, MNRAS, 373, 677
Marchant, A. B. & Shapiro, S. L. 1980, ApJ, 239, 685
Merrit, D. *et al.* 1997, ApJ, 114, 1074

Black Holes from Stars to Galaxies – Across the Range of Masses
Proceedings IAU Symposium No. 238, 2006
V. Karas & G. Matt, eds.
© 2007 International Astronomical Union
doi:10.1017/S1743921307005479

A near-infrared view of the 3CR: properties of hosts and nuclei[†]

David Floyd,[1] Marco Chiaberge,[1] Eric S. Perlman,[2] Bill Sparks,[1]
F. Duccio Macchetto,[1][‡] Juan Madrid,[1] Stefi Baum,[3] Chris O'Dea,[3]
David Axon,[3] Alice Quillen,[4] Alessandro Capetti,[5] George Miley[6]
and Stefano Tinarelli[1][¶]

[1]STScI, USA; [2]UMBC, USA; [3]RIT, USA; [4]University of Rochester, NY, USA;
[5]OA Torino, Italy; [6]Leiden Observatory, The Netherlands
email: floyd@stsci.edu

Abstract. The 3CR catalogue provides a statistical sampling of the most powerful radio galaxies out to $z \sim 0.3$. Over the decade and a half of Hubble observations we have amassed a major multi-wavelength dataset on these sources, discovering amongst other things, new jets, hotspots, dust disks, and faint point-like nuclei. We present here the results of our latest snapshot survey, a near-complete sampling of the 3CR host galaxies at $z < 0.3$ in the near-IR (H-band). This un-extinguished view of the host galaxies has provided us with an accurate measure of the stellar/spheroid masses of the sources, and an unbiased view of their morphologies. We show that they exhibit an identical Kormendy relation to nearby QSO's and the massive Elliptical population, but are distinct from the Brightest Cluster Members, and mergers. We find that while a few sources exhibit signs of a recent or impending major merger, many more sources have remnants consistent with a gas-rich minor merger in their recent history. We detect unresolved nuclear sources in most ($\sim$80%) of FRI, with their IR luminosities correlating linearly with radio core power. This implies that the IR nuclei are synchrotron radiation produced at the base of the relativistic jet, and confirms that no infrared (thermal) radiation in excess to synchrotron is present in FRIs, unlike in other classes of AGN.

Keywords. Galaxies: active – galaxies: fundamental parameters – (galaxies:) quasars: general

---

We used HST/NICMOS2 and the F160W filter (H-band) in SNAP mode to obtain IR imaging of 89 3CR RG at $z < 0.3$. Adding 11 archival targets, we have 1 ks images for 100 sources in a single infrared band, 90% of the entire sample. Observations and data reduction are described in Madrid *et al.* (2006). Many targets have additional pointings in the optical-UV, making for a rich multi-λ data-set. H-band provides a powerful diagnostic of the gravitationally-dominant stellar mass of the galaxy. We perform 2D fits to the host galaxies using Galfit (Peng, Ho, Impey, & Rix 2002), Ellipse (IRAF), and 2DM (Floyd *et al.* 2004 – for objects dominated by an unresolved nuclear point-source). This allows us to separate bulge and disc-like components in even the most strongly nuclear-dominated sources, and allows us to better characterize any residual features, such as jets, knots, and tidal or merger features. Example single-component Sérsic models are shown for 3C 288 and 3C 310. The host galaxies are generally consistent with the elliptical galaxy

† Based on observations with the NASA/ESA Hubble Space Telescope, obtained at the Space Telescope Science Institute, which is operated by the Association of Universities for Research in Astronomy, Inc. (AURA), under NASA contract NAS5-26555. We gratefully acknowledge support from HST grant STGO-10173.
‡ European Space Agency.
¶ STScI summer student.

population. The peak in the galaxy luminosity distribution is close to L^*, with steeper drop-off to high luminosity than low. But they exhibit a large spread in Sérsic index to low values. Sizes and luminosities of the host galaxies correlate with each other and with radio power. Sérsic n correlates with the radio luminosity, and the disky (low-n) wing of the sample is dominated by low-z, low-radio-power sources, merging sources, and sources with luminous companions. Additional detailed comparison with samples of merging galaxies and ellipticals appears to show a cross-over between the 2 populations (Floyd *et al.* 2006). This suggests that disky galaxies are capable of hosting powerful radio sources, providing that they are massive enough, or are undergoing major merger.

Companions and mergers. Approximately 55% of our sample exhibit compact, often unresolved companion sources too red and faint to have been detected in previous optical snapshot programs. Spectroscopy is required to determine the nature and redshifts of these sources, although typical colours of these sources are around $R - H = 2 - 3$, consistent with old stellar populations. These sources are found predominantly in elliptical host galaxies, and have 3 likely origins: Foreground stars (in cases with low galactic latitude); Synchrotron Hotspots (in a few peculiar radio sources); Merger remnants (compact cores from cannibalised small galaxies – e.g. Canalizo+03); Molecular gas clouds infalling into the galaxy (Bellamy & Tadhunter 2004). By comparison, only 10% of our sample show signs of an ongoing or recent major merger (though many more show other signs of disturbance) - typical of the elliptical galaxy population in general. If many of these sources do turn out to be galactic nuclei, it would suggest that a minor merger is sufficient to fuel or re-fuel a quiescent black hole in an elliptical galaxy into radio-loud AGN activity, while disky galaxies appear to require a more major disturbance.

NIR properties of the nuclei. Unresolved nuclear point sources (Central Compact Cores, CCC) are detected in 80% of the FRI in our sample, as well as the majority of FR II. Amongst the FR I, the IR luminosity of the CCC correlates with the radio core power and optical luminosity. The correlation over 4 decades is strong evidence that the optical-IR core emission is synchrotron in nature, and originates at the base of the jet, most likely on scales smaller than 1 pc. The lack of an IR excess implies that synchrotron radiation dominates the IR core luminosity in FRI, unlike in other classes of AGN, where thermal emission from hot dust contributes significantly.

The picture that emerges is that FRI lack a radiatively efficient accretion disk and, possibly, of a circumnuclear dusty torus (Chiaberge *et al.* in preparation). The next step in the immediate future is to combine our studies of the hosts and nuclei and in particular to examine the nature of the powerful (FRII) radio sources hosted by low-n galaxies.

References

Bellamy, M. J. & Tadhunter, C. N. 2004, MNRAS, 353, 105

Canalizo, G. Max, C. Whysong, D., Antonucci, R. & Dahm S. E. 2003, ApJ, 597, 823

Chiaberge, M., Capetti, A. & Celotti, A. 1999, A&A, 349, 77

Chiaberge, M., Capetti, A. & Macchetto, F. 2005, ApJ, 625, 716

Dunlop, J. S., McLure, R. J., Kukula, M. J., Baum, S. A., O'Dea, C. P. & Hughes, D. H. 2003, MNRAS, 340, 1095

Floyd, D. J. E., Kukula, M. J., Dunlop, J. S., McLure, R. J., Miller, L., Percival, W. J., Baum, S. A. & O'Dea, C. P. 2004, MNRAS, 355, 196

Madrid, J. P., Chiaberge, M., Floyd, D., Sparks, W. B., Macchetto, F. & Miley, G. K. *et al.* 2006, ApJS, 164, 307

McLeod, K. K. & McLeod, B. A. 2002, ApJ, 546, 782

Peng, C. Y., Ho, L. C., Impey, C. D. & Rix, H.-W. (2002), AJ, 124, 266

Taylor, G. L., Dunlop, J. S., Hughes, D. H. & Robson, E. I. (1996), MNRAS, 283, 930

Black Holes from Stars to Galaxies – Across the Range of Masses
Proceedings IAU Symposium No. 238, 2006
V. Karas & G. Matt, eds.
© 2007 International Astronomical Union
doi:10.1017/S1743921307005480

Generation of shocked hot regions in black hole magnetosphere

Keigo Fukumura,[1] Masaaki Takahashi[2] and Sachiko Tsuruta[3]

[1]NASA/Goddard Space Flight Center, Code 663, Greenbelt, MD 20771, USA

[2]Department of Physics and Astronomy, Aichi University of Education, Kariya,
Aichi 448-8542, Japan

[3]Department of Physics, Montana State University, Bozeman, MT 59717, USA

email: fukumura@milkyway.gsfc.nasa.gov, takahasi@phyas.aichi-edu.ac.jp,
uphst@gemini.msu.montana.edu

Abstract. We study magnetohydrodynamic (MHD) standing shocks in ingoing plasmas in a black hole (BH) magnetosphere. We find that low or mid latitude (non-equatorial) standing MHD shocks are both physically possible, creating very hot and/or magnetized plasma regions close to the event horizon. We also investigate the effects of the poloidal magnetic field and the BH spin on the properties of shocks and show that both effects can quantitatively affect the MHD shock solutions. MHD shock formation can be a plausible mechanism for creating high energy radiation region above an accretion disk in AGNs.

Keywords. Black hole physics – magnetohydrodynamics: MHD – shock waves

---

1. Introduction

In a series of our previous investigations in the context of general relativity (see, e.g., Takahashi *et al.* 2002, 2006) it has been shown that shock formation in MHD plasmas in-flowing onto a black hole can be a physically plausible mechanism for creating very hot (i.e., $T_i \gtrsim 10^{12}$ K for ions and $T_e \gtrsim 10^9$ K for electrons) or strongly magnetized plasma regions. These studies were conducted primarily for the equatorial accretion flows. As a natural next step we extended these studies to two dimensional, non-equatorial flows. Specifically, we investigated the polar angle dependence of various shock properties. Non-equatorial shock-heated regions would be attractive as a possible high energy radiation source above an accretion disk in the central engine of active galactic nuclei (AGN). Solid understanding of non-equatorial shock formation can be useful for comparing our theoretical implications with future observations.

2. Our models, results and conclusion

We considered the MHD shock formation around a Schwarzschild and slowly-rotating Kerr black hole. The basic equations governing plasmas consist of particle number conservation law, the equation of motion, and the ideal MHD condition. We solved these equations for wide ranges of field-aligned flow parameters, by adopting stationary and axisymmetric ideal MHD accretion flows in Kerr geometry, and applying the general relativistic Rankine–Hugoniot shock conditions in Kerr geometry – particle number conservation, energy conservation, and angular momentum conservation. The background metric is written by the Boyer-Lindquist coordinates. We systematically explored, for the first time, general relativistic MHD shock formation by employing the conserved quantities and found the allowed shocked regions.

In order to explore a general trend for allowed MHD shock formation for the vast amount of parameter sets, we sliced the entire parameter space, consisting of all of the parameters (i.e., the total plasma energy E, plasma angular momentum L, angular velocity of the magnetic field line Ω_F, particle flux per magnetic flux η, dimensionless postshock plasma temperature θ_{sh}, magnetization parameter δ, and spin parameter a), by finding the allowed *cubic* parameter space, consisting of three parameters (E, L, and Ω_F), for given black hole magnetosphere models with the other two parameters a and δ fixed. Since it is very hard to explore all the parameter space spanned by all of the parameters together, our current approach will greatly simplify and help us better understand a comprehensive picture of the resulting shock solutions. Through our search for possible non-equatorial MHD shocks we find that the allowed shock region is constrained by various physical factors – especially the regularity conditions at the magnetosonic points and the shock condition. Because the mathematical expressions for the regularity conditions for the existence of the Alfven and fast magnetosonic points are not at all simple (see, e.g., Takahashi *et al.* 2002), the topological appearance of the obtained shock regions is also complicated, but nevertheless we find it very useful. It may be emphasized that our search for the parameter space which allows MHD shock formation is carried out in a systematic manner in such a way that *our choice of the parameter sets is not arbitrary*. It is consistent, for instance, with the numbers relevant in application to, e.g., the environment of black hole magnetospheres around supermassive black holes in AGN.

Rotation and geometry of the magnetic field are among the primary factors for generating a strong shock. We find that the shocked plasma temperature has a clear anti-correlation with the shocked plasma magnetization. The more strongly magnetized plasma is formed for larger $L\Omega_F$. We also show that the energy transport between the fluid and the magnetic field can operate even more effectively across the shock front.

In summary we find that non-equatorial MHD shocks can form and could be a plausible candidate for generating a hot and/or strongly magnetized region over various latitudes in the non-equatorial plane, which could be a powerful high energy radiation source. It may be noted that in reality the actual MHD shock location in the actual astrophysical environment would probably trace a certain trajectory, rather than occupying the whole allowed shock region, because the parameters allowing the shock formation may be unique from case to case.

In various astrophysical objects, high energy activities are often associated with the magnetic fields along which the accreting plasma flows. Our investigation of MHD standing shock formation will be useful for our better understanding of complicated interactions between the plasma and the magnetic field in these situations. In the context of a strong radiation source required for the accretion-powered central engines of objects such as AGNs and GBHCs, the presence of non-equatorial shocks therefore can be attractive as a candidate for such a source.

Further details can be found in Fukumura *et al.* 2006.

References

Fukumura, F., Takahashi, M. & Tsuruta, S. 2006, ApJ, submitted

Takahashi, M., Rilett, D., Fukumura, K. & Tsuruta, S. 2002, ApJ, 572, 950

Takahashi, M., Goto, J., Fukumura, K., Rillet, D. & Tsuruta, S. 2006, ApJ, submitted; astro-ph/0511217

Black Holes from Stars to Galaxies – Across the Range of Masses
Proceedings IAU Symposium No. 238, 2006
V. Karas & G. Matt, eds.
© 2007 International Astronomical Union
doi:10.1017/S1743921307005492

The gravitational redshift in the broad line region of the active galactic nucleus Mrk 110

N. Gavrilović,[1] L. Č. Popović,[1] and W. Kollatschny[2]

[1] Astronomical Observatory, Volgina 7, Belgrade, Serbia

[2] Institut für Astrophysik, Friedrich-Hund-Platz 1, D-37077 Göttingen, Germany

email: ngavrilovic@aob.bg.ac.yu, lpopovic@aob.bg.ac.yu,
wkollat@astro.physik.uni-goettingen.de

Abstract. We used the long term spectroscopic observations of Mrk 110 (Hα and Hβ lines) to investigate the gravitational field influence on spectral line profiles. We found that effects of gravitational field can be measured and that the lines are more intense where the emission is originating close to the central black hole of Mrk 110.

Keywords. galaxies: nuclei – galaxies: active – galaxies: Seyfert

1. Introduction

Popović *et al.* (1995) showed that in a strong gravitational field broad emission lines are red-shifted and asymmetric. Müller & Wold (2006) showed that the gravitational red-shift in Mrk 110 is present. They describe shifts on larger distance from the black hole where a line is shifted as a total feature, conserving its intrinsic shape. Another mode describes the behavior of the BLR lines in strong gravity regime, where lines are strongly deformed and suppressed. Using effects of gravitational red-shift Kollatschny (2003) determined the mass of the central black hole ($M = 14\pm3 \times 10^7\ M_\odot$) in Mrk 110 and the projection of the accretion disc ($i = 21 \pm 5$ degrees). Here, we investigate the Hydrogen line shapes in order to detect gravitational field influence on the Mrk 110 broad line profiles.

2. Observations and analysis

Observations were performed with the 9.2 m Hobby-Early Telescope at Mc Donald Observatory and in a period between November 1999 and May 2000 were taken 26 spectra (Kollatschny 2003). The variable fraction of observed emission lines are probable created in distance of 1 to 50 light days from the super-massive black hole (Kollatschny 2003). Here we analyzed the Hα and Hβ lines from 24 spectra of Mrk 110. To measure the red-shift of the lines, first we subtract narrow lines and continuum.

3. Results and discussion

As one can see in Fig. 1, we obtained a linear dependence between the red-shift of lines and flux. It seems that the lines are more intensive with stronger gravitational influence, i.e. where red-shift of lines is higher. It may indicate that the lines stay more intensive in the part of the BLR closer to the black hole. The intensity of the Hβ line as a function of the width at tenth intensity maximum ($W_{1/10}$) is present in Fig. 2 (left). As one can see from Fig. 2, there is a correlation between the intensity and the width, i.e. the $W_{1/10}$

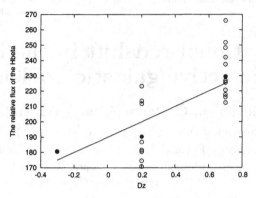

Figure 1. The relative flux of the Hβ as a function of the red-shift of the line center. White circles present measured values from each spectrum, while black circles correspond to the averaged values according to the red shift.

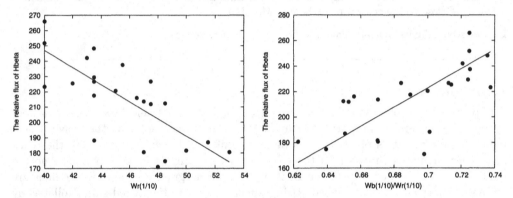

Figure 2. Left: The relative flux of the Hβ as a function of full width of the red part at tenth of intensity maximum; Right: The relative flux of the Hβ as a function of the full width ratio of blue (Wb) and red part (Wr) at tenth of intensity maximum.

is decreasing with intensity. Also, the asymmetry of the Hβ is changing (Fig. 2, right). The detailed discussion of our results will be given elsewhere (Gavrilović, *et al.* 2006).

Acknowledgements

This work was supported by the Ministry of Science and Environment Protection of Serbia through the project 146002.

References

Gavrilović, N., Popović, L. Č. & Kollatschny, W. 2006, in preparation
Kollatschny, W. 2003, A&A, 407, 461
Müller, A. & Wold, M. A&A, accepted (astro-ph/0607050)
Popović, L. Č., Vince, I., Atanacković-Vukmanović, O. & Kubičela, A. 1995, A&A, 293, 309

Black Holes from Stars to Galaxies – Across the Range of Masses
Proceedings IAU Symposium No. 238, 2006
V. Karas & G. Matt, eds.
© 2007 International Astronomical Union
doi:10.1017/S1743921307005509

Activity type of galaxies in HyperLeda

N. Gavrilović,[1,2] A. Mickaelian,[2,3] C. Petit,[2] L. Č. Popović[1] and P. Prugniel[2]

[1]Astronomical Observatory, Volgina 7, Belgrade, Serbia

[2]Université de Lyon, F-69000 Lyon; Université Lyon 1, F-69622 Villeurbanne; Centre de Recherche Astronomique de Lyon, Observatoire de Lyon, 9 avenue Charles André, F-69561 Saint-Genis Laval; Ecole Normale Supérieure, Lyon, France

[3]Byurakan Astrophysical Observatory and Isaac Newton Institute of Chile, Armenian branch, Byurakan 378433, Armenia

email: ngavrilovic@aob.bg.ac.yu, lpopovic@aob.bg.ac.yu, petit@obs.univ-lyon1.fr, prugniel@obs.univ-lyon1.fr, aregmick@apaven.am

Abstract. The HyperLeda database (http://leda.univ-lyon1.fr) is a tool to study the physics of galaxies. It is based on compilations of heterogeneous data (from large surveys and literature) which are cross-identified and homogenized to produce a uniform representation of the galaxies. We have added in the database a characterization of the nuclear and starburst activity from the Véron and Véron catalogue.

It is now possible to retrieve an *activity type* for the HyperLeda galaxies having this attribute, and to select list of objects from constraints on *activity type*. For example, one may select all Sy2 galaxies within some magnitude limits and/or redshift.

Keywords. Astronomical databases: miscellaneous – galaxies: active

1. Motivation

One of the first critical point when starting a statistical study of activity of galaxies is to construct a sample by setting constraints on various physical characteristics. Unfortunately, until present, none of the extragalactic databases, like NED or HyperLeda (http://leda.univ-lyon1.fr), was suited for this purpose because of the lack of information about nuclear activity stored in these systems. The goal of this work is to include in HyperLeda information about the active galaxies.

2. Cross-identification of the Véron-Véron catalogue with HyperLeda

The first step was to cross-identify the Véron and Véron catalogue (Véron-Cetty & Véron 2006; hereafter VCV-12) with HyperLeda, i.e. to find the VCV-12 objects in the database. HyperLeda provides a homogeneous description for over 4 millions objects from the Local Group to distant galaxies (Paturel, *et al.* 2003, Prugniel & Golev 1999), including about 3 millions galaxies and 100000 quasars. The VCV-12 is a compilation, maintained since 1984, of all active galaxies and QSOs.

The main difficulty for cross-identifying these multi-wavelength catalogues is the extremely different spatial resolution of the surveys which make automatic positional cross-identification unsafe. Fortunately, the VCV-12 already made the essential of this work and provides accurate position (typically to within 1 arcsec) for most of the sources. In many cases the redshift consistency allowed to make the proper identifications, and in some cases it was necessary to control visually the identifications using sky surveys images. In the course of this process we added in the databases objects which were missing.

3. New parameter of HyperLeda: Activity Type

After the successful cross-identification, we used for HyperLeda the activity type provided in the VCV-12:

- QSO – Quasi-stellar objects. Have very high luminosities ($M_{abs} \geqslant -23$) and broad emission lines (FWHM=5,000-30,000km/s) with a large red-shift.
- BL Lac – BL Lacertae type object. No emission or absorption lines deeper than $\sim$ 2 % are seen in any part of the optical spectrum, or only extremely week absorption and/or emission lines are observed, as a rule at minimum of their very highly variable phase. They show polarization and are strong radio sources.
- S1 – Broad-line Seyfert 1. Have broad permitted Balmer HI lines (FWHM=1000-10000km/s) and narrow forbidden lines (FWHM=300-1000km/s). Physically are the same objects as QSOs, but having smaller luminosities.
- S1n – Narrow-line Seyfert 1. Have narrow permitted lines only slightly broader than the forbidden ones.
- S1i – with a broad Paschen β line, indicating the presence of a highly reddened BLR.
- S1h – showing S1 like spectra in polarized light.
- S2 – show relatively narrow (compared to S1s) emission in both permitted Balmer and forbidden lines, with almost the same FWHM, typically in the range of 300-1000 km/s.
- S1.2, S1.5, S1.8, S1.9 – intermediate Seyfert galaxies.
- S3 – low activity AGN. Have S2-like spectra with relatively strong low-ionization lines.
- S3b – S3 with broad Balmer lines.
- S3h – S3 with broad Balmer lines seen only in polarized light.
- HII – Extragalactic HII regions. Have spectra with strong narrow (FWHM$\leqslant$300km/s) emission line spectrum but with a ratio [OIII]/Hb$\geqslant$3 and [NII]6584/Ha$<$0.5, coupled with a blue continuum.

In the new release of HyperLeda, the users can select list of objects from constraints on activity type. For example, select all Sy2 galaxies within some magnitude and/or red-shift limits and within a region of the sky.

In the future, we will continue to add more detailed information, like in particular the equivalent width of the emission lines, monochromatic fluxes and other characteristics entering in the line diagnostics and scaling relations of starbursts and active galaxies.

Acknowledgements

This work was supported by the Ministry of Science and Environment Protection of Serbia through the project 146002 and was made in a frame of French–Serbian collaboration "Pavle Savić"–CNRS. We would like to thank French Embassy in Belgrade for providing PhD fellowship for N. Gavrilović.

References

Paturel, G., Petit, C., Prugniel, P., Theureau, G. & Rousseau, J. *et al.* 2003, A&A, 412, 45P
Prugniel, P. & Golev, V. 1999, ASPC, 163, 296P
Véron-Cetty, M.-P. & Véron, P. 2006, A&A, 455, 773V

Black Holes from Stars to Galaxies – Across the Range of Masses
Proceedings IAU Symposium No. 238, 2006
V. Karas & G. Matt, eds.
© 2007 International Astronomical Union
doi:10.1017/S1743921307005510

Evidence for the AGN nature of LINERs

O. González-Martín,[1] J. Masegosa,[1] I. Márquez[1] and E. Jiménez-Bailón[2]

[1]Instituto de Astrofísica de Andalucía (CSIC), Apdo 3004, 10080 Granada, Spain
[2]Università degli Studi Roma Tre, Rome, Italy
email: omaira@iaa.es

Abstract. We report the analysis of the X-ray data for a sample of 48 LINER nuclei with available X-ray Chandra imaging. In González-Martín *et al.* 2006 21 objects had enough count rate to make the spectral analysis. Here we enlarge the sample performing the spectral analysis of the XMM-Newton observations of 7 additional galaxies. Our aim is to investigate the physical mechanisms which power the nuclear activity of LINERs. The use of multiwavelength information at radio, UV, optical HST and X-ray lead us to conclude that at least 60% of the LINERs are hosting a low luminosity AGN in their nuclei.

Keywords. Galaxies: active – X-rays: galaxies

1. Introduction

LINERs are very common in the nearby universe. Pioneering works already estimated that at least 1/3 of all the spiral galaxies are LINERs (Heckman *et al.* 1980). Nowadays, there is still an ongoing strong debate on the origin of the energy source in LINERs, with two main alternatives for the ionizing source being explored: either it is a low luminosity AGN, or it has a thermal origin related to massive star formation and/or from shock heating mechanisms resulting from the massive stars evolution. The search for a compact X-ray nucleus in LINERS is indeed one of the most convincing evidence about their AGN nature.

2. X-ray analysis

The sample has been grouped into two categories, attending to the X-ray morphology: (i) AGN–like Nuclei: unresolved point-like source in 4.5–8.0 keV band; (ii) Starburst–like Nuclei: without unresolved point-like source in 4.5–8.0 keV band. Two models (thermal and non-thermal), and their combination, were tested to account for the spectral emission of the objects with enough S/N (21 from Chandra, see González-Martín *et al.* 2006, and 7 from XMM-Newton). A combination of thermal and non-thermal components is required in most of the cases. The resulting median spectral parameters are $kT = 0.7 \pm 0.2$ keV and $\Gamma = 1.7 \pm 0.4$. Nuclear luminosities were either calculated from the best-fit model or estimated otherwise (assuming a power law of $\Gamma = 1.8$ and galactic absorption). Fig. 1 (left) shows that AGN–like nuclei tend to be more luminous than SB–like nuclei.

When the fitting was not possible (20 objects) color-color (C-C) diagrams were used to estimate Γ, kT, and n_H. Fig 1 shows the comparison between estimated and fitted values. Excepting three cases, the C–C diagrams provide a good T estimation, whereas Γ results to be slightly underestimated. The same result is obtained when XMM-Newton data are included.

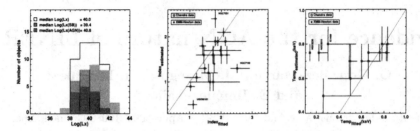

Figure 1. Left: Luminosity histogram for the whole sample (empty); objects classified as AG-N–like (grey) and Starburst–like objects (dashed). Comparison between power index (centre) and temperature (right) as estimated from C-C diagrams with those from fitted values.

3. Other evidence

In order to gain insight into the emission mechanisms in these objects, we looked for other evidence of the AGN nature in other wavelengths:

HST analysis. By Sharp-divided imaging the objects are classified into two groups: (a) compact nuclear sources (35 objects); and (b) dusty nuclear regions (8). All the galaxies classified as AGN by the X-ray imaging analysis show compact nuclei at the resolution of HST images.

Radio evidence. An unresolved nuclear radio core and flat continuum have been suggested as an evidence of AGN nature. Most of the LINERs in our sample classified by Filho *et al.* (2000) as AGNs in radio (13 objects) form an AGN–like class in X-rays (9).

UV variability. For the 7 objects in common with the sample of 17 LINER galaxies with HST/UV data by Maoz *et al.* (2005), five show time variability hinting to their AGN nature; all of them belong to our AGN–like class.

Stellar populations. High Mass X-ray Binaries are not expected to be an important ingredient for the nuclear X-ray emission. For the 14 galaxies with data available (Cid-Fernandes *et al.* 2004) the contribution of young stars is always less than 3%.

Compton thickness. L_X/L[OIII]< 1 may be used to detect Compton-thick sources. In our sample, SB–like nuclei show lower values than AGN–like (mean value of 0.96 for SB candidates and 11.0 for AGN candidates). The high percentage of SB–like in the region occupied by Compton-thick objects (50%) imply that they may host a strongly obscured AGN.

4. Conclusions

Although contributions from HMXBs and ULXs cannot be ruled out for some galaxies, we concluded in González-Martín *et al.* (2006) that 60% seems to be a lower limit for LINERs hosting an AGNs. The analysis of the Compton Thickness in our enlarged sample hints to a high percentage of our SB–like hosting strongly obscured AGNs.

This work was financed by DGICyT grants AYA2003-00128 and the Junta de Andalucía TIC114. OGM acknowledges financial support by the Ministerio de Educación y Ciencia through the Spanish grant FPI BES-2004-5044.

References

Cid Fernandes, R., Gonzalez Delgado, R. M., Schmitt, H. *et al.* 2004, ApJ, 605, 105
Filho, M. E., Barthel, P. D. & Ho, L. C. 2000, ApJS, 129, 93
González-Martín, O. *et al.* 2006, A&A in press (astro-ph/0605629)
Heckman, T. M., Crane, P. C. & Balick, B. 1980, A&A, 40, 295
Maoz, D., Nagar, N. M., Falcke, H. & Wilson, A. S. 2005, ApJ, 625, 699

Black Holes from Stars to Galaxies – Across the Range of Masses
Proceedings IAU Symposium No. 238, 2006
V. Karas & G. Matt, eds.
© 2007 International Astronomical Union
doi:10.1017/S1743921307005522

AGN polarization modeling with STOKES

René W. Goosmann,[1] **C. Martin Gaskell,**[2] **and Masatoshi Shoji**[3]

[1]Astronomical Institute, Academy of Sciences, Prague, Czech Republic

[2]Dept. Physics & Astronomy, Univ. of Nebraska, Lincoln, NE, USA

[3]Astronomy Dept., University of Texas, Austin, TX, USA

email: goosmann@astro.cas.cz, mgaskell1@unl.edu, masa1127@astro.as.utexas.edu

Abstract. We introduce a new, publicly available Monte Carlo radiative transfer code, STOKES, which has been developed to model polarization induced by scattering of free electrons and dust grains. It can be used in a wide range of astrophysical applications. Here, we apply it to model the polarization produced by the equatorial obscuring and scattering tori assumed to exist in active galactic nuclei (AGNs). We present optical/UV modeling of dusty tori with a curved inner shape and for two different dust types. The polarization spectra enable us to clearly distinguish between the two dust compositions. The STOKES code and its documentation can be freely downloaded from http://www.stokes-program.info/.

Keywords. Polarization – radiative transfer – dust, extinction – galaxies: active

1. Introduction

Spectropolarimetric data in the optical/UV range put important constraints on the emission and scattering geometry of AGNs (see e.g. Antonucci 2002). For the interpretation of spectropolarimetric data detailed modeling tools are important. Here, we use a new, publicly available radiative transfer code, STOKES (Goosmann & Gaskell 2007), to model the obscuring torus of AGNs. The code is based on the Monte-Carlo method and allows the simulation of various emission and scattering geometries. Polarization due to Thomson and dust (Mie-)scattering is included. Moreover, the code computes wavelength-dependent time delays and can thus be used to study polarization reverberation (Shoji, Gaskell & Goosmann 2005).

2. Modeling an obscuring torus

We consider a torus with an elliptical cross-section and centered on a point source. The source isotropically emits a flat continuum spectrum between 1600 Å and 8000 Å. The torus half-opening angle is set to $\theta_0 = 30°$. The inner and outer radii of the torus are fixed at 0.25 pc and 100 pc respectively. The radial optical depth in the equatorial plane is 750 for the V-band. The dust models (table 1) assume a mixture of graphite and "astronomical silicate" and a grain radii distribution $n(a) \propto a_s^\alpha$ between a_{min} and a_{max}.

The "Galactic dust" model reproduces the interstellar extinction for $R_V = 3.1$ whilst the "AGN dust" parameterization is obtained from quasar extinction curves derived by Gaskell *et al.* 2004. This latter dust type favors larger grain sizes.

Table 1. Parameterization of the dust models

Type	Graphite	Silicate	a_{min}	a_{max}	α_s
Galactic	62.5%	37.5%	0.005 μm	0.250 μm	−3.5
AGN	85%	15%	0.005 μm	0.200 μm	−2.05

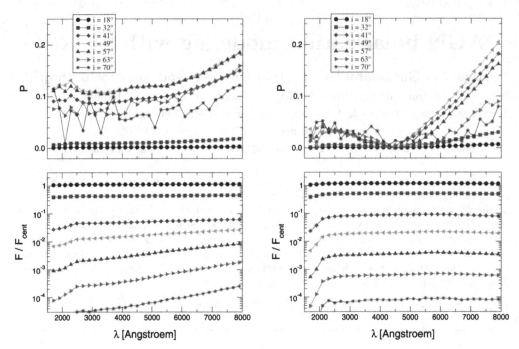

Figure 1. Polarization (top) and flux (bottom) spectra for a centrally-illuminated torus filled with "Galactic dust" (left) or "AGN dust" (right). The flux spectra are normalized to the illuminating flux and the inclination angle i is measured from the symmetry axis.

3. Results and discussion

The angular distributions of the polarization and flux spectra are shown in figure 1. In a face-on view, the central source is seen directly and the obtained polarization is low. At obscured inclinations $i > \theta_0$, the scattering properties of both dust types lead to different results. For "AGN dust", the polarization is lower in the UV but rises quickly toward longer wavelengths. Detailed spectropolarimetric observations of dust reflection thus effectively constrain the dust composition. For both dust types the polarization position angle is oriented perpendicularly to the projected symmetry axis of the object.

If the torus is filled with "AGN dust" then the total flux spectra are wavelength-independent above 2500 Å at all possible inclinations. For the "Galactic dust" model the scattered flux rises gradually toward longer wavelengths. The albedo of both dust compositions changes significantly below 2500 Å so that less radiation is scattered in the UV.

Spectropolarimetric data from AGNs contain not only information about the obscuring torus, but also include other scattering regions. To obtain a more detailed picture of AGN polarization it is therefore necessary to model several, radiatively coupled scattering regions self-consistently. While this is beyond the scope of this proceedings note, STOKES is capable of solving such problems.

References

Antonucci, R. 2002, Astrophysical Spectropolarimetry, 151 (astro-ph/0103048)

Gaskell, C. M., Goosmann, R. W., Antonucci, R. R. J. & Whysong, D. H. 2004, ApJ, 616, 147

Goosmann, R. W. & Gaskell, C. M. 2007, A&A, 465, 129

Shoji, M., Gaskell, C. M. & Goosmann, R. W. 2005, Bulletin of the AAS, 37, 1420

Black Holes from Stars to Galaxies – Across the Range of Masses
Proceedings IAU Symposium No. 238, 2006
V. Karas & G. Matt, eds.
© 2007 International Astronomical Union
doi:10.1017/S1743921307005534

The entropy increase during the black hole formation

Tetsuya Hara, Keita Sakai, Shuhei Kunitomo and Daigo Kajiura

Department of Physics, Kyoto Sangyo University, Kyoto 603-8555, Japan
email: hara@cc.kyoto-su.ac.japan

Abstract. The entropy increases enormously when a star collapses into a black hole. For the radiation dominated star, the temperature has decreased from the almost gravitational equilibrium one to the so-called black hole temperature. If we interpret this temperature decrease as the volume increase of the black hole, the entropy increase could be understood as the free expansion of radiation through the increased volume. The expected volume size is tentatively estimated. Under this interpretation, it is derived that the entropy is proportional to the horizon surface area and the microscopic states of the black hole entropy could be understood as the statistical states of the enlarged phase space.

Keywords. Uncertainty principle – black hole entropy – entropy increase

---

1. Introduction

Since the research of black hole thermodynamics was born a little over three decades ago several striking concepts have been accepted – see Bekenstein (1973); Hawking (1975); Wald (1998). One of them is that the black hole entropy is proportional to the surface area. It has been insisted that superstring theory could explain the microscopic states of the black hole entropy. However there are still questions about the black hole entropy in its microscopic interpretation. We investigate this problem through the point that the entropy has increased enormously during the black hole formation.

2. Entropy before black hole formation

The spherical massive star is considered to collapse due to gravitational instability for its radiation dominance. As the gravitational energy $(\Omega = -3GM/2r^2)$ is almost equal to the thermal energy $(U \simeq \epsilon V_\star)$ for radiation dominated star $(\Omega + U = 0)$, let take the relation $3GM^2/2r \simeq \epsilon V_\star$, where M, $V_\star$ and ϵ are the star mass, volume and radiation energy density, respectively. In the final collapse stage of the black hole formation, the temperature $T(= T_\star)$ is estimated as $T_\star \simeq \left(\frac{3GM^2}{2\tilde{a}rV_\star}\right)^{1/4} \simeq \left(\frac{3^3 \cdot 5}{2^7 \cdot \pi^3}\right)^{1/4} m_{\rm pl} \left(\frac{m_{\rm pl}}{M}\right)^{1/2}$ where $r \simeq r_{\rm g} = 2M$, $V_\star = 4\pi r_{\rm g}^3/3$, and $m_{\rm pl} = \sqrt{\hbar c/G}$ are taken and $\epsilon = \tilde{a}T_\star^4$ is assumed, being $\tilde{a} = \pi^2 k_B^4/(15\hbar^3 c^3)$. It shows $T_\star \propto M^{-1/2}$ and the total entropy $S \simeq sV_\star \simeq 4\tilde{a}T_\star^3 V_\star/3 \simeq 4Mc^2/(3T_\star)$ is expressed by $S \simeq \left(\frac{2^3}{3^7 \cdot 5}\right)^{1/4} \left(\frac{A}{\ell_{\rm pl}^2}\right)^{3/4} \simeq 3 \cdot 10^{57} \left(\frac{M}{M_\odot}\right)^{3/2}$, where $A = 4\pi r_{\rm g}^2$ and $\ell_{\rm pl} = \sqrt{G\hbar/c^3}$ are the black hole surface area and Planck length, respectively. It should be noted that entropy S is proportional to $A^{3/4}$, before black hole formation.

3. Entropy increase and the effective volume of the black hole

The Bekenstein–Hawking temperature and entropy of a black hole are $T_{\rm BH} = 1/(8\pi M)$ and $S = \int d(Mc^2)/T_{\rm BH} = M/2T_{\rm BH} = A/(4\ell_{\rm pl}^2) \simeq 10^{77} (M/M_\odot)^2$, respectively. Consequently the entropy has increased by an enormous factor $(A/\ell_{\rm pl}^2)^{1/4} \simeq 10^{20}(M/M_\odot)^{1/2}$

during the black hole formation. Here, let estimate the effective volume $\tilde{V}_{BH}$ of the black hole by assuming $S = s\tilde{V}_{BH}$, where $\tilde{V}_{BH}$ could be interpreted pedagogically to regard Bekenstein–Hawking entropy as an extensive state quantity. Then the volume is inferred as $\tilde{V}_{BH} = \frac{3A}{4 \cdot 4\tilde{a}T_{BH}^3} = 2^4 \cdot 3^3 \cdot 5 \cdot \pi \cdot V_\star \left(\frac{M}{m_{\text{pl}}}\right)^2 \propto M^5$, where $V_\star = 4\pi r_g^3/3$ is used.

The massive star is to collapse even when $r \gg r_g$ and the entropy is almost constant during the contraction before black hole formation, because the contraction of $r \gg r_g$ to $r \simeq r_g$ is almost adiabatic. From the relation $S \simeq M/T$ before and after the black hole formation, this entropy increase is mainly due to the temperature decrease. Before the black hole formation, the temperature is $T_\star \propto m_{\text{pl}} (m_{\text{pl}}/M)^{1/2}$ and, after the black hole formation, it becomes $T_{BH} \propto m_{\text{pl}} (m_{\text{pl}}/M)$.

If the process is free expansion, the temperature changes appropriately as $T \propto (1/V)^{1/4}$, because the internal energy is constant ($U \propto T^4 V = const$), and/or, thermodynamically, the following relation is satisfied:

$$\left(\frac{\partial T}{\partial V}\right)_U = -\left(\frac{\partial U}{\partial V}\right)_T \bigg/ \left(\frac{\partial U}{\partial T}\right)_V = \frac{1}{c_V V}\left(p - T\left(\frac{\partial p}{\partial T}\right)_V\right) = -\frac{T}{4V},$$

where p and c_V are pressure and the specific heat at constant volume, respectively. Then the entropy increase could be interpreted as the free expansion of the radiation due to the volume increase by the black hole formation. The expected volume V_{BH} for the temperature decrease is given by $V_{\text{BH}} \simeq V_\star \left(\frac{T_\star}{T_{\text{BH}}}\right)^4 \simeq 2^3 \cdot 3^3 \cdot 5 \cdot \pi \cdot V_\star \left(\frac{M}{m_{\text{pl}}}\right)^2 \propto M^5$, where V_{BH} is the effective black hole volume or the effective volume of the gravitationally collapsed object. It is noted that V_{BH} is almost equal to $\tilde{V}_{\text{BH}}$, so we use $V_{\text{BH}} = \tilde{V}_{\text{BH}}$ to denote the assumed effective volume of the black hole.

The interpretation of the volume increase $V_{\text{BH}} \simeq V_\star (M/m_{\text{pl}})^2$ is supplemented by the following consideration. The uncertainty principle $\Delta x \Delta p \geqslant \hbar$ shows one massless particle energy Δe should satisfy the relation $\Delta e \simeq \Delta p c \geqslant \hbar c/\Delta x$. Taking $\Delta x \simeq r_g = 2M$, the upper limit of the particle number is given by $N_{\text{BH}} \simeq M/(\Delta e) \leqslant GM^2/(\hbar c) = (M/m_{\text{pl}})^2$. On the other hand, the number of single particle quantum states in phase space of a black hole is given by $n_{\text{BH}} \simeq V_{\text{BH}}(\Delta p)^3/\hbar^3 \simeq (M/m_{\text{pl}})^2$, where $V_{\text{BH}} \simeq V_\star (M/m_{\text{pl}})^2$ and $V_\star \simeq (\Delta x)^3$ are used. Then the number of ways W to distribute the N_{BH} particles among the n_{BH} states is $W = (N_{\text{BH}} + n_{\text{BH}} - 1)!/(N_{\text{BH}}!(n_{\text{BH}} - 1)!)$. Taking Stirling's formula $\ln n! \simeq n \ln n - n$, the entropy becomes as $S = \ln W \simeq N_{BH}((1 + \alpha)\ln(1 + \alpha) - \alpha \ln \alpha) \propto N_{BH} \simeq (M/m_{\text{pl}})^2 \simeq A/l_{\text{pl}}^2$, where $N_{BH} \simeq \alpha n_{BH} (\gg 1)$ is taken.

4. Discussion and Conclusions

If we accept the increase of the effective black hole volume, the microscopic states of black hole entropy could be interpreted as the statistical state number in the phase space and the entropy could be taken as the extensive state quantity. The enormous entropy increase and temperature decrease in the black hole formation could be understandable as the free expansion of composed particles to this increased black hole volume.

References

Bekenstein, J. D. 1973, Phys. Rev., D7, 2333
Hawking, S. W. 1975, Commun. Math. Phys., 43, 199
Wald, R. M. 1998, Black Holes and Relativistic Stars (Chicago Univ. Press), gr-qc/9912119
Frolov, V. P. & Novikov, I. D. 1998, Black Hole Physics (Kluwer: Dordrecht)
Mukohyama, S. 1998, PhD Thesis (Kyoto University), gr-qc/9812079
Page, D. N. 2004, hep-th/0409024
Kiefer, C. 2002, astro-ph/0202032

Black Holes from Stars to Galaxies – Across the Range of Masses
Proceedings IAU Symposium No. 238, 2006
V. Karas & G. Matt, eds.

Uncertainty principle for the entropy of black holes, de Sitter and Rindler spaces

Tetsuya Hara, Keita Sakai, Shuhei Kunitomo and Daigo Kajiura

Department of Physics, Kyoto Sangyo University, Kyoto 603-8555, Japan
email: hara@cc.kyoto-su.ac.japan

Abstract. By a simple physical consideration and uncertain principle, we derive that temperature is proportional to the surface gravity and entropy is proportional to the surface area of the black hole. We apply the same consideration to de Sitter space and estimate the temperature and entropy of the space, then we deduce that the entropy is proportional to the boundary surface area. By the same consideration, we estimate the temperature and entropy in the uniformly accelerated system (Rindler space). The cases in higher dimensions are considered.

Keywords. Uncertainty principle – black hole entropy – de Sitter space – Rindler space

---

1. Introduction

Although it has passed almost 30 years since the discussion of the thermodynamics of black hole and event horizon began, there are still many problems about the fundamental concepts. One of them is that the entropy of the black hole is proportional to its surface. The other is that the temperature of the black hole and event horizon is related to the acceleration strength. These results could be understood if heuristic assumptions are adopted. One of the key concept is the uncertainty principle. We apply it to the black hole, de Sitter space and Rindler space to derive the characteristic features about the entropy in these spaces.

2. Black hole

As the black hole of mass M has the gravitational radius $r_g = 2GM/c^2$, there is the corresponding momentum $\Delta p = \hbar/(2c_1 r_g)$ from the uncertainty principle $\Delta p \Delta x \simeq \hbar/2$, putting Δx as $\Delta x \simeq c_1 r_g$. If the black hole is formed by photons corresponding to this momentum, it could be speculated that the temperature of the black hole is proportional to the surface gravity $kT \simeq \Delta pc \simeq 1/(4c_1 GM) \simeq M/r_g^2$ and it could be explained that the entropy of the black hole is proportional to the surface of the black hole.

If the black hole is formed by the black body radiation of temperature T, the volume V, total photon number N and total entropy S are given by

$$V = \frac{Mc^2}{\epsilon} = \frac{15}{\pi^2}\frac{(8\pi G)^4 M^5}{\hbar c^7}, \quad N = nV = \frac{30\zeta(3)}{\pi^4}\frac{8\pi G}{\hbar c}M^2 \simeq \frac{240}{\pi^3}\frac{G}{\hbar c}M^2,$$

$S/k_B = sV/k_B = \frac{32\pi}{3}\left(GM^2/\hbar c\right)$, where the following radiation density ϵ, number density n, and entropy density s of the radiation temperature T are used ($\epsilon = \tilde{a}T^4$, $n = \frac{2\zeta(3)}{\pi^2}\left(\frac{k_B T}{\hbar c}\right)^3 = 0.244\left(\frac{k_B T}{\hbar c}\right)^3$, $s = \frac{4}{3}\tilde{a}T^3$, $\frac{n}{\frac{s}{k_B}} = \frac{\frac{2}{\pi^2}\zeta(3)}{\frac{\pi^2}{15}\times\frac{4}{3}} = \frac{45\zeta(3)}{2\pi^4} \simeq 0.2776$, being $\tilde{a} = \pi^2 k_B^4/(15\hbar^3 c^3)$ and $\zeta(3) = 1.202$ the radiation constant and zeta function).

Using the surface of the black hole $A = 4\pi r_g^2 = 16\pi G^2 M^2/c^4$, and taking $c_1 = \frac{3\pi}{4}$, the following relation is derived, $S/k_B \simeq c^3 A/(4G\hbar) \simeq A/(4\ell_p^2)$, where $\ell_p = \sqrt{G\hbar/c^3}$ is the Planck length.

3. De Sitter space

The SNe Ia and WMAP observations confirm that our universe is now accelerating. For simplicity, we consider the universe with Λ term as de Sitter space with metric
$$ds^2 = -\left(1 - \tfrac{\Lambda}{3}r^2\right)c^2dt^2 + \left(1 - \tfrac{\Lambda}{3}r^2\right)^{-1}dr^2 + r^2\left(d\theta^2 + \sin^2\theta d\phi^2\right).$$
The characteristic point of this space is that there is the horizon with the radius of $\ell_\Lambda = \sqrt{3/\Lambda}$ (de Sitter horizon). Applying the uncertainty principle for this length, the energy ΔE for the photon or particle is estimated as $\Delta E \simeq \Delta pc \simeq \frac{\hbar c}{2\Delta x} \simeq k_B T$. Taking $\Delta x = c_3\ell_\Lambda$, the temperature T is given by $k_B T = \frac{\hbar c}{2c_3\ell_\Lambda} \simeq \frac{\hbar c}{2c_3\sqrt{3}}\sqrt{\Lambda}$. When we put $c_3 = \pi$ and the acceleration of the space as $a = c^2\sqrt{\Lambda/3}$, it becomes the one $k_B T = \hbar c\sqrt{\Lambda}/(2\pi\sqrt{3}) = a\hbar/(2\pi c)$ what Gibbons and Hawking have derived.

Because the cosmological constant is related to the vacuum energy density ρ_Λ as $\rho_\Lambda = \frac{\Lambda c^2}{8\pi G}$ the number density of the particle is given by $n_\Lambda \frac{\Delta E}{c^2} = \rho_\Lambda$. Assuming that particle energy is given by $\Delta E = \hbar c/2c_3\ell_\Lambda$, the number density n_Λ becomes as $n_\Lambda = \frac{\rho_\Lambda c^2}{\Delta E} = \frac{\Lambda c^4}{8\pi G}\frac{2c_3\ell_\Lambda}{\hbar c} = \frac{3c_3}{4\pi G}\frac{c^3}{\hbar\ell_\Lambda}$. Taking the volume of the universe V as $V = \frac{4}{3}\pi\ell_\Lambda^3$, the total number is given by $N_\Lambda = n_\Lambda V = \frac{3c_3}{4\pi G}\frac{c^3}{\hbar\ell_\Lambda}\frac{4}{3}\pi\ell_\Lambda^3 = c_3\left(\frac{c^3}{G\hbar}\right)\ell_\Lambda^2 = c_3\frac{\ell_\Lambda^2}{\ell_p^2}$.

If we assume the particle as Bose particle such as photon, the total entropy is proportional to the total number as ($N/(S/k_B) = 0.2776$) $\frac{S}{k_B} \sim \frac{N}{0.2776} \sim 4c_3\frac{\ell_\Lambda^2}{\ell_p^2}$. Using the area of the de Sitter horizon $A = 4\pi\ell_\Lambda^2$, it is expressed as $\frac{S}{k_B} \sim \frac{c_3}{\pi}\frac{A}{\ell_p^2}$, where the entropy is proportional to the horizon area A. If we take $c_3 = \pi/4$, it becomes $S/k_B = A/(4\ell_p^2)$ which is derived by Gibbons and Hawking.

4. Rindler space

Unruh Effect has been said that in the uniformly accelerating coordinate (Rindler coordinate) he or she (observer) seems to be in a bath of blackbody radiation at the temperature T which is related to the acceleration $\kappa(= a)$ as $k_B T = \hbar\kappa/(2\pi c)$.

There is a characteristic length $\ell_\kappa = c^2/\kappa$ due to the acceleration κ. Applying the uncertainty principle to this length, the energy ΔE of the particle is given by $\Delta E \simeq \Delta pc \simeq \frac{\hbar c}{2\Delta x} \simeq \frac{\hbar c}{2c_4\ell_\kappa} \simeq \frac{\hbar\kappa}{2c_4c} \simeq k_B T$, where we put $\Delta x = c_4\ell_\kappa$. The relation between the temperature and the acceleration κ is given by $k_B T \simeq \frac{\hbar\kappa}{2c_4c}$. If we take $c_4 = \pi$, the result is the same derived by Unruh.

In the following we consider the relation of this temperature to the entropy as $k_B S = \frac{A}{4\ell_p^2}$. One way of the derivation is to assume that the acceleration κ is the gravitational acceleration by the mass M, which is composed of the blackbody radiation of temperature T. If we take the volume of the considering region as V, the mass is given as $M = \frac{\tilde{a}T^4}{c^2}V$. The acceleration κ due to this mass is given by $\kappa = \frac{GM}{V^{\frac{2}{3}}} = \frac{G\tilde{a}T^4}{c^2}\frac{V}{V^{\frac{2}{3}}} = \frac{G\pi^2}{15}\frac{k_B^4 T^4}{\hbar^3 c^3}\frac{V^{\frac{1}{3}}}{c^2}$. Here we use the relations $k_B T = \hbar c/(2c_4\ell_\kappa)$, the above equation becomes $\kappa = \frac{c^2}{\ell_\kappa} = \frac{\pi^2}{15}\frac{1}{16c_4^4}\frac{G\hbar c}{c^2}\frac{V^{\frac{1}{3}}}{\ell_\kappa^4}$. Then the region size is given by $V^{\frac{1}{3}} = \frac{15\times16c_4^4}{\pi^2}\frac{\ell_\kappa^3}{\ell_p^2}$.

As the total entropy is $S = \frac{4}{3}\tilde{a}T^3V$, the entropy per surface $S/V^{\frac{2}{3}}$ is given by
$$\frac{S/k_B}{V^{\frac{2}{3}}} = \frac{4}{3}\frac{\pi^2}{15}\frac{k_B^3 T^3}{\hbar^3 c^3}V^{\frac{1}{3}} = \frac{8}{3}\frac{c_4}{\ell_p^2}.$$

If we take $c_4 = \frac{3}{32}$ and $A = V^{\frac{2}{3}}$, the relation $\frac{\frac{S}{k_B}}{A} = \frac{1}{4\ell_p^2}$ is derived, which means that the entropy per surface is $1/(4\ell_p^2)$. Finally, we derived the relation $S/(k_B) = A/(4\ell_p^2)$, which is common for the black hole and de Sitter space.

Black Holes from Stars to Galaxies – Across the Range of Masses
Proceedings IAU Symposium No. 238, 2006
V. Karas & G. Matt, eds.
© 2007 International Astronomical Union
doi:10.1017/S1743921307005558

Intensity and polarization light-curves from radiatively-driven clouds

Jiří Horák and Vladimír Karas

Astronomical Institute, Academy of Sciences, Boční II, CZ-140 31 Prague, Czech Republic
email: horak@astro.cas.cz, vladimir.karas@cuni.cz

Abstract. We study linear polarization from scattering of light on a cloud of relativistic electrons. We assume that the cloud is hovering above an accretion disc, or it is in an accelerated motion under the combined influence of the radiation and gravitational fields near an accreting black hole. At first we derive simple and general analytical formulae for the Stokes parameters. These formulae are then used in calculations of the temporal evolution of the observed signal.

We find that higher-order images can significantly enhance the observed flux. Possible targets where the effect should be searched are accreting super-massive black holes and Galactic microquasars exhibiting episodic accretion/ejection events.

Keywords. Polarization – Thomson scattering – radiative acceleration – gravitational lensing

1. Model

We consider a cloud of particles moving through the radiation field of a standard thin accretion disc. Primary photons from the disc are scattered by electrons in the cloud, they are beamed in the direction of the cloud motion, and polarized by Thomson mechanism. We adopt the first-scattering approximation and we restrict the cloud motion to the symmetry axis. The cloud moves in the radiation field of the disk that acts on the cloud together with gravity of the central black hole. The light rays of primary and scattered photons are also influenced. Direct and indirect (retro-lensed) light rays exhibit different degree of linear polarization and they experience different amplification and the Doppler boosting. The indirect photons contribute most intensely to the total signal if the cloud moves toward the black hole and the observer inclination i is small.

2. Local intensity and polarization

The electron distribution is considered isotropic in the cloud comoving frame. We derived simple formulae for the non-vanishing frequency-integrated Stokes parameters I, Q and U of the scattered radiation (Horák & Karas 2006a,b):

$$I = A\left[(1 + \mathcal{A})\left(T^{tt} + T^{ZZ}\right) + \mathcal{B}\left(T^{tt} - 3T^{ZZ}\right) - 2\mathcal{A}T^{tZ}\right],$$

$$Q = A\left(T^{YY} - T^{XX}\right), \qquad U = -2\,A\,T^{XY},$$

where $\mathcal{A} \equiv \frac{4}{3}\langle\gamma_e^2\beta_e^2\rangle$ and $\mathcal{B} \equiv 1 - \langle\ln[\gamma_e(1 + \beta_e)]/(\beta_e\gamma_e^2)\rangle$. $\mathcal{A}$ is a constant factor proportional to the cloud optical depth; β_e and γ_e are electron individual velocity and the Lorentz factor; $\langle\ldots\rangle$ denotes the averaging over the particle distribution in the cloud comoving frame. The fourth Stokes parameter, V, vanishes (linear polarization). The Stokes parameters are evaluated in the polarization frame comoving with the cloud, whose one basis vector is pointed along the direction of the scattered radiation and the two other basis vectors lie in the perpendicular observation plane. The incident unpolarized radiation comes into the formulae as components of the stress-energy tensor $T^{\alpha\beta}$ with respect to this reference frame.

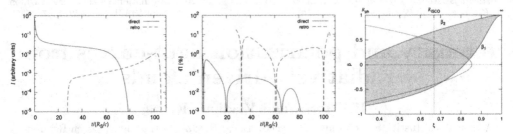

Figure 1. Examples of intensity (left panel) and polarization (middle panel) lightcurves. Contributions of the retro-lensing images have been summed together and plotted (by a dashed line); they are clearly distinguished from the signal produced by the direct-image photons (solid line). Polarization vanishes at the moment when the cloud crosses one of the critical velocities $\beta_1(\xi)$, $\beta_2(\xi)$, where $x \equiv 1 - R_S/r$ is dimensionless radius. In both cases the view angle was $i = 5$ deg from the symmetry axis. Regions of different polarization are shown together with the velocity curve $\beta(\xi)$ (red curve; see the right panel).

The total four-force acting on the cloud is a superposition of the radiation and inertial terms. We solved the equation of motion in the spacetime of a static black hole (Schwarzschild radius $R_S \equiv 2GM_{\bullet}/c^2$). The radiation field influences both components of the model – the bulk motion of the cloud as well as the electron distribution of the cloud. For polarization we obtain two critical velocities at which the polarization vector changes its orientation between transversal and longitudinal one. Similar effect of polarization direction changing with velocity of the scattering medium was studied by Beloborodov (1998) for the model of a wind from the disc. Below, by solving the equations of motion, we find how the observed polarization depends on time.

3. Retro-lensing lightcurves and polarization

When determining the temporal evolution of observed intensity and polarization we consider the first three images of the observed radiation – the direct one and two retro-lensed images. The latter are formed by rays making a single round about the black hole by the angle $2\pi \pm i$. The effect is usually small but for small inclinations the images take the form of Einstein arcs and may be quite significant (the flux level is up to several percent of the total signal). The retro-lensed photons give rise to peaks in the observed signal occurring with a characteristic mutual time lag after the direct-image photons. Duration of these features is very short and comparable to the light crossing time.

4. Conclusions

Our calculation is self-consistent in the sense that the motion of the blob and of photons is connected with the resulting polarization properties of the emerging signal. We concentrated ourselves on gravitational effects and compared the polarization magnitudes of direct and retrolensing images. We have estimated the mutual delay between the signal peaks formed by photons of different orders. The delay is characteristic to the effect and has a value proportional to the black hole mass.

We thank the Grant Agency of the Academy of Sciences (ref. 300030510) for support.

References

Beloborodov, A. M. 1998, ApJ, 496, L105
Horák, J. & Karas, V. 2006a, MNRAS, 365, 813
Horák, J. & Karas, V. 2006b, PASJ, 58, 204

Black Holes from Stars to Galaxies – Across the Range of Masses
Proceedings IAU Symposium No. 238, 2006
V. Karas & G. Matt, eds.
© 2007 International Astronomical Union
doi:10.1017/S174392130700556X

Physical properties of emitting plasma near massive black holes: the Broad Line Region

D. Ilić,[1] G. La Mura,[2] L. Č. Popović,[3] A. I. Shapovalova,[4]
S. Ciroi,[2] V. H. Chavushyan,[5,6] P. Rafanelli,[2] A. N. Burenkov[4]
and A. Marcado[7]

[1]Department of Astronomy, Faculty of Mathematics, University of Belgrade, Belgrade, Serbia
[2]Dipartimento di Astronomia, Università di Padova, Padova, Italy
[3]Astronomical Observatory, Belgrade, Serbia
[4]Special Astrophysical Observatory of the Russian AS, Nizhnij Arkhyz, Russia
[5]Instituto Nacional de Astrofísica, Optica y Electrónica, Puebla, Pue. México
[6]Instituto de Astronomía, Universidad Nacional Autónoma de México, Ensenada, México
[7]Observatorio Astronomico Nacional, Instituto de Astronomía, Universidad Nacional Autónoma de México, Ensenada, México

Abstract. We apply the Boltzmann-Plot (BP) method to the Balmer lines to estimate the physical properties in the Broad Line Region of Active Galactic Nuclei. We study the Balmer lines of a sample of 90 AGN from Sloan Digital Sky Survey database, as well as the time variability of the BP parameter A of NGC 5548.

Keywords. Galaxies: active – (galaxies:) quasars: emission lines – galaxies: Seyfert

---

1. Introduction

The Broad Emission Lines (BELs), which originate in the Broad Line Region (BLR) of the Active Galactic Nuclei (AGN), could be used to probe the properties of the massive Black Hole (BH) that is assumed to be in the center of an AGN. The understanding of the physics and kinematics of the BLR is crucial because: (i) kinematics of the BLR is probably determined by the massive BH, with the competing effects of gravity and radiation pressure, therefore the parameters of the BLR should be connected with the general characteristics of an AGN (e.g. BH mass); (ii) the BLR reprocesses the UV energy emitted by the continuum source, consequently BELs can provide indirect information about the continuum source. In order to connect the physical and kinematical parameters of the BLR, we study the parameters of Balmer lines of the 90 AGN from Sloan Digital Sky Survey (SDSS), as well as the variation of the parameters of Balmer lines of NGC 5548 observed from 1996 till 2004. We apply the Boltzmann-Plot (BP) method, as described by Popović (2003, 2006), to the Balmer lines to estimate the physics of a typical BLR.

2. Application of the BP to the Balmer lines

The Boltzmann Plot Method. We start from the assumptions, that the BLR has a significant optically thin component and that the Balmer line series is emitted from a region with the same physical properties. In such a case, it has been demonstrated that the BP method, commonly used for laboratory plasma diagnostics – see Griem (1997), can be used for average temperature estimates of the line emitting medium; La Mura *et al.* (2006), Ilić *et al.* (2006), Popović (2003, 2006), Popović *et al.* (2006).

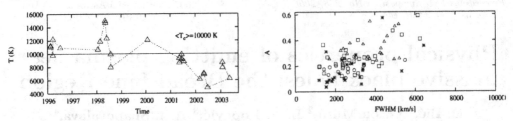

Figure 1. Left: the temperature variation from 1996 till 2004 of NGC 5548. Right: temperature parameter A as a function of the FWHM that we determined from the line profiles of the SDSS sample. Open circles are for the cases when BP is working and $T_e < 20000$ K; open squares are for the cases when the BP is working, but one point of the plot has too small error-bars that do not cover the BP line; open triangles are for the cases when the BP may work, but two points are not covering the BP line (with their estimated error-bars) or the BP works, but $T_e > 20000$ K; and asterisks are for the cases when BP does not work at all.

The BLR of NGC 5548. We study the variability of physical parameters in the BLR of NGC 5548 using the BP method. We apply the method on Balmer lines observed from 1996 till 2004 with the 6m and 1m telescopes of SAO (Russia, 1996–2004) and at INAOE's 2.1m telescope at the Guillermo Haro Observatory (GHO) at Cananea, Sonora, Mexico (1998–2004); Shapovalova *et al.* (2004). We found that variability seen in lines is also present in the electron temperature (Fig. 1, left panel). The average electron temperature for the considered period was T≈10000 K, and that varies from 6000 K (in 2002) till 15000 K (in 1998). The more detailed discussion is given in Popović *et al.* (2006).

The SDSS results. We apply the BP method to the Balmer series and we discuss the physical parameters of the emitting plasma for an AGN sample collected at the 3rd data release from the SDSS database, and we study their correlations with other BLR and AGN parameters (La Mura *et al.* (2006)). Moreover, we perform line flux and profile analysis, as well as continuum luminosity measurements at 5100 Å, estimating some parameters of the sources, including the mass of central BH, the velocity fields of the emitting gas and the size of the BLR. We use these results to discuss a model that can explain the observed effects. We found that: (i) PLTE is a suitable approximation to describe the physical conditions of the BLR emitting gas in a number of AGN (in our sample ∼ 30%) and this greatly simplifies the task of gathering even general information about the physics of these sources, since most of the standard methods used in astrophysics cannot deal with them; (ii) there is a general trend, for AGN showing broader line profiles, to be associated with averagely colder BLR (Fig. 1, right panel).

This work was supported by the Ministry of Science and Environment Protection of Serbia through the project 146002, as well as by grants: CONACYT (N39560F, Mexico) and RFBR (06-02-16843, Russia).

References

Griem, H. R. 1997, Principles of Plasma Spectroscopy (Cambridge University Press)

Ilić, D., Popović, L. Č., Bon, E., Mediavilla, E. G. & Chavushyan, V. H. 2006, MNRAS, 371, 1610

La Mura, G., Popović, L. Č., Ciroi, S., Rafanelli, P. & Ilić, D. 2006, ApJ submitted

Popović, L. Č. 2003, ApJ, 599, 140; 2006, ApJ, 650, 1217

Popović, L. Č., Shapovalova, A. I., Chavushyan, V. H., Ilić, D., Burenkov, A. N. & Marcado, A. 2006, astro-ph/0511676

Shapovalova, A. I. *et al.* 2004, A&A, 422, 925

Black Holes from Stars to Galaxies – Across the Range of Masses
Proceedings IAU Symposium No. 238, 2006
V. Karas & G. Matt, eds.
© 2007 International Astronomical Union
doi:10.1017/S1743921307005571

Fe K line profile in PG quasars: the averaged shape and Eddington ratio dependence

H. Inoue,[1] Y. Terashima,[2] and L. C. Ho[3]

[1]Institute of Space and Astronautical Science, 3-1-1 Yoshinodai, Sagamihara 229-8510, Japan

[2]Ehime Univesity, 2-5 Bunkyo, Matsuyama 790-8577, Japan.

[3]The Observatories of the Carnegie Institution of Washington, 813 Santa Barbara Street, Pasadena, CA 91101-1292

Abstract. Fe K emission line is a powerful probe of the inner part of an accretion disk. We analyze X-ray spectra of 43 Palomar-Green (PG) quasars taken from Boroson & Green (1992) observed with *XMM-Newton* and make an averaged Fe K line profile. We study the Eddington ratio dependence of the Fe K line profile. The width of the Fe line becomes broader ($\sigma = 0.1$ to 0.7 keV) and its peak energy becomes higher (6.4 to 6.8 keV) as the Eddington ratio gets larger.

These results indicate that the physical state of the accretion disk, such as the geometrical structure and/or ionization state, changes with the Eddington ratio.

Keywords. Quasars – X-rays – accretion, accretion disks

---

1. The sample

We selected 43 PG quasars (Boroson & Green 1992) observed with EPIC-PN onboard *XMM-Newton*. Their redshifts, central black hole masses, and Eddington ratios are well known from optical spectra. The quasars in our sample have relatively low redshifts ($z \leqslant 0.5$), and the central black hole masses range widely from 10^6 to 10^9 $M_\odot$. The Eddington ratio, which is one of the most important parameters to characterize an accretion disk, ranges from 0.05 to 4.50. This wide range enables us to study the Eddington ratio dependence of the Fe line profile.

2. Results

We fitted the spectra of the 43 quasars systematically by using a canonical model consisting of a power-law, a blackbody, an edge for objects showing signature of warm absorbers, and a Gaussian, all modified by the Galactic absorption. The spectra of 38 objects are well reproduced by this model. The spectra of the remaining five objects require an additional absorbed power law with a column density of 10^{23} cm^{-2}.

In order to characterize the mean Fe K line profile quantitatively, we fitted all spectra simultaneously with a model consisting of the best-fit continuum model and a Gaussian. The Fe K line in each quasar has a different peak energy in the observed frame according to their redshifts. In order to properly treat the detector response, which is energy dependent, we applied simultaneous fits rather than fits to a composite spectrum. A common peak energy and a width were used for the Gaussian and the normalizations were left free individually. The best-fit parameters are $E = 6.48^{+0.05}_{-0.04}$ keV, $\sigma = 0.36^{+0.08}_{-0.08}$ keV, and EW=248±168 eV, where the EW is the mean equivalent width and its error is a standard deviation of the distribution of the EW.

The sample was then divided into four groups according to the Eddington ratio and averaged spectra were derived in each group. The co-added spectra are shown in Figure 1

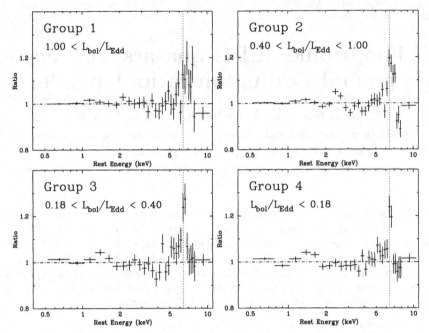

Figure 1. Eddington ratio dependence of the Fe-K line profile.
The vertical dotted line is at the energy 6.4 keV.

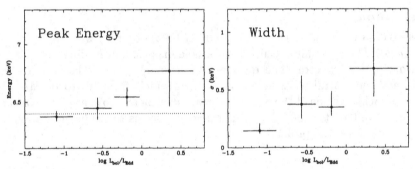

Figure 2. Eddington ratio dependence of the peak energy (left) and the width (right).
The horizontal dotted line is at the energy 6.4 keV.

for a presentation purpose. We fitted the spectra in each group simultaneously to quantify the Fe line shape. The best-fit parameters are shown in Figure 2. The peak energy of the Fe K line becomes higher (6.4 to 6.8 keV) and its width becomes broader (σ=0.1 to 0.7 keV) as the Eddington ratio gets larger. The mean equivalent width is EW$\approx$280 eV except for the lowest Eddington ratio group (EW$\approx$130 eV). These results indicate that the physical state of the accretion disk, such as the inner radius of the optically thick disk and/or ionization state, changes with the Eddington ratio.

References

Boroson, T. A. & Green, R. F. 1992, ApJS, 80, 109
Jiménez-Bailón, E. *et al.* 2005, A&A, 435, 449
Page, K. L. *et al.* 2004, MNRAS, 347, 316
Piconcelli, E. *et al.* 2005, A&A, 432, 15
Streblyanska, A. *et al.* 2005, A&A, 432, 395
Tanaka, Y. *et al.* 1995, Nature, 375, 659

Black Holes from Stars to Galaxies – Across the Range of Masses
Proceedings IAU Symposium No. 238, 2006
V. Karas & G. Matt, eds.
© 2007 International Astronomical Union
doi:10.1017/S1743921307005583

Radiation in models with cosmological constant

Hedvika Kadlecová and Jiří Podolský

Institute of Theoretical Physics, Faculty of Mathematics and Physics, Charles University,
Prague, Czech Republic
email: Hedvika.Kadlecova@centrum.cz, podolsky@mbox.troja.mff.cuni.cz

Abstract. We analyze the asymptotic directional structure and properties of gravitational field of specific exact solutions belonging to the large family of black hole spacetimes of type D found by Plebański and Demiański. We discuss the structure in the case of de Sitter ($\Lambda > 0$) and anti-de Sitter ($\Lambda < 0$) conformal infinity. With the aim of further interpretation, we have found the relation between the structure of the sources (the mass m, the electric e and magnetic g charges, the NUT parameter l, the rotational parameter a, the acceleration α) and the properties of radiation generated by them. In particular, we studied the amplitude of radiation and illustrated it in several figures.

Keywords. Plebański–Demiański family of solutions – asymptotic directional structure of radiation – de Sitter, anti-de Sitter

1. Introduction

The investigation of properties of gravitational radiation is an important problem of general relativity. In last few years, the theory was further developed for asymptotically de Sitter ($\Lambda > 0$) and anti-de Sitter ($\Lambda < 0$) spacetimes in Krtouš & Podolský (2004). The work assumed, in analogy with the asymptotically flat situation, that the radiative field components are decaying as η^{-1}, where η is an affine parameter along the null geodesics. We apply the new approach on particular exact model spacetimes of type D from the Plebański and Demiański family of solutions, concretely on the Kerr–Newman–de Sitter metric and then on the general form of the metric which contains not only the mass, the charge and the rotation, but also the NUT parameter, non-zero acceleration and the cosmological constant, for details see Plebański & Demiański (1976), Griffiths & Podolský (2006). In both cases, we found the explicit dependence on the parameters of the metric in the final formulae of the directional structure of radiation.

2. Results

The general form of the metric has the form (Plebański & Demiański 1976)

$$\mathbf{g}_{ab} = \frac{1}{\Omega^2} \left[-\frac{Q}{\rho^2}(\mathrm{d}\tilde{t} - (a\sin^2\vartheta + 4l\sin^2\tfrac{\vartheta}{2})\mathrm{d}\tilde{\phi})^2 + \frac{\rho^2}{\hat{P}}\mathrm{d}\vartheta^2 + \frac{\rho^2}{Q}\,\mathrm{d}r^2 \right.$$
$$\left. + \frac{\hat{P}\sin^2\vartheta}{\rho^2}(\mathrm{d}\tilde{t} - (r^2 + (a+l)^2)\mathrm{d}\tilde{\phi})^2 \right],$$

(2.1)

where $\Omega = 1 - \frac{\alpha}{\omega}(l + a\cos\vartheta)$, $\rho^2 = r^2 + (l + a\cos\vartheta)^2$, $\hat{P} = 1 - a_3\cos\vartheta - a_4\cos^2\vartheta$, $Q = (\omega^2 k + e^2 + g^2) - 2mr + \epsilon r^2 - 2\alpha\omega^{-1}nr^3 - (\alpha^2 k + \Lambda/3)r^4$.

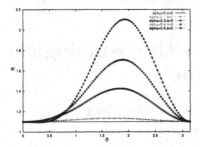
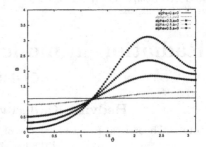

Figure 1. Left panel: the magnitude of radiation $B(\vartheta)$ on spacelike $\mathcal{I}$ for C–metric without charges, where $a = 0$ and α varies. Right panel: the magnitude of radiation $B(\vartheta)$ on spacelike $\mathcal{I}$ for charged C–metric, where $a = 0$ and α varies.

The directional pattern for $\Lambda > 0$ is determined by two principal null directions (PNDs) oriented outwards with respect to the future spacelike conformal infinity $\mathcal{I}^{+}$, explicitly

$$|\Psi_4^i| \approx \frac{3}{2}\frac{1}{|\eta|}B(\vartheta)\mathcal{A}(\theta, \phi, \theta_s), \tag{2.2}$$

where $B(\vartheta)$ is the amplitude of radiation, which contain parameters of the metric. $\mathcal{A}(\theta, \phi, \theta_s)$ is the directional structure, where θ and ϕ parametrize the direction along geodesic which approaches the conformal infinity, θ_s is a parameter between two tetrads. In particular, for the C–metric the amplitude $B(\vartheta)$ has the form

$$B(\vartheta) = (1 + \tfrac{3}{\Lambda}\alpha^2\mathcal{P}\sin^2\vartheta)\sqrt{m^2 - 4m\alpha(e^2 + g^2)\cos\vartheta + 4\alpha^2(e^2 + g^2)^2\cos^2\vartheta}$$

where $\mathcal{P} = 1 - 2\alpha m\cos\vartheta + \alpha^2(e^2 + g^2)\cos^2\vartheta$. The dependence on parameters is illustrated in two graphs in Fig. 1. We verified results for C-metric obtained in Krtouš & Podolský (2003) and in Podolský, Ortaggio & Krtouš (2003). For Kerr–Newman–de Sitter, only the mass parameter occurs in the amplitude $B(\vartheta) = m$. We also investigated the structure near timelike $\mathcal{I}$, for details see Kadlecová (2006).

3. Conclusions

We investigated influence of the physical parameters of the sources, namely α, a, m, e, g and Λ on the amplitude of the radiation $B(\vartheta)$ which appears at spacelike and timelike conformal infinity in the subcase $l = 0$ of the general metric (2.1) and on its special subcases: the Kerr–Newman–de Sitter solution and the C-metric. Generally, we observe from Fig. 1 and other figures that the acceleration parameter α and the cosmological constant Λ have *dominant* influence on the amplitude. The acceleration parameter α has stronger influence than the rotational parameter a. The charges e and g have a smaller influence than acceleration but their influence is quite stronger when the rotational parameter a is small or vanishes. The asymptotic structure depends substantially on *rotation* when the acceleration is non-vanishing. We hope that these results will provide a deeper insight into the general theory of radiation in general relativity.

References

Plebański, J. F. & Demiański, M. 1976, Ann. Phys. (NY), 98
Griffiths, J. B., Podolský, J. 2006, Int. J. Mod. Phys. D, 15, 335
Krtouš, P., Podolský, J. 2004, Class. Quantum Grav., 21 , R233–273
Krtouš, P., Podolský, J. 2003, Phys. Rev. D, 68, 024005
Podolský, J., Ortaggio, M. & Krtouš, P. 2003, Phys. Rev. D, 68, 124004
Kadlecová, H. 2006, Diploma Thesis, Charles Univesity, Prague

Black Holes from Stars to Galaxies – Across the Range of Masses
Proceedings IAU Symposium No. 238, 2006
V. Karas & G. Matt, eds.
© 2007 International Astronomical Union
doi:10.1017/S1743921307005595

The results of X-ray binary Cyg X-1 investigations based on the optical photometry and high-resolution spectroscopy

Eugenia A. Karitskaya

Astronomical Institute of RAS, 48 Pyatnitskaya str., Moscow 119017, Russia

email: karitsk@sai.msu.ru

Abstract. Selected results of 1994-2004 optical photometric and high-resolution spectral observations including those obtained in the frame of a coordinated CIS countries campaign "Optical Monitoring of Unique Astrophysical Objects" and their comparison with ASM/RXTE X-ray data are briefly reviewed.

Keywords. X-rays: binaries – black hole physics – accretion, accretion disks – stars: early-type – stars: fundamental parameters (colors, luminosities, masses, radii, temperatures)

1. Introduction

Cyg X-1/HDE226868 is an X-ray binary system (the orbital period $P = 5.6^d$) whose relativistic component is the first candidate to black hole. The optical component, an O9.7 Iab supergiant, is responsible for about 95% of the system optical luminosity. The remaining 5% are due to the accretion structure (disc and surrounding gas) near BH. In spite of the fact that the investigations of Cyg X-1 are being carried on for almost 40 years and have resulted in about 1000 publications, a lot of phenomena in the system, its geometrical and physical parameters remain unclear.

2. The main results of photometric researches

In the frame of 1994–1998 international campaign "Optical Monitoring of Unique Astrophysical Objects" (Georgia, Kazakhstan, Russia, Uzbekistan, and Ukraine) 2258 Cyg X-1 UBVR observations were made during 407 nights. The main results are reported in papers Karitskaya, *et al.* (2000), Karitskaya, *et al.* (2001), Karitskaya (2002) and briefly described below. By comparing photoelectric (UBVR) and X-ray ASM/RXTE (3–12 keV) flux variations we found different kinds of variability (orbital variations, different kinds of flares, dips, so-called precession period 147/294 days) and a correspondence between optical and X-ray variations. Cross-correlation analysis revealed lags between X-ray (2–10 keV) and optical long-term variations equaling 7^d in 1996 and 12^d in 1997–1998.

One item in our list of accretion instability evidence is the existence of several-day-long optical flares followed by X-ray ones coinciding with X-ray dips. X-ray flare lag lengths in respect to the optical ones are the same as for slow variations (see above). We proposed the scenario for such X-ray flares in which the matter flows in separate portions from the supergiant to the accretion structure.

The characteristic time of the matter transfer through the accretion disk (the time delay of X-ray responses) was about 7 days in Summer and Autumn 1996 and 12 days in 1997-98. These times are too short for α – disk model. We may suggest for example that the most accreting matter goes through a high latitude of the accretion structure.

3. The main results of high-resolution spectral researches

We report the results of Cyg X-1 spectral monitoring over 2002-2004 (Karitskaya (2003), Karitskaya *et al.* (2005), Karitskaya, *et al.* (2006)). The observations were carried out with the echelle-spectrographs of the Peak Terskol Observatory (altitude 3100 m, North Caucasus) 2 m telescope (70 spectra, spectral resolution R = 13000 or 45000) and BOAO (Korea) 1.8 m telescope (5 spectra, R = 30000) covering most of the optical range. Signal-to-noise ratio is S/N> 100–200 near H_α. The observations fell at different states of Cyg X-1 X-ray spectrum ("hard", "soft", and transitional). We used RXTE/ASM X-ray data. The spectra contain the supergiant absorption lines and emission components of H_α and HeIIλ4686Å lines with complicated profiles showing variation with the orbital period 5.6^d.

Optical spectral line profile variations were found during X-ray flare. The comparison of the observed and non-LTE model HI, HeI and MgII profiles is given. Tidal distortion of Cyg X-1 optical component and its illumination by X-ray emission are taken into account. We set limits on the optical component main characteristics $T_{\rm eff}$ = 30400±500 K, log g = 3.31±0.07, and overabundance of He and Mg: [He/H]= 0.43±0.06 dex, [Mg/H]= 0.4 − 0.6 dex by using 2003–2004 spectra.

For Doppler tomography we used the method developed by Agafonov (2004), on the basis of the radioastronomical approach. The main features are deconvolution with introduction of a synthesized beam and the removal of distortions by sidelobes' impact on the summary image (after back projecting) using CLEAN algorithm. This method is developed specially for a small number of irregularly distributed observations. Tomography maps of Cyg X-1 using HeIIλ4686Å profiles were constructed (Karitskaya *et al.* (2005)). A new method of parameter determination through Doppler tomograms was proposed and tested. The Doppler images and Roche lobe model allowed to put a limitation on the black hole / supergiant mass ratio $1/4 < q < 1/3$.

4. Long-term supergiant variations in Cyg X-1 binary system

The photometric and spectral variations point to the supergiant parameter changes on a time scale of tens of years. Line profile non-LTE simulations lead to the conclusion that the star radius has grown about 1–4% from 1997 to 2003–2004 and the temperature decreased by 1300–2400 K (Karitskaya, *et al.* (2006)). This agrees with the X-ray activity growth in that period. The increasing of the degree of the Roche lobe filling go to intensification and instability of the matter outlet toward the X-ray source.

Acknowledgements

The work is supported by RFBR grants 04-02-16924 and 06-02-16234.

References

Agafonov, M. I. 2004, AN, 325, 259; and ibid., 325, 263
Karitskaya, E. A., Goranskij, V. P., Grankin, E. N. *et al.* 2000, Astron. Lett., 26, 22
Karitskaya, E. A., Voloshina, I. B., Goranskij, V. P., *et al.* 2001, Astron. Rep., 45, 350
Karitskaya, E. A. 2002, in: R. K. Manchanda & B. Paul (eds.), Multi Colour Universe, Proc. Conf. TIFR, (Ebenezer Printing House: Mumbai), p. 45
Karitskaya, E. A. 2003, Kinematika i Fizika Nebesnykh Tel, Suppl., No. 4, 230
Karitskaya, E. A., Agafonov, M. I., Bochkarev, N. G. *et al.* 2005, Astron. Astrophys. Trans., 24, 383
Karitskaya, E. A., Lyuty, V. M., Bochkarev, N. G. *et al.* 2006, IBVS, No. 5678, 1

Black Holes from Stars to Galaxies – Across the Range of Masses
Proceedings IAU Symposium No. 238, 2006
V. Karas & G. Matt, eds.
© 2007 International Astronomical Union
doi:10.1017/S1743921307005601

Observational manifestations of accretion onto isolated black holes of different masses

Sergey Karpov and Gregory Beskin

Special Astrophysical Observatory of Russian Academy of Sciences, Nizhniy Arkhyz,
Karachaevo-Cherkessia, Russia

email: karpov@sao.ru

Abstract. The process of accretion onto the isolated black holes under the various conditions of the ISM and its observational manifestations are discussed. For the majority of the Galaxy volume the accretion rate is as low as 10^{-6}–10^{-9} of Eddington one, and the accretion is spherically-symmetric. Such objects manifest itself as a weak optical and x-ray sources with featureless spectra and the significant variability of the emission.

For the BH located inside the dense molecular cloud the regime of accretion depends on its mass and velocity. A massive (100–$1000 M_\odot$) BH born in the cloud or having low relative velocity, may manifest itself as a ultra-luminous x-ray source (ULX).

Keywords. Accretion – black hole physics – ISM: clouds – ISM: general

---

For the majority of Galaxy, filled with hot and warm ionized hydrogen, the regime of accretion onto the stellar mass isolated black holes is spherical, as the captured specific angular momentum is much smaller than one on the last stable orbit. The accretion rate is also small, $\dot{m} \dot{M}/\dot{M}_{edd} \sim 10^{-10}$–$10^{-5}$ for the velocity of 50–100 km/s (Shvartsman 1971, Ipser & Price 1982). So, the emission of the accretion flow is dominated by the non-thermal synchrotron component due to accelerated particles(Beskin & Karpov 2005). The total luminosity of such object is then $L = 9.6 \cdot 10^{33} M_{10}^3 n_1^2 (V^2 + c_s^2)_{16}^{-3} \sim 10^{29} - 10^{34}$ erg/s.

For typical interstellar medium parameters, a $10 M_\odot$ black hole at 100 pc distance will be a 16–25$^{\rm m}$ optical source coinciding with the highly variable bright X-ray counterpart and a very faint gamma-ray one. The hard emission consists of a distinct flares (Beskin & Karpov 2005; see Fig. 1) carrying the information on a structure of space-time near the horizon. Such black hole mimics the optical appearance of known classes of objects with featureless optical spectra, such as DC-dwarfs and blazars (see the sample spectrum in Fig. 2).

The intermediate mass black holes (IMBH, 100–$1000 M_\odot$) may form as a result of the collapse of first massive stars or due to the mergers of the stars in the centers of globular clusters, and is assumed to populate the halo of the galaxies. They may be observable while crossing the galactic plane. Typical velocities for them are 50–100 km/s, the accretion rate is again $\dot{m} \sim 10^{-10}$–10^{-6}, and the luminosity is 10^{30}–10^{35} erg/s. Such objects appear as a strong IR sources coincided with optical and UV/X-ray counterparts. The example of computed spectrum for such object is shown in Fig. 2.

The accreting stellar mass black hole may be located in a dense molecular cloud (Mapelli *et al.* 2006). The time scale of the cloud crossing is about 10^5–10^6 years for the velocity of 50–100 km/s. The black hole moving slower ionizes and heats the cloud before it may start to accrete it, and falls in the class of low-density accretion models. The fast-moving black hole also ionizes the cloud while flying through it, as the Strömgren radius is always larger than the Bondi one. The ionization time scale is 10–100 years – this is

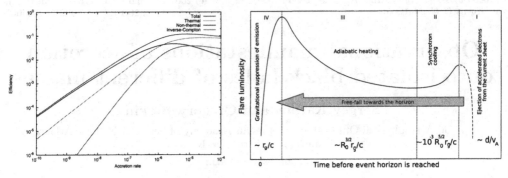

Figure 1. Left panel: efficiencies of the synchrotron emission of thermal and non-thermal electron components of the accretion flow. Right panel: internal structure of a flare as a reflection of the electron cloud evolution. The prevailing physical mechanisms defining the observed emission are denoted and typical durations of the stages are shown.

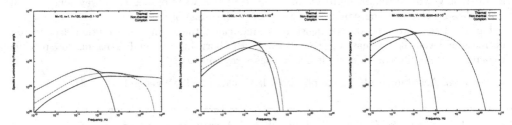

Figure 2. Spectra of accreting isolated black holes. The left panel corresponds to the stellar-mass black hole in low-density ISM, the middle – to the IMBH in ISM, and the right one – to the IMBH in dense molecular cloud.

the "active phase" of the black hole inside the dense molecular cloud accretion. After this time the luminosity drops down several orders of magnitude. During the "active phase" the accretion rate is high, so the inverse Compton emission is important. Strong gamma-ray counterpart appears along with the radio, infrared, optical and X-ray one. The luminosity is of order 10^{34}–10^{36} erg/s, and the sample spectrum is shown in Fig. 2.

In the rare case of IMBH born in the cloud or having low relative velocity (less than 10 km/s), it may appear as ULX, with luminosity $L \sim 10^{39}$–10^{40} erg/s for either spherical or disk regime of accretion.

In the case of the accretion disk formation (due to the large gradients of density and velocity in molecular clouds) the picture basically remains the same. The difference is in the spectral shape – the X-ray part becomes larger while the optical and IR fainter.

Acknowledgements

This work has been supported by the Russian Foundation for Basic Research (grant No 04-02-17555) and by the Russian Science Support Foundation.

References

Beskin, G. M. & Karpov, S. V. 2005, A&A, 440, 223
Ipser, J. R. & Price, R. H. 1982, ApJ, 255, 654
Mapelli, M. *et al.* 2006, MNRAS, 368, 1340
Shvartsman, V. F. 1971, AZh, 48, 479

Black Holes from Stars to Galaxies – Across the Range of Masses
Proceedings IAU Symposium No. 238, 2006   © 2007 International Astronomical Union
V. Karas & G. Matt, eds.   doi:10.1017/S1743921307005613

AGN types as different states of massive black holes

Boris V. Komberg

Astrospace Center, Physical Institute P. N. Lebedev, Moscow, Russian Federation

email: bkomberg@asc.rssi.ru

Abstract. After discovering microquasars – active stellar systems with relativistic outflows (Mirabel *et al.* 1992) – many papers bringing the observational arguments on similar nature of accreting systems with different BH masses (from stellar to galactic nuclei) have appeared. It became clear that a level of activity in such systems could depend on few parameters: BH mass and angular momentum; accretion rate and magnetization parameter of accretion disk (AD); relative orientation of BH and AD angular momentum. The bolometric luminosity and characteristic variation time in such systems is proportional to BH masses, which span an interval in 5–8 decades.

Keywords. Black hole physics – galaxies: nuclei – galaxies: active

1. Introduction

Active systems with different masses can be placed on a single sequence (Falcke & Biermann 1996) that can be expressed using known data about Seyfert nuclei: $\log L_5(\text{erg/s}) \approx 1.67 \log L_{\text{BBB}}(\text{erg/s}) - 33.4$. Here L_5 – radio luminosity at 5 GHz, L_{BBB} – disk luminosity in the big blue bump range. The more strict sequence connecting active stellar systems with different masses is traced on so called "fundamental plane" (Merloni *et al.* 2003): $\log L_r(\text{erg/s}) \approx 0,6 \log L_{\text{XR}}(\text{erg/s}) + 0.78(M_{\text{BH}}/M_\odot) + 7.33$. Monitoring data on the brightest microQSOs in X-rays showed that light curves of such objects strongly vary in time which can be described by the existence of some characteristic states occasionally changing each other (Yadav 2001).

In the best-studied microQSO GRS 1915+105 sharp bursts of the soft X-ray radiation (spikes) occasionally appears in low/hard state followed by ejecting of radio outflows with relativistically moving components. A counter-rotation of the central BH and accreting material can also support jet formation (e.g. Koide *et al.* 2000).

Similar case is observed in some AGN (e.g. ON 231, 3C345; Belokon & Babadzhanyats 2000) with VLBI jets appearing after bursts in the optical or UV range. But while in microQSOs X-ray–radio time lag is about hundreds of seconds, the observed time lag in the case of AGN is of order of years. Furthermore, it should be noted that mean AD temperature is connected with BH mass: $T_D = 2 \cdot 10^7 (M_{\text{BH}}/M_\odot)^{-1.4}$ K. So X-ray bursts from disks of microQSOs correspond to UV bursts from AGN disks.

2. AGN types as different states of massive black holes

X-ray monitoring of microQSOs shows that they can be in the strong radio state for no more than 0.1 of time (e.g. Yadav 2001). It means that strong radio emission could be observed at 0.1 of microQSOs. This ratio looks similar to that of quasars – the fraction of the radio loud QSOs is ~ 0.1 from the radio quiet QSOs. This also refers to the ratio of the radio galaxies to the luminous E-galaxies (e.g. Komberg 2003; Nipoti *et al.* 2005). If these coincidences are not random, they can be the evidence of strong radio emission of AGN being relatively short state of its activity. And if radio quiet- radio loud ratio transition in microquasars lasts 10^3 s, in AGN it corresponds to 10^2–10^3 yrs. Such transition has been confirmed by the detection of the extended steep spectrum radio structures at the long radio wavelengths (90 cm) of the radio quiet QSOs – the remains of the past nuclear activity (Blundel et $al.$ 2000). Different AGN states are well observed in BL Lacertae objects. They have a typical "two-hump" spectral energy distribution (SED) $\log \nu f_\nu - \log \nu$ (Mei et $al.$ 2002). A peak with a maximum on the lower frequency ν_{P1} (synchrotron peak) and that one with a maximum on the higher frequency ν_{P2} (inverse Compton peak) can move in the frequency axes depending on the radio luminosity: $\log L_r(\text{erg/s}) = -0.67 \log \nu_{P1} + 53.13$ (Di Mateo & Psaltis 1999). Radio luminosity depends on the accretion rate increasing in low/hard state when $m < m_{\text{crit}}$. This shows a similarity in the accreting systems with very different BH masses.

Another common phenomena for the systems considered can be so called quasiperiodic oscillations (QPO) observed in microquasars in X-rays. Low-frequency QPO with $\nu_{\text{QPO}} < 10$ Hz are frequently observed in the low/hard state, rarely in the intermediate state and never in the high/soft state. One point of view on QPO nature lies in the assumption of instabilities on the inner edge of the accretion disk ($r_{\text{in}} \sim 10 R_g$) characterized by Keplerian frequency (e.g. Belanger et $al.$ 2006): $\nu_k = \frac{1}{2}\pi(GM_{\text{BH}}/R^3)^{1/2} (M_{\text{BH}})^{-1}$. Observations give $\nu_k \sim 0.1$–10 Hz for microquasars. So for AGN with $M_{\text{BH}} \sim 10^{6-9} M_\odot$ QPO frequencies correspond to typical oscillation times of order of days to months.

3. Conclusions

Observational arguments favoring to the similarity of different active system's properties listed above argue that different types of AGN are in fact the different states of massive accreting BH in the galactic centers. Conclusions following such assumption could promote a deeper understanding of AGN evolution. We can list some of them:

1. By the analogy with unstable "intermediate" state of microquasars, BL Lacertae objects can be thought to have a strongly variable intermediate state connecting the radio loud and radio quiet QSOs (RLQ and RQQ). In that case two types of BL Lacertae objects are possible. When transition is from RLQ to RQQ state, then RL BL Lac would be observed, and RQ BL Lac could be observed in opposite case.

2. One can expect higher accretion rates of BHs in the nuclei of the distant massive galaxies on large redshifts. That in turn leads to a smaller probability of the radio loud QSOs detection on higher z. So one should expect a reduction of RL QSO to RQ QSO number ratio on large z. Respectively, the RL BL Lac to RQ BL Lac number ratio, also should decrease on large z.

References

Mirabel, I. F., Rodrigez, L. F. & Cordier, B. 1992, Nature, 358, 215

Falcke, H. & Biermann, P. L. 1996, A&A, 308, 321

Merloni, A., Heinz, S. & Di Matteo, T. 2003, MNRAS, 345, 1057

Yadav, J. S. 2001, AJ, 548, 876

Koide, S. et $al.$ 2000, AJ, 536, 668

Belokon, E. T. & Babadzhanyats, M. K. 2000, A&A, 356 , L21

Komberg, B. V. 2003, http://www.asc.rssi.ru (Educational Center, in Russian)

Nipoti, C., Blundell, K. M. & Binney, J. 2005 MNRAS, 361, 633

Blundell, K. M. & Rawkings, S. 2000, AJ, 562, L5

Mei, D. C., Zhang, L., & Jiang, Z. J. 2002, A&A, 391, 917

Di Mateo, T. & Psaltis, D. 1999, AJ, 526, L101

Belanger, G., Terrier, R., De Jager, O., Goldwurm A. & Melia, F. 2006, ApJL, submitted (astro-ph/0604337)

Black Holes from Stars to Galaxies – Across the Range of Masses
Proceedings IAU Symposium No. 238, 2006
V. Karas & G. Matt, eds.
© 2007 International Astronomical Union
doi:10.1017/S1743921307005625

Relativistic jets and non-thermal radiation from collapse of stars to black holes

V. Kryvdyk and A. Agapitov

Depart. of Astronomy, Kyiv National University, av. Glushkova 2/1, Kyiv 03022, Ukraine
email: kryvdyk@univ.kiev.ua

Abstract. The formation of the relativistic jets and a non-thermal emission from the collapsing magnetized stars with dipole magnetic fields and the heterogeneous particles distribution are investigated. These polar jets are formed when the stellar magnetosphere compress during collapse its magnetic field increases considerable. The electric field is produced in magnetosphere, which the charged particles will be accelerated. As follow from the calculation, the jets can be formed from collapsing stars already the explosion of supernova stars without shock waves. These jets will generate the non-thermal radiation. The radiation flux depends on the distance to the star, its magnetic field and the particle spectrum in the magnetosphere. This flux can be observed near Earth by means of modern telescopes in the form of the radiation pulse with duration equal to time collapse.

Keywords. Collapsing stars – particle acceleration – non-thermal emission – relativistic jets

1. Introduction

In our previous papers (Kryvdyk 1999, 2004, 2005) we investigated the particles acceleration and the non-thermal emission from the collapsing magnetized stars. It was showed that the collapsing stars can be powerful sources of the non-thermal radiation, which can be observed by means of modern telescopes. In this paper we consider the formation of the relativistic jets by collapse of stars with heterogeneous magnetospheres to black holes. These jets formed when the stellar magnetosphere compress during the collapse and its magnetic field increases considerably. A cyclic electric field is produced and the charged particles are accelerated in polar caps of collapsing stars.

2. Magnetosphere of collapsing star

We examine particles in magnetospheres of collapsing stars with dipolar magnetic fields. We consider three cases of initial heterogeneous distributions of the particles – power-series distribution, the relativistic Maxwell one, and the Boltzmann distribution,

$$N_P^i(E, R) = (r_*)^3 K_P E_*^{-\gamma} R_*^{-\beta_P} \tag{2.1}$$

$$N_M^{'}(E, R) = (r_*)^3 K_M E_*^2 R_*^{-\beta_M} \exp(-E/kT), \tag{2.2}$$

$$N_B^{'}(E, R) = (r_*)^3 K_B R_*^{-\beta_B} \exp(-E/kT), \tag{2.3}$$

Here, $E_* = E/E_0$; $r_* = r_0/r$; $\beta_P = a_1(\gamma - 1)$; $\beta_M = a_1(E/kT \ln E_* - 3)$; $\beta_B = a_1(E/kT \ln E_* - 1)$; $a_1 = (5k_1/3)(3\cos^4\theta + 1.2\cos^2\theta - 1)(1 + 3\cos^2\theta)^{-2}$; K_C, K_M, K_B are the spectral coefficients, k is the Boltzmann constant, and T is temperature.

Figure 1 shows the transformation of the stellar magnetosphere during collapse. We can see that the initial stellar magnetosphere transforms during the collapse. The polar jets are formed in the magnetosphere already at the initial stage of collapse. Particle

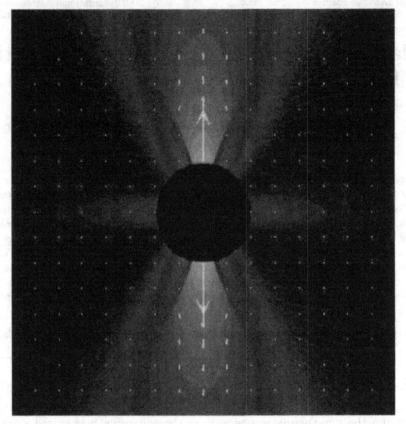

Figure 1. Relativistic jets emerging from a collapsing star. This simulation shows the structure of the stellar magnetosphere during collapse.

density and energy in polar jets grow during the process of collapse. Thus, the jets from collapsing stars can be formed already at the moment of the explosion of supernovae without shock waves.

3. Conclusions

As follows from our results, the charged particles are accelerated in the magnetosphere of collapsing stars to relativistic energy; the relativistic polar jets are formed. These jets emit the electromagnetic waves in all frequency bands. The radiation flux grows during collapse and reaches maximum at the final stage of collapse. This radiation can be observed as bursts in all frequency bands, from radio to gamma ray.

References

Kryvdyk, V. 1999, MNRAS, 309, 593
Kryvdyk, V. 2004, Adv. Sp. Res., 33, 484
Kryvdyk, V. & Agapitov, A. 2005, ASP CS, 330, 415

Black Holes from Stars to Galaxies – Across the Range of Masses
Proceedings IAU Symposium No. 238, 2006
V. Karas & G. Matt, eds.
© 2007 International Astronomical Union
doi:10.1017/S1743921307005637

Long-term photometry of blazars at Abastumani Observatory

Omar M. Kurtanidze, Maria G. Nikolashvili, Givi N. Kimeridze, Lorand A. Sigua, Bidzina Z. Kapanadze and Roman Z. Ivanidze

Abastumani Astrophysical Observatory, 383792 Abastumani, Georgia

Abstract. We give a brief summary of the ongoing Abastumani Active Galactic Nuclei Monitoring Program started in the May 1997. More than 110 000 frames are obtained during more than 1300 nights of observations for about 50 target objects, among them gamma-ray, X-ray and optical blazars. All observations were done in the BVRI bands using ST-6 CCD based photometer attached to the Newtonian focus of 70-cm meniscus telescope.

Keywords. BL Lacertae objects – variability

1. Introduction

Among active galactic nuclei (AGNs), blazars are objects most dominated by nonthermal continuum emission, which extends from radio to gamma-rays, and whose properties are best explained by emitting plasma in relativistic motion towards the observer, closely aligned with the line of sight (Urry 1995). One of the distinguishing characteristics of the blazars which include BL Lacertae type objects, high polarization quasars (HPQ) and optical violently variable (OVV) quasars is that their flux densities are highly variable at all wavelength from radio to gamma-rays. The optical multiband monitoring along with other ones gives unique clues into the size and structure of the radiating region.

Variability timescales have been derived for many blazars from monitoring programs which attain a time resolution of days and to years. Unfortunately, existing multi-wavelength data are not adequate yet to permit definite conclusions to be drawn about the nature of blazars due to the optical coverage in the previous campaigns have been much to sparse. We started systematic multi-wavelength optical monitoring of blazars at Abastumani Observatory in the may of 1997. In the late of October 1997 we joined the Whole Earth Blazar Telescope (WEBT, http://www.to.astro.it/blazars/webt). The aim of the program is to study short-term and long-term variability of blazars and their correlations with that in radio, X-ray and gamma-ray bands.

2. Observation and data reduction

Abastumani Observatory is located in the South-Western part of Georgia at a latitude of 41.8 grad and a longitude of 42.8 grad on the top of the Mt. Kanobili at 1700 m above mean sea level. The weather and seeing are very good in Abastumani (150 nights per year, one third with seeing less than 1 arcsec). The mean values of the night sky brightness are B=22.0 mag, V=21.2 mag, R=20.6 mag, and I=19.8 mag. Blazar Monitoring Program at Abastumani Observatory was started in the May 1997 and is carried out with Petlier cooled CCD Imaging Camera attached to the Newtonian focus of the 70-cm meniscus telescope (1:3). The pointing accuracy of the meniscus telescope is one-two arcminutes and it is good enough to locate target object inside of the full frame field of view 14.9×10.7 square arcminute.

The ST-6 Imaging Camera uses the TC241 chip (375x242, 23x27 sq. micron) with a maximum quantum efficiency 0.7 at 675 nm. All observations were performed using combined filters of glasses which match the standard B, V (Johson) and Rc, Ic (Cousins) bands well. Reference sequences in the blazar fields are calibrated using the Landolt's equatorial standard stars (Landolt 1992). In photometric nights at least one equatorial field is observed with different exposures. Because of the scaling of CCD and resolution f the meniscus telescope are equal to 2.3x2.7sq. arcsec per pixel and 1.5 arcsec respectively, the images are under-sampled, therefore to improve sampling it is needed to defocus frames slightly. Unfortunately, the high dark current limits the exposure time to 900 sec. The images are reduced using Daophot-II (Stetson 1987). To eliminate the effects of seeing induced spurious IDV and IHV (Cellone 2000) the apertures are taken to include the whole host galaxy. The highest differential photometrical accuracy reached in R-band is 0.005 mag (rms) magnitude at m(R)=14.00 mag during 180 sec.

3. Results and conclusions

List of target objects was compiled using two catalogues of AGNs (Padovani 1996, Perlman 1996). In the period from May 1997 to September 2005 (about 1450 observing nights) more than 130 000 frames were obtained. In the table the list of the target objects along with the number of nights observed and frames obtained in every of the BVRI bands in excess of one hundred fifty in R band are given. Last column shows the number of frames obtained to study the intra-day and intra-hour variability. Among most frequent observed objects are BL Lacertae, S5 0716+714, AO 0235+164 and Mkr 421.

So far we took part in many international multi-wavelength campaigns conducted during outburst and post-outburst era of BL Lacertae, S5 0716+714, AO 0235+164, Mkr 421, 3C 279, 4C 29.45, OJ 287, S4 0954+650 and ON 231. The main part of the results are already published (Ostorero 2006; Osterman 2006; Villata 2006; Krawczynski 2004; Jorstad 2004; Raiteri 2005; Bottcher 2005). Most objects under study show light variations over one magnitude in the optical band. Largest one was observed for AO 0235+164 and equals to 4.0 mag in R-band. A few faint variable stars (B 16.5 mag) with amplitude 0.3–0.4 mag and periods 3–5 hours were also identified.

Acknowledgements

O.M.K. gratefully acknowledge financial support and the kind collaboration with Dr. G. M. Richter without which this Programme would never have been conducted. We thank European Astronomical Society for donation of the ST-6 CCD camera.

References

Bottcher, M. *et al.* 2005, ApJ, 631, 169
Celone, S. A. *et al.* 2000, AJ, 121, 90
Jorstad, S. *et al.* 2004, AIPC CS, 714, 202
Krawczynski, H. *et al.* 2004, ApJ, 601, 151
Landolt, A. U. 1992, AJ, 104, 340
Osterman, A. M. *et al.* 2006, A&A, 132, 873
Ostorero, L. *et al.* 2006, A&A, 451, 797
Padovani, P., Giommi, P. *et al.* 1996, MNRAS, 277, 1477
Perlman, E. *et al.* 1996, A&A, 424, 497
Stetson, P. B. 1987, PASP, 99, 19
Villata, M. *et al.* 2004, A&A, 453, 817
Raiteri, C. M., Villata, M. *et al.* 2005, A&A, 438, 39
Urry, C. M., Padovani, P. *et al.* 1995, PASP, 107, 803

Black Holes from Stars to Galaxies – Across the Range of Masses
Proceedings IAU Symposium No. 238, 2006
© 2007 International Astronomical Union
V. Karas & G. Matt, eds.
doi:10.1017/S1743921307005649

Optical variability of X-ray selected blazars

Omar M. Kurtanidze, Maria G. Nikolashvili, Givi N. Kimeridze, Lorand A. Sigua and Bidzina Z. Kapanadze

Abastumani Astrophysical Observatory, 383792 Abastumani, Georgia

Abstract. We present optical R band photometry of nine X-ray selected BL Lac objects: 1ES 0229+200, 1ES 0323+022, 1ES 502+675, 1ES 0647+250, 1ES 0806+524, 1ES0927+500, 1ES 1028+511, 1ES 1959+650, 1ES2344+514.

Variability on long time scales within one magnitude in R band was found for all of the observed objects, except 1ES 0229+200 and 1ES0927+500. Largest variation was detected for 1ES 0502+675 and equals to 1.07 mag. Only few objects show statistically significant variation on intra-day scale.

Keywords. BL Lacertae objects – X-ray blazars – variability

---

1. Observation and data reduction

Blazar Monitoring Program at Abastumani Observatory was started in the May 1997 and is carried out with ST-6 CCD camera attached to the Newtonian focus of the 70 cm meniscus telescope (1:3, 14.9×10.7 square arcmin). All observations are performed using combined filters of glasses that match the standard B, V (Johnson) and Rc, Ic (Cousins) bands well (Kurtanidze 1999). Reference sequences in the blazar fields are calibrated using the equatorial standard stars (Landolt 1992). List of target objects was compiled from Einstein Slew Survey Sample of BL Lacertae objects (Perlman 1996). During more than 200 nights about 1400 CCD frames were obtained in R band to study long-term and intra-day variability of selected objects.

The most frequently observed object in our sample (on long-term as well as on intra-day scales) is 1ES 1959+650. The duration of observational runs varied from two hours to six hours and exposure times varied from 60 to 180 sec depending on the brightness of the object and the filter used. The images were reduced using Daophot-II (Stetson 1987). To eliminate the effects of seeing induced spurious IDV and IHV variability (Cellone 2000) the apertures are taken into account to include the whole host galaxy.

2. Results and conclusions

2.1. *1ES 0323+022 and 1ES 0502+675*

The brightest state $R = 15.62$ was detected in Dec 1982 (Feigelson 1986). Largest amplitude 0.67 mag in R band was detected during three years (23.10.1996–23.01.1999) of observation by the Torino group with a maximum of $R = 16.60$ mag (Villata 2000). Our observations include the period from 04 Oct 1997 to 04 Feb 2002. There were two dramatic changes of brightness: first, up to 0.43 mag from 31 Aug 1998 to 23 Nov 1998 and the second one from 12 Sept 1999 to 23 July 2000 about 0.26 mag, while the maximum amplitude was 0.45 mag. Early observations of 1ES 0502+675 (31.10.1996–22.02.1997) show that maximum amplitude in R band equals to 0.58 (Raiteri 1998). Our observations include the period from 09 Nov 1997 to 12 Feb 2000. Dramatic changes $R = 15.67$–16.74 was detected before 23 Nov 1998 with an amplitude 1.07 mag. After the minimum it rapidly increases again and reach a mean state characterised by $R = 16.40$ and with a maximum amplitude of variation 0.36 mag.

2.2. *ES 0647+250, 1ES 0806+524 and 1ES 1028+511*

The previous observation of 1ES 1028+511 during December 3, 1996 – May 8, 1997 revealed variation $dR = 0.18$ and maximum brightness $R = 16.53$ (Villata 2000). Our observations of these objects include the period from 25 Nov 1997 to 25 Jan 2002. All three objects show significant light variations that are equal to 0.37 mag (25 Nov 1997-13 Dec 1998), 0.88 (28 Dec 1997-06 June 2000) and 0.60 (28 Jan 1998-25 Jan 2002), respectively.

2.3. *1ES 1959+650 and 1ES 2344+514*

Observations of 1ES 1959+650 from February 29,1996 to May 30, 1997 shows that the light curve in the R band is characterized by rapid flickering, a decrease of 0.28 mag in 4 days (Villata 2000). Both objects during our observations show light variations bellow 0.4 mag in R band. Largest one is observed for 1ES 1959+650 (Kurtanidze 2001). The 1ES 2344+514 show obvious long-term variability trend over the observing period at 0.1 mag level (Fan 2004). Consequently, the intra-day variability is very week bellow 0.05 mag and may only be detected in exceptional cases of very high photometric accuracy. More higher level activity of 1ES 1959+650 relative to 1ES 2344+514 may be attributed to its higher radio luminosity (Raiteri 1998).

2.4. *Conclusion*

Seven of the nine X-Ray BL Lacertae object studied show variability on long-term scale. Three of them show variation over 0.5 mag (1ES 0502+675, 1ES 0806+524 and 1ES 1028+511), while other four bellow 0.5 (1ES 0323+022, 1ES 0647+250, 1ES 1959+650 and 1ES 2344+514). Long-term variability was not detected for two BL Lacertae 1ES 0229+200 and 1ES 0927+500. Intra-day variability of 1ES 1959+650 and 1ES 2344+514 is bellow 0.05 mag. In general, X-ray selected blazars show week optical variability in comparison with radio selected blazars.

Acknowledgements

O.M.K. gratefully acknowledge invaluable financial support and the kind collaboration with Dr. G. M. Richter without which this Programme would have never been conducted. We thank European Astronomical Society for donation of the ST-6 CCD camera.

References

Celone, S. A. *et al.* 2000, AJ, 121, 90
Fan, J. H., Kurtanidze, O. M. *et al.* 2004, ChAA, 4, 133
Feigelson, S. D. *et al.* 1997, ApJ, 302, 337
Landolt, A. U. 1992, AJ, 104, 340
Nikolashvili, M. G. *et al.* 2001, in: S. Ritz, *et al.* (eds.) 6th Compton Symp. (AIP CP), 587, 333
Kurtanidze, O. M. *et al.* 1999, in: C. M. Raiteri *et al.* (eds.), Blazar Monitoring Towards the 3rd millenium (Osserv. Astron. di Torino: Torino), p. 29
Perlman, E. *et al.* 1996, A&A, 424, 497
Raiteri, C. M. *et al.* 1998, A&A, 144, 481
Stetson, P. B. 1987, PASP, 99, 19
Villata, M. *et al.* 2000, A&A, 144, 481

Black Holes from Stars to Galaxies – Across the Range of Masses
Proceedings IAU Symposium No. 238, 2006
V. Karas & G. Matt, eds.
© 2007 International Astronomical Union
doi:10.1017/S1743921307005650

Rotation curves, dark matter and general relativity

Patricio S. Letelier

Departamento de Matemática Aplicada-IMECC, Universidade Estadualde Campinas,
13083-970 Campinas, S. P., Brazil

Abstract. We study the possibility of pure general relativistic models without exotic matter to describe the observed flattening of the rotation curves for stars moving in circular orbits in a galaxy disk. In particular we consider the dragging of inertial frames (rotation of the source) and the presence of a Taub–NUT (Newman–Unti–Tamburino) "charge", the gravitational equivalent to a magnetic monopole in electrodynamics.

Keywords. Thin disks – rotation curves – dragging of inertial frames – Taub–NUT parameter

1. Introduction

Thin disk models in General Relativity can be constructed using the "Displace, Cut and Reflect method" (image method). This method amount to take a solution to the Einstein vacuum field equations in cylindrical coordinates and perform the transformation $z \to |z| + d$, where d is a constant. From the Einstein field equations we find $G_{\mu\nu} = -T_{\mu\nu}\delta(z)$, where $\delta(z)$ is the usual Dirac distribution. See for instance González & Letelier (2000).

The rotation curves associated to a particular disk are found using the geodesic equation computed from the particular metric that represents the gravitational field of the disk: $ds^2 = g_{ij}dx^i dx^j + f(dr^2 + dz^2)$, were $(x^i) = (t, \varphi)$. For a circular geodesic we find that the tangential velocity is $v^2 = -det(g_{ij})\frac{d\varphi}{dt}/(g_{tt} + g_{t\varphi}\frac{d\varphi}{dt})^2$.

2. Rotation curves dragging of inertial frames and Taub–NUT mass.

The Kerr metric that represents a rotating black hole has two parameters the mass m and the rotation parameter a, once we apply the image method me add the new parameter d, so for the Kerr disk we have (m, a, d) as parameters. Thin Kerr disks are studied in some detail in González & Letelier (2000), see also Vogt & Letelier (2005).

For these disk we have two classes of circular orbits the ones co-rotating with the disk and the ones moving in counter rotation. For Newtonian disks both rotation curves are the same, but in the general relativistic context we have different behaviour due to the fact that the rotating bodies drag the spacetime around them. In Fig. 1 we show the profiles of the rotations curves for a Kerr thin disk with $m = 1$, $d = 3$ and $a = 0, 0.5, 9.98$, (co-rotation) and $a = 0, -0.5, -0.98$ (counter-rotation).

We find the the counter-rotation does help a little the flattening of the rotation curves co-rotation does not. So dragging of inertial frames has a very little effect on the flattening of the rotation curves. Note that the limit value of the parameter a is one.

The Taub–NUT metric has also two parameters the mass, m and the Taub–NUT parameter b. Loosely speaking we can associate mass to the electric part of the Riemann–Christoffel curvature tensor and the Taub–NUT parameter to its the magnetic part. So one can say that this parameter represents a kind of "magnetic mass". In Fig. 2 we

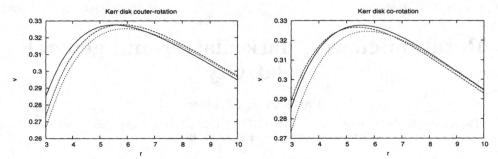

Figure 1. Rotation curves for a Kerr disk with parameters $m = 1$, $d = 3$ and $a = 0$, 0.5, 9.98, (co-rotation) and $a = 0$, -0.5, -0.98 (counter-rotation).

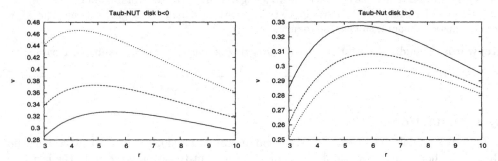

Figure 2. Rotation curves for a Taub-NUT disk with parameters $m = 1, d = 3$ and $b = 0, 0.5, 1$ and $b = 0, -0.5, -1$.

show the the rotation curves for a Taub-NUT disk with parameters $m = 1, d = 3$ and $b = 0, 0.5, 1$ and $b = 0$, -0.5, -1. We see that the Tab-NUT charge $b > 0$ helps a little the flattening of the rotation curves, but when $b < 0$ we have the opposite effect; see also González & Letelier (2000).

The disk constructed from the Kerr metric can have a perfect fluid as a source. The Taub–NUT disk has heat flow and tensions, it has a rather not astrophysical source.

One can fit actual galaxy rotation curves with disks constructed from the van Stockum solution Cooperstock & Tieu (2005), but again the source has hoop tensions and heat flow Vogt & Letelier (2005). Also it is possible to construct disks with more general solutions like Kerr–NUT or Tomimatsu–Sato solutions. We do not foresee a different behaviour of the rotation curves as the one presented above.

In summary, pure relativistic effects like dragging of inertial frames and Taub-NUT charges do not have a big effect on the flattening of rotation curves. To flatten the rotation curves within GR we need exotic matter, not a very appealing solution.

Acknowledgements

The author thanks CNPq and FAPESP for partial financial support.

References

González, G. A. & Letelier, P. S. 2000, Phys. Rev. D, 62, 064025
Vogt, D. & Letelier, P. S. 2005, MNRAS, 363, 268
Cooperstock, F. I. & Tieu, S. 2005, astro-ph/051204.
Vogt, D. & Letelier, P. S. 2005, astro-ph/0510750.

Black Holes from Stars to Galaxies – Across the Range of Masses
Proceedings IAU Symposium No. 238, 2006
V. Karas & G. Matt, eds.
© 2007 International Astronomical Union
doi:10.1017/S1743921307005662

The capture of main sequence stars and giant stars by a massive black hole

Y. Lu,[1] Y. F Huang,[2] Z. Zheng[1] and S. N. Zhang[3]

[1]National Astronomical Observatories, Chinese Academy of Sciences, Beijing 100012, China

[2]Department of Astronomy, Nanjing University, Nanjing 210093, China

[3]Physics Department and Center for Astrophysics, Tsinghua University, Beijing 100084, China

email: ly@bao.ac.cn;

Abstract. Since the mass-radius relation is quite different for a main sequence (MS) star and a giant (G) star, we find that the radiation efficiencies in the star capture processes by a black hole (BH) are also very different. This may provide a useful way to distinguish the capture of MS and G stars. Comparing with observations of the very high energy (VHE) gamma-ray emissions, we argue the event that triggers the gamma-ray emission in the energy range 4–40 TeV should be a G star capture. On the other hand, the capture of MS stars by the massive BH is required when the measured spectrum of VHE gamma-rays extends from 10^9 to 10^{15} eV.

Keywords. Galaxy: center – galaxies: active – galaxies: jets – galaxies: gamma rays – accretion, accretion disks – black holes

---

When a star with a given mass of M_* and radius of R_* passes by a BH with a mass of M_{bh}, the star could be tidally disrupted and captured. This picture is described in the left panel of Fig.1. In the phase 2 (the fallback stage), a fraction of the material in the disrupted stars remains gravitationally bound to the BH at the pericenter. The timescale and the accretion rate (Phinney 1989) for the debris to return to the pericenter are $t_{min} \approx 0.11 m_*^{-1} r_*^{3/2} M_6^{1/2}$ yr and $\dot{M} = 1.29 f_{0.3} m_*^2 r_*^{-3/2} M_6^{-1/2} (\frac{t}{t_{min}})^{-5/3} M_\odot$ yr^{-1}, respectively, where $m_* = M_*/M_\odot$, $M_6 = M_{bh}/10^6 M_\odot$, $r_* = R_*/R_\odot$, f is the fraction of the stellar material falling back to the periastron, and $f_{0.3} = f/0.3$. Assuming that the fallback material radiates the energy release promptly, the radiation efficiency ϵ (Li *et al.* 2002) is $\epsilon \approx 5.38 \times 10^{-3} r_*^{-1} m_*^{1/3} M_6^{2/3}$. Immediately, the total kinetic energy of the jet can be estimated as (Lu *et al.* 2006), $E_{jtot}^{proton} = \int_{t_{peak}}^{t_{crit}} q_j \epsilon \dot{M} c^2 dt$, where q_j is the efficiency that transfers the accretion energy into the jet power, $t_{peak} \sim 0.157 m_*^{-1} r_*^{3/2} M_6^{1/2}$ yr and $t_{crit} \sim 1.67 \times 10^2 f_{0.3}^{3/5} m_*^{2/5} M_6^{3/5}$ yr are the time since the radiation reaches the peak luminosity and the time that the jet exists, respectively. The viscous accreting timescale (Lu *et al.* 2006) is $t_{acc} \approx 2.08 \times 10^2 (h/r)_{-2}^{-2} \alpha_{0.1}^{-1} m_*^{-1/2} r_*^{3/2}$ yr, where $(h/r)_{-2} = (h/r)/10^{-2}$ and $\alpha_{0.1} = \alpha/0.1$. For typical MS stars with masses $0.08 M_\odot \leqslant M_* \leqslant 1 M_\odot$, the radius and mass could be related by $r_* = m_*$. And for G stars, the radius-mass relation (Joss *et al.* 1987) is $r_* \approx \frac{3.7 \times 10^3 m_*^4}{1 + m_*^3 + 1.75 m_*^4}$, where $m_* \equiv M_c/M_\odot$ for a giant star, M_c is a mass of its core, and $0.17 M_\odot \lesssim M_c \lesssim 0.45 M_\odot$.

With these mass-radius relations of stars and $q_j = 0.1$, the radiation efficiency and the total injection energy of protons captured by the BH with mass of $10^6 M_\odot$, $10^7 M_\odot$ and $10^8 M_\odot$ are plotted in Fig. 2. Paying attention to the injection energy required by the hadronic model(Aharonian & Neronov 2005) to produce the VHE gamma-rays (Lu *et al.* 2006), one can find that the G star capture by the BH is favored for the

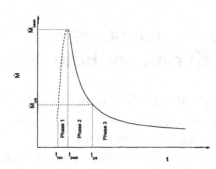 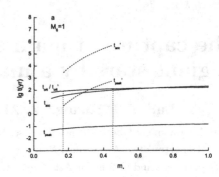

Figure 1. The left panel shows the behavior of the accretion rate versus time ($\dot{M} - t$) when a black hole captures a star, where $t_{circ} = 2t_{min}$. The right panel shows the various timescales involved in the capture processes change as function of the stellar mass: t_{acc}, t_{peak} and t_{crit} are the capture of the MS star, while those primes are for the capture of the G star.

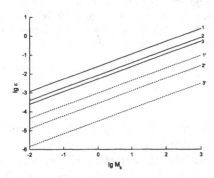 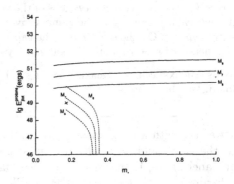

Figure 2. Lines of 1, 2, 3 in the left panel and the light lines in the right panel correspond to the capture of the MS stars, respectively. While lines of 1', 2', 3' in the left panel and the black lines in the right panel are for the capture of G stars. In the right panel, the cross represents the capture of a G star fit to the injection energy of protons required by the hadronic model with the parameters of $M_6 = 3$, $\epsilon = 2.17 \times 10^{-3}$ and $m_* = 0.17$; while the filled dot stands for the capture of a MS star fitting with $M_6 = 3$, $\epsilon = 1.12 \times 10^{-2}$ and $m_* = 1$.

gamma-rays emitted in the energy range of 4-40 TeV, and the MS case is reliable for the VHE gamma-rays with the radiated energy extended to 10^{15} TeV.

Acknowledgements

Supported by the National Natural Science Foundation of China (Grants 10273011, 10573021, 10433010), and by the Special Funds for Major State Basic Research Projects.

References

Aharonian, F. A. & Neronov, A. 2005, ApJ, 619, 306
Joss, P. C., Rappaport, S. & Lewis, W. 1987, ApJ, 319, 180
Li, L. X., Narayan, R. & Menou, K. 2002, ApJ, 576, 753
Lu, Y., Cheng, K. S. & Huang, Y. F. 2006, ApJ, 641, 288
Phinney, E. S. 1989, Nature, 340, 595

Black Holes from Stars to Galaxies – Across the Range of Masses
Proceedings IAU Symposium No. 238, 2006
V. Karas & G. Matt, eds.
© 2007 International Astronomical Union
doi:10.1017/S1743921307005674

Low-frequency, one-armed oscillations in black hole accretion flows obtained from direct 3D MHD simulations

Mami Machida[1] and Ryoji Matsumoto[2]

[1]Division of Theoretical Astronomy, National Astronomical Observatory of Japan, 2-21-1, Osawa, Mitaka, Tokyo 181-8588, Japan

[2]Department of Physics, Faculty of Science, Chiba University 1-33, Yayoi-cho, Inage-ku, Chiba, 263-8522 Japan

email: mami@th.nao.ac.jp, matumoto@astro.s.chiba-u.ac.jp

Abstract. We present the results of global 3D MHD simulations of optically thin black hole accretion flows. The initial disk is embedded in a low density, spherical, isothermal halo and threaded by weak ($\beta \equiv P_{\mathrm{gas}}/P_{\mathrm{mag}} = 100$) toroidal magnetic field. General relativistic effects are simulated by using the pseudo-Newtonian potential. When the Maxwell stress in the innermost region of the disk is reduced due to the loss of magnetic flux or by decrease of disk temperature, inner torus is created around $4 - 10r_{\mathrm{s}}$. We found that in such an inner torus, one-armed ($m = 1$) density enhancement grows and that the inner torus oscillates quasi-periodically. The oscillation period is about 0.1s when we assume a $10M_\odot$ black hole. This frequency agrees with the low-frequency QPOs observed in low/hard state of black hole candidates. The disk ejects winds whose opening angle is about 30 degree. The maximum velocity of the wind is about $0.05c$.

Keywords. accretion, accretion disks – MHD – QPO

1. Introduction

Low/hard state in black hole X-ray binaries is characterized by the large amplitude rapid variability of X-ray flux, power law X-ray spectrum extending to hard X-rays, and the radio excess indicating the existence of outflows. In the low/hard state, low-frequency QPOs are sometimes observed (e.g., Belloni *et al.* 2006). In this paper, we present the results of three-dimensional magneto-hydrodynamic (MHD) simulations of radiatively inefficient, optically thin accretion flows. Such simulations reproduced rapid time variability, low-frequency QPOs and outflows.

2. Numerical Methods and Results

We numerically solved resistive MHD equations by using a 3D resistive MHD code in cylindrical coordinates (ϖ, ϕ, z) assuming the anomalous resistivity. General relativistic effects are simulated by using the pseudo-Newtonian potential. The initial state is a disk with angular momentum $L \propto \varpi^{0.43}$ threaded by weak ($\beta = P_{\mathrm{gas}}/P_{\mathrm{mag}} = 100$) toroidal magnetic field. The initial disk is embedded in a low density, spherical, isothermal halo. The unit of length is the Schwarzschild radius r_{s}, and the unit time is $t_0 = r_{\mathrm{s}}/c = 10^{-4}$sec, when we assume a $10M_\odot$ black hole.

Left top panel of figure 1 shows the time evolution of mass accretion rate at $\varpi = 2.5r_{\mathrm{s}}$ computed by integrating in $|z| < 8r_{\mathrm{s}}$. The mass accretion takes place due to the efficient angular momentum transport by Maxwell stress generated by the magneto-rotational instability. The mass accretion rate shows large amplitude time variations similar to the

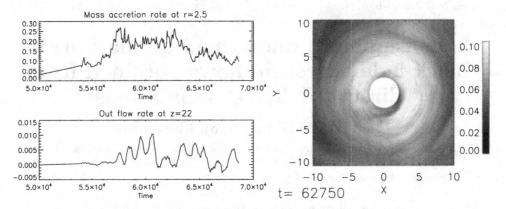

Figure 1. Left panels: The time evolution of mass accretion rate and outflow rate. top) Mass accretion rate is calculated at $\varpi = 2.5$. bottom) Outflow rate measured at $z = 22$. Right panel: Density distribution in $\varpi - \phi$ plane averaged in $|z| < 1$.

X-ray time variations in low/hard states of black hole candidates. Left bottom panel shows the time variation of the mass outflow rate measured at $z = 22r_s$. The mass outflow rate oscillates quasi-periodically with period $\sim 1000t_0$ and correlates with the mass accretion rate with time lag around $3000t_0$. The maximum speed of the outflow is $0.05c$. Numerical results indicate that the outflow emerging from the inner torus is created intermittently by magnetic reconnections taking place in the torus. The driving mechanism of the outflow is similar to that in proto-stellar flares (Hayashi et al. 1996). In addition to the intermittent outflow, quasi-steady funnel wall wind appears around $\varpi = 2.5r_s$. The funnel wall wind is driven by gas pressure gradient in the plunging region.

Right panel of figure 1 shows the vertically averaged density distribution of the disk. The inner torus is deformed into a crescent shape with azimuthal mode number $m = 1$.

3. Discussion

Global 3D MHD simulations revealed that in optically thin black hole accretion flows, crescent shaped, non-axisymmetric density distribution develops in the inner torus created around $\varpi \sim 8r_s$. Magnetic energy release generates outflows emerging from the inner torus. The mass outflow rate shows quasi-periodic oscillation with period 0.1s when the mass of the black hole is $10M_\odot$. This period is comparable to the period of low-frequency QPOs observed in low/hard states.

Let us discuss whether we can observe the time variation of magnetic fields in Galactic center black hole Sgr A*. Recently, polarization of Sgr A* has been measured by sub-mm observation. These observations show the time evolution of the polarization angle. By using the formula for synchrotron self-absorption, we computed the photosphere where the optical depth $\tau = 1$ at 43GHz and 690GHz. Numerical results indicate that the direction of azimuthal magnetic field at 690GHz oscillates with period about $1000t_0$. Lower frequency observation at 43GHz can reveal the structure of the outflowing region.

References

Belloni, T., Parolin, I. et al. 2006, MNRAS, 367, 1113
Hayashi, M., R., Shibata, K. & Matsumoto, R. 1996, ApJ, 468, L37

Black Holes from Stars to Galaxies – Across the Range of Masses
Proceedings IAU Symposium No. 238, 2006
V. Karas & G. Matt, eds.

© 2007 International Astronomical Union
doi:10.1017/S1743921307005686

The orbiting spot model gives constraints on the parameters of the supermassive black hole in the Galactic Center

Leonhard Meyer,[1] Andreas Eckart,[1] Rainer Schödel,[1] Michal Dovčiak,[2] Vladimír Karas[2] and Wolfgang J. Duschl[3,4]

[1]I. Physikalisches Institut, Universität zu Köln, Zülpicher Str. 77, 50937 Köln, Germany

[2]Astronomical Institute, Academy of Sciences, Boční II, CZ-14131 Prague, Czech Republic

[3]Institut für Theoretische Physik und Astrophysik, Universität zu Kiel, 24098 Kiel, Germany

[4]Steward Observatory, The University of Arizona, 933 N. Cherry Ave. Tucson, AZ 85721, USA

email: leo@ph1.uni-koeln.de

Abstract. We report on recent polarimetric observations of the 18 ± 3 min quasi-periodicity present in near-infrared flares from Sagittarius A*. Observations in the K-band allow us a detailed investigation of the flares and their interpretation within the hot spot model. The interplay of relativistic effects plays a major role. By simultaneous fitting of the lightcurve fluctuations and the time-variable polarization angle, we give constraints to the parameters of the hot spot model, in particular, the dimensionless spin parameter of the black hole and its inclination. We consider all general relativistic effects that influence the polarization lightcurves. The synchrotron mechanism is most likely responsible for the intrinsic polarization. We consider two different magnetic field configurations as approximations to the complex structure of the magnetic field in the accretion flow. Considering the quality of the fit, we suggest that the spot model is a good description of the origin for the QPOs in NIR flares.

Sagittarius A* (Sgr A*), the supermassive black hole at the Galactic center, exhibits frequent radiation outbursts, so-called flares. Near-infrared (NIR) observations with the ESO VLT in 2005 and 2006 (see Eckart *et al.* 2006b; Meyer *et al.* 2006a,2006b, and the contribution by Andreas Eckart in these proceedings) strongly support the previous evidence of 17 ± 2 min quasi-periodicity (Genzel *et al.* 2003). The emission shows significant polarization during the flares.

These QPOs manifest themselves as individual 'sub-flares'. They are seen in some NIR lightcurves with very short rise and fall timescales, suggesting a compact emission region with a size of only a few Schwarzschild radii (Eckart *et al.* 2006a). The value of ~ 17 min suggest to identify the QPOs with Keplerian motion around a Kerr black hole (BH), as this timescale is comparable to the orbital timescale near the innermost stable circular orbit (ISCO) of a spinning BH (Bardeen *et al.* 1972). Indeed, calculating the emission of confined hot spots in a circular orbit around a BH taking all relativistic effects into account (i.e. red- and blue-shifts, lensing, change of polarization angle) leads to lightcurves consistent with those that are observed (Broderick & Loeb 2005,2006; Dovčiak *et al.* 2004). Adding an underlying truncated disk/ring that accounts for the overall flare can then result in good-quality fits (Meyer *et al.* 2006a,2006b). Here we present constraints on the parameters of the Galactic BH that can be inferred from fitting the hot spot model to the observations. Details of the model and the fitting procedure can be found in Meyer *et al.* (2006a).

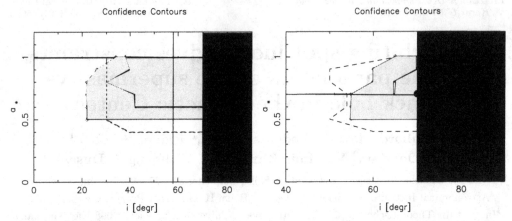

Figure 1. Confidence contours within the plane of black hole angular momentum versus inclination. We assumed two cases with different orientation of the local electric and magnetic fields: firstly, E-vector is kept constant, perpendicular to the disc plane (left), and secondly, a strictly azimuthal B vector is assumed (right). See Meyer *et al.* 2006a for details). The red (green) lines represent the analysis of the 2006 data (Meyer *et al.* 2006b) and are chosen in such a way that the projection onto one of the parameter axes gives the 1σ (3σ) limit for the corresponding parameter. The blue dashed lines indicate the 3σ contour for the 2005 data (Meyer *et al.* 2006a). The χ^2-minimum for the 2006 data is marked by the circle. The χ^2-minimum of the 2005 epoch lies at $a_\star = 1$, $i = 70°$ (left) and $a_\star = 0.5$, $i = 70°$ (right). Our analysis is limited to $i \lesssim 70°$.

Figure 1 shows the confidence contours in the dimensionless spin parameter – inclination plane ($0°$ corresponds to face on). The left panel shows the results that assumed a magnetic field scenario that leads to a constant E-vector along the orbit of the hot spot. The results in the right panel assumed an azimuthal, i.e. toroidal, magnetic field configuration. Following the χ^2 values, the former case leads to slightly better fits. The contours show that a big part of the parameter space can be ruled out already by present data. The inclination i of the blob orbit must be $i \gtrsim 30°$ on a 3σ level, and the dimensionless spin parameter of the black hole is derived to be $a_\star > 0.5$, i.e. the BH is spinning. Future observations will allow to further tighten the possible parameter regions.

While the QPO observations seems to favour the interpretation in terms of orbiting spots, the model of relativistic plasma clouds expanding in a cone-jet geometry can also account for the flares and sub-flares of Sgr A*. The inclusion of both models offers an exciting possibility of connecting the NIR/X-ray activity to the sub-mm/radio regime.

Acknowledgements

L.M. is supported by the International Max Planck Research School (IMPRS) for Radio and Infrared Astronomy at the Universities of Bonn and Cologne. V.K. and M.D. acknowledge support from the Academy of Sciences of the Czech Republic (IAA 300030510).

References

Bardeen, J. M., Press, W. H. & Teukolsky, S. A. 1972, ApJ, 178, 347
Broderick, A. E. & Loeb, A. 2006, MNRAS, 367, 905
Broderick, A. E. & Loebm A. 2005, MNRAS, 363, 353
Dovčiak, M., Karas, V. & Yaqoob, T. 2004, ApJS, 153, 205
Eckart, A., Baganoff, F. K., Schödel, R. *et al.* 2006a, A&A, 450, 535
Eckart, A., Schödel R., Meyer, L. *et al.* 2006b, A&A, 455, 1
Genzel, R., Schödel, R., Ott, T. *et al.* 2003, Nature, 425, 934
Meyer, L., Eckart, A., Schödel, R. *et al.* 2006a, A&A, 460, 15
Meyer, L., Schödel, R., Eckart, A. *et al.* 2006b, A&A, 458, L25

Black Holes from Stars to Galaxies – Across the Range of Masses
Proceedings IAU Symposium No. 238, 2006
V. Karas & G. Matt, eds.

© 2007 International Astronomical Union
doi:10.1017/S1743921307005698

Masses of radiation pressure supported stars in extreme relativistic realm

Abhas Mitra

Theoretical Astrophysics Section, BARC, Mumbai-400085, India
email: abhasmitra@rediffmail.com

Abstract. We discuss that in the extreme relativistic limit, i.e., when $z \gg 1$, where z is the surface gravitational redshift, there could be radiation pressure supported and dominated stars with arbitrary gravitational mass, high or low. Such objects are called Eternally Collapsing Objects (ECOs). ECOs are practically as compact as Schwarzschild black holes (BH) and, observationally, are likely to be mistaken as BHs. Further since any object undergoing continued collapse must first become an ECO before becoming a true BH state characterized by $M = 0$, the observed BH Candidates are ECOs.

Keywords. Gravitation – black holes

1. Introduction

Let the ratio of radiation pressure to gas pressure in a star be $x = p_r/p_g$, and let the ratio of densities of radiation and rest masses be $y = \rho_r/\rho_0$. Then it follows that, within a factor of 2, $y \sim \alpha z$, where $\alpha = L/L_{ed}$ (Mitra 2006a). Here L is the luminosity of the star and L_{ed} is the maximal or Eddington luminosity of the same. It is also seen that α increase with z and saturates to a maximal value of $\alpha \approx 1$. The conventional known "stars" have $z \ll 1$, i.e., they are Newtonian objects. In such a case $x \approx 0.2(M/M_\odot)^{1/2} \mu^{-1}$ where μ is the chemical composition and $M_\odot$ is solar mass (Mitra 2006b). In order to have a "radiation pressure supported star", one must have $x \gg 1$. Then the above equation demands that $M \gg M_\odot$. Such stars are known as (Newtonian) Supermassive Stars.

This picture changes dramatically when we allow for the likely occurrence of extreme relativistic situations with $z \gg 1$ when (Mitra 2006b)

$$x \approx 0.25\, z \left(\frac{M}{M_\odot}\right)^{1/2} \mu^{-1}. \tag{1.1}$$

Because of the presence of z on the RHS, now a radiation supported quasi-static state ($x \gg 1$) can occur for arbitrary values of M. For instance, an object with $M = 10^{10} M_\odot$ with a value of $z = 100$ will have $x \gg 1$ and, on the other hand, a fireball of energy 1 TeV created by two colliding elementary particles in a lab collider with $M = 10^{-43} M_\odot$ could attain a radiation dominated quasi-static state with $z > 10^{25}$! In both the cases, the radius of the object would be practically same as the radius of a supposed BH. Such a large value of z may sound unphysical. However, it is not so because the continued collapse is supposed to proceed to the true BH state with $z = \infty$ (and all finite numbers including 10^{25} are of course infinitely smaller than ∞).

It has been shown that, as physical gravitational collapse must be radiative (Mitra 2006c) and z keeps on increasing relentlessly during continued collapse, quanta of collapse generated photons and neutrinos must eventually move in *almost* closed circular orbits and be quasi-trapped even before an *absolute* capture would occur by the formation of trapped surfaces or Event Horizon (EH). It is this quasi-trapping of radiation which causes $x \gg 1$ and also $y \gg 1$. Since such states correspond to $\alpha \approx 1$, ECOs are formed much before a true BH state ($z = \infty$) would be formed. Since the life time of this relativistic radiation dominated state is $t = 5.10^8 (1 + z)$ yr and $t \to \infty$ as $z \to \infty$ ECOs are *eternally collapsing* (Mitra 2006d).

2. Discussion

As the ECOs evolve, their EOS $\rho \to 3p$ and the gravitational mass $M = \int (\rho - 3p) d\Omega \to 0$ where $d\Omega$ is element of proper volume (Mitra 2006e). Consequently, the eventual BH mass $M = 0$ and thus the observed BHs with $M > 0$, either discussed in this conferences or elsewhere, cannot be true BHs and must be ECOs. This conclusion that true BHs have $M \equiv 0$ has been also obtained by using the basic differential geometry property that proper spacetime 4-volume is coordinate independent (Mitra 2005, 2006f) and non-occurrence of trapped surfaces and EH (Mitra 2002, 2004).

Since astrophysical plasma is endowed with intrinsic magnetic field the super compact ECOs must have strong intrinsic magnetic fields and hence they are also called Magnetospheric ECOs or MECOs. In contrast a true uncharged BH with an EH has no intrinsic magnetic field. Thus MECOs can be observationally distinguished from true BHs by using this physical property. And there are strong circumstantial evidences that the BH Candidates in (i) X-ray binaries (Robertson & Leiter 2002, 2003, 2004), or the (ii) the compact object at the galactic centre (Robertson & Leiter 2006) are MECOs with physical surfaces rather than true BHs. Most importantly, detailed microlensing aided mapping of the central engine of the one of the most well studied quasar has directly revealed that it is a MECO and not a BH (Schild, Leiter & Robertson 2006). The observed magnetic field of MECOs will however be smaller by factor of $(1 + z)$ and could be well below pulsar values. Because of joint effect of gravitational redshift and time dilatation, the observed luminosity of the ECOs would be also also extremely small, $L = 1.3 \times 10^{38} (M/M_\odot) (1 + z)^{-1}$ erg/s even though they are shining at their respective Eddington values. This feeble quiescent ECO radiation could be in millimeter, microwave or radio range and is yet to be detected for any individual case.

Since the ECO formation is a generic effect due to inevitable quasi-trapping of collapse generated radiation, continued collapse does not give rise to any mysterious phase transition resulting in negative pressure. Since Chandrasekhar mass limit pertains to *cold* objects, it is irrelevant for *hot* ECOs whose local mean temperature is reaches the value of $T = 600(M/M_\odot)^{-1/2}$ MeV.

References

Mitra, A. 2006a, MNRAS, 367, L66, gr-qc/0601025
Mitra, A. 2006b, MNRAS, 369, 492, gr-qc/0603055
Mitra, A. 2006c, PRD, 74, 024010, gr-qc/0605066
Mitra, A. 2006d, New Astronomy, 12, 146, astro-ph/0608178
Mitra, A. 2006e, PRD, in press, gr-qc/0607087
Mitra, A. 2006f, Adv. Sp. Res., in press, astro-ph/0510162
Mitra, A. 2005, physics/0504076
Mitra, A. 2002, Found. Phys. Lett., 15, 439, astro-ph/0207056
Mitra, A. 2004, astro-ph/0408323
Robertson, S. L. & Leiter, D. J. 2002, ApJ, 565, 447, astro-ph/0102381
Robertson, S. L. & Leiter, D. J. 2003, ApJL, 596, L203, astro-ph/0310078
Robertson, S. L. & Leiter, D. J. 2004, MNRAS, 350, 1391, astro-ph/0402445
Robertson, S. L. & Leiter, D. J. 2006, astro-ph/0603746
Schild, R. E., Leiter, D. J. & Robertson, S. L. 2006, AJ, 132, 420, astro-ph/0505518

Black Holes from Stars to Galaxies – Across the Range of Masses
Proceedings IAU Symposium No. 238, 2006
V. Karas & G. Matt, eds.
© 2007 International Astronomical Union
doi:10.1017/S1743921307005704

Radiation spectra from MHD simulations of low angular momentum flows

Monika A. Moscibrodzka,[1] **Daniel Proga,**[2] **Bożena Czerny**[1] **and Aneta Siemiginowska**[3]

[1]N. Copernicus Astronomical Center, Bartycka 18, 00-716, Warsaw, Poland [2]University of Nevada, Las Vegas, Las Vegas, United States
[3]Harvard-Smithsonian Center for Astrophysics, 60 Garden Street, Cambridge, MA 02138, USA

email: mmosc@camk.edu.pl

Abstract. We perform Monte Carlo calculation to determine the radiation spectra from magnetized, low angular momentum accretion flow. Magneto-hydrodynamical simulation (with angular momentum parameter lambda of about 2 Rg c) was performed by Proga & Begelman 2003. Because simulation neglect radiative cooling, we compute electron temperature separately, including ohmic heating, parametrized with small δ coefficient, ion-electron coupling, radiative cooling and advection. Radiation spectra are obtained taking into account, thermal synchrotron and bremsstrahlung radiation, self absorption and Comptonization processes.

We show which parts of the flow are responsible for characteristic spectral features and how the spectrum changes in different accretion states. We compare our results with Galactic Center black hole radiation spectra, where low angular momentum accretion is suggested. Accretion state changes seems to be a promising model for the flaring behavior of Galactic Center black hole. We also show the radiation intensity maps in radio and X-rays energy band for viewing angle i=90°.

Keywords. Accretion, accretion disks – magnetic fields – (magnetohydrodynamics:) MHD – Galaxy: center

---

1. Modeling

To perform Monte Carlo simulation, we follow the algorithms given by Pozdnyakov *et al.* 1983 and Gorecki & Wilczewski 1984. We have chosen two characteristic accretion states in late time of evolution of the accretion flow. First one, when evolved torus (extending to the last stable orbit) accrete (T1), and the second one, when the accretion is stopped (T2) due to strong magnetic field that forms a magnetized polar cylinder around the black hole (Fig. 1, on the left). Moments of time in simulation: T1 and T2 are measured in units of Keplerian orbital time at $r = R_B$ (Bondi radius) (for Sgr A* $T2-T1 = 0.02 \times A$ [s], $A = 2\pi R_B / \sqrt{GM/R_B} \sim 0.32$ [year], thus $T2-T1 \sim 2.3$ [days]).

We rescaled density and magnetic field in purpose to test the model in case of Sgr A*. Scaling factor was chosen to fit the mass accretion rate at the outer and inner boundary of the flow (limit for mass accretion rate at inner boundary is superimposed by polarization measurements). Our calculations show that synchrotron photons are created mainly in a magnetized corona of the torus very close to the black hole. Bremsstrahlung radiation comes out from outer regions (Fig. 1, on the right). Because only inner flow changes, the variability can be seen just in synchrotron and Comptonization energy range. We computed broad band radiation spectra (Fig. 2, on the left) and intensity maps for two cases T1 and T2 (in radio and X-ray bands), when the observer is situated near the equator ($i = 90°$) (Fig. 2, on the right).

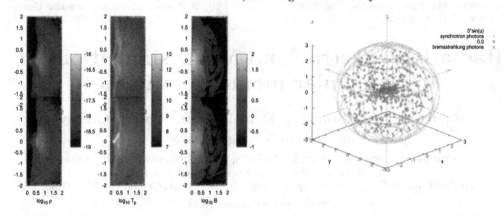

Figure 1. Density,temperature and mean magnetic field maps for two-accretion states T1 (upper panels) and T2 (lower panels) (on the left). 3-D map of synchrotron and bremsstrahlung photons birth place for one of the accretion states (radius is given in $\log R_g$ units) (on the right).

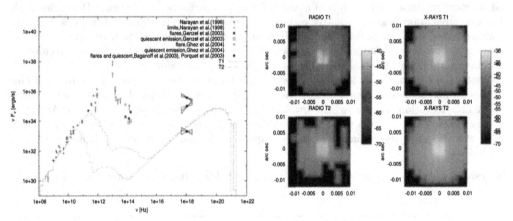

Figure 2. Broadband radiation spectra for two accretion states T1 and T2 calculated for Galactic Center black hole (on the left) Radio (10^{12} Hz) and X-ray(10^{17} Hz) pictures for two accretion states for observer at $i = 90°$ (on the right).

2. The case of Sgr A*

'Quasi-stationary' Sgr A* radiation spectrum cannot be fitted satisfactory by neither of accretion states due to a weak magnetic field. Changes in radio-IR level of radiation can indicate that similar process can be responsible for radio-IR flaring in Sgr A*, however emission level, predicted by the model, is much too weak at these energies. Low energy radio tail is fitted very well to the observed points. At higher energies (X-ray band) the spectrum does not vary because bremsstrahlung photons are born mostly in outer parts of the flow. X-ray flaring must be created by non-thermal process, or another accretion state (see Proga & Begelman 2003).

References

Gorecki, A. & Wilczewski, W. 1984, AcA, 34, 141
Pozdnyakov, L. A., Sobol, I. M. & Syunyaev, R. A. 1983, ASPR, 2, 189
Proga, D. & Begelman, M. C. 2003, ApJ, 592, 767

Black Holes from Stars to Galaxies – Across the Range of Masses
Proceedings IAU Symposium No. 238, 2006
V. Karas & G. Matt, eds.
© 2007 International Astronomical Union
doi:10.1017/S1743921307005716

The K-correction for LMXBs: an application to X1822-371 and GX339-4

T. Muñoz-Darias,[1] J. Casares[1] and I. G. Martínez-Pais[1,2]

[1]Instituto de Astrofísica de Canarias, 38200, La Laguna, Tenerife, Spain
[2]Departamento de Astrofísica, Universidad de La Laguna, 38206 La Laguna, Tenerife, Spain

Abstract. We model the K-correction for emission lines formed on the irradiated face of companion stars in compact binaries. We compute this K-correction in a general approach as function of the mass ratio and the disc flaring angle. Our results, combined with the detection of high excitation emission lines arising from the donor star, can be used to set constraints to the masses of neutron stars and black holes in persistent Low Mass X-ray Binaries (LMXBs).

The application of our model to X1822-371 (Muñoz-Darias, Casares & Martínez-Pais 2005) lends strong support to the presence of a massive neutron star in this LMXB (i.e. $M_{NS} > 1.6 M_\odot$). Here, we also present the K-correction for the Black Hole binary GX339-4, where we obtain a solid lower limit to the black hole mass of $M_{BH} > 5.8 M_\odot$.

Keywords. Accretion, accretion disks – binaries: close – X-rays: stars

1. The K-Correction for LMXBs

Low mass X-ray binaries (LMXBs) are interacting binaries where a low mass donor star transfers mass onto a Neutron Star (NS) or Black Hole (BH). If the accretion rate is high enough the X-ray luminosity remains above $\sim 10^{36}$ erg s^{-1} resulting in persistent systems. In these sources the companion star is usually undetectable and thus dynamical studies have been restricted to transients where the companion dominates the optical spectrum of the source. The situation has recently changed thanks to the discovery of narrow emission components within the Bowen blend at $\lambda\lambda$ 4634-50 arising from the donor in the prototypical LMXB Sco X-1 (Steeghs & Casares 2002). Fortunately, this property is not peculiar to Sco X-1 and this technique has successfully been applied also to others LMXBs (e. g. X1822-371: Casares *et al.* 2003, GX339-4: Hynes *et al.* 2003; hereafter H03, V801 Ara and V926 Sco: Casares *et al.* 2006). However, these high energy emission lines are excited on the inner hemisphere of the donor and only provide a lower limit to the true K_2 velocity which corresponds to the center of mass of the companion. In Muñoz-Darias *et al.* (2005; hereafter MCM05) we model the deviation between the reprocessed light center and the center of mass of a Roche lobe-filling star (i.e. the so-called K-correction). To do this, we integrate a Gaussian profile over the Roche lobe geometry which is divided in $\sim 10^5$ resolution elements. We find that the K-correction depends on:

(i) Mass ratio, $q = M_2/M_1$;
(ii) Disc flaring angle, α, which obscures the companion;
(iii) Orbital inclination, i.

Therefore, we have computed the K-correction as function of q and α for the cases of low and high orbital inclination. The K-correction is well fitted by fourth-order polynomials and we report the coefficients of the fits in MCM05.

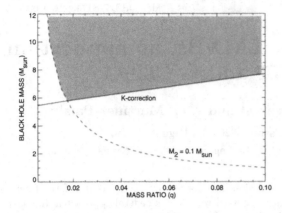

Figure 1. Black Hole mass determination in GX339-4. Dashed line represents the constraint $M_2 = 0.1 M_\odot$. Solid line shows the lower limit to M_{BH} obtained through the K-correction by using $\alpha = \alpha_M$. The possible BH mass range is restricted by the grey region.

2. The mass of the neutron star in X1822-371

To date, the accretion disc corona X1822-371 is the best candidate to apply the K-correction since we know with high accuracy the following system parameters:

(i) Orbital period = 5.57 hr (Hellier & Mason 1989);

(ii) Orbital inclination = $82.5 \pm 1.5°$ (Heinz & Nowak 2001);

(iii) Radial velocity of the NS, $K_1 = 94.5 \pm 0.5$ km s^{-1} (Jonker & van der Klis 2001);

(iv) α constrained between 12–14° (Hellier & Mason 1989, Heinz & Nowak 2001);

(v) Doppler tomography of the NIII emission line at $\lambda 4640$ reveals a compact spot at the position of the companion with $K_{em} = 300 \pm 15$ km s^{-1} (Casares et al. 2003).

As we show in MCM05, the application of the K-correction strongly suggest that X1822-371 harbours a heavy neutron star of 1.6–2.3$M_\odot$.

3. The Black Hole binary GX339-4

GX339-4 was proposed as a BH candidate by Samini et al. (1979). However, this transient LMXBs has a high level of X-ray activity and the features of the companion are undetectable even during quiescent epochs. Only during the 2002 outburst H03 discovered narrow NIII emission likely arising from the companion star. These authors obtained a best fit orbital period of $P_{orb} = 1.7557 \pm 0.0004$ days and measured a NIII radial velocity of $K_{em} = 317 \pm 10$ km s^{-1}. Moreover, the small motion in the wings of the HeII $\lambda 4686$ emission line suggests $q \lesssim 0.08$. Here, we apply the K-correction for $0.05 < q < 0.1$ by assuming $\alpha = \alpha_M$ for each value of q, where α_M is the highest value of the flaring angle for the companion to be irradiated. The absence of X-ray eclipses sets an upper limit to the orbital inclination for each q. We use these constraints to derive a lower limit to the mass of the BH. Assuming that $M_2 \geqslant 0.1 M_\odot$ we obtain $M_{BH} > 5.8 M_\odot$ (see figure 1).

References

Casares, J., Steeghs, D., Hynes, R. I., Charles, P. A. & O'Brien K. 2003, ApJ, 590, 1041

Casares, J., Cornelisse, R., Steeghs, D., Charles, P. A., Hynes, R. I., O'Brien, K. & Strohmayer, T. E. 2006, MNRAS, 373, 1235

Hynes, R. I., Steeghs, D., Casares, J., Charles, P. A. & O'Brien, K. 2003, ApJ, 583, L95

Heinz, S. & Nowak, M. A. 2001, MNRAS, 320, 249

Hellier, C. & Mason, K. O. 1989, MNRAS, 239, 715

Jonker, P. G. & van der Klis, M. 2001, ApJ, 553, L43

Muñoz-Darias, T., Casares, J. & Martinez-Pais, I. G. 2005, ApJ, 635, 502

Samimi, J. et al. 1979, Nature, 278, 434

Steeghs, D. & Casares, J. 2002, ApJ, 568, 273

Black Holes from Stars to Galaxies – Across the Range of Masses
Proceedings IAU Symposium No. 238, 2006
V. Karas & G. Matt, eds.
© 2007 International Astronomical Union
doi:10.1017/S1743921307005728

Proper motions of thin filaments at the Galactic Center

K. Mužić,[1] A. Eckart,[1,2] R. Schödel,[1] L. Meyer[1] and A. Zensus[2,1]

[1] I. Physikalisches Institut, Universität zu Köln, Zülpicher Str. 77, 50937 Köln, Germany

[2] Max-Planck-Institut für Radioastronomie, Auf dem Hügel 69, 53121 Bonn, Germany

email: muzic@ph1.uni-koeln.de

Abstract. We present the proper motion study of the thin filaments observed in L'-band (3.8 μm) adaptive optics images of the central parsec of the Milky Way. Observed filaments are associated with the mini-spiral and, in some cases, with stars. They can be interpreted as shock fronts formed by the interaction of a central wind with the mini-spiral or extended dusty stellar envelopes.

Keywords. Galaxy: center – infrared: ISM

1. Introduction

The central cavity of the Galactic Center is surrounded by a dense clumpy molecular ring (also called circum-nuclear disk, CND) of warm dust (Zylka *et al.* 1995) and neutral gas (e.g., Güsten *et al.* 1987). The central cavity itself has a much lower mean gas density and contains the so called mini-spiral, which consists of mostly ionized gas and dust, and connects the CND to the center of the stellar cluster. Adaptive optics L'-band (3.8μm; Fig. 1, left) images of the central parsec of the Milky Way show large amounts of gas and dust belonging to the mini-spiral. Additionally, one can distinguish between a large number of thin filaments (some have already been reported by Clénet *et al.* 2004) Here we identify them and present proper motion measurements. We argue that they are most likely shock fronts formed in interaction of the strong winds with an ambient ISM.

Observations were performed using the NAOS/CONICA adaptive optics system at the ESO VLT. The data set includes L'-band images from 5 epochs (2002 to 2006). Data reduction and image transformations were performed using the DPUSER software for astronomical image analysis (T. Ott) and IDL routines. Proper motions of the thin filaments were determined using the cross-correlation technique.

2. Identification and proper motions of the thin filaments

In this report we present only the thin filaments for which we were able to determine the proper motions. The results obtained in our study are given in Table 1. and shown in Fig. 1. Filaments labeled NE are located along the inner rim of the northern arm and are curved with their convex sides eastwards, same as the stellar bow-shock X1. Filaments to the west and southwest of Sgr A* are curved with their convex sides westward and are positioned almost perpendicular to the Bar (those labeled SW). Feature X7 shows a cometary structure and points approximately to Sgr A*.

Our results support the model of a collimated outflow originating from the disk of young mass-losing stars around Sgr A* (e.g., Paumard *et al.* 2006) or, alternatively, from the massive black hole itself (for more details see Mužić *et al.* 2006).

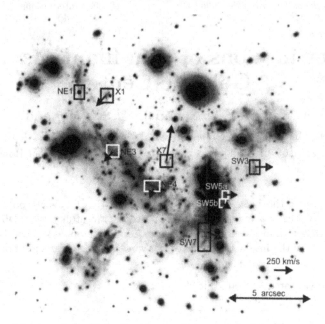

Figure 1. L'-band image of the Galactic Center. Boxes mark thin filaments with measurable proper motions. Note that boxes in this image are larger than those used for measurements (stated in Tab. 1). Arrows show proper motions of the thin filaments obtained in our study. Insignificant motion of the Northern Arm filament NE1 is marked with a circle.

Table 1. Proper motions of the thin filaments

Feature	$\Delta\alpha$ (arcsec)	$\Delta\delta$ † (arcsec)	size ($\Delta x \times \Delta y$) (arcsec×arcsec)	v ‡ (km/s)	Δv (km/s)
NE1	4.76	3.42	0.32×0.46	-29	16
NE3	2.62	-0.09	0.38×0.41	167	25
NE4	0.32	-1.95	0.43×0.43	130	18
SW3	-6.08	-0.77	0.35×0.35	-229	19
SW5a	-4.20	-2.31	0.35×0.11	-165	15
SW5b	-4.13	-2.75	0.16×0.24	-123	22
SW7 (R.A.)	-3.07	-5.10	0.73×1.22	-186	17
SW7 (Dec.)	-3.07	-5.10	0.73×1.22	-80	21
X1 (R.A.)	3.22	3.40	0.30×0.38	113	16
X1 (Dec)	3.22	3.40	0.30×0.38	-78	21
X7 (R.A.)	-0.56	0.55	0.54×0.41	-71	17
X7 (Dec)	-0.56	0.55	0.54×0.41	458	21

† Relative to Sgr A*; position of the center of the box
‡ Velocities of all NE and SW features, except that of SW7, are in a direction perpendicular to the shock feature. Other proper motions are given in both R.A. and Dec. direction.

References

Clénet, Y., Rouan, D., Gendron, E. *et al.* 2004, A&A, 417, L15
Güsten, R., Genzel, R., Wright, M. C. H. *et al.* 1987, ApJ, 318, 124
Mužić, K., Eckart, A., Schödel, R. *et al.* 2006, submitted to A&A
Paumard, T., Genzel, R., Martins, F. *et al.* 2006, ApJ, 643, 1011
Zylka, R., Mezger, P. G., Ward-Thompson, D. *et al.* 1995, A&A, 297, 83

Black Holes from Stars to Galaxies – Across the Range of Masses
Proceedings IAU Symposium No. 238, 2006
V. Karas & G. Matt, eds.

Two-dimensional MHD simulations of accretion disk evaporation

Kenji E. Nakamura

Department of Sciences, Matsue National College of Technology
14-4, Nishiikuma, Matsue, Shimane, 690-8518, Japan
email: nakamrkn@matsue-ct.jp

Abstract. We simulate the accretion disk evaporation to study the changes of accretion disk structure during the state transition from the soft state to the hard state. We performed 2 dimensional MHD simulations by including the heat conduction process. We assume the axisymmetric accretion and put a cold rotating gas torus in a hot halo in hydrostatic equilibrium initially. Weak magnetic fields are threaded vertically. Heat conduction equation and MHD equations are solved separately according to the time splitting method. We obtained the result that accretion disk is heated by the hot corona and the hot gas evaporates from the accretion disk surface. We found that magnetic fields lines bended by disk rotation restrict the energy transport vertically and make disk evaporation ineffective.

Keywords. Accretion, accretion disks – black hole physics – X-rays: binaries

1. Introduction

The accretion disk evaporation model (Meyer-Hofmeister & Meyer 1999) is one of attractive models which can explain the state changes of black hole accretion flows. The hot corona heats the surface of cold accretion disk. The gas evaporates from the disk into the corona. Geometrically thin accretion disks change to geometrically thick hot accretion flows. The evaporation model can explain that the X-ray spectrum from accretion flows changes from the soft state to the hard state. We simulate the accretion disk evaporation by taking account of a heat conduction process to study the state transitions.

2. Numerical Models

Basic equations. We solve resistive MHD equations and heat conduction equation due to the time splitting method (Hayashi, Shibata, Matsumoto 1996, Yokoyama, Shibata 1997). We adopt the cylindrical coordinate and assume the axial symmetry.

The heat conduction equation is written as

$$\frac{1}{\gamma - 1}\frac{k}{m}\rho\frac{dT}{dt} = \nabla \cdot \left(\kappa_0 T^{5/2} \frac{\boldsymbol{B}}{B^2}(\boldsymbol{B} \cdot \nabla) T \right). \tag{2.1}$$

Energy is transfered along to the magnetic fields lines. We adopt the heat conductivity, $\kappa_0 = 6.0 \times 10^{-6} [\text{g cm s}^{-3} \text{ K}^{-7/2}]$. Other symbols have their usual meanings.

Conditions. An isothermal halo is in hydrostatic equilibrium initially. A rotating gas torus is dynamically balanced (Abramowicz *et al.* 1978). Magnetic fields are vertically threaded. Plasma β is 100. Black hole mass is 10 solar mass. An absorbing inner boundary condition is imposed.

Table 1. Units

Quantity	Symbol	Numerical unit
Length	r_0	3.0×10^{10} cm
Density	ρ_0	8.3×10^{-7} gcm^{-3}
Time	t_0	1.4×10^2 s
Temperature	T_0	5.4×10^8 K

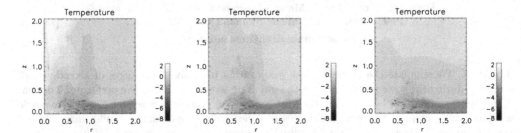

Figure 1. Temperature $\log(T/T_0)$ at $t = 8.0\,t_0$. Left: No heat conduction. Middle: Anisotropic heat conduction. Right: Isotropic heat conduction.

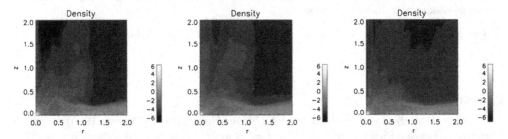

Figure 2. Density $\log(\rho/\rho_0)$ at $t = 8.0\,t_0$. Left: No heat conduction. Middle: Anisotropic heat conduction. Right: Isotropic heat conduction.

3. Results

Anisotropic conduction. Thermal energy is transfered mildly from a corona to a disk along to the magnetic field lines. The azimuthal component of magnetic fields become large by differential rotation, therefore the vertical energy transfer from a corona to a disk become inefficient due to the magnetic field line bending. The flow evolves almost same as the case of no conduction.

Isotropic conduction. Energy transfer from a hot corona to a cool disk continues regardless of the disk evolution. The structure of corona and accretion disk change wholly in short time scale. A corona become cool and dense and restrains the jet formation.

References

Abramowicz, M., Jaroszynski, M. & Sikora, M. 1978, A&A, 63, 221
Hayashi, M. R., Shibata, K. & Matsumoto, R. 1996, ApJ, 468, L37
Yokoyama, T. & Shibata, K. 1997, ApJ, 474, L61
Meyer-Hofmeister, E. & Meyer, F. 1999, A&A, 348, 154

Black Holes from Stars to Galaxies – Across the Range of Masses
Proceedings IAU Symposium No. 238, 2006
V. Karas & G. Matt, eds.
© 2007 International Astronomical Union
doi:10.1017/S1743921307005741

Long-term and intra-day variability of BL Lacertae since the last great outburst

Maria G. Nikolashvili and Omar M. Kurtanidze

Abastumani Observatory, 383792 Abastumani, Georgia

Abstract. We present the results of optical photometry of BL Lacertae carried out using ST-6 CCD camera attached to the Newtonian focus of the 70 cm meniscus telescope of Abastumani Observatory. On the basis of observations conducted since August 1997 during more than 550 nights about 17 000 frames were collected. They have been reduced using Daophot II.

It has been shown that optical variability of BL Lacertae is very complex. The maximum variation was observed at long-term scale and is equal to equals to 3.0 mag (rms=0.03) in B band, while the variation in V and R bands are within 2.71 mag (0.02) and 2.53 mag (0.01), respectively. This means that variations are larger at shorter wavelength or the object become bluer in the active phase. It were also demonstrated that BL Lacertae shows intra-day variability within 0.30 mag (0.02), while intra-hour variability within 0.10 mag (0.01) magnitudes.

Keywords. BL Lacertae objects – CCD photometry – variability

1. Introduction

BL Lacertae is a prototype of one of the most extreme subclass of AGNs. It was discovered in 1929 by Cuno Hoffmeister, who found it to vary by more than a factor two in one week and classified it as a short period variable star (Hoffmeister 1990). Since its identification as an extragalactic source it was the subject of numerous studies in many frequency bands. Historically, BL Lac is known to show 5.0 mag variation in optical band with episodic outbursts (Maessano 1997). Maximum variation in the infrared K band is 3.0 mag (Fan 1999).

During the summer 1997 outburst BL Lac exhibited a very strong activity (Clements 2001). The strong activity was also detected in the radio, X-ray and gamma-ray bands (Sambruna 1999; Madejski 1999; Tanihata 2000; Bottcher 2000).

2. Observation and data reduction

We have been intensively monitoring BL Lacertae at Abastumani Observatory since 1997, when it remained in a high state for more than two months. Rapid and large amplitude flux variations characterized the source during this period. Here we present observations carried out from August 1997 to November 2001. All observations were carried out with 70-cm meniscus telescope and CCD camera ST-6 attached to the Newtonian focus (1/3). To study the long-term variability we observed BL Lacertae during 317 nights, collected 320 frames in each of the BVI bands, and 465 frames in the R band. More than 16 000 frames were obtained in R band during 259 nights to study intra-day variability (Wagner 2001) as well as intra-hour variability (Miller 1989). The duration of observational runs varied from two hours to six hours. The exposure times varied from 60 to 180 sec depending on the brightness of the object and the filter used. Instrumental differential magnitudes were calculated relative comparison stars C and H, that have nearly the same colours as the object under study (Smith 1985). The images were reduced using

Daophot-II (Stetson 1987). The highest differential photometric accuracy reached in R band is 0.005 (rms) magnitude at $m(r) = 14.00$ mag during 180 sec. Magnitudes were calculated relative comparison stars C and H, that have nearly the same colours as the object under study. To eliminate the effects of seeing induced spurious IDV and IHV (Cellone 2000) the apertures are taken to include the whole host galaxy.

3. Results and conclusions

The constructed long-term variability lightcurves have shown that maximum B-band variation was recorded during August 1997 and it amounts to 3.0 mag (rms=0.03), while the maximum amplitude in R band equals 2.53 mag (0.01). The amplitudes of variation in R band of the other observing seasons are presented in the table. The results of optical observations of BL Lacertae during great summer 1997 outburst have been presented by different blazar monitoring groups (Speziali 2000; Fan 2001; Clements 2001; Villata 2002, Villata 2004). On the basis of observations of BL Lacertae during the period from August 1997 to November 2001 it was clearly demonstrated that variations are larger in B band or the object become bluer in the active phase (Nikolashvili 1999; Kurtanidze 2001; Nesci 2001), that were also confirmed by other groups (Clements 2001; Fan 2001).

The significant statistical evidences of intra-day and intra-hour variabilities are found during many nights of observations. The typical intra-day and intra-hour variability amplitudes in R band are within 0.30 (0.02) and 0.10 (0.01) magnitudes, respectively. Detailed study of BL Lacertae during multi-wavelength campains carried out in the frame of WEBT collaboration are most extensive study of this prototype source ever conducted (Villata 2004).

Acknowledgements

O.M.K. gratefully acknowledges invaluable financial support and the kind collaboration with Dr. G. M. Richter without which this Programme would have never been conducted.

References

Bottcher, M. & Bloom, S. D. 2000, AJ, 119, 469
Celone, S. A. et al. 2000, AJ, 121, 90
Clements, S. D. & Carini, M. T. 2001, AJ, 119, 534
Fan, J. H. et al. 1999, AAS, 133, 217
Fan, J. H. et al. 2001, A&A, 369, 758
Hoffmeister, G. et al. 1990, Veranderlichte Sterne, Leipzig
Kurtanidze, O. M. et al. 2001, in: S. Ritz et al. (eds.), 6th Compton Symp. (AIP CP), 587, 333
Madejski, G. M. et al. 1999, ApJ, 521, 145
Maessano, M.,et al. 1997, AAS, 122, 267
Miller, H. R. et al. 1995, Nature, 337, 627
Nikolashvili, M. G. et al. 1999, in: Blazar Monitoring Towards the 3rd millenium, C. M. Raiteri, et al. (eds.), (Osserv. Astron. di Torino: Torino), p. 33
Sambruna, R. M. et al. 1999, ApJ, 515, 140
Smith, P. S. et al. 1985, AJ, 90, 1184
Speziali, R. & Natali, G. 2000, AAS, 339, 382
Stetson, P. B. 1987, PASP, 99, 191
Tanihata, S. et al. 2000, ApJ, 543, 124
Villata, M. et al. 2002, A&A, 390, 407
Villata, M. et al. 2004, A&A, 424, 497
Wagner, S. J. 2001, ASP CS, 227, 112

Black Holes from Stars to Galaxies – Across the Range of Masses
Proceedings IAU Symposium No. 238, 2006
V. Karas & G. Matt, eds.
© 2007 International Astronomical Union
doi:10.1017/S1743921307005753

Steady models of optically thin, magnetically supported two-temperature accretion disks around a black hole

H. Oda,[1] K. E. Nakamura,[2] M. Machida[3] and R. Matsumoto[4]

[1]Graduate School of Science and Technology, Chiba University, 1-33 Yayoi-Cho, Inage-ku, Chiba 263-8522, Japan

[2]Department of Sciences, Matsue National College of Technology, 14-4 Nishiikuma-Cho, Matsue, Shimane 690-8518, Japan

[3]Division of Theoretical Astronomy, National Astronomical Observatory of Japan, 2-21-1 Osawa, Mitaka, Tokyo 181-8588, Japan

[4]Department of Physics, Faculty of Science, Chiba University, 1-33 Yayoi-Cho, Inage-ku, Chiba 263-8522, Japan

email: oda@astro.s.chiba-u.ac.jp, nakamrkn@matsue-ct.jp, mami@th.nao.ac.jp, matumoto@astro.s.chiba-u.ac.jp

Abstract. We obtained steady solutions of optically thin two-temperature magnetized accretion disks around a black hole. We included relativistic bremsstrahlung cooling, synchrotron cooling and inverse Compton effects and assumed that the disk is threaded by toroidal magnetic field. We found that a magnetic pressure dominated new branch, which we call a 'low-β branch', appears in the thermal equilibrium curves. The luminosity of the optically thin, magnetically supported disk can exceed 10% of the Eddington luminosity ($0.1L_{\rm Edd}$).

Keywords. Accretion: accretion disks – black holes – state transitions – MHD

1. Introduction

X-ray hard state (low/hard state) of black hole candidates (BHCs) can be modeled by optically thin advection dominated accretion flows (ADAFs). Abramowicz *et al.* (1995) showed that when the accretion rate $\dot{M}$ exceeds a critical accretion rate $\dot{M}_{\rm c}$, optically thin solution disappears. Above this critical accretion rate, the hard X-ray emitting optically thin hot accretion disk undergoes a transition to an optically thick cold disk which emit soft X-rays. RXTE observations of galactic BHCs revealed that some BHCs (e.g., XTE J1550-564) stay in X-ray hard state even when $L > 0.1L_{\rm Edd}$ (e.g., Gierliński, Newton 2006). This luminosity is higher than the critical luminosity above which thermal instability takes place. Machida *et al.* (2006) carried out global 3D MHD simulations of this cooling instability and showed that magnetically supported disk is created.

2. Models

We extended the basic equations for 1D steady, axisymmetric, optically thin black hole accretion flows (e.g., Kato *et al.* 1998) by incorporating the azimuthal magnetic field. General relativistic effects are simulated by using pseudo-Newtonian potential (Paczyńsky & Wiita 1980). We prescribe the vertically integrated $\varpi\varphi$-component of the stress tensor based on the results of local and global 3D MHD simulations (e.g., Machida *et al.* 2006; Hirose *et al.* 2006) as $T_{\varpi\varphi} = -\alpha_{\rm B}W$ and we set $\alpha_{\rm B} = 0.05$. Machida *et al.* (2006) showed that $\int_{-\infty}^{\infty} v_{\varpi}\langle B_{\varphi}\rangle dz \propto \varpi^{-1}$ where $\langle B_{\varphi}\rangle$ is the azimuthally averaged

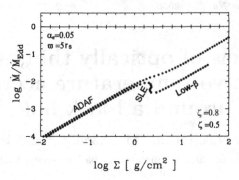

Figure 1. Local thermal equilibrium curves of optically thin disks at $\varpi = 5r_s$ when $M = 10M_\odot$, $\alpha_B = 0.05$ and $\zeta = 0.5$ (black) and $\zeta = 0.8$ (red). We obtained a low-β branch in addition to the ADAF and SLE branches. $\dot{M}_{Edd} = 4\pi GM/(\eta_e \kappa_{es} c)$ is the Eddington accretion rate where η_e is the energy conversion efficiency taken to be $\eta_e = 0.1$.

toroidal field. Here we assume the radial advection rate of the toroidal magnetic field $\dot{\Phi}$ as

$$\dot{\Phi} = -\int_{-\infty}^{\infty} v_\varpi \langle B_\varphi \rangle dz = -v_\varpi B_0(\varpi)\sqrt{4\pi}H \equiv \dot{\Phi}_{out}\left(\frac{\varpi}{\varpi_{out}}\right)^{-\zeta}$$

where $B_0(\varpi) = 2^{5/4}\pi^{1/4}(\Re T/\mu)^{1/2}\Sigma^{1/2}H^{-1/2}\beta^{-1/2}$ is the strength of equatorial toroidal magnetic field and $\dot{\Phi}_{out}$ is the radial advection rate of the toroidal magnetic field at the outer boundary.

3. Results

We obtained local thermal equilibrium curves of optically thin, two-temperature black hole accretion flows at $\varpi = 5r_s$ (figure 1). We found a low-β branch in which the magnetic heating balances with the radiative cooling. The low-β branch exists even when $\dot{M} \sim 0.5\dot{M}_{Edd}$ and explains the luminous, X-ray hard state in BHCs observed during the transition from a low/hard state to a high/soft state. Since $T_e < 10^9$K when $\dot{M} > \dot{M}_{Edd}$ in low-β branch, the power-law X-ray spectrum in such disks becomes steeper and can explain the steep power law spectra observed in very high/steep power law state.

Acknowledgements

The authors thank Drs. M.A. Abramowicz, C. Akizuki, C. Done, J. Fukue, S. Hirose, Y. Kato, S. Mineshige, K. Ohsuga, R.A. Remillard, and K. Watarai for their valuable comments and discussions. This work was supported in part by the Grants-in-Aid for Scientific Research of the Ministry of Education, Culture, Sports, Science and Technology (RM: 17030003), Japan Society for the Promotion of Science (JSPS) Research Fellowships for Young Scientists (MM: 17-1907, 18-1907).

References

Abramowicz, M. A., Chen, X., Kato, S., Lasota, J.-P. & Regev, O. 1995, ApJ, 438, L37
Gierliński, M. & Newton, J. 2006, MNRAS, 370, 837
Hirose, S., Krolik, J. H. & Stone, J. M. 2006, ApJ, 640, 901
Kato, S., Fukue, J. & Mineshige, S. 1998, Black hole accretion disks (Kyoto: Kyoto University Press)
Machida, M., Nakamura, K. E. & Matsumoto, R. 2006, PASJ, 58, 193
Paczyński, B. & Wiita, P. J. 1980, A&A, 88, 23

Black Holes from Stars to Galaxies – Across the Range of Masses
Proceedings IAU Symposium No. 238, 2006   © 2007 International Astronomical Union
V. Karas & G. Matt, eds.   doi:10.1017/S1743921307005765

QPOs expected in rotating accretion flows around a supermassive black hole

T. Okuda,[1] V. Teresi[2] and D. Molteni[2]

[1]Hokkaido University of Education, Hachiman-Cho 1-2, 040-0429 Hakodate, Japan

[2]Dipartimento di Fisica e Tecnologie Relative, Università di Palermo, Viale delle Scienza, 90128 Palermo, Italy

email: okuda@cc.hokkyodai.ac.jp, molteni@unipa.it

Abstract. It is well known that rotating inviscid accretion flows with adequate injection parameters around black holes could form shock waves close to the black holes, after the flow passes through the outer sonic point and can be virtually stopped by the centrifugal force. We numerically examine such shock waves in 2D accretion flows with 10^{-5} to 10^{6} Eddington critical accretion rates around a supermassive black hole with $10^{6} M_{\odot}$. As the results, the luminosities show QPO phenomena with modulations of a factor 2–3 and with quasi-periods of a few to several hours.

Keywords. Accretion, accretion disks – black hole physics – galaxies: active – hydrodynamics – radiation mechanisms: thermal – shock waves

---

1. Introduction

Since the pioneer works of transonic problems of accretion and wind by Fukue(1987) and Chakrabarti and his collaborators (Chakrabarti 1989, Abramowicz & Chakrabarti 1990, and Chakrabarti & Molteni 1993), it has been shown that these generalized accretion flows could be responsible for the quasi-periodic oscillation (QPO) from the black hole candidates (Molteni et al. 1996.; Ryu et al. 1997.; Lanzafame et al. 1998.). Here, following the recent numerical 2D simulations of the shocks (Okuda et al. 2004.; Chakrabarti et al. 2004.), we examine the QPOs phenomena due to the centrifugally supported shocks around a supermassive black hole with $10^{6} M_{\odot}$, while taking account of the cooling and heating of the gas and the radiation transport. .

2. Model Parameters and Numerical Methods

For the central black hole, we consider a supermassive black hole with $10^{6} M_{\odot}$. The basic equations and the numerical methods used here are given in Okuda et al.(2004). A typical set of injection parameters, such as the specific angular momentum, λ_{out}, the radial velocity v_{out}, the sound velocity a_{out}, the ambient density ρ_{out}, and the accretion rate $\dot{m}$ normalized to the Eddington critical accretion rate $\dot{M}_{E}(= 2.7 \times 10^{23} \text{ g s}^{-1})$ at an outer boundary radius R_{out}, are given in table 1. Here, the velocities and distances are

Table 1. Injection flow parameters

λ_{out}	v_{out}	a_{out}	$\rho_{out}(\text{g cm}^{-3})$	$\dot{m}$	ϕ	R_{out}/R_{g}
1.875	0.0751	0.0654	$10^{-17} - 10^{-6}$	$10^{-5} - 10^{6}$	29°	30.0

given in units of the speed of light, c, and the Schwarzschild radius, R_g, respectively. ϕ is the subtended angle of the central black hole to the initial disk at $r = R_{out}$, that is, $\tan \phi = (h/r)_{out}$, where h is the inflow thickness.

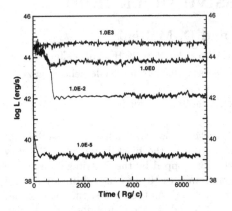

Figure 1. Luminosity L as a function of time in units of R_g/c for cases of $\dot{m} = 10^{-5}, 10^{-2}$, 1, and 10^3.

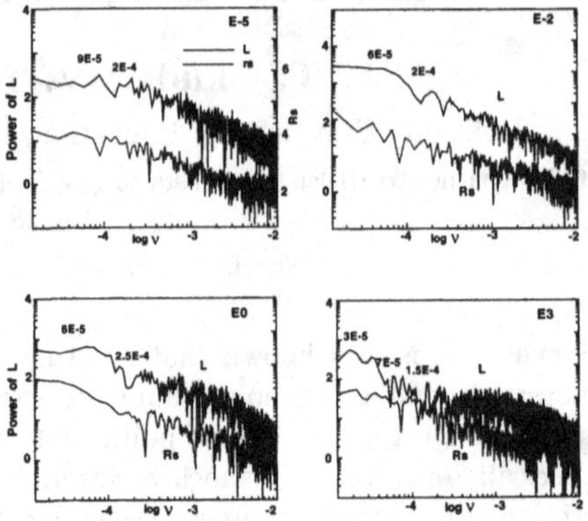

Figure 2. Power density spectra of luminosity L (solid line) and shock positions R_s (dotted line) for cases of figure 1.

3. Results

In all of the above models, the centrifugally supported shocks are formed at the region of $5 - 12\ R_g$ around the black hole, depending weakly on the accretion rate, and oscillate quasi-periodically around the shock position. Figure 1 shows the luminosity curves for cases of $\dot{m} = 10^{-5}, 10^{-2}, 1$, and 10^3, respectively. The luminosity increases in proportion to the accretion rate when it is low, but it tends to a saturated value of ~ 3 Eddington luminosity when it exceeds considerably the Eddington critical rate. The power density spectra for the luminosity and the shock position on the equatorial plane are given in figure 2. Here we find the QPO frequency $\nu_{qpo} \sim 3 \times 10^{-5} - 3 \times 10^{-4}$, that is, the period $P \sim 1 - 10$ hours. The luminosities show QPO behaviour with modulations of a factor $2 - 3$ and with quasi-periods of a few to several hours. These results may suggest the existence of QPOs with the time scale of hours in AGNs .

References

Fukue, J. 1987, PASJ, 39, 309

Chakrabarti, S. K. 1989, ApJ, 347, 365

Abramowicz, M. A. & Chakrabarti, S. K. 1990, ApJ, 350, 281

Chakrabarti, S. K. & Molteni, D. 1993, ApJ, 417, 671

Molteni, D., Sponholz, H. & Chakrabati, S. K. 1996, ApJ, 457, 805

Ryu, D., Chakrabarti, S. K. & Molteni, D. 1997, ApJ, 474, 378

Lanzafame, G., Molteni, D. & Chakrabarti, S. K. 1998,MNRAS, 299, 799

Okuda, T., Teresi, V. & Molteni, D. 2004, PASJ, 56, 547

Chakrabarti, S. K., Acharyya, K. & Molteni, D. 2004, A&A, 421, 1

Black Holes from Stars to Galaxies – Across the Range of Masses
Proceedings IAU Symposium No. 238, 2006
V. Karas & G. Matt, eds.
© 2007 International Astronomical Union
doi:10.1017/S1743921307005777

Power spectra from spotted accretion discs

Tomáš Pecháček,[1,2] Michal Dovčiak[2] and Vladimír Karas[2]

[1]Charles University, Faculty of Mathematics and Physics, Prague, Czech Republic
[2]Astronomical Institute, Academy of Sciences, Prague, Czech Republic
email: pechacek_t@seznam.cz

Abstract. Some aspects of power-spectral densities (PSD) of active galactic nuclei are similar to those of galactic black hole X-ray binary systems (McHardy *et al.* 2005). The signal originates near a black hole and its modulation by general-relativistic effects should be taken into account (Życki & Nedźwiecki 2005). We modified the previous calculations of these effects, assuming a model of spots which occur on the disc surface and decay with a certain lifetime.

Keywords. Accretion disks, black hole physics, methods: analytical, X-rays: general

1. Introduction

Let us consider an infinitesimal surface element, $r \, dr \, d\phi$, residing a thin accretion disc near a non-rotating black hole. This element of area is assumed to be orbiting together with the disc material with Keplerian orbital frequency, $\Omega \equiv \Omega(r)$. The redshift factor $g(\phi, r, \theta_o)$ (which defines the change of photon energy ν_o/ν_s with an observer inclination θ_o) and the total flux received by a detector $F(\phi, r, \theta_o)$ can be approximated by (Pecháček *et al.* 2005, 2006)

$$g = \frac{\sqrt{r(r-3)}}{r + \sin\phi \sin\theta_o \sqrt{r - 2 + 4(1 + \cos\phi \sin\theta_o)^{-1}}}, \quad (1.1)$$

$$F \equiv \frac{f}{I} = g^4 \left[1 + \frac{1}{r} \frac{1 - \sin\theta_o \cos\phi}{1 + \sin\theta_o \cos\phi} \right] \cos\theta_o. \quad (1.2)$$

The exact calculation of the time delay in the curved spacetime leads to elliptic integrals (e.g. Čadež & Kostić 2005). Using the power series expansion method several simple approximative formulae can be found. According to Beloborodov (2002) the light bending in Schwarzschild spacetime can be with high accuracy described by the relation $1 - \cos\alpha = (1 - \cos\psi)(1 - u)$ where $u = 2/r$, α is the angle of emission with respect to the radial direction in the local static frame and ψ is the angle measured in the orbital plane of the photon from the emission point to the direction of the observer. By expanding in $(1 - \cos\alpha)/(1 - u)$ up to the second order, Poutanen & Beloborodov (2006) give

$$u\Delta t = (1 - \cos\psi)\left(1 + \tfrac{1}{8}u(1 - \cos\psi)\left[1 + (1 - \cos\psi)(1/3 - u/14)\right]\right), \quad (1.3)$$

where Δt is the delay between signal emitted from some ψ and signal emitted radially toward a distant observer from the same initial distance u as the first signal.

Collecting all terms depending linearly on u we found a slightly different expression,

$$u\Delta t = (1 - \cos\psi) - u\left[\tfrac{1}{2}(1 - \cos\psi) + \ln\left(1 - (1 - \cos\psi)/2\right)\right]. \quad (1.4)$$

Accuracy of different approximations is assessed in figure 1.

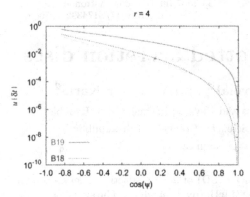
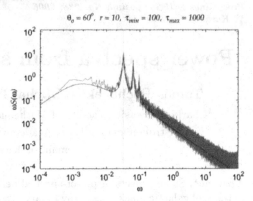

Figure 1. Left: Comparison of three different approximations for the time delay. The difference of the approximation and exact numerical solution is plotted. The solid line "B18" refers to (1.3), "B19" corresponds to $(1 - \cos\psi)$ (i.e. the first term of (1.3)) and "P" stands for (1.4). Right: Power spectrum from the spot model calculated for a narrow ring $r = 10$, observer's inclinations $\theta_o = 60°$, and life-time intervals $\langle \tau_{\min}, \tau_{\max}\rangle$. The red curve is a result of direct numerical simulation. Dashed blue curve is an approximation calculated from formula (2.1) assuming probability density function $\rho(\tau) \propto 1/\tau$. The vertical lines give the Keplerian orbital frequency $\Omega(r)$ for the corresponding radius.

2. Effects of decaying spots on the disc

As a simple model of the X-ray flux variability we consider a stochastic process consisting of independent exponentially decaying events, with uniformly distributed ignition times. After averaging we find PSD in the form

$$S(\omega) = n \sum_{k=-\infty}^{\infty} |c_k|^2 \int_{\tau_{\min}}^{\tau_{\max}} \rho(\tau) \frac{\tau^2}{1 + \tau^2 (\omega - k\Omega)^2} d\tau, \qquad (2.1)$$

where Ω and c_k are the frequency and Fourier series coefficients of $F(t)$, n is a mean rate of events and $\rho(\tau)$ is the probability density function of the random variable τ. See Pecháček et al. (2006) for details.

3. Results & Conclusions

We have developed approximative analytical formulae useful to describe relativistic effects acting on photons that originate from spots on a geometrically thin accretion disc near the Schwarzschild black hole. We used these formulae to derive the theoretical variability power spectra and compared their shapes with the corresponding power spectra generated by a numerical routine.

We thank the Czech Science Foundation (ref. 205/03/H144) and the Grant Agency of the Academy of Sciences (ref. 300030510) for support.

References

Beloborodov, A. M. 2002, ApJ 586, L85
Čadež, A. & Kostić, U. 2005, Phys. Rev. D 72, 104024
McHardy, I. M., Gunn, K. F., Uttley, P. & Goad, M. R. 2005, MNRAS 359, 1469
Misner, C. W., Thorne, K. S. & Wheeler, J. A. 1973, Gravitation (Freeman: San Francisco)
Pecháček, T., Dovčiak, M., Karas, V. & Matt, G. 2005, A&A 441, 855
Pecháček, T., Dovčiak, M., Karas, V. & Matt, G. 2005, AN 327, 957
Poutanen, J. & Beloborodov, A. M. 2002, MNRAS 373, 836
Życki, P. T. & Nedźwiecki, A. 2005, MNRAS 359, 308

Black Holes from Stars to Galaxies – Across the Range of Masses
Proceedings IAU Symposium No. 238, 2006
V. Karas & G. Matt, eds.
© 2007 International Astronomical Union
doi:10.1017/S1743921307005789

Brownian motion of black holes in stellar systems with non-Maxwellian distribution of the field stars

Isabel Tamara Pedron[1] and Carlos H. Coimbra-Araújo[2]

[1]Universidade Estadual do Oeste do Paraná, 85960-000, Marechal Cândido Rondon, PR, Brasil

[2]Instituto de Física Gleb Wataghin, Universidade Estadual de Campinas, 13083-970, Campinas, SP, Brasil

Abstract. A massive black hole at the center of a dense stellar system, such as a globular cluster or a galactic nucleus, is subject to a random walk due gravitational encounters with nearby stars. It behaves as a Brownian particle, since it is much more massive than the surrounding stars and moves much more slowly than they do. If the distribution function for the stellar velocities is Maxwellian, there is a exact equipartition of kinetic energy between the black hole and the stars in the stationary state. However, if the distribution function deviates from a Maxwellian form, the strict equipartition cannot be achieved.

The deviation from equipartition is quantified in this work by applying the Tsallis q-distribution for the stellar velocities in a q-isothermal stellar system and in a generalized King model.

Keywords. black holes, equipartition, Tsallis distribution, King models

1. Introduction

A massive black hole as a Brownian particle conducts to a equipartition of kinetic energy at the equilibrium, when the distribution function of stellar velocities is Maxwellian. When it is non-Maxwellian is not surprising to find a deviation from equipartition. This deviation is defined as $\eta = (M\langle V^2 \rangle)/(m\langle v^2 \rangle)$, where M and V represent the mass and velocity of the black hole, and m and v masses and velocities of stars. Following Chatterjee *et al.* (2002) we have $\eta = [3 \int_0^\infty f(r,v)v dv \int_0^\infty f(r,v)v^2 dv]/[f(r,0) \int_0^\infty f(r,v)v^4 dv]$, with $f(r,v)$ representing the distribution function for stars field. Evidently, for Maxwellian distribution $\eta = 1$. However, astrophysical systems are non extensive since they are subject to long range interactions. Boltzmann-Gibbs (BG) thermostatistics is able to predictions in extensive systems, in the sense that microscopic interactions are short or ignored, temporal or spatial memory effects are short range or does not exist and is valid ergodicity in the phase space. BG domain is enlarged by non extensive statistical mechanics, based on Tsallis entropy (Tsallis 1988) and in the q-distribution function: $p(x) \propto [1 - (1 - q)\beta x]^{1/(1-q)}$. Boltzmann distribution is recovered in the $q \to 1$ limit, and all the usual statistical mechanics as well. A summary about mathematical properties of these functions can be found in Umarov *et al.* (2006).

2. Extended stellar models, deviation from equipartition and results

The isothermal sphere is the simplest model for spherical systems. It was generalized by Lima & Souza (2005) based on the distribution function $f_q(v) \propto \left[1 - (1 - q)v^2/(2\sigma^2)\right]^{\frac{1}{1-q}}$ and the deviation results in $\eta(q) = (7 - 5q)/[2(2 - q)]$. Most commonly used are King models, based on truncated isothermal spheres. Extended King models were presented in

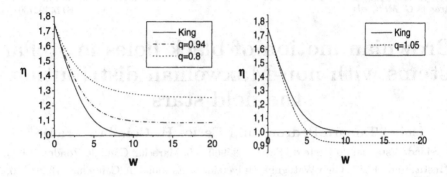

Figure 1. At the left, deviation for q-King model with $q = 0.94$, and $q = 0.8$. The limit value is $\eta = (7 - 5q)/2(2 - q)$. At the right, η for $q = 1.05$. King means $q = 1$, the usual King model. Strict equipartition implies $\eta = 1$.

Fa & Pedron (2001) and applied to fit surface brightness of the NGC3379 and 47TUC with excellent results. The distribution function in q-King models is $f_{k_q}(\epsilon) = \rho_1(2\pi\sigma^2)^{-3/2}\{[1 + (1 - q)\epsilon/\sigma^2]^{1/(1-q)} - 1\}$ for $\epsilon > 0$ and $f_{k_q}(\epsilon) = 0$ for $\epsilon \leqslant 0$. Here ϵ is the relative energy $\epsilon = \psi - v^2/2 > 0$ and $\psi = -\phi(r) + \phi_0$ the relative potential. The central potential is $W = \psi(0)/\sigma^2$ and in the $\psi(0)/\sigma^2 \to \infty$ limit the isothermal sphere is recovered. For such model the η value for $q < 1$ and $w = -\phi(r)$ is

$$\eta(q, w) = \frac{3}{2} \frac{\left[a^{-1}A^{\frac{2-q}{a}}\beta(y, 1, \frac{2-q}{1-q}) - w\right]\left[a^{\frac{-3}{2}}A^{\frac{5-3q}{2a}}\beta(y, \frac{3}{2}, \frac{2-q}{1-q}) - \frac{2}{3}w^{3/2}\right]}{(A^{1/a} - 1)\left[a^{-5/2}A^{(7-5q)/2a}\beta(y, \frac{5}{2}, \frac{2-q}{1-q}) - \frac{2}{5}w^{5/2}\right]} \quad (2.1)$$

where β is the incomplete Beta function, $a = 1 - q$, $y = \frac{w}{1/a+w}$, and $A = 1 + aw$. For $q > 1$ we obtain

$$\eta(q, w) = \frac{3}{2} \frac{\left[b^{-1}\tilde{A}^{(2-q)/b}I_1 - w\right]\left[b^{-\frac{3}{2}}\tilde{A}^{(5-3q)/2b}I_2 - \frac{2}{3}w^{3/2}\right]}{(\tilde{A}^{-1/b} - 1)\left[b^{-\frac{5}{2}}\tilde{A}^{(7-5q)/2b}I_3 - \frac{2}{5}w^{5/2}\right]} \quad (2.2)$$

where $b = q - 1$, $\tilde{A} = 1 - bw$, and $xm = \frac{w}{1/b-w}$. Furthermore, $I_1 = \int_0^{xm}(1 + x)^{-1/b}dx$, $I_2 = \int_0^{xm}(1 + x)^{-1/b}x^{1/2}dx$, and $I_3 = \int_0^{xm}(1 + x)^{-1/b}x^{3/2}dx$ with the conditions $\tilde{A} = 1 - bw \geqslant 0$ and $q \leqslant 1 + \frac{1}{w}$.

In the Figure 1 results indicate that there is a deviation from equipartition *a priori*, even at very long time-scales. The equipartition is never achieved in both cases ($q < 1$ and $q > 1$). The q parameter will be, in some way, dictated by the system.

References

Chatterjee, P., Hernquist, L. & Loeb, A. 2002, Phys. Rev. Lett., 88, 121103
Fa, K. S. & Pedron, I. T. 2001, astro-ph/0108370
Lima, J. A. S. & Souza, R. E. 2005, Physica A, 350, 303
Tsallis, C. 1988, J. Stat. Phys., 52, 479
Umarov, S., Tsallis, C., Gell-Mann, M. & Steinberg, S. 2006, cond-mat/0606038

Black Holes from Stars to Galaxies – Across the Range of Masses
Proceedings IAU Symposium No. 238, 2006
V. Karas & G. Matt, eds.
© 2007 International Astronomical Union
doi:10.1017/S1743921307005790

XMM-Newton study of the spectral variability in NLS1 galaxies

G. Ponti,[1,2] M. Cappi,[2] B. Czerny,[3] R. W. Goosmann[4] and V. Karas[4]

[1]Dipartimento di Astronomia, Università degli Studi di Bologna, Via Ranzani 1, I–40127 Bologna, Italy

[2]INAF-IASF Bologna, Via Gobetti 101, I–40129 Bologna, Italy

[3]Copernicus Astronomical Centre, Bartycka 18, P–00 716 Warsaw, Poland

[4]Astronomical Institute, Academy of Sciences, Boční II, CZ–14131 Prague, Czech Republic

email: ponti@iasfbo.inaf.it

Abstract. Preliminary results of the study of the X–ray spectral variability of 12 Narrow Line Seyfert 1 (NLS1) galaxies are presented. Rms spectra are calculated and compared for the whole sample to search for possible variations with black hole mass. A larger sample of AGN is under investigation.

Keywords. X-rays: galaxies – galaxies: Seyfert – methods: data analysis – black hole physics

We have been studying Root Mean Square (rms) variability of 12 NLS1 galaxies. This sample comprises all NLS1s (at $|b| \geqslant 15°$) detected in the RASS with a PSPC count rate higher than 0.2 c/s that have been observed with *XMM-Newton* for more than 30 ks (public data up to July 2006). Figure 1 shows the rms spectra calculated with time bins of a few ks (Ponti *et al.* 2004). All the sources exhibit significant degree of variability.

Rms spectra have been ordered with increasing black hole mass M from left to right and from top to bottom. All sources with $M < 10^7 M_\odot$ show a peak of variability between about 0.5 and 2 keV. The decrease of variability at low energy is likely associated to the strong soft excess present in all these sources, the origin of which is highly debated (e.g. Gierliński & Done 2004). The lower variability at high energy could be due to a pivoting power law (Markowitz & Edelson 2004; Haardt *et al.* 1997). Alternatively the whole shape of the rms spectrum could also be due to a variable power law with the presence of either an almost constant ionized disc reflection component (Crummy *et al.* 2006; Ponti *et al.* 2006), or a variable absorbing medium (Gierliński & Done 2006; Chevallier *et al.* 2006).

For the objects of higher black hole mass the rms spectra become flatter and the variability lower, except for I Zw 1 which shows the maximum of variability at low energy.

References

Chevallier, L., Collin, S., Dumont, A.-M., Czerny, B., Mouchet, M., Gonçalves, A. C. & Goosmann, R. 2006, A&A, 449, 493

Crummy, J., Fabian, A. C., Gallo, L. & Ross, R. R. 2006, MNRAS, 365, 1067

Gierliński, M. & Done, C. 2004, MNRAS, 349, L7

Gierliński, M. & Done, C. 2006, MNRAS, 371, L16

Haardt, F., Maraschi, L. & Ghisellini, G. 1997, ApJ, 476, 620

Markowitz, A. & Edelson, R. 2004, ApJ, 617, 939

O'Neill, P. M., Nandra, K., Papadakis, I. E. & Turner, T. J. 2005, MNRAS, 358, 1405

Ponti, G., Cappi, M., Dadina, M. & Malaguti, G. 2004, A&A, 417, 451

Ponti, G., Miniutti, G., Cappi, M., Maraschi, L., Fabian, A. C. & Iwasawa, K. 2006, MNRAS, 368, 903

Ponti *et al.* 2006

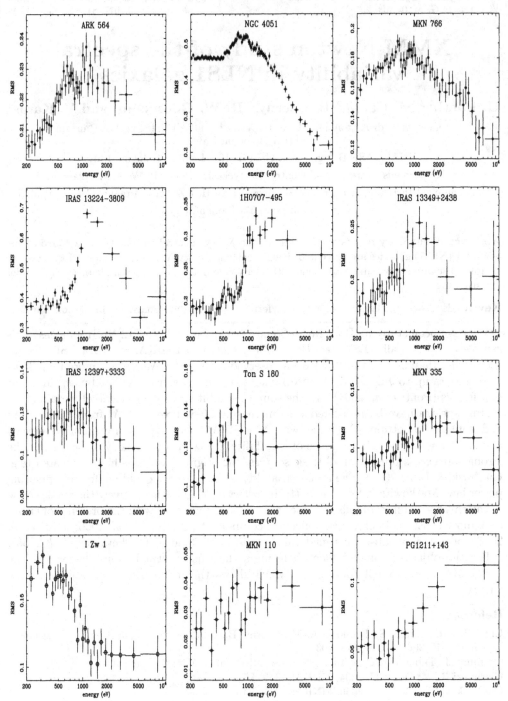

Figure 1. RMS spectra of all the NLS1 of the *ROSAT* All Sky survey, with a PSPC count rate higher than 0.2 s^{-1} and a Galactic latitude $|b| \geqslant 15°$, observed with *XMM-Newton* for more than 30 ks. The sources are sorted according to the black hole masses: filled squares, open circles, and stars indicate objects with the black hole mass lower than $10^7 M_\odot$, between 10^7 and $10^8 M_\odot$, and higher than $10^8 M_\odot$, respectively. Note: Black hole masses have been all taken from literature (see O'Neill *et al.* 2005) except for IRAS13224-3809 and 1H0707-495 for which these were assumed based on their strong and fast variability.

Black Holes from Stars to Galaxies – Across the Range of Masses
Proceedings IAU Symposium No. 238, 2006
V. Karas & G. Matt, eds.

© 2007 International Astronomical Union
doi:10.1017/S1743921307005807

Can gravitational microlensing be used to probe geometry of a massive black-hole?

Luka Č. Popović and Predrag Jovanović

Astronomical Observatory, Volgina 7, 11160 Belgrade, Serbia
email: lpopovic@aob.bg.ac.yu

Abstract. Here we discuss the possibility to use gravitational microlensing in order to probe the geometry around a massive black hole. Taking into account that lensed quasars are emitting X-rays which come from the heart of these objects, we investigated the influence of microlensing on the Fe Kα line shape originated in Schwarzschild and Kerr metrics.

Keywords. Accretion, accretion disks – gravitational lensing – galaxies: active

---

1. Introduction

The X-Ray emission of AGN is assumed to be generated in the innermost region of an accretion disk around a central super-massive Black Hole (BH). An emission line from iron Kα (Fe Kα) has been observed at 6–7 keV in the vast majority of AGNs (see e.g. Nandra *et al.* 1997). This line is probably produced in a very compact region near the BH and can bring essential information about the space geometry around the BH. Thus it seems clear that the Fe Kα line can be strongly affected by microlensing and recent observations of three lens systems supported this idea (Popović et al. 2006).

The aim of this paper is to investigate the influence of microlensing (ML) on the AGN Fe Kα line shape originated in a compact accretion disk around a non-rotating (Schwarzschild) and rotating (Kerr) BH and discuss the possibility that this effect can be used in probing of the massive black-hole geometry.

2. Microlensing of an accretion disk around massive black hole

To model an accretion disk around a massive BH, we use the ray tracing method (Popović *et al.* 2003) considering only those photon trajectories which reach the sky plane at a given observer's angle θ_{obs}. If X and Y are the impact parameters which describe the apparent position of each point of the accretion disk image on the celestial sphere as seen by an observer at infinity, the amplified line intensity is given by

$$I_p = \varepsilon(r)g^4(X,Y)\delta(x - g(X,Y))A(X,Y) \tag{2.1}$$

where $x = \nu_{obs}/\nu_0$ (ν_0 and ν_{obs} are the transition and observed frequencies, respectively); $g = \nu_{em}/\nu_{obs}$ (ν_{em} is the emitted frequency from the disk), and $\varepsilon(r)$ is the emissivity in the disk; $\varepsilon(r) = \varepsilon_0 \cdot r^{-q}$. $A(X,Y)$ is the amplification caused by microlensing. To calculate $A(X,Y)$ we use three approximation for microlens: point-like, caustic and pattern microlens (see Fig. 1, and Popović et al. 2003,2006).

Here we consider an accretion disk with following parameters: $i = 30°$ and $q = 2.5$, the inner radius is taken to be $R_{in}=R_{ms}$. (R_{ms} is the radius of the marginal stability orbit; $R_{ms} = 6R_g$ in the Schwarzschild and $R_{ms} = 1.23R_g$ in the the Kerr metric with angular momentum $a = 0.998$.) We adopt for the outer radius $R_{out} = 20R_g$.

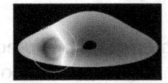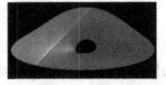

Figure 1. Simulations of the accretion disk microlensing by point-like microlens (in the Kerr metric, left) and straight-fold caustic (in the Schwarzschild metric, right).

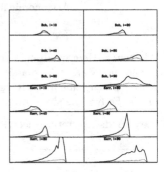

Figure 2. The amplification and deformation of the Fe Kα line in the case of Schwarzschild (left panel) and Kerr (middle panel) metrics. The different shapes and amplification of the Fe Kα line for different inclinations (right panel).

Some of our results are presented in Fig. 2 (also see Popović *et al.* 2003, 2006) where the difference of the Fe Kα line amplification in Schwarzschild (first panel) and Kerr (second panel) metrics are given for different positions of the point-like lens (X,Y in R_g) across the accretion disk (first two panels). Third panel of Fig. 2 shows the shape of Fe Kα line (dashed line) for different inclination and its deformation induced by ML (solid line) for a point-like microlens of EER=10 R_g, located at position of $X_0 = -5R_g$, $Y_0 = 5R_g$ in both metrics; Schwarzschild (Sch) and Kerr (Kerr).

Several outstanding changes of the line shape with the location of the microlens, and consequently with the transit of a microlens across the disk, can be inferred from Fig. 2 in both metrics: (i) the number of peaks, their relative separation and the peak velocity could change (this also affects to the velocity centroid), (ii) an asymmetrical enhancement of the line profile, (iii) amplification has a maximum for negative values of X_0 that correspond to the approaching part of the rotating disk.

On the other hand, as one can see from Fig. 2, the amplification and deformation of the Fe Kα line originated in the Kerr metric are more prominent that in the Schwarzschild one. Detailed discussion of our investigations can be found in Popović *et al.* (2003,2006) and Jovanović (2006). The obtained results show that microlensing can be used in investigations of an unresolved X-ray emitting region geometry around massive black holes.

Acknowledgements

This work was supported by the Ministry of Science and Environment Protection of Serbia through the project 146002.

References

Jovanović, P. 2006, PASP, 118, 656
Nandra, K., George, I. M., Mushotzky, R. F., Turner, T. J. & Yaqoob, T. 1997, ApJ, 477, 602
Popović, L. Č., Jovanović, P., Mediavilla, E. *et al.* 2006, ApJ, 637, 620
Popović, L. Č., Mediavilla, E. G., Jovanović, P. & Muñoz, J. A. 2003, A&A, 398, 975

Black Holes from Stars to Galaxies – Across the Range of Masses
Proceedings IAU Symposium No. 238, 2006
V. Karas & G. Matt, eds.
© 2007 International Astronomical Union
doi:10.1017/S1743921307005819

Exact solutions for discs around stationary black holes

N. Požár, O. Semerák, J. Šácha, M. Žáček and T. Zellerin

Inst. of Theor. Physics, Faculty of Maths and Phys., Charles Univ., Prague, Czech Republic
email: oldrich.semerak@mff.cuni.cz

Abstract. Black holes surrounded by axisymmetric structures prosper in some of the most interesting sources in the universe. However, a consistent exact description of the gravitational field of these systems is still lacking. In a static case, the task reduces to Laplace equation and the fields of multiple sources follow by mere superposition. In a rotating case, non-linearity of the Einstein equations resists simple grasp, but even then the theory of completely integrable systems seems to verge on satisfactory solutions. It seeks them in terms of θ-functions on special manifolds connected – symptomatically – with the names of Riemann and Hilbert.

Keywords. Accretion disks – black hole physics – gravitation – relativity

---

All the evidence for black holes is based on their interaction with the surrounding matter, but in theoretical models the gravity of this matter is neglected: the field is fully due to the hole as given by general relativity. Certainly, real accretion discs are probably much lighter than their central holes, but (i) this is not always the case (black hole with neutron torus as a temporary stage of a compact-binary collapse), and (ii) some important properties of the disc flow were found sensitive to the details of the gravitational field. However, general relativity being non-linear, it is not simple to incorporate the disc effect.

Einstein equations can be treated numerically and for light additional sources a perturbative approach may also be adequate, but only an exact solution can convey all the richness of the theory with full accuracy and generality. In the case of stationarity and axial symmetry, the field outside of sources can be described by the metric

$$ds^2 = -f(dt + A d\phi)^2 + f^{-1} \left[\rho^2 d\phi^2 + e^{2k} (d\rho^2 + dz^2) \right]$$

in the Weyl–Lewis–Papapetrou cylindrical coordinates (ρ, ϕ, z). The unknown functions $f(\rho, z)$, $A(\rho, z)$ must satisfy the Ernst equation $f \vec{\nabla}^2 \mathcal{E} = (\vec{\nabla}\mathcal{E})^2$ for the complex (Ernst) potential $\mathcal{E} \equiv f + i\psi$ with ψ introduced by $\rho \psi_{,z} = f^2 A_{,\rho}$, $\rho \psi_{,\rho} = -f^2 A_{,z}$; index-posed commas denote partial derivatives, $\vec{\nabla} \equiv (\partial_\rho, \partial_z)$ and $\vec{\nabla}^2 = \partial_{\rho\rho} + \frac{1}{\rho}\partial_\rho + \partial_{zz}$.

Up to date, multi-body exact solutions have only been studied in more detail in a *static case* $(A = 0)$ when the equation reduces to Laplace equation for f and compound fields follow by linear superposition plus a line integration for $k(\rho, z)$. In particular, we explored the Schwarzschild black hole surrounded, in the equatorial plane, by several types of sources – thin discs (i) of the inverted Morgan-Morgan family (Lemos & Letelier 1994; Semerák *et al.* 1999; Semerák & Žáček 2000; Žáček & Semerák 2002; Letelier 2003; Semerák 2003) and (ii) of the inverted family with power-law density profiles (Semerák 2004), (iii) Bach-Weyl thin ring (Chakrabarti 1988; Semerák *et al.* 1999), and (iv) thick toroid (Šácha & Semerák 2005). The total fields were e.g. studied on field lines, geodesics, equatorial circular motion or redshift behaviour. We observed that

• the disc's gravity affects Keplerian motion (in terms of which the disc itself is interpreted), the density profile being the most important "input"; the real accretion discs may thus be sensitive to their own field, even if their mass is not big relative to the centre

• this mainly applies to the disc's stability; for some density profiles, instability first occurs *inside* the disc (not at the rim) when its rim is shifted closer and closer to the hole, which "in practice" could imply radial fragmentation of the disc.

Such "ad-hoc" static superpositions are far from describing real systems: they are not linked with any accretion model and do not involve rotation. However, some of them possess physically acceptable properties within a reasonable range of parameters, so they can at least indicate some effects that might occur. (See Karas *et al.* 2004 for a review.)

In stationary setting, the problem gets fully non-linear and much more involved. Wide classes of new generic solutions have still been found – by "generating techniques" developed in 1950s–1980s. Unfortunately, it is almost impossible to impose physical prerequirements on these mathematical procedures and most of their outcomes have hardly any relevance. We made just minor attempt in this direction (Zellerin & Semerák 2000), to derive a solution for a rotating hole with a thin disc by a certain version of the "inverse-scattering method". The result was not physical (Semerák 2002).

The "generating" algorithms translate the original, non-linear equations into a pair of linear equations. In the last decade, the linear problem equivalent to the Ernst equation has been addressed using the ideas of algebraic geometry (e.g. Ansorg *et al.* 2002; Korotkin 2004; Klein & Richter 2005; and references therein). Instead of the boundary-value problem for $\mathcal{E}$, the Riemann-Hilbert problem is tackled to find certain functions having prescribed jump(s) across prescribed contour(s) on a specific Riemann surface. The solution is not known in general, but in some cases it can be given in terms of Riemann θ functions. The method seems to verge on providing results that can have physical relevance (Frauendiener & Klein 2001; Klein 2003); in particular, numerical codes were created for evaluation of theta functions (Frauendiener & Klein 2004; Frauendiener & Klein 2006; Požár 2006). Now it should mainly be clarified what properties of the Riemann surfaces correspond to particular spacetime features and locations in order that the scheme might be supplemented by a detailed and running physical control.

We thank V. Karas and colleagues from the Prague Relativity Group.

References

Ansorg, M., Kleinwächter, A., Meinel, R. & Neugebauer, G. 2002, Phys. Rev. D, 65, 044006
Chakrabarti, S. K. 1988, J. Astrophys. Astron., 9, 49
Frauendiener, J. & Klein, C. 2001, Phys. Rev. D, 63, 084025
Frauendiener, J. & Klein, C. 2004, J. Comp. Appl. Math., 167, 193
Frauendiener, J. & Klein, C. 2006, Lett. Math. Phys., 76, 249
Karas, V., Huré, J.-M. & Semerák, O. 2004, Class. Quantum Grav., 21, R1
Klein, C. 2003, Phys. Rev. D, 68, 027501
Klein, C. & Richter, O. 2005, Ernst equation on hyperelliptic Riemann surfaces (Springer, Berlin)
Korotkin, D. A. 2004, Math. Ann., 329, 335
Lemos, J. P. S. & Leterier, P. S. 1994, Phys. Rev. D, 49, 5135
Letelier, P. S. 2003, Phys. Rev. D, 68, 104002
Požár, N. 2006, diploma thesis (Charles Univ., Faculty Math. & Phys., Prague)
Šácha, J. & Semerák, O. 2005, Czech. J. Phys., 55, 139
Semerák, O. 2002, Class. Quantum Grav., 19, 3829
Semerák, O. 2003, Class. Quantum Grav, 20, 1613
Semerák, O. 2004, Class. Quantum Grav., 21, 2203
Semerák, O. & Žáček, M. 2000, Class. Quantum Grav., 17, 1613
Semerák, O., Zellerin, T. & Žáček, M. 1999, MNRAS, 308, 691 (Erratum: 2001 322 207) and 705
Žáček, M. & Semerák, O. 2002, Czech. J. Phys., 52, 19
Zellerin, T. & Semerák, O. 2000, Class. Quantum Grav., 17, 5103

Black Holes from Stars to Galaxies – Across the Range of Masses
Proceedings IAU Symposium No. 238, 2006
V. Karas & G. Matt, eds.
© 2007 International Astronomical Union
doi:10.1017/S1743921307005820

IR contamination in quiescence X-Ray Novae

Mark T. Reynolds, Paul J. Callanan and Colm Kelleher

Department of Physics, University College Cork, Ireland
email: m.reynolds@ucc.ie

Abstract. We present archival optical-IR observations of a number of short period X-ray novae (XRNe). We use these to show that in contrast to the current paradigm, there is likely significant non stellar flux in the infra-red (IR).

Keywords. Stars: individual (XTE J1118+480, GRO J0422+32, A0620-00, GS2000+25, GS1142-68 & Cen X-4) – X-rays: binaries

1. Introduction

Observations of X-ray novae (XRNe) in quiescence are an important means by which it has been possible to constrain the mass of the compact object in accreting binaries (see Charles & Coe (2006) for a thorough review). Such observations are easiest to perform in the optical; however, contamination from the accretion disc can be significant here, particularly for measurements of the ellipsoidal modulation. As such observations in the IR are to be preferred, where the flux from the secondary star is thought to be dominant (Charles & Coe 2006).

Previous quiescent IR observation of the XRNe V404 Cyg and A0620-00 (Shahbaz *et al.* 1996, 1999) have been used to limit the IR contamination from the accretion disc to be $\leqslant 14\%$ and 27% respectively. However, recent K-band observations of GRO J0422+32 by Reynolds *et al.* (2006) appear to show a light curve dominated by emission from the accretion disc and not the ellipsoidal modulation of the secondary star, see Fig. 1. Here, we take a photometric approach to constraining any accretion disk contribution to the IR flux of quiescent XRNe: we use the measured secondary star contribution in the optical to predict its contribution to the overall flux at longer wavelengths. We will show that there appears to be a significant excess of flux, above that expected from the secondary alone, present in the IR.

The data consist of the optical-IR (including *Spitzer* data where possible) spectral energy distribution (SED) of a number of XRNe consisting of data taken from the literature, namely : XTE J1118+480, GRO J0422+32, A0620-00, GS2000+25, GS1142-68 & Cen X-4.

2. Spectral Energy Distributions

In contrast to previous authors who normalised to the K-band flux (Muno *et al.* 2006; Gelino, et al. 2003; Gelino *et al.* 2001), where the IR contamination from the accretion disc is often ill-unconstrained, we choose to normalise the model atmospheres of the appropriate spectral type to the measured secondary contribution in the R-band (which has been directly measured for these systems). We fit a NextGen model atmosphere (Hauschildt *et al.* 1999a; Hauschildt *et al.* 1999b) corresponding to the spectral type of the secondary star in each system.

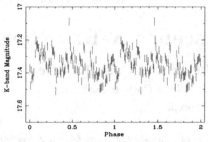

Figure 1. 2 minute resolution K-band light curve of GRO J0422+32. Two orbital phases are displayed for added clarity.

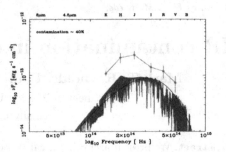

Figure 2. The optical-IR SED of GRO J0422+32.

3. Discussion

It is immediately apparent, when normalising to the flux in this way, that there is significant non-stellar flux in the IR, see Fig. 2. We note that there is a systematic excess of flux in the IR in all 6 systems. In each system the excess flux at IR wavelengths is similar to that observed in the optical. It is unclear why such systematic differences should occur for all 6 systems. However, we cannot rule out the following possibilities;

- There is a systematic under-estimation of the extinction towards the XRNe.
- The optical veiling is systematically over estimated.

The data suggest an alternative explanation: that the IR flux for these systems may have a significant component that does not originate from the secondary. The spectral shape of the IR excess we observe is consistent with having a blackbody origin. Hence, the observed excess is unlikely to be synchrotron in nature and is instead more likely to originate in a circumbinary disc or the cool outer regions of the accretion disc.

4. Conclusions

The current paradigm assumes that the observed IR flux from XRNe is dominated by emission from the secondary; however, the analysis presented here may cast some doubt on this assumption. If confirmed, the existence of a significant IR excess in these systems may have important consequences for any mass estimates of the compact object that use IR ellipsoidal modulation measurements.

Acknowledgements

This research made use of the SIMBAD database, operated at CDS, Strasbourg, France, and NASA's Astrophysics Data System. The authors wish to acknowledge financial support from Science Foundation Ireland.

References

Charles, P. A. & Coe, M. J., 2006, in Lewin W. H. G., van der Klis M., eds, Compact Stellar X-Ray Sources (Cambridge University Press, Cambridge)
Gelino, D. M., Harrison, T. E. & Orosz, J. A., 2001, AJ, 122, 2668
Gelino D. M. & Harrison T. E., 2003, ApJ, 599, 1254
Hauschildt, P. H., Allard, F. & Baron, E., 1999, ApJ, 512, 377
Hauschildt, P. H., Allard, F., Ferguson, J., Baron, E. & Alexander, D. R., 1999, ApJ, 525, 871
Muno, M. P. & Mauerhan, J., 2006, ApJ, 648, 135
Reynolds, M. T., Callanan, P. J. & Filippenko, A. V., 2006, MNRAS submitted (astro-ph/0610272)
Shahbaz, T., Bandyopadhyay, R., Charles, P. A. & Naylor, T., 1996, MNRAS, 282, 977
Shahbaz, T., Bandyopadhyay, R. & Charles, P. A., 1999, A&A, 346, 82

Black Holes from Stars to Galaxies – Across the Range of Masses
Proceedings IAU Symposium No. 238, 2006
V. Karas & G. Matt, eds.
© 2007 International Astronomical Union
doi:10.1017/S1743921307005832

First e-VLBI observations of GRS 1915+105

A. Rushton,[1] R. E. Spencer,[1] M. Strong,[1] R. M. Campbell,[2]
S. Casey,[1] R. P. Fender,[3,4] M. A. Garrett,[2] J. C. A. Miller-Jones,[4]
G. G. Pooley,[5] C. Reynolds,[2] and A. Szomoru,[2] V. Tudose[4,6]
and Z. Paragi[2]

[1]The University of Manchester, Jodrell Bank Observatory, Cheshire SK11 9DA, UK

[2]Joint Institute for VLBI in Europe, Postbus 2, 7990 A A Dwingeloo, The Netherlands

[3]School of Physics and Astronomy, University of Southampton, Southampton, UK

[4]"Anton Pannekoek" Astronomical Inst., Univ. of Amsterdam, Amsterdam, The Netherlands

[5]University of Cambridge, Mullard Radio Astronomy Observatory, CB3 0HE Cambridge, UK

[6]Astronomical Institute of the Romanian Academy, Cutitul de Argint 5, Bucharest, Romania

email: arushton@jb.man.ac.uk

Abstract. We present results from the first successful open call e-VLBI science run, observing the X-ray binary GRS 1915+105. e-VLBI science allows the rapid production of VLBI radio maps, within hours of an observation rather than weeks. A total of 6 telescopes observing at 5 GHz across the European VLBI Network (EVN) were correlated in real time at the Joint Institute for VLBI in Europe (JIVE). Throughout this, GRS 1915+105 was observed for a total of 5.5 hours, producing 2.8 GB of visibilities of correlated data. The peak brightness was 10.2 mJy per beam, with a total integrated radio flux of 11.1 mJy.

The use of the Internet for VLBI data transfer offers a number of advantages over conventional recorded VLBI, including improved reliability due to real time operation and the possibility of a rapid response to new and transient phenomena. Decisions on follow-up observations can be made immediately after the observation rather than delayed by potentially weeks due to problems in shipment of tapes/discs to the correlator. One aim of the project was also to develop a strategy for rapid response (ToO) e-VLBI observations for when this technique is more mature.

In this e-VLBI experiment, the data were transferred from the telescope to the correlator using Mark 5A disk-based VLBI data systems. These units have been fitted with 1 Gbps Network Interface Cards which allow the units to transfer the telescope data to the correlator over the Internet and private optical networks at rates exceeding 100 Mbps. Production Internet connections for institutions within each participating country are provided and controlled by the local and national network providers. Most of the telescopes connect to the national networks, and then are connected to the GÉANT 2 network allowing pan-European multi-gigabit connectivity.

On 2006 April 20–21 the e-EVN observed GRS 1915+105 at 4.994 GHz using phase-referencing for a total time of 5.5 hours. Each station sustained a transfer rate of 128 Mbps across the e-VLBI network. The radio image of GRS 1915+105 on 2006 April 20 – 21 is shown in Fig. 1 (left) using a uv weighting robustness parameter of 0. The source had a position of R.A. 19^h 15^m $(11.548 \pm 0.001)^s$ and Dec. $10° 56' (44.71 \pm 0.01)''$ (J2000). The detected source has a major and minor axis of 9.8×6.9 mas, observed with a beam size of 9.6×6.5 mas, respectively. This was de-convolved with the beam which revealed an extended component of 2.7×1.2 mas at a position angle of 140 $(\pm 2)°$. This is similar

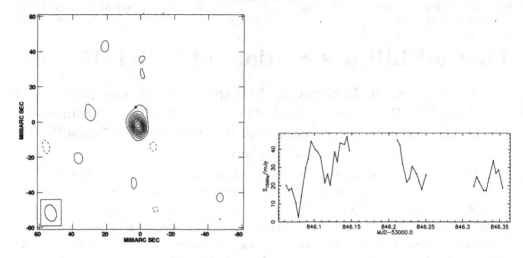

Figure 1. Left: e-EVN map of GRS 1915+105 at 5 GHz using 6 telescopes on 21 April 2006. Contour levels are (-1, 1, 2, 4, 6, 8, 10) times 1 mJy per beam, with an rms of 0.3 mJy. Right: Ryle Radio Telescope 15 GHz flux monitoring of GRS 1915+105, observed at 2006 April 21 01:27–08:32 UT.

to the P.A. of the large scale jets previously observed (Fender *et al.* 1999). The total integrated radio flux density was 11.1 ($\pm$0.6) mJy.

Fig. 1 (right) shows the Ryle Radio Telescope data on 2006 April 21 between 01:27–08:32 UT. A flare of 40 mJy was detected, which quickly decayed to $\sim$ 20 mJy within 4.5 hours. Assuming equipartition and the flare expands isotropically the minimum energy in the magnetic field for a distance of 11 kpc (Fender *et al.* 1999) is 2×10^{41} ergs. Using the de-convolved size from the image rather than assuming spherical expansion, we find a minimum energy of 1×10^{40} ergs, a lower value due to the source being collimated. This is unlike the major flares studied by the VLA and MERLIN (Mirabel & Rodriguez 1994) where the decay is over several days and the ejecta can be followed for up to 2 months after the flare. The behaviour of the strong flare is consistent with the shock-in-jet model (Miller-Jones *et al.* 2005); however the short flares seem to show the characteristic of an expanding source without continuous ejection of relativistic electrons.

The use of e-VLBI enabled us to obtain images within approximately a day of the VLBI run, rather than the many weeks needed for conventional recording based observations. This work clearly shows the ability of the e-EVN to produce high resolution radio maps in real time, hence eliminating the need of tape / disc recording. In the future, e-VLBI transmission rates will keep increasing with network development, yielding higher sensitivities and longer baselines will be achieved with the addition of more telescopes to the network.

References

Fender, R. P., Garrington, S. T., McKay, D. J., Muxlow, T. W. B., Pooley, G. G., Spencer, R. E. *et al.* 1999, mnras, 304, 865

Mirabel, I. F. & Rodriguez, L. F. 1994, Nature, 371, 46

Miller-Jones, J. C. A., McCormick, D. G., Fender, R. P., Spencer, R. E., Muxlow, T. W. B. & Pooley, G. G. 2005, MNRAS, 363, 867

Black Holes from Stars to Galaxies – Across the Range of Masses
Proceedings IAU Symposium No. 238, 2006
V. Karas & G. Matt, eds.
© 2007 International Astronomical Union
doi:10.1017/S1743921307005844

Detection of IMBHs from microlensing in globular clusters

Margarita Safonova[1] and Sohrab Rahvar[2]

[1]Indian Institute of Astrophysics, Koramangala, Bangalore, 560034, India
[2]School of Physics, Institute for Theoretical Studies in Physics and Mathematics,
Farmanieh Building, Tehran, Iran
email: rita@iiap.res.in, rahvar@physics.ipm.ir

Abstract. Globular clusters have been long predicted to host intermediate-mass black holes (IMBHs) in their centres. The growing evidence that some/all Galactic globular clusters (GCs) could harbour middle range ($10^2 - 10^4 \, M_\odot$) black holes, just as galaxies do, stimulates the searches and the development of new methods for proving their existence. Here we propose another method of detection – the microlensing of the cluster stars by the central BH.

Keywords. Gravitational lensing, (Galaxy:) globular clusters: general

1. Introduction

The BH in a GC centre acts as a telescope with a lens of effective diameter ~ 40 AU; we propose to consider the microlensing events expected when globular cluster star passes behind the central BH, which acts as a lens. Globular clusters are very advantageous for microlensing search of IMBHs, since the location of both potential lens and sources and their relative motions are well constrained, removing the ambiguities usually presented in the microlensing events detected towards the bulge and the Magellanic Clouds. The Galaxy has ~ 150 globular clusters, each containing $\sim 10^4 - 10^5$ stars, and though a significant fraction of stars in the cores of GCs will be unresolved with the ground-based telescope, one can use the pixel lensing technique.

2. Microlensing of stars in globular clusters

As an example we show the details of calculations for the M15 globular cluster, a very promising cluster with the possible mass of the central BH $3.2 \times 10^3 \, M_\odot$ (Gerssen *et al.* 2002). In a globular cluster, $D_{LS} \ll D_L \approx D_S$, and the Einstein radius reduces to $R_E \approx (2.8 \, \text{AU}) \, (M_{bh}/M_\odot)^{1/2} \, (D_{LS}/1 \, \text{Kpc})^{1/2}$. The optical depth is the number of stars to be lensed by a central BH, considering a cone with the cross-section $\pi R_E^2(r)$, $dN_{event} = (\rho(r)/M_*) \, \pi R_E^2(r) dr$. Using the Plummer density profile, we estimate the number of microlensing events for M15 as $N \approx 1.3 \times 10^{-4}$, which would mean that if we monitor the centres of about 10^4 globular clusters we have a chance to see a microlensing event already in progress. To estimate the mean duration of events, we divide the mean Einstein radius by the central velocity dispersion σ, $\langle t_E \rangle = \langle R_E \rangle / \sigma$. The mean Einstein radius is $\langle R_E \rangle = \left(\int_0^\infty P(r) R_E(r) dr \right) / \left(\int_0^\infty P(r) dr \right)$, where $P(r) = dN_{event}/dr$ is the probability of a star being inside the Einstein ring of the central BH. For M15, $\langle R_E \rangle = 2.07$ AU, and with $\sigma = 12$ km/sec, $t_E = 300$ days.

We calculate these quantities for a specific list of globular clusters, assuming the mass of the central BH as $10^3 \, M_\odot$ for all candidates except M15 (see above) and G1 (Gebhardt *et al.* 2002), which is mainly motivated by the idea based on $M_{bh} - \sigma$ correlation existing for galaxies. A possible source of contamination is the self-lensing of the GC objects by themselves. We estimate the self-lensing integrated optical depth to be $\tau \simeq 10^{-3}$, and the self-lensing rate for a typical GC as 0.05 events per one year of observation. The time scale of self-lensing is estimated to be less than a year.

3. Choice of globular clusters and observational strategy

We have included in our sample all Galactic proposed core-collapsed (CC) clusters (Trager *et al.* 1995), due to current belief that central BHs can only reside in centrally concentrated clusters (e.g., M15, a proto-typical CC cluster (Djorgovski & King 1986; Lugger *et al.* 1997)). However, recently Baumgardt *et al.* (2004) claimed that, on the contrary, one has to look for medium-concentration King-profile clusters with nearly flat cores. We have included in our sample the candidates of Baumgardt as well.

Following the globular clusters classification scheme (Mackey & van der Bergh 2005), we found that majority of the clusters candidates belong to the old halo/bulge–disk (OH/BD) group. We noticed that as far as cluster luminosity versus half-light radius is concerned, there is no considerable difference between the Baumgardt and CC sets. We noticed the clustering of our candidates in a small area of the plot, indicating the region where lie the clusters that are both tight and bright. It was noticed before (Gebhardt *et al.* 2002) that to the extent that a massive, bound cluster can be viewed as a 'mini-bulge', it may be that every dense stellar system (small or large) hosts a central black hole. We observe that, most probably, dense and high-luminosity clusters are better candidates for central BH search than diffuse and low-luminosity ones. Ideally, we would monitor all $\sim$ 150 Galactic globular clusters. However, it just may be useless to look for central BHs in diffuse and faint clusters. Moreover, in CC clusters, if we take into account the mass segregation effect or more concentrated mass distribution law, a $r^{-7/4}$ profile, the lensing rate increases. For example, for nearby 47 Tuc, NGC 6397, and NGC 6752 it would nearly double.

4. Conclusions

Currently we are writing a proposal for a 2-m class telescope to look at the clusters for the central BH using the pixel lensing technique. We came to the tentative conclusion that some classes of GCs possess characteristics which indicate that they are more likely to be the ones to look for the central BH. The OH, BD and possible ex-spheroidals (like ω Cen) clusters are most likely ones. Since the average event duration is up to years, we can monitor clusters cores with the frequency once or twice a month, which gives the advantage of easily differentiating other possible sources of contamination, like self-lensing or lensing of the stars belonging to the Galactic buldge or Magellanic Clouds backgrounds. Besides, no far-lying background stars can be detected within the highly crowded core radius, so any discovered event will be due to the stars within the cluster.

References

Gerssen, J., van der Marel, R. P., Gerbhardt, K., Guhathukurta, P., Peterson, R. C. & Pryor, C. 2002, AJ, 124, 3270
Gerssen, J., van der Marel, R. P., Gerbhardt, K., Guhathukurta, P., Peterson, R. C. & Pryor, C. 2003, AJ 125, 376
Gerbhardt, K., Rich, R. M. & Ho, L. C. 2002, ApJ 578, L41
Djorgovski, S. & King, I. 1986, ApJ, 305, 61
Lugger, P. M., Cohn, H., Grindlay, J. E., Baylin, C. D. & Hertz, P. 1987, ApJ, 320, 482
Trager, S. C., King, I. & Djorgovski, S. 1995, AJ, 109, 218
Baumgardt, H., Makino, J. & Hut, P. 2004, astro-ph/0410597
Mackey, A. D. & van den Bergh, S. 2005, astro-ph/0504142

Black Holes from Stars to Galaxies – Across the Range of Masses
Proceedings IAU Symposium No. 238, 2006
V. Karas & G. Matt, eds.
© 2007 International Astronomical Union
doi:10.1017/S1743921307005856

Episodic ejection from super-massive black holes

Lakshmi Saripalli,[1]† Ravi Subrahmanyan[1] and Richard W. Hunstead[2]

[1]Raman Research Institute, Bangalore, India

[2]School of Physics, University of Sydney, Australia

email: lsaripal@rri.res.in, rsubrahm@rri.res.in, rwh@physics.usyd.edu.au

Abstract. Episodic activity in super-massive black holes is shown by radio galaxies exhibiting 'double-double' radio morphologies (Subrahmanyan *et al.* 1996, Schoenmakers *et al.* 2000). Spectacular examples showing a renewal of beam activity in the form of new beams emerging within relic radio lobes of previous activity have placed the phenomenon of recurrence in AGN outflows on a firm footing.

By using the SUMSS and WENSS GRG samples, we infer that on timescales of order a few million years, low luminosity radio sources are more likely to exhibit episodic behaviour in the accretion on to their supermassive black holes as compared to the more powerful radio galaxies.

Keywords. Galaxies: active – galaxies: nuclei – accretion

---

1. Introduction

Extended radio galaxies provide a historical record of the central activity: the extended radio lobes were built up over the source lifetime and represent a history of the beam power from the supermassive black hole at the centre of the galaxy. Radio galaxies of the largest linear sizes, the giant radio galaxies (GRGs; linear size > 0.7 Mpc) are particularly suitable for such studies. Their large ages imply that there is a higher probability that they were witness to changes in activity at the centres during their lifetime. Additionally, abrupt changes in the activity of the central engine would take longer times to propagate over the giant radio lobes, thus increasing the likelihood of observing such interruptions and restarting in these sources.

2. Recurrence in nuclear activity in GRGs

There are four samples of GRGs available today that form a resource from which compilations of re-starting radio galaxies may be made. They are the WENSS 325 MHz sample (Schoenmakers *et al.* 2001), two NVSS 1.4 GHz samples that have been formed using different angular size criteria (Lara *et al.* 2001 and Machalski *et al.* 2001) and the SUMSS 843 MHz sample (Saripalli *et al.* 2005).

In this work we use two samples of GRGs: the WENSS and the SUMSS samples, to explore the phenomenon of episodic ejection from supermassive black holes. WENSS (47 GRGs over 8100 sq. deg.) and SUMSS (18 GRGs over 2100 sq. deg.) have complete redshift information and are matched in their selection criteria. Given that these samples have been made using a lower limit of 5′ for the angular extent, and a lower limit of 0.7 Mpc for the linear size, the two samples are expected to contain all giants with redshift $z \leqslant 0.13$ that have radio luminosities above the survey sensitivity limits.

† Present address: Raman Research Institute, C. V. Raman Avenue, Sadashivanagar, Bangalore 560 080, India.

3. The sample of low redshift restarting GRGs

The SUMSS GRG sample has four radio galaxies with $z \leqslant 0.13$ and in the WENSS sample there are 25 such sources. Together, there are 29 $z \leqslant 0.13$ GRGs, and their radio powers have a clear bimodal distribution. 19 GRGs form a distinct population at lower radio power with a median $\sim 2 \times 10^{24}$ W Hz^{-1}; the remaining 10 GRGs have relatively higher power with a median $\sim 2 \times 10^{25}$ W Hz^{-1}.

We have used the following morphological indicators for recognizing a restarting of activity: (i) Twin pairs of nested lobes or sources with 'double-double' morphology, or alternately (ii) a pair of outer lobes together with inner emission features that are observed to terminate at hot/warm spots well recessed from the ends of the lobes. Among the 10 powerful GRGs as many as 9 are in the ON state (currently active) of which only 1 may be classified as restarting. There is only one source among the 10 that appears to be in the OFF state, i.e. in a relic phase with neither hot-spots at the lobe ends nor inner double structure. In contrast, among the low power GRGs we find that 15 of the 19 sources are in the ON state; however, of the 15 as many as 7 appear to be in the restarting phase. Additionally, 4 of the 19 low power GRGs appear to be in the relic phase and show no signs of jets or hot-spots.

4. Discussion

The fraction of visible relics is small. This must imply that the time for which these sources remain visible after the central beams switch OFF is small as compared to their active lifetimes. We infer that that the low power GRG source population is more vulnerable to restarting, since a significantly greater fraction of low power sources are observed to be in the restarting phase as compared to the higher power source population. The low power GRGs, with radio power below the FR-I/II dividing line, are inferred to have relatively short duty cycles in their central activity, alternating between ON and OFF states. In contrast, in the high power GRGs, with radio powers above the FR-I/II dividing line, restarting on timescales of a few Myr is rare.

If giant radio sources, in which recurrence is more easily recognized, are the tail end of the linear size distribution in the radio source population and are not a distinct peculiar population, then our results imply that low and high power radio sources may have very different recurrence rates in their AGN activity: low luminosity sources may have more frequent episodes of accretion on to their supermassive black holes (Marchesini *et al.* 2004) with timescales of a few Myr.

References

Lara, L., Cotton, W. D., Feretti, L., Giovannini, G., Marcaide, J. M., Marquez, I. & Venturi, T. 2001, A&A, 370, 409

Machalski, J., Jamrozy, M. & Zola, S. 2001, A&A, 371, 445

Marchesini, D., Celotti, A. & Ferrarese, L. 2004, MNRAS, 351, 733

Saripalli, L., Hunstead, R. W., Subrahmanyan, R. & Boyce, E. 2005, AJ, 130, 896

Schoenmakers, A. P., de Bruyn, A. G., Roettgering, H. J. A. & van der Laan, H. 2001, A&A, 374, 861

Schoenmakers, A. P., de Bruyn, A. G., Roettgering, H. J. A., van der Laan, H. & Kaiser, C. R. 2000, MNRAS, 315, 371

Subrahmanyan, R., Saripalli, L. & Hunstead, R. W. 1996, MNRAS, 279, 257

Black Holes from Stars to Galaxies – Across the Range of Masses
Proceedings IAU Symposium No. 238, 2006
V. Karas & G. Matt, eds.
© 2007 International Astronomical Union
doi:10.1017/S1743921307005868

Detailed structure of the X-ray jet in 4C 19.44 (PKS1354+195)

D. A. Schwartz,[1] D. E. Harris,[1] H. Landt,[1] A. Siemiginowska,[1]
E. S. Perlman,[2] C. C. Cheung,[3] J. M. Gelbord,[4] D. M. Worrall,[5]
M. Birkinshaw,[5] S. G. Jorstad,[6] A. P. Marscher[6] and L. Stawarz[7]

[1]Smithsonian Astrophysical Observatory, Cambridge MA 02138, USA
[2]University of Maryland, Baltimore County, Baltimore MD 21250, USA
[3]NRAO/Stanford University, Palo Alto CA 94305, USA
[4]Massachusetts Institute of Technology, Cambridge MA 02139, USA
[5]Physics Department, University of Bristol, Bristol BS8 1TL UK
[6]Institute for Astrophysical Research, Boston University, Boston MA 02215, USA
[7]Kavli Inst. for Particle Astroph. and Cosmology, Stanford Univ., Stanford CA 94305, USA
email: das@cfa.harvard.edu

Abstract. We investigate the variations of the magnetic field, Doppler factor, and relativistic particle density along the jet of a quasar at $z = 0.72$. We chose 4C 19.44 for this study because of its length and straight morphology. The 18 arcsec length of the jet provides many independent resolution elements in the Chandra X-ray image. The straightness suggests that geometry factors, although uncertain, are almost constant along the jet. We assume the X-ray emission is from inverse Compton scattering of the cosmic microwave background. With the aid of assumptions about jet alignment, equipartition between magnetic-field and relativistic-particle energy, and filling factors, we find that the jet is in bulk relativistic motion with a Doppler factor $\approx$ 6 at an angle no more than $10°$ to the line of sight over de-projected distances $\approx$ 150–600 kpc from the quasar, and with a magnetic field $\approx$10 μGauss.

Keywords. (Galaxies:) quasars: individual (4C 19.44, PKS 1354+135) – galaxies: jets – X-rays

1. Analysis of the jet

We present preliminary results from a 200 ks observation of the quasar 4C 19.44 (=PKS 1354+135), using *Chandra* observation identifications (OBSID's) 6903, 6904, 7302, 7303. We analyze the jet assuming X-ray production by inverse Compton scattering on the cosmic microwave background . This is motivated in the middle panel of Figure 1 since the optical flux for two knots (from Sambruna *et al.* (2004)) do not allow an extrapolation of the synchrotron radio spectrum to the X-ray region. We follow Tavecchio *et al.* (2000), and Celotti *et al.* (2001) to calculate the rest frame magnetic field B and the Doppler factor $\delta = [\Gamma(1 - \beta \cos(\theta))]^{-1}$. The formalism requires many assumptions on the physics and on unknown values of parameters. Consistent with our previous work (Schwartz *et al.* 2006) we assume a minimum total energy in magnetic field and relativistic particles, filling factor $=1$, an electron spectral index m=2.4 with cutoffs at low and high energies tuned to give radio emission from 10^6 Hz to 10^{12} Hz, equal energy in electrons and protons, and that the bulk Lorentz factor $\Gamma = \delta$ so that the angle to the line of sight takes on the maximum value, $\approx 1/\delta$, for a given δ. We take the volume of regions 4 through 14 to be

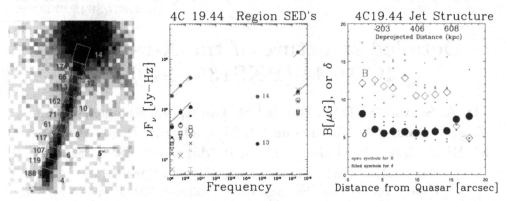

Figure 1. Left panel shows the *Chandra* 0.5 to 7 keV band X-ray image of the 18 arcsec long jet. The boxes numbered 4 through 14 are an arbitrary division of the jet into distinct regions for independent analysis. Numbers to the left of each box show how many X-ray photons were detected in that region. The contours are of the 4.8 GHz emission. The middle panel shows the spectral energy distributions for these 11 regions. Optical emission is detected only from the regions 14 and 13 (knots A and B of Sambruna *et al.* 2004). The summed X-ray spectrum of all the regions is consistent with $f_\nu \propto \nu^{-0.7}$, and the corresponding νf_ν is shown as the solid lines. The optical detection from region 13 can serve as an upper limit to any region, and along with the radio indices prohibits a simple synchrotron spectrum being extrapolated from the radio to X-ray region. The right hand panel shows the results for the magnetic field (large diamonds) and Doppler factor (filled circles) as a function of distance along the jet. Alternate assumptions can lead to systematic changes in these parameters as follows: small crosses use the best-fit power law radio spectral index; small triangles and circles assume conical jets starting at the quasar and 1 kpc from the quasar, respectively, reaching 1" diameter at the end of the jet in each case; small diamonds use the 1.4 to 4.86 GHz spectral index instead of the assumed slope of 0.7.

cylinders of the (resolved) lengths shown in the left panel of Figure 1, and (unresolved) radii assumed to be $0''.5 = 3.62$ kpc at the redshift $z = 0.72$ of this quasar.

2. Structure of the Jet

The results for B and δ are presented in the rightmost panel of the figure. With the above assumptions the jet shows similar structure along its length. We give the de-projected distance on the top axis by assuming $\delta = 5.5$. We note that if the radii were all 1 kpc, the values for B and δ would be a factor of 1.44 higher. The systematic effects of our assumptions allow alternate profiles, but the numerical values are constrained within a factor roughly 2. In particular, the increase in δ at the end of the jet would be peculiar in the IC/CMB scenario. For the baseline values of B and δ, we can derive the kinetic power, the minimum electron Lorentz factor, and the relativistic electron density as a function of distance along the jet. For de-projected distances 200 to 600 kpc from the quasar, regions 13 to 6, we require a cutoff to the electron spectrum at energies below $80mc^2$ and we estimate a kinetic power of approximately $(1-2) \times 10^{46}$ erg s^{-1}, and a total relativistic electron density of roughly $(2-3) \times 10^{-8}$ cm^{-3}.

We acknowledge support of NASA contract NAS8 39073 to SAO, CXC grants GO5-6116B, GO6-7111A, GO6-7111B, and HST grant HST-GO-10762.01-A.

References

Celotti, A., Ghisellini, G. & Chiaberge, M. 2001, MNRAS, 321, L1
Sambruna, R. M. *et al.* 2004, ApJ, 608, 698
Schwartz, D. A. *et al.* 2006, ApJ, 640, 592
Tavecchio, F., Maraschi, L., Sambruna, R. M. & Urry, C. M. 2000, ApJ, 544, L23

Black Holes from Stars to Galaxies – Across the Range of Masses
Proceedings IAU Symposium No. 238, 2006
V. Karas & G. Matt, eds.
© 2007 International Astronomical Union
doi:10.1017/S174392130700587X

Color variations of gravitationally lensed quasars as a tool for accretion disk size estimation

V. N. Shalyapin

Institute for Radiophysics and Electronics, Kharkov, Ukraine
email: vshal@ire.kharkov.ua

Abstract. Since the amplitude of microlensing variability depends on a source size the monitoring of gravitationally lensed quasars can produce the valuable information about the accretion disk size. Comparison of standard deviations in two wavebands yields useful constraints from quite moderate number of multicolor observations.

Existing optical VRI monitoring data for the quadruple gravitationally lensed quasar Q2237+0305 allow to obtain the source size ratio in two bands, V and R or V and I, as a function of size in the V band. Agreement with the standard accretion disk theory which predicts the dependency of a source size on wavelength $\sim \lambda^{4/3}$ can be achieved only if the quasar size in V band is comparable with the Einstein radius ($\simeq 10^{17}$ cm).

Keywords. Gravitational lensing – accretion, accretion disks – quasars: individual (Q2237+0305)

---

1. Multiband monitoring of Q2237+0305

Monitoring of microlensing variability of Q2237+0305 continues for more than 20 years, however, most of the observations have been carried out in a single band. In this study we limit ourselves by only 42 time moments when simultaneous observations in three wavebands V, R and I are available. VR color diagram for all 4 components A-D is shown in Fig. 1a. Variations of all images on color diagrams follow along straight lines with regression slopes $0.91 \pm 0.03, 0.93 \pm 0.02, 0.85 \pm 0.04, 0.61 \pm 0.06$ for A, B, C, D components respectively in VR bands and $0.86 \pm 0.04, 0.83 \pm 0.02, 0.71 \pm 0.04, 0.56 \pm 0.04$ in VI bands.

2. Numerical simulations and results

We generated magnification maps by the ray-shooting method (Kayser, Refsdal & Stabell 1986) taking into account microlensing parameters of Q2237+ 0305A,B images the convergence $k = 0.4$ and the shear $\gamma = 0.4$. Observed light curves are produced by moving of a source across the magnification pattern(Mortonson, Schechter & Wambsganss 2005). In order to estimate the ratio of two variation amplitudes and build color diagrams like Fig. 1a 42 random situated points from microlensing patterns were taken at random positions (Goicoechea, Gil-Merino, Ullan *et al.* 2005). Variations of two sources with radii $0.1R_e$ (R_e is the Einstein radius) and $0.2R_e$ are correlated quite closely with the regression slope 0.91 ± 0.03 (Fig. 1b). 10 000 Monte Carlo repetitions confirm robustness of the slope estimation as 0.92 ± 0.03.

In order to estimate the ratio of variation dispersions for two sources magnification maps were generated for source sizes from $0.01R_e$ to $2.0R_e$. The dependency of the dispersion on a source size is presented in Fig. 2a. The dispersion ratio is obtained by

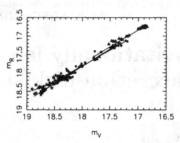 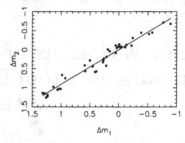

Figure 1. Left: VR color diagram of Q2237+0305. Triangle marks – A image, circles – B, squares – C, and stars – D image. Right: a sample of color variations for two sources with different radii.

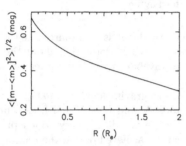 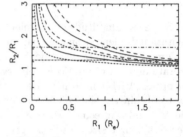

Figure 2. Left: standard deviation of microlensing fluctuations as the function of a source radius. Right: source sizes ratio for R and V bands (thin lines) and I and V bands (thick lines). Horizontal lines show the predictions from the standard accretion disk theory.

dividing two ordinate values for corresponding source sizes. This ratio cannot be presented as a function of only the source sizes ratio R_2/R_1 but it depends on both R_1 and R_2 in nonlinear way. So, to obtain the dispersion ratio about 0.92 as for Q2237+0305A,B values in VR diagram with a small source $R_1 = 0.01R_e$ in V band we need choose $R_2 = 0.1R_e$, i.e. in 10 times larger. For $R_1 = 0.1R_e$ we need increasing of the second source only nearly in two times till $R_2 = 0.2R_e$ and for very large sources an asymptotic inverse proportional of the source size (Refsdal & Stabell 1991) is getting corrected $R_2/R_1 = 1/0.92 \sim 1.09$. Fig. 2b illustrates the above statement for two band couples VR and VI. For the last pair the dispersion ratio value 0.84 for Q2237+0305A,B was used.

It is interesting to compare the obtained results with predictions of the standard accretion disk theory which claims the power-law dependency $R(\lambda) \sim \lambda^{4/3}$ (Shakura & Sunyaev 1973) (horizontal lines in Fig. 2b). Concordance of microlensing calculations with the standard accretion disk theory is reached in crossing place of the two curves (solid and dashed lines). Therefore, the RV color diagram yields the estimation of source radius in V band as $\sim 1.1R_e$ and VI relation gives $\sim 0.9R_e$, i.e. both estimations $\sim R_e$ that for this gravitational lens system equals $\simeq 10^{17}cm$ for solar mass microlenses.

The work was partially support by Spanish Department of Education and Science (project no. AYA2004-08243-C03-02).

References

Goicoechea, L. J., Gil-Merino, R., Ullan, A. *et al.* 2005, ApJ, 619, 29
Mortonson, M. J., Schechter, P. L. & Wambsganss, J. 2005, ApJ, 628, 594
Kayser, R., Refsdal, S. & Stabell, R. 1986, A&A, 166, 36
Refsdal, S. & Stabell, R. 1991, A&A, 250, 62
Shakura, N. I. & Sunyaev, R. A. 1973, A&A, 24, 337

Black Holes from Stars to Galaxies – Across the Range of Masses
Proceedings IAU Symposium No. 238, 2006
V. Karas & G. Matt, eds.
© 2007 International Astronomical Union
doi:10.1017/S1743921307005881

Outburst of the unique X-ray transient CI Cam and its impact on the system

V. Šimon,[1] C. Bartolini,[2] A. Guarnieri,[2] A. Piccioni[2] and D. Hanžl[3]

[1]Astronomical Institute, Academy of Sciences, 25165 Ondřejov, Czech Republic

[2]Dipartimento di Astronomia, Università di Bologna, via Ranzani 1, 40127 Bologna, Italy

[3]N. Copernicus Observatory and Planetarium, Kraví hora 2, 61600 Brno, Czech Republic

Abstract. We argue that the outburst of CI Cam (XTE J0421+560), probably containing a black hole, was caused by the thermal instability of the accretion disk. Applying the model of King & Ritter (1998), we obtain a realistic disk mass and radius. The differences from soft X-ray transients belonging to the low-mass X-ray binaries can be explained if the disk in CI Cam heats up an extended envelope and/or a strong jet is formed. We resolve several spectral components in the optical colors in quiescence after the outburst; they vary in a complicated way during a possible 1350 d cycle.

We find indications that the variations of the source of the optical light can be related to those of the X-ray source in quiescence. The accretion disk seems to refill at present. Nowadays, only non-periodic intra-night optical ($\sim$0.02 mag) fluctuations are present. As regards the absence of coherent changes in the optical band, we point out the similarities in the situation of CI Cam and the microquasar LS5039/RX J1826–1450.

Keywords. Accretion, accretion disks – instabilities – stars: variables: other – X-rays: binaries

---

1. Introduction and results

CI Cam, an optical counterpart of the unique X-ray transient XTE J0421+560 (e.g. Frontera *et al.* 1998), is a remarkable system consisting of a B[e] supergiant and probably a black hole (BH, e.g. Belloni *et al.* 1999).

The 1998 outburst. We investigate various models and find that the thermal instability of an accretion disk is the best interpretation of the outburst. This event is of type FRED in accordance with the conditions of the model of King & Ritter (1998), in which the irradiation by X-rays is strong enough to ionize the entire disk out to its outer edge. The luminosity at the outburst peak is $L_{\rm peak} \approx 3 \times 10^{38}$ erg/s (0.8 $L_{\rm Edd}$) ($d = 5$ kpc (Robinson *et al.* 2002)), the disk mass being $M_{\rm h}(0) \approx 1.5 \times 10^{23}$ g and the disk radius being $R_{\rm h}(0) \approx 2.5 \times 10^{10}$ cm. These disk parameters are consistent with those of soft X-ray transients (SXTs). More details are given in Šimon *et al.* (2006). The small $M_{\rm h}(0)$ and $R_{\rm h}(0)$ speak in favor of a small, wind-fed disk, in variance with the stream-fed disks in SXTs. A comparison of the absolute magnitude of CI Cam with those of SXTs puts an important constraint on the emission mechanism. The absolute peak magnitude of outburst is $M_{\rm Vout} \approx -7$ ($d = 5$ kpc). $M_{\rm Vout}$ of SXTs tends to brighten with the orbital period (Shahbaz & Kuulkers 1998). A very large disk radius would be needed if it were the site of luminosity in outburst of CI Cam, in contradiction with the small $R_{\rm h}(0)$. Most luminosity thus comes from a different site, which is supported also by the reddening of the colors in outburst. This can be explained by heating up an extended envelope by a small disk and/or jet formation.

Post-outburst activity of CI Cam. The light curve displays smooth waves on the time scale of hundreds of days, which is in striking variance with the pre-outburst fluctuations (Bergner et al. 1995). The color variations are not explicable by the changes of the reddening intrinsic to CI Cam. Significant variations of the continuum must play a role; dominant line(s) changes would lead to rather independent variations of indices. Several superposed spectral components appear to be present. We find the division of the dominant contributions of the spectral components near $\lambda = 550$ nm: f-f/f-b emission from the wind and/or envelope (in the red and near-IR passbands; Clark et al. 2000), and another component – (pseudo)photospheric emission – in the blue region. Hα changes are resolved in the colors and suggest that the Hα region evolves on the time scale of hundreds of days. We detect two maxima of a possible 1350 d cycle: (pseudo)photospheric emission, f-f/f-b emission, and Hα emission are involved in a complicated way in this cycle.

X-ray and optical emission in quiescence. We find indication that even if the donor and its wind dominate in the optical in quiescence, they still affect the X-ray flux and X-ray absorption N_H, measured e.g. by Parmar et al. (2000). Using the Predehl & Schmitt (1995) relation, the changes of N_H should lead to an increase of $E(B - V)$ by several mag, which is not reflected in the optical colors. Changes of N_H are thus confined to the region hotter than the temperature of the dust condensation and may imply a refilling of the disk embedding the BH. This speaks in favor of the origin of the dominant X-ray emission from the close vicinity of the BH, not from the donor.

Rapid variations in quiescence. A chaotic profile prior to the outburst (Bergner et al. 1995) may suggest under-sampled rapid variations. In any case, the amplitude of the residuals of the night-to-night variations significantly decreased after the outburst. We detect only low-amplitude (~ 0.02 mag) intra-night variations in four nights; they have a form of waves on the time scale of about an hour. Such rapid changes in B[e] stars may not be related to the presence of the compact object; those in FS CMa were interpreted by de Winter & van den Ancker (1997) as either accretion events or inhomogeneities in the circumstellar (CM) matter. Since in CI Cam the dust envelope is located too far from the B[e] component (13–52 AU; Robinson et al. 2002), only the inhomogeneities in the CM matter are a viable explanation. As for the non-detection of coherent variations from the compact object, the situation of CI Cam can be similar to the microquasar LS 5039/RX J1826–1450 (Martí et al. 2004). Its absolute mag $M_V = -5.0 \pm 0.3$ (Ribó et al. 2002) is close to CI Cam in quiescence. The optical luminosity of the disk embedding the compact object can be out shined by the donor in CI Cam.

This analysis has been supported by the grant 205/05/2167 of the Grant Agency of the Czech Republic.

References

Belloni, T., Dieters, S., van den Ancker, M. E., Fender, R. P. et al. 1999, ApJ, 527, 345
Bergner, Y. K., Miroshnichenko, A. S., Yudin, R. V., Kuratov, K. S. et al. 1995, A&AS, 112, 221
Clark, J. S., Miroshnichenko, A. S., Larionov, V. M., Lyuty, V. M. et al. 2000, A&A, 356, 50
de Winter, D. & van den Ancker, M. E. 1997, A&AS, 121, 275
Frontera, F., Orlandini, M., Amati, L., dal Fiume, D. et al. 1998, A&A, 339, L69
King, A. R. & Ritter, H. 1998, MNRAS, 293, L42
Martí, J., Luque-Escamilla, P., Garrido, J. L., Paredes, J. M. et al. 2004, A&A, 418, 271
Parmar, A. N., Belloni, T., Orlandini, M., Dal Fiume, D. et al. 2000, A&A, 360, L31
Predehl, P. & Schmitt, J. H. M. M. 1995, A&A, 293, 889
Ribó, M., Paredes, J. M., Romero, G. E., Benaglia, P. et al. 2002, A&A, 384, 954
Robinson, E. L., Ivans, I. I. & Welsh, W. F. 2002, ApJ, 565, 1169
Shahbaz, T. & Kuulkers, E. 1998, MNRAS, 295, L1
Šimon, V., Bartolini, C., Piccioni, A. & Guarnieri, A. 2006, MNRAS, 369, 355

Black Holes from Stars to Galaxies – Across the Range of Masses
Proceedings IAU Symposium No. 238, 2006
V. Karas & G. Matt, eds.

Marginally stable thick discs orbiting the Kerr–de Sitter black holes: the mass estimates

Petr Slaný and Zdeněk Stuchlík

Institute of Physics, Faculty of Philosophy and Science, Silesian University in Opava,
Bezručovo nám. 13, CZ-746 01 Opava, Czech Republic
email: zdenek.stuchlik@fpf.slu.cz, petr.slany@fpf.slu.cz

Abstract. Basic properties of equipotential surfaces in test perfect fluid tori with uniform distribution of the specific angular momentum orbiting KdS black holes are summarized. The central mass-densities of adiabatic non-relativistic tori, for which the approximation of test fluid is adequate, are given and compared with the typical densities of Giant Molecular Clouds.

Keywords. Accretion, accretion disks – black hole physics – cosmological parameters

---

Presence of a repulsive cosmological constant, $\Lambda > 0$, changes the asymptotic structure of the black-hole backgrounds and implies strong consequences in the structure of thin and thick discs around Kerr–de Sitter (KdS) black holes, see the review of Stuchlík (2005).

KdS spacetimes are characterized by three parameters, M, a, y, representing the mass and spin of the black hole, and the dimensionless "cosmological parameter" defined in geometrical units ($c = G = 1$) as $y = \Lambda M^2/3$. For simplicity, we put $M = 1$ hereafter. Gravitational attraction of the black hole is just compensated by the cosmic repulsion at the "static radius", the only radius where the static geodesic observer resides: $r_s = y^{-1/3}$ in the Boyer–Lindquist coordinates. Stationary discs exist only in the spacetimes admitting stable circular geodesics. Beside the inner marginally stable (ms) and the inner marginally bound (mb) circular geodesics located near the black-hole horizon, there are the outer ms and mb circular geodesics near the static radius (Stuchlík & Slaný (2004)).

Basic properties of geometrically thick discs are determined by the equilibrium configurations of test perfect fluid orbiting a black hole. Marginally stable thick discs are characterized by uniform distribution of the specific angular momentum, $\ell(r, \theta) \equiv -U_\varphi/U_t = $ const. Solving the Euler equation for the fluid with 4-velocity $U^\mu = (U^t, 0, 0, U^\varphi)$, see, e.g., AJS (1978), we obtain the relation for the potential W in the form

$$W(r, \theta) = \ln U_t(r, \theta) = \ln \left[\frac{\Sigma}{I^2} \cdot \frac{\Delta_r \Delta_\theta \sin^2 \theta}{\Delta_\theta \sin^2 \theta \left(r^2 + a^2 - a\ell\right)^2 - \Delta_r \left(\ell - a \sin^2 \theta\right)^2} \right]^{1/2}$$

$$\Delta_r = r^2 - 2r + a^2 - yr^2 \left(r^2 + a^2\right), \quad \Delta_\theta = 1 + ya^2 \cos^2 \theta, \quad I = 1 + ya^2, \quad \Sigma = r^2 + a^2 \cos^2 \theta.$$

The boundary of any torus is given by any closed equipotential surface, however the last closed surface is critical enabling the outflows of matter from the disc through the cusp(s) due to the violation of hydrostatic equilibrium. We can distinguish three kinds of discs: accretion discs, marginally bound accretion discs and the excretion discs–completely new kind of toroidal structures from which the outflows into the outer space are only possible, see Fig. 1. Repulsive cosmological constant causes the existence of the outer cusp corresponding to the outer edge of the disc being located close to but bellow the static radius, and the strong collimation of open equipotential surfaces along the rotational axis

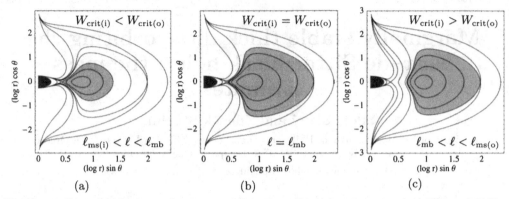

Figure 1. Typical behaviour of equipotential surfaces (meridional sections) in SdS and KdS black-hole spacetimes. Light gray region contains closed equipotential surfaces. The last closed surface is self-crossing in the cusp(s). Possible toroidal configurations correspond to: (a) accretion discs, (b) marginally bound accretion discs and (c) excretion discs.

indicating some role of $\Lambda > 0$ for collimation of jets far away from the maternal galaxy. For the current value of the cosmological constant, $\Lambda_0 \approx 1.3 \times 10^{-56}$ cm^{-2}, the maximal dimensions of test disc-like structures around supermassive black holes are smaller but comparable with the dimensions of large galaxies, see dimensions of r_s in Table 1.

Relevance of test-fluid approximation can be tested by determining the central-mass density ϱ_c of the torus for which the masses of the disc and central black hole are comparable. For an adiabatic non-relativistic perfect fluid the total mass-energy of the torus is given by the relation bellow. The results are presented in Table 1. Comparing with the typical densities of Giant Molecular Clouds (GMC), $\varrho_{GMC} \sim 10^{-18}$ kg m^{-3}, we see that for diatomic GMC the limiting $\varrho_c \gg \varrho_{GMC}$.

Table 1. Mass parameter (M), the static radius (r_s) and the central mass-density (ϱ) of an adiabatic non-relativistic perfect fluid torus corresponding to two values of the adiabatic index $\gamma = (5/3, 7/5)$, for which $m_{disc} \approx M_{BH}$, in SdS and KdS $(a = 0.9)$ black-hole spacetimes with $\Lambda_0 \approx 1.3 \times 10^{-56}$ cm^{-2}.

M [$M_\odot$]	r_s [kpc]	$\varrho_{5/3}$ [kg m^{-3}]	$\varrho_{7/5}$ [kg m^{-3}]	$\varrho_{5/3}$ [kg m^{-3}]	$\varrho_{7/5}$ [kg m^{-3}]
10^6	10	10^{-22}	10^{-13}		
10^7	22	10^{-22}	10^{-14}	10^{-21}	10^{-12}
10^8	50	10^{-22}	10^{-15}	10^{-22}	10^{-13}
10^9	110	10^{-23}	10^{-16}	10^{-22}	10^{-14}

$$m = 2\pi\varrho_c \int_{\mathrm{disc}} \left[\frac{1 + \ell\Omega(r, \theta)}{1 - \ell\Omega(r, \theta)} \right] \left[\frac{\exp\{W_{in} - W(r, \theta)\} - 1}{\exp\{W_{in} - W_c\} - 1} \right]^n (r^2 + a^2 \cos^2\theta) \sin\theta \, dr \, d\theta$$

This work has been done as a part of the research project MSM 4781305903.

References

Abramowicz, M. A., Jaroszyński, M. & Sikora, M. 1978, A&A, 63, 221

Slaný, P. & Stuchlík, Z. 2005, Classical Quantum Gravity, 22, 3623

Stuchlík, Z. 2005, Mod. Phys. Lett. A, 20, 561

Stuchlík, Z. & Slaný, P. 2004, Phys. Rev. D, 69, 064001

Black Holes from Stars to Galaxies – Across the Range of Masses
Proceedings IAU Symposium No. 238, 2006 © 2007 International Astronomical Union
V. Karas & G. Matt, eds. doi:10.1017/S174392130700590X

An $f(R)$ gravitation for galactic environments

Yousef Sobouti

Institute for Advanced Studies in Basic Sciences, P.O. Box 45195-1159, Zanjan, Iran
email: sobouti@iasbs.ac.ir

Abstract. We propose an action-based $f(R)$ modification of Einstein's gravity which admits of a modified Schwarzschild-deSitter metric. In the weak field limit this amounts to adding a small logarithmic correction to the Newtonian potential. A test star moving in such a spacetime acquires a constant asymptotic speed at large distances. This speed turns out to be proportional to the fourth root of the mass of the central body in compliance with the Tully-Fisher relation. A variance of MOND's gravity emerges as an inevitable consequence of the proposed formalism.

1. Introduction

Convinced of cosmic speed up and finding that the dark energy hypotheses is not a compelling explanation, some cosmologists have looked for alternatives to Einstein's gravitation. There is a parallel situation in galactic studies. Dark matter hypotheses, intended to explain the flat rotation curves of spirals or the large velocity dispersions in ellipticals and clusters of galaxies have raised more questions than answers. Alternatives to Newtonian dynamics have been proposed but have had their own critics. The foremost among such theories, the Modified Newtonian Dynamics (MOND) of Milgrom (1983 a,b,c) is capable of explaining the flat rotation curves of spirals (Sandres *et al.*, 1998 and 2002) and of justifying the Tully-Fisher relation with considerable success. But it is often criticized for the lack of an axiomatic foundation.

Here we are concerned with galactic problems. We suggest to follow cosmologists and look for a modified Einstein gravity tailored to suit galactic environments.

2. A modified field equation

The model we consider is an isolated point mass embedded in an asymptotically flat spacetime. As an alternative to the Einstein-Hilbert action we assume the following $S = \frac{1}{2} \int f(R)\sqrt{-g}d^4x$, where R is the Ricci scalar and $f(R)$ is an, as yet, unspecified but differentiable function of R. Variations of S with respect to the metric tensor leads to the following field equation (Capozziello *et al.*, 2003)

$$R_{\mu\nu} - \frac{1}{2}g_{\mu\nu}\frac{f}{h} = \left(h_{;\mu\nu} - h^{\lambda}_{;\lambda}\, g_{\mu\nu}\right)\frac{1}{h}, \tag{2.1}$$

where $h = df/dR$. The case $f(R) = R + constant$ and $h = 1$ gives the Einstein field equations. For the purpose of galactic studies we envisage a spherically symmetric static metric

$$ds^2 = -B(r)dt^2 + A(r)dr^2 + r^2\left(d\theta^2 + \sin^2\theta d\varphi^2\right). \tag{2.2}$$

3. Kinship with MOND

The features of this model are also shared by Milgrom's MOND. Below we show that some version of MOND can actually be derived from the present formalism. We recall that in the weak field approximation, Newton's dynamics is derived from that of Einstein by writing $B = \left(1 + 2\phi/c^2\right)$, $\phi = GM/r$, and expanding all relevant quantities and equations up to first order in ϕ/c^2. In a similar way for our modified Newtonian dynamics we find

$$B(r) = 1 + \alpha + \alpha \ln(r/s) - s/r = 1 + 2\phi(r)/c^2, \qquad (3.1)$$

where the second equality defines $\phi(r)$. Let us write $\alpha = \alpha_0 (GM/GM_\odot)^{1/2}$ and find the gravitational acceleration

$$g = |d\phi/dr| = (a_0 g_n)^{1/2} + g_n, \quad a_0 = \alpha_0^2 c^4 / 4GM_\odot, \quad \text{and} \quad g_n = GM/r^2. \qquad (3.2)$$

The limiting behaviors of g are the same as those of MOND. One may comfortably identify a_0 as MOND's characteristic acceleration and calculate α_0 anew from Eq (3.2). For $a_0 = 1.2 \times 10^{-8} cm/sec^2$, one finds

$$\alpha = 2.8 \times 10^{-12} \left(M/M_\odot\right)^{1/2}. \qquad (3.3)$$

It is gratifying how close this value of α is to the one obtained above, and how similar MOND and the present formalism are, in spite of their totally independent starting points.

4. Concluding remarks

We have developed an $f(R) \approx R^{1-\alpha/2}$ gravitation which is essentially a logarithmic modification of Einstein- Hilbert action. In spherically symmetric static situations the theory admits of a modified Schwarzschild-deSitter metric. The latter in the limit of weak fields gives a logarithmic correction to the Newtonian potential. From the observed asymptotic speeds of galaxies we learn that the correction is proportional to the square root of the mass of the central body. Flat rotation curves, the Tully-Fisher relation and a version of MOND emerge as natural consequences of the theory.

There are two practices to obtain the field equations of $f(R)$ gravity, the metric approach, where $g_{\mu\nu}$'s are considered as dynamical variables and that of Palatini, where the metric together with the affine connections are treated as such (see Magnano, 1995 for a review). Unless $f(R)$ is linear in R, the resulting field equations are not identical (see Ferraris et al., 1994). The metric approach is often shied away from, for its leading to fourth order differential equations. In the present paper we do not initially specify $f(R)$. Instead, at some intermediate stage in the analysis we adopt an ansatz for $df(R)/dR$ as a function of r and work forth to obtain the metric, R, and eventually $f(R)$. This enables us to avoid the fourth order equations. The trick should work in other contexts (cosmological, say). Further details can be found in Sobouti (2005).

References

Capozziello, S., Cardone, V. F., Carloni, S. & Troisi, A. 2003, astro-ph/0307018
Ferraris, M., Francaviglia, M. & Volvich, I. 1994, Class. Quantum Grav., 11, 1505
Magnano G. 1995, gr-qc/9511027
Milgrom, M. 1983, a, b, c, ApJ271, 365, 371, 384
Sanders, R. H. & Verheijen, M. A. W. 1998, astro-ph/9802240
Sanders, R. H. & Mc Gough, S. S. 2002, astro-ph/0204521
Sobouti, Y. 2005, astro-ph/0503302

Black Holes from Stars to Galaxies – Across the Range of Masses
Proceedings IAU Symposium No. 238, 2006
V. Karas & G. Matt, eds.
© 2007 International Astronomical Union
doi:10.1017/S1743921307005911

Chilled disks in ultraluminous X-ray sources

Roberto Soria,[1] Zdenka Kuncic,[2] and Anabela C. Gonçalves[3,4]

[1]Harvard-Smithsonian CfA, 60 Garden st, Cambridge, MA 02138, USA

[2]School of Physics, University of Sydney, NSW 2006, Australia

[3]LUTH, Observatoire de Paris-Meudon, 5 Place Jules Janssen, 92195 Meudon, France

[4]CAAUL, Observatorio Astronomico de Lisboa, Tapada da Ajuda, 1349-018 Lisboa, Portugal

email: rsoria@cfa.harvard.edu, z.kuncic@physics.usyd.edu.au, anabela.darbon@obspm.fr

Abstract. If the standard disk-blackbody approximation is used to estimate black hole (BH) masses in ultraluminous X-ray sources (ULXs), the inferred masses are $\sim 1000 M_\odot$. However, we argue that such an approximation cannot be applied to ULXs, because their disks are only radiating a small fraction of the accretion power, and are therefore cooler than they would be in a thermal-dominant state, for a given BH mass. Instead, we suggest that a different phenomenological approximation should be used, based on three observable parameters: disk luminosity, peak temperature, and ratio between thermal and non-thermal emission. This method naturally predicts masses $\sim 50 M_\odot$, more consistent with other theoretical and observational constraints.

Keywords. X-rays: binaries, black hole physics, accretion, accretion disks

1. Mass estimates from thermal disk spectra

The masses of the accreting BHs in ULXs, and hence their nature and physical origin, are still unknown. In the absence of direct kinematic measurements, indirect methods based on X-ray spectral modelling have been used, by analogy with stellar-mass BH X-ray binaries in our Galaxy. For stellar-mass BHs in the high/soft state, most of the accretion power is radiated by the disk, and the X-ray spectrum is well fitted by a multicolour blackbody (Shakura & Sunyaev 1973; Frank, King & Raine 2004). The fitted peak temperature T_0 and the integrated disk-blackbody luminosity L_0 are simply related to the mass accretion rate $\dot{M}$ and the size of the inner disk $R_{\rm in}$ (Makishima *et al.* 1986):

$$L_0 \approx 4\pi\sigma T_0^4 R_{\rm in}^2 \tag{1.1}$$

$$L_0 = \eta\dot{M}c^2 \tag{1.2}$$

$$\sigma T_0^4 \approx \frac{3GM\dot{M}}{8\pi R_{\rm in}^3}, \tag{1.3}$$

where we have ignored a factor related to the no-torque condition at $R_{\rm in}$ and a hardening factor, whose combined effect is ≈ 1.35 (Fabian, Ross & Miller 2004). The radiative efficiency $\eta \approx 0.1$–0.3 for a source in a high/soft state.

¿From (1.1), (1.2), and (1.3), one can estimate the BH mass:

$$M \approx \frac{c^2\eta L_0^{1/2}T_0^{-2}}{3G(\sigma\pi)^{1/2}} \approx 5.6 \left(\frac{\eta}{0.2}\right)\left(\frac{L_0}{5\times 10^{38}\ {\rm erg\ s^{-1}}}\right)^{1/2}\left(\frac{T_0}{1\ {\rm keV}}\right)^{-2} M_\odot. \tag{1.4}$$

Studies of Galactic stellar-mass BHs show that the spectroscopic mass estimate (1.4) is in agreement (within a factor of 2) with the kinematic mass.

2. Spectroscopic mass estimates for power-law-dominated ULXs

Let us assume that the "soft-excess" component in ULXs is indeed thermal emission from a disk (see Gonçalves & Soria 2006 for an alternative scenario). Direct application to (1.4) of the observed values of their X-ray luminosity ($L_0 \sim 10^{40}$ erg s^{-1}) and colour temperature ($kT_0 \approx 0.15$ keV) has led to the suggestion that ULXs may contain intermediate-mass BHs with masses $\sim 1000 M_\odot$ (Miller, Fabian & Miller 2004).

However, the thermal component is only a small fraction ($\sim 10\%$) of the X-ray emission in ULX spectra. In the inner region, at $R_{\rm in} \leqslant R \leqslant R_{\rm c}$, most of the accretion power is released via non-thermal processes. This implies that the inner disk is *cooler than a standard disk*, because it radiates only a flux

$$\sigma T(R)^4 \approx \frac{3GM\dot{M}}{8\pi R^3} - F_{\rm nr}(R) < \frac{3GM\dot{M}}{8\pi R^3} \qquad (2.1)$$

where $F_{\rm nr}(R)$ is the energy flux released via non-radiative processes, for example transferred to a corona or outflow via magnetic stresses (Kuncic & Bicknell 2004). If $T(R)$ increases more slowly that $R^{-1/2}$ for $R \to R_{\rm in}$, the maximum contribution to the disk emission occurs at $R \approx R_{\rm c}$, $T \approx T(R_{\rm c})$ (peak in the fitted spectrum). For simplicity, here we assume that $T = T(R_{\rm c}) = $ constant inside $R_{\rm c}$; only a fraction $\beta < 1$ of the disk emission from the inner region is directly visible, depending on the optical depth of the scattering region. We can now re-write (1.1), (1.2), (1.3) as:

$$L_0 = 4\pi R_{\rm c}^2 \sigma T_0^4 + 2\pi (R_{\rm c}^2 - R_{\rm in}^2)\sigma T_0^4 \times \beta \approx 4\pi R_{\rm c}^2 \sigma T_0^4 \times (1 + \beta/2) \qquad (2.2)$$

$$L_0 = f\eta \dot{M} c^2 \qquad (2.3)$$

$$\sigma T_0^4 \equiv \sigma T(R_{\rm c})^4 \approx \frac{3GM\dot{M}}{8\pi R_{\rm c}^3}. \qquad (2.4)$$

where L_0 is the total radiative luminosity of the disk, $T_0 \equiv T(R_{\rm c})$ is the peak temperature, and $f \sim 0.1$ is the fraction of accretion power radiated by the disk; f cannot be directly measured, but we can estimate it based on the fitted ratio of soft thermal emission over total X-ray luminosity. We can then solve (2.1), (2.2), (2.3) for M, $\dot{M}$ and $R_{\rm c}$ as a function of the observable quantities f, T_0 and L_0. In particular, for the BH mass we obtain:

$$M \approx \frac{49.8}{(1 + \beta/2)^{3/2}} \left(\frac{\eta}{0.2}\right) \left(\frac{f}{0.1}\right) \left(\frac{L_0}{2 \times 10^{39} \text{ erg s}^{-1}}\right)^{1/2} \left(\frac{T_0}{0.15 \text{ keV}}\right)^{-2} M_\odot \qquad (2.5)$$

We conclude that the fitted spectral features of ULXs (X-ray luminosity, temperature and ratio of thermal/non-thermal contribution) suggest masses $\sim 50 M_\odot$. This is at the extreme end of, but still consistent with models of stellar evolution. If that is the case, the emitted luminosity is a few times the Eddington luminosity $L_{\rm Edd}$, but the disk radiative contribution alone is $\lesssim L_{\rm Edd}$. The rest is generated outside the disk by non-thermal processes, which dominate at radii $\lesssim R_{\rm c} \sim 100$ gravitational radii.

References

Fabian, A. C., Ross, R. R. & Miller, J. M. 2004, MNRAS, 355, 359
Frank, J., King, A. & Raine, D. 2002, Accretion Power in Astrophysics (Cambridge University Press)
Gonçalves, A. C. & Soria, R. 2006, MNRAS, 371, 673
Kuncic, Z. & Bicknell, G. V. 2004, AJ, 616, 669
Makishima, K. *et al.* 1986, AJ, 308, 635
Miller, J. M., Fabian, A. C. & Miller, M. C. 2004, ApJ, 614, L117
Shakura, N. I. & Sunyaev, R. A. 1973, A&A, 24, 337.

Black Holes from Stars to Galaxies – Across the Range of Masses
Proceedings IAU Symposium No. 238, 2006
V. Karas & G. Matt, eds.
© 2007 International Astronomical Union
doi:10.1017/S1743921307005923

A ULX and a giant cloud collision in M 99

Roberto Soria[1] and Diane Sonya Wong[2]

[1]Harvard-Smithsonian CfA, 60 Garden st, Cambridge, MA 02138, USA

[2]Astronomy Dept, 601 Campbell Hall, Univ. of Cal. at Berkeley, CA 94720-3411, USA

email: rsoria@cfa.harvard.edu, dianew@astron.berkeley.edu

Abstract. The Sc galaxy M 99 in the Virgo Cluster has been strongly affected by recent tidal interactions, responsible for an asymmetric spiral pattern and a high star formation rate ($\sim 10 M_\odot$ yr^{-1}). We studied the galaxy with *XMM-Newton*, Keck and the Very Large Array (VLA). The inner disk is dominated by hot plasma with a total X-ray luminosity $\approx 10^{41}$ erg s^{-1}. At the outskirts of the galaxy, away from the main star-forming regions, there is an ultra-luminous X-ray source (ULX) with an X-ray luminosity $\approx 2 \times 10^{40}$ erg s^{-1} and a hard spectrum (power-law photon index $\Gamma \approx 1.7$). This source is close to the location where a massive H I cloud appears to be falling onto the M 99 disk at a relative speed > 100 km s^{-1}. The infalling gas may have been stripped from the nearby "dark galaxy" candidate VIRGOHI 21. We speculate there may be a relation between collisional events, infall of metal-poor gas clouds, and ULX formation.

Keywords. X-rays: galaxies – radio lines: galaxies – galaxies: individual (NGC 4254) – X-rays: binaries – black hole physics

1. Why M 99 is an interesting galaxy

The Sc galaxy M 99 (NGC 4254; $d \approx 17$ Mpc) is the brightest spiral in the Virgo Cluster ($M_B = -20.8$). It has a kinematic mass $\approx 10^{11} M_\odot$ (Vollmer *et al.* 2005; Phookun *et al.* 1993). It shows unusual features such as a lopsided spiral-arm structure, an H I gas tail and a string of H I clouds around the stellar disk; the total gas mass in those extra-disk gas structures is $\approx 2 \times 10^8 M_\odot$ (Phookun *et al.* 1993). This suggests tidal disruptions over the last ~ 300 Myr, due to a close encounter with another Virgo Cluster galaxy, or simply to the cluster potential. The gas clouds around M 99 could have been ram-pressure or tidally stripped from the edge of its own gas disk; or they could come from an interacting galaxy. One such nearby object is the "dark galaxy" candidate VIRGOHI 21, a large H I cloud, with a gas mass $\approx 2 \times 10^8 M_\odot$ but without any associated stars, ≈ 120 kpc to the north-west of M 99 (Davies *et al.* 2004).

Gas-rich galaxies with recent tidal interactions tend to have very high star formation rates (SFRs), and M 99 is no exception, with a rate $\sim 10 M_\odot$ yr^{-1} (Kennicutt *et al.* 2003). Another phenomenon often associated with tidal interactions and high SFR is the presence of ultraluminous X-ray sources (ULXs). We have used *XMM-Newton*, VLA and optical data to investigate the X-ray properties of M 99, and the possible connection between ULXs, star formation and collisional events.

2. Main results of our study

We summarize here our main results (see Soria & Wong 2006 for details):
- In the inner galactic disk (at $\lesssim 5$ kpc from the nucleus), the X-ray emission is dominated by a soft thermal-plasma component with $kT \lesssim 0.30$ keV. There is no starburst core. The total unabsorbed luminosity (not including the bright ULX) inside the D_{25}

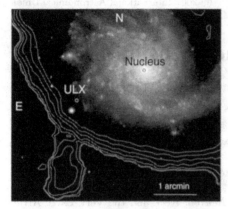 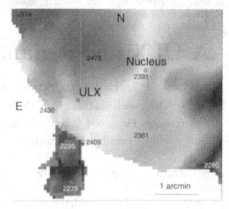

Figure 1. Left panel: H I 21-cm radio flux (from VLA archival data), represented as contours, marking the outer edge of the gas disk; the underlying greyscale image is from *VLT*/FORS1, in the *R*-band. The projected position of the ULX is well inside the gas disk, but at the edge of the stellar disk, near a large H I cloud. Right panel: velocity map of the H I 21-cm radio emission. Characteristic radial velocities (in km s^{-1}) are also marked at various locations on the plot. The gas cloud is consistent with tidal debris infalling onto the galactic disk from behind (with respect to our viewpoint), at a projected speed > 100 km s^{-1}. It appears as though it is impacting or merging with the disk near the ULX position.

ellipse is $\approx (1.2 \pm 0.2) \times 10^{41}$ erg s^{-1} in the 0.3–12 keV band; about 15% of this is due to resolved or unresolved discrete sources.

● The optical nucleus consists of a bright nuclear star cluster ($M_V \approx -13.5$ mag) with an old stellar population. Around it, there are a few smaller, much bluer clusters of young stars (typical $M_V \approx -11$ mag), much brighter than the nucleus itself in the UV band.

● The bright ULX at the outer edge of the stellar disk has an unabsorbed luminosity of $\approx 2 \times 10^{40}$ erg s^{-1} in the 0.3–12 keV band. Its spectrum is well fitted by a power-law with photon index $\Gamma \approx 1.7$. Such high/hard states are not uncommon in ULXs but are not generally seen in Galactic BHs. It suggests that we are not seeing the accretion disk directly, and that most of the gravitational power is efficiently transferred to a comptonizing region before being radiated via non-thermal processes.

● There is a massive gas cloud (H I mass $\sim 10^7 M_\odot$) seen in projection very close to the ULX. The cloud seems to be falling onto the galactic disk (from behind) with a projected radial velocity $\gtrsim 100$ km s^{-1}, and perhaps impacting it near the ULX location.

If ULXs are simply a statistical by-product of normal star formation at a sufficiently high rate, it is difficult to explain why there are no ULXs in the inner disk of M 99 despite an SFR $\sim 10 M_\odot$ yr^{-1}; instead, the bright ULX is at the outer edge of the stellar disk, where the SFR is orders of magnitude lower. Intriguingly, this ULX seems to be associated with a collisional event between the disk and an infalling gas cloud. The cloud might be made of metal-poor gas stripped from the nearby dark galaxy candidate VIRGOHI 21. We cannot rule out that all this is simply a chance coincidence, but we speculate that there is a physical link between triggered star formation in metal-poor environments and the creation of massive black hole remnants (~ 30–$100 M_\odot$) powering ULXs.

References

Davies, J. *et al.* 2004, MNRAS, 349, 922
Kennicutt, R. C., *et al.* 2003, PASP, 115, 928
Phookun, B., Vogel, S. N. & Mundi, L. G. 1993, ApJ, 418, 113
Soria, R. & Wong, D. S. 2006, MNRAS, in press (astro-ph/0608648)
Vollmer, B., Huchtmeier, W. & van Driel, W. 2005, A&A, 439, 921

Black Holes from Stars to Galaxies – Across the Range of Masses
Proceedings IAU Symposium No. 238, 2006
V. Karas & G. Matt, eds.
© 2007 International Astronomical Union
doi:10.1017/S1743921307005935

Near-IR integral field spectroscopy of the NLR of ESO428-G14: the role of the radio jet

Thaisa Storchi-Bergmann,[1] Rogemar A. Riffel,[1] Fausto K. B. Barbosa[1] and Cláudia Winge[2]

[1]Instituto de Física, UFRGS, CP 15051, Porto Alegre 91501-970, RS, Brazil

[2]Gemini Observatory, c/o AURA Inc., Casilla 603, La Serena, Chile

email: thaisa@ufrgs.br, rogemar@ufrgs.br, faustokb@if.ufrgs.br, cwinge@gemini.edu

Abstract. We present two-dimensional (2D) gas kinematics and excitation of the inner 300 pc of the Seyfert galaxy ESO428-G14 at a sampling of 14 pc^2, from near-infrared spectroscopic observations at R=5900 obtained with the Integral Field Unit of the Gemini Near-Infrared Spectrograph. Blue-shifts of up to $400\,\mathrm{km\,s^{-1}}$ and velocity dispersions of up to $150\,\mathrm{km\,s^{-1}}$, are observed in association with the radio jet running from SE to NW along position angle 129°. Both X-rays emitted by the active galactic nucleus and shocks produced by the radio jet can excite the H$_2$ and [Fe II] emission lines. We use the 2D velocity dispersion maps we estimate upper limits of 90% to the contribution of the radio jet to the excitation of [Fe II]λ1.257μm, and of 80% to the excitation of H$_2$$\lambda$2.121$\mu$m in the jet region.

Keywords. Galaxies: individual (ESO 428-G14) – galaxies, active – galaxies: nuclei – galaxies: ISM – galaxies: kinematics and dynamics – galaxies: jets

1. Introduction

The Narrow-Line Region (NLR) of Seyfert galaxies is one of the best probes of the mechanisms in operation in the surrounding of accreting supermassive black holes in galaxies. The excitation and dynamics of the inner NLR gas can reveal how radiation and mass outflows from the nucleus interact with circumnuclear gas. In this work we use the IFU of the Gemini Near-Infrared Spectrograph which provides two dimensional (hereafter 2D) mapping in the near-IR (thus minimizing the effect of reddening) to study the circumnuclear emitting gas of the Seyfert galaxy ESO 428-G14. Further results of this study can be found in Riffel *et al.* (2006).

2. Results and conclusions

In Fig. 1 we present 2D maps for the flux distribution and kinematics of the H$_2$, H I and [Fe II] emitting gas. The intensity maps show that all emission lines are most extended along the position angle PA~129°, which is the orientation of the radio jet and approximately coincides with the photometric major axis of the galaxy. The line emission distributions present a bibolar structure extended to both sides of the nucleus and show a detailed correspondence between with the radio map: the strongest line-emission is observed to the NW, and approximately coincides with the strongest emission in radio; the emission-line distribution to the SE bends to NE, as also observed in the radio contours. The strong influence of the radio jet is also observed in the gas kinematics. In particular, the location of a radio hot spot at $\approx 0.8''$ NW from the nucleus corresponds

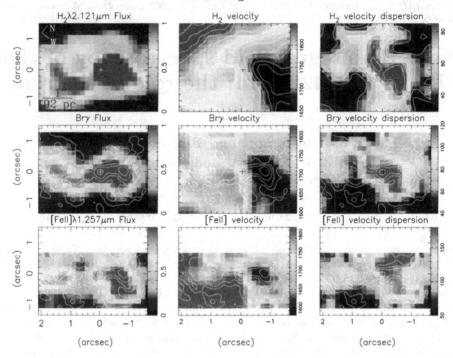

Figure 1. 2D maps for the flux distribution and kinematics of the $H_2\lambda 2.121$, $Br\gamma$ and [Fe II] emission lines. Left panels: intensity maps in arbitrary units; middle: radial velocity fields; right: velocity dispersions. Radio contours at wavelength 2 cm (heavy white lines) from Falcke, Wilson & Simpson (1998) are overlaid on the $Br\gamma$ and [Fe II] maps. The cross marks the position of the nucleus, defined as the peak of the K band continuum emission.

to the highest blueshifts observed in the gas. The highest velocity dispersion (σ) values are observed between the nucleus and the radio hot spot, and do not coincide with the emission line flux and velocity peaks, being shifted towards the nucleus. We interpret this result as due to the kinetic energy deposited by the radio jet in the circumnuclear ISM, producing a compression in the gas and emission enhancement just beyond this compressed region. Velocities slices along each emission line profile are presented in Riffel *et al.* (2006). Blueshifts of up to $400\,\mathrm{km\,s^{-1}}$ and velocity dispersions of up to $150\,\mathrm{km\,s^{-1}}$, are observed in association with the radio jet.

From the observed kinematics, we conclude that the radio jet has a fundamental role in shaping the emission line region as it interacts with the galaxy ISM surrounding the galaxy nucleus. We have used the 2D velocity dispersion maps to estimate the kinetic energy deposited in the circumnuclear ISM by the radio jet relative to regions away from the jet. Assuming that the H_2 and [Fe II] excitation in the latter regions is dominated by X-rays, and that the X-rays excitation is the same in the jet region, we obtain contributions by shocks of up to 80 - 90% for the [Fe II] and up to 70 - 80% for H_2 in the jet region. These are however, upper limits due to the fact that the X-rays contribution along the jet axis may be larger than away from it. The stronger association of the [Fe II] emission and kinematics with the radio structure supports a larger contribution of the radio jet to the excitation of [Fe II] than to that of H_2.

References

Falcke, H., Wilson, A. S. & Simpson, C. 1998, ApJ, 502, 199
Riffel, R. A., Storchi-Bergmann, T., Winge, C. & Barbosa, F. K. B. 2006, MNRAS 373, 2

Black Holes from Stars to Galaxies – Across the Range of Masses
Proceedings IAU Symposium No. 238, 2006
V. Karas & G. Matt, eds.
© 2007 International Astronomical Union
doi:10.1017/S1743921307005947

Humpy LNRF-velocity profiles in accretion discs orbiting rapidly rotating Kerr black holes: a possible relation to epicyclic oscillations

Zdeněk Stuchlík, Petr Slaný and Gabriel Török

Institute of Physics, Faculty of Philosophy and Science, Silesian University in Opava,
Bezručovo nám. 13, CZ-746 01 Opava, Czech Republic
email: zdenek.stuchlik@fpf.slu.cz, petr.slany@fpf.slu.cz, terek@volny.cz

Abstract. Coordinate-independent definition of the characteristic, so-called "humpy" frequency related to the positive gradient of the orbital velocity in locally non-rotating frames around nearly extreme Kerr black holes is given and compared with the epicyclic and orbital frequencies for both Keplerian thin discs and limiting marginally stable thick discs.

Keywords. accretion disks – black hole physics – relativity – instabilities

---

Gradient sign change of the orbital velocity $\mathcal{V}^{(\varphi)}$ related to the locally non-rotating frames (LNRF) in Kerr backgrounds (see, e.g., Bardeen, Press, Teukolsky 1972) has been found for accretion discs orbiting rapidly rotating Kerr black holes with spin $a > 0.9953$ for thin (Keplerian) discs (Aschenbach 2004) and $a > 0.99979$ for marginally stable thick discs characterized by the uniform distribution of the specific angular momentum, $\ell(r, \theta) = \text{const}$ (Stuchlík *et al.* 2005). Such a "humpy" orbital velocity profiles occur close to but above the marginally stable circular geodesic of the black hole spacetimes.

Non-monotonic behaviour of the orbital velocity in the marginally stable ($\ell = \text{const}$) tori is characterized by topology change of the von Zeipel/equivelocity surfaces, $\mathcal{R}(r, \theta) \equiv \ell/\mathcal{V}^{(\varphi)} = \text{const}$. In addition to the open surface crossing itself under the inner edge of the torus and existing for all values of the rotational parameter a, for $a > 0.99979$ the second self-crossing (marginally closed) surface together with toroidal surfaces occur. Toroidal von Zeipel surfaces exist under the newly developing cusp, being centered around the circle corresponding to the minimum of the equatorial LNRF velocity profile. The whole effect, elucidated by the toroidal von Zeipel surfaces, is located inside the ergosphere of a given Kerr spacetime.

The maximal positive rate of change of the orbital velocity in terms of the proper radial distance $\tilde{R}$,

$$\nu_{\text{crit}}^{\tilde{R}} = \frac{\partial \mathcal{V}^{(\varphi)}}{\partial \tilde{R}}\bigg|_{\max}, \quad d\tilde{R} = \sqrt{g_{rr}}dr, \tag{1.1}$$

where r is the Boyer-Lindquist radial coordinate, introduces a locally defined critical frequency characterizing possible disc oscillations connected with the velocity hump. Comparing the "humpy frequency" related to distant observers,

$$\nu_{\text{h}} = \sqrt{-(g_{tt} + 2\omega g_{t\varphi} + \omega^2 g_{\varphi\varphi})}\, \nu_{\text{crit}}^{\tilde{R}}, \tag{1.2}$$

where $\omega = -g_{t\varphi}/g_{\varphi\varphi}$ is the angular velocity of the LNRF, with the radial and vertical epicyclic frequencies, ν_{r}, ν_{v} (their analytic expressions can be found, e.g., in Aliev &

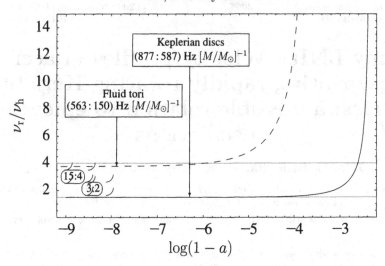

Figure 1. Spin dependence of the ratio of the radial epicyclic frequency and the "humpy frequency" related to distant observers for Keplerian discs and $\ell = \ell_{\mathrm{ms}}$ fluid tori. The ratio is given in the radius of definition of the humpy frequency. Both the ratios rapidly falls down to the asymptotic values of $3:2$ for Keplerian discs and $15:4$ for fluid tori. Then an exact $1/M$ scaling holds with frequencies depicted in the figure.

Galtsov 1981, Nowak & Lehr 1998), we can show that in the Keplerian discs orbiting extremely rapid Kerr holes ($1 - a < 10^{-4}$) the ratio of the epicyclic frequencies and the humpy frequency is nearly constant, i.e., almost independent of a, being $\sim 3:2$ for the radial epicyclic frequency and $\sim 11:2$ for the vertical epicyclic frequency. For black holes with $a \sim 0.996$, i.e., when the resonant phenomena with ratio $3:1$ between the vertical and radial epicyclic oscillations occur near the radius of the critical humpy frequency, there is ratio of the radial epicyclic and the humpy frequency $\sim 12:1$, i.e., the critical frequency is close to the low-frequency QPOs related to the high-frequency QPOs in such spacetimes. For $a > 0.996$ the resonant orbit $r_{4:1}$ (with the ratio $4:1$ between the vertical and radial epicyclic oscillations) occurs in the region of the hump. In the case of thick discs, the situation is more complex due to the dependence on the distribution of the specific angular momentum ℓ characterizing the disc rotation. For $\ell = \mathrm{const}$ tori and $1 - a < 10^{-6}$ the ratios of the orbital and epicyclic frequencies to the humpy frequency are again almost constant and independent of both a and ℓ. Moreover, the resonant orbit $r_{4:1}$ lays very close to the maximum rate of change of the velocity gradient and the ratio of the radial epicyclic and the humpy frequency is close to $4:1$ there. Spin dependence of the ratio $\nu_{\mathrm{r}} : \nu_{\mathrm{h}}$ for both Keplerian discs and fluid tori is illustrated in Fig. 1.

Acknowledgements

The work has been done as a part of the research project MSM 4781305903.

References

Aliev, A. N. & Galtsov, D. V. 1981, General Relativity and Gravitation, 13, 899
Aschenbach, B. 2004, A&A, 425, 1075
Bardeen, J. M., Press, W. H. & Teukolsky, S. A. 1972, ApJ, 178, 347
Nowak, M. A. & Lehr, D. E. 1998, in: M. A. Abramowicz, G. Björnsson, and J.E. Pringle (eds.), Theory of Black Hole Accretion Disks, (Cambridge: Cambridge University Press), p. 233
Stuchlík, Z., Slaný, P., Török, G. & Abramowicz, M. A. 2005, Phys. Rev. D, 71, 024037

Black Holes from Stars to Galaxies – Across the Range of Masses
Proceedings IAU Symposium No. 238, 2006
V. Karas & G. Matt, eds.
© 2007 International Astronomical Union
doi:10.1017/S1743921307005959

The X-ray–radio association in RATAN and RXTE monitoring of microquasar GRS 1915+105

S. A. Trushkin,[1] T. Kotani,[2] N. Kawai,[3] M. Namiki,[4] and S. N. Fabrika[5]

[1]Special Astrophysical Observatory RAS, Nizhnij Arkhyz, 369167, Russia

[2]Tokyo Tech, 2-12-1 O-okayama, Tokyo 152-8551, Japan,

[3]Tokyo Tech, 2-12-1 O-okayama, Tokyo 152-8551, Japan

[4]Osaka University, 1-1 Machikaneyama, Toyonaka, Osaka 560-0043, Japan

[5]Special Astrophysical Observatory RAS, Nizhnij Arkhyz, 369167, Russia

email: satr@sao.ru, kotani@hp.phys.titech.ac.jp, kawai@hp.phys.titech.ac.jp, namiki@ess.sci.osaka-u.ac.jp, email:fabrika@sao.ru

Abstract. In the daily RATAN-600 monitoring of the radio variability of the microquasar GRS1915+105 we detected a clear correlation of the flaring radio emission and X-rays "spikes" at 2–12 keV emission (1–2 Crab) detected with RXTE (ASM data) during nine bright (200–600 mJy) radio flares in October 2005. The spectra of these flares in maximum were optically thick at ν <2.3 GHz and optically thin at $\nu \geqslant$2.3 GHz. During radio flares the spectra of the X-ray spikes became definitely softer than those of a quiescent radio state. Thus these data indicated transitions from very high/hard states to high/soft ones during which massive ejections are probably happened, and the ejections are detected as the radio flares.

Keywords. X-rays: binaries – radio continuum: stars – radiation mechanisms: nonthermal

---

The X-ray transient source GRS 1910+105 was discovered by Castro-Tirado *et al.* (1992) with WATCH instrument on board GRANAT. In 1994, a superluminal motion of the radio jets had been detected from GRS 1915+105 (Mirabel F. & Rodriguez 1994). Since then a new class of astrophysical objects 'microquasars' was established. Nearly 20 microquasars were distinguished from near 300 Galactic X-ray binaries. We are far form the full understanding the jet phenomena in microquasars, however, from radio monitoring data a jet activity and state of the source can be diagnosed and even predicted.

We have carried out the 250-day almost daily monitoring observations of the microquasar GRS 1915+105 with RATAN-600 radio telescope at 1–22 GHz from September 2005 to May 2006. We have used a standard continuum radiometer complex of cryo-receivers at 4.8, 7.7, 11.2, and 21.7 GHz and the low-noise HEMT-based radiometers at 1 and 2.3 GHz. Observations were carried out using the 'Northern sector' antenna of the RATAN-600 radio telescope at the upper culmination of the sources. The measured multi-frequency light curves were directly compared with series from the X-ray observatory RXTE (ASM, Levine *et al.* 1996).

During the total set we have detected nine radio flares from 200 to 600 mJy, and all of them had counterparts in the soft X-rays emission (ASM RXTE). In Fig. 1 the radio and X-ray light curves in October 2005 with the first five flares are showed. The X-ray emission was defined as high (1–2.5 crabs) and hard (HR2 = 1–1.5). Radio spectra were optically thin in the first two flares, and optically thick in third one or were so-called "gigahertz-peaked" spectra, often measured from AGNs.

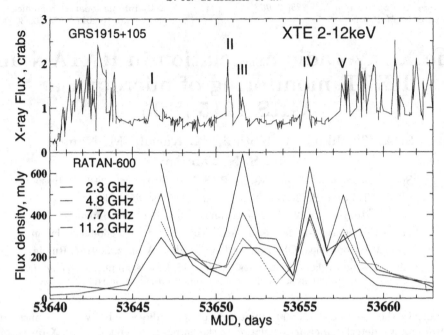

Figure 1. The radio and X-ray light curves of GRS1915+105 in October 2005.

The profiles of the X-ray spikes during the radio flares are clearly distinguishable from other spikes because of its shape, it shows the fast rise and the exponential decay. The other X-ray spikes, which reflect the activity of an accretion disk, exhibit an irregular pattern. During the radio flare, the spectra of the X-ray spikes become softer than those of the quiescent phase by a fraction of $\sim30\%$ in the hardness ratio (1.3–5 keV/5–12 keV). Thus a X-ray spike associated with a radio flare has a specified characteristics, and probably only such ones could produce the massive ejections (Namiki *et al.* 2007).

Miller-Jones *et al.* (2006) have detected a large-scale radio jet with VLBA mapping during the ninth radio outburst on 23 February 2006 (MJD53789.258), associated with a X-ray spike. Then the optically thin flare with fluxes 340, 340, 342, 285, 206, and 153 mJy was detected at frequencies 1, 2.3, 4.8, 7.7, 11,2 and 21.7 GHz.

Acknowledgments

These studies were supported by Russian Foundation Base Research (RFBR) grant N 05-02-17556 and mutual RFBR and Japan Society for the Promotion of Science (JSPS) grant N 05-02-19710.

References

Levine, A., Bradt, H. & Cui, W. 1996, ApJ, 469, L33
Castro-Tirado, A. J., Brandt, S. & Lund, N. 1992, IAUC, 5590
Mirabel, I. F. & Rodriguez, L. F. 1994, Nature, 371, 46
Miller-Jones, J. C. A., Rupen, M. P. & Trushkin, S. 2006, ATel, 758, 1
Namiki, M., Trushkin, S. & Bursov, N. 2007, in VI microquasars workshop, Komo, Italy, PoS,
    (in press)

Black Holes from Stars to Galaxies - Across the Range of Masses
Proceedings IAU Symposium No. 238, 2006
V. Karas & G. Matt, eds.
© 2007 International Astronomical Union
doi:10.1017/S1743921307005960

Flaring activity of microquasars from multi-frequency daily monitoring program with RATAN-600 radio telescope

S. A. Trushkin, N. A. Nizhelskij, N. N. Bursov and E. K. Majorova

Special Astrophysical Observatory RAS, Nizhnij Arkhyz, 369167, Russia

email: satr@sao.ru

Abstract. We report about the multi-frequency (1–30 GHz) daily monitoring of the radio flux variability of the three microquasars: SS433, GRS1915+105 and Cyg X-3 during 2005–2006. After a quiescent radio emission we have detected a drop down of the fluxes (~20 mJy) from Cyg X-3, a sign of the following bright flare, and indeed a 1 Jy flare was detected on 2 February 2006 after 18 days of quenched radio emission. The daily spectra of the flare in the maximum was found flat from 2 to 110 GHz, using the quasi-simultaneous observations at 109 GHz with the RT45m telescope and the NMA millimeter array of Nobeyama Radio Observatory in Japan. Several bright radio flaring events (1–15 Jy) followed during the state of highly variable and intense 1–12 keV X-ray emission (~0.5 Crab), monitored in the RXTE ASM program. We discussed various spectral and temporal characteristics of the detected 180 day light curves from three microquasars in comparison with the Rossi XTE ASM data.

Keywords. X-rays: binaries, radio continuum: stars, radiation mechanisms: nonthermal

---

The two-side relativistic jets, collimated high-velocity outflows, ejected from polar regions of accretion disks around black holes or neutron stars in the microquasars are efficient sources of variable synchrotron emission. Light curves and spectra variations are a key probe to test models of cosmic jets. We have carried out the 250-day monitoring of the microquasars Cyg X-3, GRS 1915+105, and SS433, with RATAN-600 radio telescope at 1–30 GHz from September 2005 to May 2006. Trushkin *et al.* (2007a) discussed the detected X-ray/radio light curves correlation from GRS 1915+105.

We have used a continuum radiometer complex: the four cryo-receivers at 3.9, 7.7, 11.2, and 21.7 GHz and three low-noise transistor radiometers at 1, 2.3 and 30 GHz. The observations were carried out with the 'Northern sector' antenna of RATAN-600.

During 100 days, Cyg X-3 was in a quiescent state of ~100 mJy (Fig. 1). In December 2005 its X-ray and radio fluxes began to increase. Then the flux density at 4.8 GHz was found to drop from 103 mJy on Jan 14.4 (UT) to 22 mJy on Jan 17.4 (UT). The source is known to exhibit the radio flares typically with a few peaks exceeding 1–5 Jy following such a quenched state as Waltman *et al.* (1994) have showed in the intensive monitoring of Cyg X-3 at 2.25 and 8.7 GHz. The source has been monitored from Jan 25 (UT) with the Nobeyama Radio Observatory 45m Telescope and the Nobeyama Millimeter Array (NMA). On Feb 2.2 (UT), about 18 days after it entered the quenched state, the rise of a first peak is detected with the NRO45m Telescope. On Feb 3.2 (UT), the flux densities reached to the first peak at all the sampling frequencies from 2.25 GHz to 110.10 GHz (Tsuboi *et al.* 2006). The spectrum on Feb 03 of the flare was flat as measured by RATAN, NRO RT45m and NMA from 2 to 110 GHz. The next peak of the active events on 10 February have a flat spectrum at a level of 1Jy again. Later three short-duration flares have happened during a week. A flare on Feb 18 had the inverted power-law spectrum with the spectral index $\alpha=+0.75$ from 2.3 to 22 GHz.

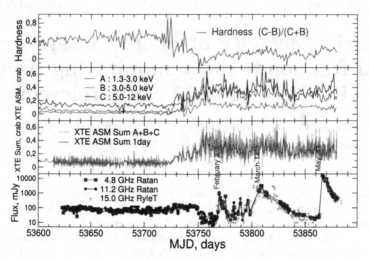

Figure 1. The RATAN and RXTE ASM light curves of Cyg X-3, September 2005 – May 2006.

We detected three powerful flares: on March 14 to 5 Jy, on May 11 to 16 Jy and on July 26 to 13 Jy . In the May flare fluxes have grown up by a factor ∼1000 during a ∼ one day. The change of the spectrum during the flare on May 11–19 followed to model of single ejection of the relativistic electrons, moving in thermal matter in a intense WR star wind. It was in a optically thin mode at frequencies 2–22 GHz, meanwhile at 614 MHz (Pal *et al.* 2006) and 1 GHz Cyg X-3 was in a hard absorption.

The first microquasar SS433, a bright variable emission star, supercritical accreting X-ray binary, identified with a radio source 1909+048 has an unresolved compact radio core and 1 arcsec long aligned jets, as discovered with the MERLIN system. Kotani *et al.* (2006) detected the fast variation in the X-ray emission of SS433 during the radio flares, probably even QPOs of 0.11 Hz. In Trushkin *et al.* (2007b) the daily RATAN light curves are given. In the bright flare of Feb 16, 2006 we detected a delay of the maximum flux at 1 GHz about 2 days, and 1 day at 2.3 GHz, respectively, relatively the maxima at $nu > 4$ GHz. Some flares happened just close to the multi-band program of the studies of SS433 in 3–6 April 2006 (Kotani *et al.* 2007).

Acknowledgements

These studies were supported by the Russian Foundation Basic Research (RFBR) grant N 05-02-17556 and by the mutual RFBR and Japan Society for the Promotion of Science (JSPS) grant N 05-02-19710.

References

Kotani, T., Trushkin, S. A., Valiullin, R. K. *et al.* 2006, ApJ, 637, 486
Kotani, T., Trushkin, S. *et al.* 2007, VI microquasars workshop, Komo, Italy, PoS, (in press)
Levine, A., Bradt, H., Cui, W. *et al.* 1996, ApJ, 469, L33
Pal, S., Ishwara-Chandra, C. H. & Pramesh, A. 2006, ATel, #809
Tsuboi, M., Kuno, N., Umemoto, T. *et al.* 2006, ATel, #727, 1
Trushkin, S., Kotani, T. *et al.* 2007a, in this Volume
Trushkin, S., Bursov, N. *et al.* 2007b, VI microquasars workshop, Komo, Italy, PoS, in press
Waltman, E. B., Fiedler, R. L., Johnston, K. L. & Ghigo, F. D. 1994, AJ, 108, 179

Black Holes from Stars to Galaxies – Across the Range of Masses
Proceedings IAU Symposium No. 238, 2006
V. Karas & G. Matt, eds.
© 2007 International Astronomical Union
doi:10.1017/S1743921307005972

Peculiar outbursts of a black hole X-ray transient, V4641 Sgr

M. Uemura,[1]† T. Kato,[2] D. Nogami,[2] A. Imada,[2] and R. Ishioka[3]

[1]Astrophysical Science Center, Hiroshima Univ., Kagamiyama, Higashi-Hiroshima, Japan

[2]Department of Astronomy, Kyoto University, Kitashirakawa, Kyoto, Japan

[3]National Astronomical Observatory of Japan, Mitaka, Japan

uemuram@hiroshima-u.ac.jp

Abstract. We have kept optical monitoring of a peculiar black hole X-ray binary, V4641 Sgr. Based on our observations, we show that its unprecedented activity can be divided into 5 phases. In this paper, we report their observational properties.

Keywords. Accretion, accretion disks – black hole physics – binaries: close – stars: individual (V4641 Sgr)

1. Introduction

V4641 Sgr is a black hole X-ray binary (BHXB), which has shown mysterious outbursts almost once a year (Uemura *et al.* 2002; Uemura *et al.* 2004a; Uemura *et al.* 2004b; Uemura *et al.* 2005). We pay attention to this object, not only because of its peculiar behaviour, but also because it provides us with a number of information about topics which have recently received much attention in BHXBs, such as the jet-disk interaction (e.g. Mirabel *et al.* 1998), the synchrotron emission in the optical and IR regions (e.g. Fender 2001), and rapid optical variations (e.g. Kanbach *et al.* 2001). We keep on monitoring V4641 Sgr since 1999, and observed 5 optical outbursts. Here, we summarize optical properties of V4641 Sgr outbursts.

2. Result and Discussion

Its outbursts are quite unique in the points of short durations and rapid time-evolutions. Due to these characteristics, the number of data during outbursts is so small that it had been difficult to give a unified description for them even phenomenologically. Based on our long monitoring, however, we have finally found common properties in outbursts in 1999, 2002, 2003, 2004, and 2005. Photometric observations were performed using 30-cm class telescopes at Kyoto University, Universidad de Concepcion, and a number of observatories under collaborations by VSNET.

Whole light curves of the previous five outbursts are shown in the left panel of Fig. 1. The outburst duration is typically one week, while the object keeps active even after the main outbursts. The state of V4641 Sgr can be divided in 5 phases based on its optical variability; i) the initial rising phase, ii) the high variability phase, iii) the super-Eddington flare, iv) the post-outburst active phase, and v) the quiescence. During the phases ii) and iv), the object exhibits rapid optical variations with time-scales of a few seconds. No periodic or quasi-periodic signal has been found in these phases (Uemura *et al.* 2004a; Uemura *et al.* 2004b). While X-ray luminosities were low in the phases ii)

† Present address: Kagamiyama 1-3-1, Higashi-Hiroshima, Japan.

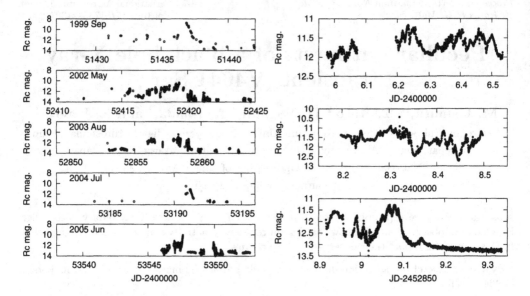

Figure 1. Left: 5 outbursts of V4641 Sgr. Right: Short-term variations during outbursts.

and iv) $(L_X \sim 10^{35-36}$ erg s^{-1}; Uemura *et al.* 2004b; Swank *et al.* 2005), it has reached super-Eddington levels during the final flares, at least in 1999 and 2005, and possibly in 2002 (Hjellming *et al.* 2000; Shimokawabe *et al.* 2005). Relativistic jets were associated to those flares (Hjellming *et al.* 2000).

Between the high variability phase and the super-Eddington flare, we may see a transition of accretion flows from the radiatively inefficient flow to the slim disk. We can expect a strong disk wind at the forming of the slim disk, which might have been observed as quite strong and broad emission lines observed during the final flare in 2004 (Lindstrom *et al.* 2005).

Acknowledgements

This research was partially supported by the Ministry of Education, Science, Sports and Culture, Grant-in-Aid for Young Scientists, 2006.

References

Fender, R. P. 2001, MNRAS, 322, 31
Hjellming, R. M. *et al.* 2000, ApJ, 544, 977
Kanbach, G., Straubmeier, C. & Spruit, H. 2001, Nature, 414, 180
Lindstrom, C. *et al.* 2005, MNRAS, 363, 882
Mirabel, I. F. *et al.* 1998, A&A, 330, L9
Shimokawabe, T. *et al.* 2005, Autumn Meeting of Astronomical Society of Japan
Swank, J. 2003, ATel, 303
Swank, J. 2005, ATel, 536
Uemura, M. *et al.* 2002, PASJ, 54, 95
Uemura, M. *et al.* 2004, PASJ, 56, 823
Uemura, M. *et al.* 2004, PASJ, 56, S61
Uemura, M. *et al.* 2005, IBVS, 5626

Black Holes from Stars to Galaxies – Across the Range of Masses
Proceedings IAU Symposium No. 238, 2006
V. Karas & G. Matt, eds.
© 2007 International Astronomical Union
doi:10.1017/S1743921307005984

Self-gravitating warped disks around nuclear black holes

Ayse Ulubay-Siddiki,[1] Ortwin Gerhard[1] and Magda Arnaboldi[2]

[1]Max Planck Institute for Extraterrestrial Physics, D-85748 Garching, Germany

[2]ESO, Karl-Schwarzschildstr. 2, D-85748 Garching, Germany

email: siddiki@exgal.mpe.mpg.de, gerhard@exgal.mpe.mpg.de, marnabol@eso.org

Abstract. Many galactic nuclei contain disks of gas and possibly stars surrounding a supermassive black hole. These disks may play a key role in the evolution of galactic centers. Here we address the problem of finding stable warped equilibrium configurations for such disks, considering the attraction by the black hole and the disk self-gravity as the only acting forces. We model these disks as a collection of concentric, circular rings.

We find the equilibria of such systems of rings, and determine how they scale with the ring parameters and the mass of the central black hole. We show that in some cases these disk equilibria may be highly warped. We then analyze the stability of these disks, using both direct time integration and linear stability analysis. This shows that the warped disks are stable for a range of disk-to-black hole mass ratios, when the rings extend over a limited range of radii.

Keywords. Galaxies: nuclei, dynamics, black hole physics.

---

1. Introduction

Nuclear disks are known to exist in many galaxies (see for example Herrnstein *et al.* (1996); Genzel *et al.* (2003)). We address the problem of finding stable warped configurations for such disks.

2. Stable Warped Disks

The disk is modeled as a collection of concentric, self-gravitating circular rings surrounding a black hole. The mutual torque between two rings (i, j) has the following dependency (see Arnaboldi & Sparke (1994)):

$$\frac{\partial V_m(\alpha_{ij})}{\partial \alpha_{ij}} \propto \frac{m_i m_j r_i r_j \sin 2\alpha_{ij}}{\pi^2 (r_i^2 + r_j^2)^{3/2}}. \tag{2.1}$$

Here, m and r are the ring masses and radii respectively, and α denotes the inclination angle between the rings. The equilibrium is evaluated by setting all the torques around line of nodes (LONs) to zero. The solutions give the inclination angles of all rings, for a given precession rate. After some algebra, the latter reduces to:

$$\frac{\dot{\phi}}{\Omega_j} = -\frac{D_i}{2A_i} \frac{m_i}{M_{bh}}, \tag{2.2}$$

where D and A are constants specific to each equilibrium.

The stability of the system is studied using standard perturbational method. Figure 2 depicts the equilibrium configurations for the least and most massive disks of one stable family. One sees clearly that the higher mass disk has a more pronounced warp.

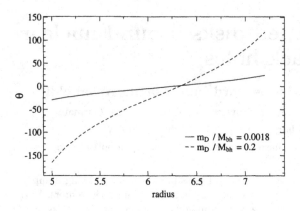

Figure 1. Stable equilibria for a disk of constant surface density.

3. Time Evolution of The Disk

Time integration of the equations of motion is carried out to cross-check the disk stability. Rings still sharing a common LONs after many orbital periods are considered to form a stable disk. Figures 2 and 3 show disks with masses $0.25M_{bh}$ and $0.05M_{bh}$ respectively. The effect of changing the disk mass on the stability is obvious. The higher mass disk dissolves after 10 orbital periods, where the lower mass disk is stable.

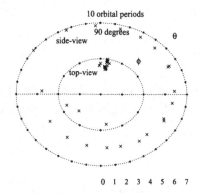

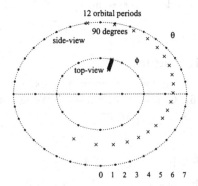

Figure 2. Time evolution of a disk with mass $M_d = 0.25M_{bh}$. After 10 orbital periods the disk dissolves into subrings, hence is unstable.

Figure 3. As in figure 2, but for a disk having $M_d = 0.05M_{bh}$. The rings still lie on the same LONs after 12 orbital periods, hence the disk is stable.

4. Conclusions

We found that more massive nuclear disks in galaxies can have more pronounced warps. These warps are stable in a mass range $0.002 \lesssim M_d/M_{bh} \lesssim 0.2$.

References

Arnaboldi, M. & Sparke, L. S. 1994, AJ, 107, 958
Herrnstein, J. R., Greenhill, L. J. & Moran, J. M. 1996, ApJ, 468, 17
Genzel, R., Schödel, R., Ott, T., Eisenhauer, F., Hofmann, R., Lehnert, M. *et al.* 2003, ApJ, 594, 812

Black Holes from Stars to Galaxies – Across the Range of Masses
Proceedings IAU Symposium No. 238, 2006
V. Karas & G. Matt, eds.
© 2007 International Astronomical Union
doi:10.1017/S1743921307005996

The modelling of the line-locking effect in broad absorption line QSOs

Emmanuil Y. Vilkoviskij and Sergey N. Yefimov

Fessenkov Astrophysical Institute, Observatory, 050020 Almaty, Kazakhstan
email: vilk@aphi.kz

Abstract. We present a unification model of the matter outflow from AGN. The model includes calculations of the hot gas dynamics, the dynamics of the cold clouds in the hot gas flow, and the radiation transfer in this two-phase medium. We used the model for calculation of the ultraviolet spectrum of the quasar q1303+308. The spectrum of this object is a well-know example of the line-locking effect. Our model calculations permits for the first time to obtain some thin details seen in the observed absorption spectrum, which confirms the validity of the main issues of our theory and model.

Keywords. Galaxies: active – (galaxies:) quasars: absorption lines – (galaxies:) quasars: individual (q1303+308) – galaxies: evolution

1. Introduction

Our theoretical approach to the problem of matter outflow from AGN was presented in Vilkoviskij & Karpova (1996), and in complete it was described in Vilkoviskij *et al.* (1999, 2006).

We suppose that AGN consists of three main subsystems: the central super-massive black hole (SMBH), compact stellar cluster (CSC), and gas subsystem. The AGN evolution, driven by intergalactic interactions and merging (e.g., Sanders *et al.* (1988); Menci *et al.* (2003) and the references therein) leads naturally to this structure. The corresponding evolution sequence of an AGN during a "duty cycle" is very well compatible with the standard geometrical unification model (Antonucci 1993).

2. The unifying model and the line-locking effect

From the evolution scheme a unifying outflow model follows, which includes both the "classical" AGN1–AGN2 unification and the absorbing outflow models. We suppose that the main absorbing outflow is placed at the internal surface of the obscuring torus. We take into account the dynamics of the hot gas outflow in the summary gravitation field of the massive black hole and compact stellar cluster, the cold cloud dynamics in the hot gas flow (including radiation pressure and drag forces calculations), and the radiation transfer in the clouded media. Our model was tested with the calculations of spectra of different objects, including in particularly the ultraviolet spectrum of the quasar q1303+308.

The line-locking effect appears in our model as the result of non-linear interaction of the radiation pressure and drag force, which leads to a "ladder" velocity structure $V(r)$. The stability of the narrow spectral details in this case means stability of the velocity structure.

The resulting calculated spectrum in comparison to the observed one (Vilkoviskij & Irwin 2001) is shown in Fig. 1. One can see that it is rather good coincidence in many

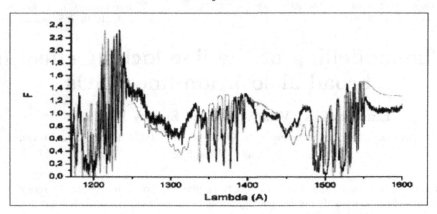

Figure 1. Calculated (gray) and observed (black) spectrum of QSO q1303+308.

details of the calculated and observed spectra (taking into account the model simplifications such as equal masses of cloudlets, some uncertainty of the supposed non-absorbed spectrum, and so on). This proves that our dynamical model explains the line-locking effect.

3. Conclusions

Though the numerical dynamical model, which we used for the q1303+308 spectrum simulation, includes some simplifications, it is the only model explaining the observed spectra with the effect of line-locking seen in the UV band of BALQSOs. We call to those observing in X-rays to obtain the spectrum of q1303+308 in this spectral band. The variability of the UV spectrum of the object is of great interest (Foltz *et al.* 1987, Vilkoviskij & Irwin 2001). Further observations of this unique object in the optical, UV and X-ray bands are desirable for the sake of deeper understanding of the AGN mass outflow physics.

References

Antonucci, R. 1993, ARAA, 31, 473

Foltz, C. B., Weymann, R. J., Morris, S .L. & Turnshek, D. A. 1987, ApJ, 317, 450

Menci, N., Cavaliere, A., Fontana, A., Giallongo, E., Poli, F. & Vittonini, V. 2003, ApJ, 587, L63

Sanders, D. B., Soifer, B .T., Elias, J. H., Neugebauer, G. & Matthews, K. 1988, ApJ, 328, L35

Vilkoviskij, E. Y., Efimov, S. N., Karpova, O. G. & Pavlova, L. A. 1999, MNRAS, 309, 80

Vilkoviskij, E. Y. & Irwin, M. J. 2001, MNRAS, 321, 4

Vilkoviskij, E. Y. & Karpova, O. G. 1996, PAZ, 22, 148

Vilkoviskij, E. Y., Lovelace, R. V. E., Pavlova, L. A., Romanova, M. M. & Yefimov, S. N. 2006, Ap&SS, 306, 129

Black Holes from Stars to Galaxies – Across the Range of Masses
Proceedings IAU Symposium No. 238, 2006
V. Karas & G. Matt, eds.
© 2007 International Astronomical Union
doi:10.1017/S174392130700600X

Observational properties of relativistic black hole winds

Ken-ya Watarai†

Department of Education, Osaka Kyoiku University, 582-8582, Osaka, Japan
email: watarai@cc.osaka-kyoiku.ac.jp

Abstract. We examine observational properties of relativistic black hole winds as an origin of high luminosity sources such as microquasars and ultra-luminous X-ray sources (ULXs). When strong relativistic wind/outflow happens in the vicinity of the black hole, the wind might form the optically-thick photosphere. Therefore the emission observed in ULXs might come from the photosphere of the wind, not from the accretion disk.

We found that the location of the photosphere is larger than the disk thickness for super-Eddington mass-outflow rates and sub-relativistic wind velocities ($v \sim 0.1$–$0.2\,c$). To understand the radiative structure in the high luminosity sources, we should take into account not only the emission from the accretion disk but also the emission from the outflow at the same time.

Keywords. Black hole physics – stars: winds, outflows – X-rays: binaries

1. Introduction

As for observed thermal spectral components in Galactic black hole candidates, it generally seems that accretion disk is the origin. However, if strong, radiation-pressure driven wind blows in a luminous black hole candidate, how do we observe the wind? King & Pounds (2003) have recently proposed the black hole wind scenario in order to explain the origin of the high luminosity in ULXs and a few PG quasars. Here we will discuss the observational properties using a simple relativistic wind model for a stellar mass black hole.

2. Model for Relativistic Black Hole Winds

Our model for relativistic black hole wind is based on Abramowicz *et al.* (1991). Assumptions of the model are steady, spherical symmetry, special/general relativity. From mass conservation equation, the density distribution in co-moving frame is given by

$$\rho = \frac{\dot{M}_0}{4\pi R^2 v Y^*},\tag{2.1}$$

where $\dot{M}_0$ is the mass-outflow rate, and v is the wind velocity, and R is the distance from a black hole, $R = \sqrt{r^2 + z^2}$, respectively. The relativistic corrections are included in $Y^* = \gamma g_{00}^{1/2}$, where $\gamma = (1 - \beta^2)^{-1/2}$, and $g_{00} = (1 - r_g/R)$. Here β is the wind velocity normalized by the speed of light, $\beta \equiv v/c$, and r_g is the Schwarzschild radius, $r_g = 2GM/c^2$. We assume that the wind velocity β is constant, and a 10 solar mass black hole.

---

† Research fellow of the Japan Society for the Promotion of Science.

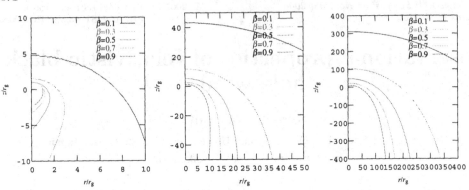

Figure 1. Location of the photosphere for various wind velocities β. Mass loss rates, $\dot{m}_0 = \dot{M}_0/\dot{M}_{\rm Edd}$, equal to 1, 10, and 100 from left to right.

3. Calculation Method & Location of the Photosphere

To calculate flux around a Schwarzschild black hole, we adopt the "Ray-Tracing Method" including special/general relativity (Luminet 1979). Total optical depth of moving media from an observer at infinity to the location of the "photosphere" is given by

$$\tau_{\rm ph} = \int_{R_{\rm ph}}^{\infty} \kappa \rho \gamma (1 - \beta \cos \theta) ds = 1 \tag{3.1}$$

where θ, κ, ds are the viewing angle to the observer, opacity for electron scattering, and small distance along the null geodesics. We obtain the location of the photosphere, $R_{\rm ph}$, from the $\tau_{\rm ph} = 1$ condition. The location of the photosphere is shown in figure 1.

Increasing of $\dot{m}_0$ ($\gtrsim 10$) leads to high-density wind, and the $R_{\rm ph}$ also be large. On the other hand, when the wind velocity closes to the light speed for high-$\dot{m}_0$, the density of the wind decreases because of the mass conservation. Thus we obtain small emitting region. The $R_{\rm ph}$ is given by the combination of these parameters. We note that the location of the photosphere is larger than the disk thickness in high-$\dot{m}_0$ and low-β cases.

4. Conclusions

According to Begelman *et al.* (2006), mass-loss rate in SS433 is extremely high, $dM/dt = 5 \times 10^{-4} M_\odot yr^{-1}$. In this surroundings we can no longer see the inner accretion disk, the emission from the photosphere of the outflow will be observed. As for ULXs, if the super-Eddington mass outflow occurs, the size of the photosphere increases, but the temperature of the outflow decreases as increasing $R_{\rm ph}$. Thus this feature conflicts with the observations in several ULXs. Comparing time-dependent spectral properties with various models will be important in the future, and it will help to understand real nature of ULXs.

Acknowledgements

This work was supported in part by the Grants-in-Aid of the Ministry of Education, Science, Sports, and Culture of Japan (16004706; K.W.).

References

Abramowicz, M. A., Novikov, I. D. & Paczyński, B. 1991, ApJ, 369, 175
King, A. R. & Pounds, K. A. 2003, MNRAS, 345, 657
Luminet, J.-P. 1979, A&A, 75, 228

Black Holes from Stars to Galaxies – Across the Range of Masses
Proceedings IAU Symposium No. 238, 2006
V. Karas & G. Matt, eds.
© 2007 International Astronomical Union
doi:10.1017/S1743921307006011

The relation between the properties of the NLR in narrow-line Seyfert 1 galaxies and the accretion rate

Dawei Xu[1] and Stefanie Komossa[2]

[1]National Astronomical Observatories, Chinese Academy of Sciences, Beijing 100012, China
[2]Max-Planck-Institut für Extraterrestrische Physik, Postfach 1312, 85741 Garching, Germany
email: dwxu@bao.ac.cn, skomossa@mpe.mpg.de

Abstract. We present a systematic study of the properties of the narrow-line region (NLR) of narrow-line Seyfert 1 galaxies (NLS1s) using Sloan Digital Sky Survey (SDSS) spectroscopy. Various correlations between the observed parameters and physical properties of NLS1s and broad-line Seyfert 1 galaxies (BLS1s) are detected. We search for possible origins of these trends by employing correlation analyses. We further investigate the relationship between black hole mass, Eddington ratio ($L/L_{\rm Edd}$) and physical parameters of the NLR.

1. Introduction

NLS1s are intriguing due to their extreme emission line and continuum properties. They lie at one extreme end of the Boroson & Green (1992) eigenvector 1 (EV1) which governs the correlations between AGN properties. Boroson (2002) and Sulentic *et al.* (2003) suggested that EV1 is an indicator of $L/L_{\rm Edd}$. Density of an out-flowing wind was speculated to be among the other main drivers underlying EV1 (e.g. Lawrence *et al.* 1997). However, there has been little investigation on NLR density for NLS1s. We present for the first time the NLR density measurements for one of the largest homogeneously analyzed NLS1 sample to date and compare it with that of BLS1s.

2. Sample selection

We homogeneously analyze and compare optical emission line properties of a large number of NLS1s with BLS1s based on spectra obtained in the course of the SDSS Data Release 3 (Abazajian *et al.* 2005). Only objects with redshift z less than 0.3 are taken into account. In order to get an accurate measurement, we require the density-sensitive ratio [S II] $\lambda\lambda6716,6731$ to have a signal-to-noise ratio (S/N) greater than 5. A total of 61 objects with FWHM of the broad component of Hβ (hereafter H$\beta_{\rm b}$) $\leqslant 2000\,{\rm km\,s^{-1}}$ are included in the NLS1 sample, while 40 with FWHM(H$\beta_{\rm b}$) > $2000\,{\rm km\,s^{-1}}$ are included in the BLS1 sample.

3. The density of the NLR

One of our main goals is to examine whether or not there is a difference in electron density n_e between NLS1s and BLS1s, in order to test different NLS1 models. We use the density diagnostic [S II] $\lambda6716/\lambda6731$ to measure the NLR electron density. We find the sources do not homogeneously populate the n_e – FWHM(H$\beta_{\rm b}$) diagram (Figure 1). Our key finding is the detection of a '*zone of avoidance*' in the n_e – FWHM(H$\beta_{\rm b}$) diagram: BLS1s (FWHM H$\beta_{\rm b}$ > $2000\,{\rm km\,s^{-1}}$) avoid low average densities, and all show n_e >

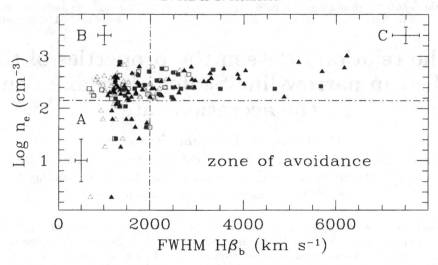

Figure 1. Electron density in dependence of FWHM of $H\beta_b$. Filled and open symbols represent $H\beta_b$ modeled by Gaussian and Lorentzian profiles, respectively. Squares are QSOs; triangles are Seyfert 1s. The dot-dashed lines distinguish areas populated by: (A) NLS1s with small $H\beta_b$ width and low density; (B) NLS1s small $H\beta_b$ width and high density; and (C) BLS1s with large $H\beta_b$ width and high density. Median error bars of each regime are given.

$140 \, \mathrm{cm}^{-3}$. On the other hand, NLS1s show a larger scatter in densities in the range $n_e = 2 \sim 770 \, \mathrm{cm}^{-3}$, including a significant number of objects with low densities (see details in Xu *et al.* 2006).

In order to search for possible origins of this peculiar trend, we investigate the relationship between black hole mass, L/L_{Edd} and physical parameters of the NLR. The correlation analyses show that the L/L_{Edd} anti-correlates with density. We investigate a number of different models for the '*zone of avoidance*' in density. We tentatively favor winds/outflows and/or different host galaxy (ISM) properties.

Acknowledgements

We thank the Chinese National Science Foundation (NSF) for support under grant NSFC-10503005. DX acknowledges Max-Planck-Institut für extraterrestrische Physik and Max-Planck-Gesellschaft for financial support. This research made use of the SDSS archives and the Catalogue of Quasars and Active Nuclei.

References

Abazajian, K., Adelman-McCarthy, J. K., Agüeros, M. A. *et al.* 2005, AJ, 129, 1755
Boroson, T. A. & Green, R. F. 1992, ApJS, 80, 109
Boroson, T. A. 2002, ApJ, 565, 78
Sulentic, J. W., Zamfir, S., Marziani, P., Bachev, R., Calvani, M. & Dultzin-Hacyan, D. 2003, ApJ, 597, L17
Xu, D., Komossa, S., Zhou, H., Wang, T. & Wei, J. 2006, ApJ, to be submitted

Black Holes from Stars to Galaxies – Across the Range of Masses
Proceedings IAU Symposium No. 238, 2006
V. Karas & G. Matt, eds.
© 2007 International Astronomical Union
doi:10.1017/S1743921307006023

Measuring the supermassive black hole parameters with space missions

Alexander F. Zakharov

National Astronomical Observatories of Chinese Academy of Sciences, Beijing 100012, China

ITEP, 25, B. Cheremushkinskaya st., Moscow 117259, Russia

BLTP, JINR, Dubna, Russia

email: zakharov@itep.ru

Abstract. Recent X-ray observations of microquasars and Seyfert galaxies reveal broad emission lines in their spectra, which can arise in the innermost parts of accretion disks. Recently Müller & Camenzind (2004) classified different types of spectral line shapes and described their origin. Zakharov (2006b) clarified their conclusions about an origin of doubled peaked and double horned line shapes in the framework of a radiating annulus model and discussed s possibility to evaluate black hole parameters analyzing spectral line shapes.

Keywords. Black holes – X-ray astronomy – supermassive black holes

---

There are a lot of papers discussing theoretical aspects of possible scenarios for generation of broad iron lines in AGNs, see for example, reviews by Fabian *et al.* (2000); Matt (2006). Moreover, an influence of microlensing on Fe K_α line shapes and spectra was discussed by Popovic *et al.* (2006); optical depths for these phenomena were calculated by Zakharov *et al.* (2004, 2005a,b). Formation of shadows (mirages) is another example when general relativistic effects are extremely important and in principle they could be detected with forthcoming interferometrical facilities such as Radioastron, Millimetron, MAXIM, as was shown by Zakharov *et al.* (2005c,d,e,f,g,h) (perspective studies of microlensing with Radioastron facilities were discussed (Zakharov (2006a)). Observations of shadows could give a real chance to observe "faces" of black holes of black holes and confirm general relativity predictions in the strong gravitational field, and to obtain new constraints on alternative theories of gravity.

Müller & Camenzind (2004) classified different types of spectral line shapes and described their origin. Zakharov (2006b) and Zakharov & Repin (2005, 2006) clarified their conclusions about an origin of doubled peaked and double horned line shapes. Based on results of numerical simulations we showed using a radiating annulus model that double peaked spectral lines arise for *almost any* locations of narrow emission rings (annuli) (except closest orbits as we could see below) although Müller & Camenzind (2004) concluded that such profiles arise for relatively flat spacetimes and typical radii for emission region about 25 r_g. We did not impose assumptions about an emissivity law; we only assume that the radiating region is a narrow circular ring (annulus). We used an approach which was discussed in details by Zakharov (1991a); Zakharov (1994a); Zakharov (1995, 2003, 2004, 2005); Zakharov & Repin (1999, 2002, 2003a,b,c, 2004a); Zakharov *et al.* (2003, 2004). The model is based on results of qualitative analysis done earlier by Zakharov (1986, 1988, 1989, 1991b).

Acknowledgements

AFZ is grateful to the National Natural Science Foundation of China (NNSFC) (Grant 10233050) and National Key Basic Research Foundation of China (Grant TG 2000078404) for a partial financial support of the work.

References

Fabian, A. C., Iwazawa, K., Reynolds, C. S. & Young, A. J. 2000, PASP, 112, 1145

Matt, G. 2006, Astronomische Nachrichten, 327, 949

Müller, A. & Camenzind, M. 2004, A&A, 413, 861

Popović, L. C., Jovanović, P., Mediavilla, E., Zakharov, A. F., Abajas, C. *et al.* 2006, AJ, 637, 630

Zakharov, A. F. 1986, Sov. Phys. - J. Exper. Theor. Phys., 64, 1

Zakharov, A. F. 1988, Soviet Astron., 32, 456

Zakharov, A. F. 1989, Sov. Phys. – J. Exper. Theor. Phys. 68, 217

Zakharov, A. F. 1991a, Soviet Astron., 35, 30

Zakharov, A. F. 1991b, Soviet Astron., 35, 147

Zakharov, A. F. 1994a, MNRAS, 269, 283

Zakharov, A. F. 1994b, Classical & Quantum Gravity,, 11, 1027

Zakharov, A. F. 1995, in H. Böhringer, G. E. Morfill, J. E. Trümper (eds.), Proc. of the 17th Texas Symposium on Relativistic Astrophysics, (Ann. NY Academy of Sciences), p. 550

Zakharov, A. F. 2003, Publ. of the Astron. Observatory of Belgrade, 76, 147

Zakharov, A. F. 2004, in: L. Hadzievski, T. Gvozdanov, and N. Bibić (eds.) The Physics of Ionized Gases (AIP Conf. Proc.), Vol. 740, p. 398

Zakharov, A. F. 2005, Intern. J. Mod. Phys. A, 20, 2321

Zakharov, A. F. 2006a, Astronomy Reports, 50, 79

Zakharov, A. F. 2006b, Physics of Atomic Nuclei, 37, 1

Zakharov, A. F., Kardashev, N. S., Lukash, V. N. & Repin, S. V. 2003, MNRAS, 342, 1325

Zakharov, A. F., Ma, Z. & Bao, Y. 2004, New Astronomy, 9, 663

Zakharov, A. F., Nucita, A. A. & De Paolis, F., Ingrosso G. 2005c, New Astronomy, 10, 479

Zakharov, A. F., Nucita, A. A., De Paolis, F. & Ingrosso, G. 2005d, in: G. Vilasi, G. Espositio, G. Lambiase, G. Marmo, G. Scarpetta (eds.), Proc. of the 16th SIGRAV Conference on General Relativity (AIP Conf. Proceedings), Vol. 751, p. 227

Zakharov, A. F., Nucita, A. A., De Paolis, F. & Ingrosso, G. 2005e, in: P. Chen, E. Bloom, G. Madejski, V. Petrosian (eds.), Proc. of the 22nd Texas Symposium on Relativistic Astrophysics (Stanford University: Stanford), PSN 1226

Zakharov, A. F., Nucita, A. A., De Paolis, F. & Ingrosso, G. 2005f, in: H. V. Klapdor-Kleingrothaus and D. Arnowitt (eds.), Proc. of "Dark Matter in Astro- and Particle Physics", (Springer: Heidelberg), p. 77

Zakharov, A. F., Nucita, A. A., De Paolis, F. & Ingrosso, G. 2005j, A&A, 442, 795

Zakharov, A. F., Nucita, A. A., De Paolis, F. & Ingrosso, G. 2005h, in: J. Tran Thanh Van and J. Dumarchez (eds.) Proc. of XXXXth Rencontres de Moriond "Very High Energy Phenomena in the Universe", (The GIOI Publishers), p. 223

Zakharov, A. F. Popović, L. C. & Jovanović, P. 2004, A&A, 420, 881

Zakharov, A. F., Popović, L. C. & Jovanović, P. 2005a, in: Y. Mellier & G. Meylan (eds.), Proc. of IAU Symposium, "Gravitational Lensing Impact on Cosmology", (Cambridge University Press: Cambridge), vol. 225, p. 363

Zakharov, A. F., Popović, L. C. & Jovanović, P. 2005b, in: Y. Giraud-Heraud and J. Tran Thanh Van and J. Dumarchez (eds.) Proc. of XXXIXth Rencontres de Moriond "Exproring The Universe", (The GIOI Publishers), p. 41

Zakharov, A. F. & Repin, S. V. 1999, Astronomy Reports, 43, 705

Zakharov, A. F. & Repin, S. V. 2002, Astronomy Reports, 46, 360

Zakharov, A. F. & Repin, S. V. 2003a, A&A, 406, 7

Zakharov, A. F. & Repin, S. V. 2003b, Astronomy Reports, 47, 733

Zakharov, A. F. & Repin, S. V. 2003c, Nuovo Cimento, 118B, 1193

Zakharov, A. F. & Repin, S. V. 2004, Adv. Space Res., 34, 1837

Zakharov, A. F. & Repin, S. V. 2005, Mem. S. A. It. Suppl., 7, 60

Zakharov, A. F. & Repin, S. V. 2006 New Astronomy, 11, 405

Black Holes from Stars to Galaxies – Across the Range of Masses
Proceedings IAU Symposium No. 238, 2006
V. Karas & G. Matt, eds.

© 2007 International Astronomical Union
doi:10.1017/S1743921307006035

Doppler boosting and de-boosting effects in relativistic jets of AGNs and GRBs

Jianfeng Zhou[1] and Yan Su[2]

[1]Department of Engineering Physics, Center for Astrophysics, Tsinghua University, Beijing, 100084, China

[2]National Astronomical Observatories, Chinese Academy of Sciences, Chaoyang District, Datun Road 20A, Beijing, 100012, China

email: zhoujf@tsinghua.edu.cn, suyan@bao.ac.cn

Abstract. It is widely accepted that the Doppler de-boosting effects exist in counter relativistic jets. However, people often neglect another important fact that both Doppler boosting and de-boosting effects could happen in forward relativistic jets. Such effects might be used to explain some strange phenomena, such as the invisible gaps between the inner and outer jets of AGNs, and the rapid initial decays and re-brightening bumps in the light curves of GRBs.

Keywords. Relativity – acceleration of particles – galaxies: jets – gamma rays: bursts

1. Doppler factors in relativistic jets

In the relativistic jets of AGNs or GRBs, the observed flux is related to their intrinsic flux by $F_{\rm obs} = \delta^{3+\alpha} F$, where δ is Doppler factor, $F_{\rm obs}$ and F are the observed and intrinsic flux respectively , and α is the spectral index (Blandford & Königl 1979). If δ is greater than 1, then the observed flux will be enhanced, which is called Doppler boosting effect. On the other hand, if *delta* is less than 1, the observed flux is attenuated, which is named to Doppler de-boosting effect.

The Doppler factor of a jet is

$$\delta = \frac{1}{\gamma(1 - \beta \cos\theta)} \tag{1.1}$$

where β is the velocity, γ is the Lorentz factor, and $0 \leqslant \theta \leqslant \pi$ is the viewing angle.

For a counter relativistic jet where $\theta > \pi/2$, the Doppler factor δ is always less than 1, which can easily be derived from equation 1.1. So a counter jet is always Doppler de-boosted. For a forward relativistic jet, the situation is more complex. Provided that the viewing angle of jet $\theta > 0$ (when $\theta = 0$, δ is always greater that 1), the Doppler factor can be greater than 1 as well as less than 1 (see figure 1 for detail). Therefore, both Doppler boosting and de-boosting effects could happen in forward jets.

2. Application to AGNs and GRBs

Some observational facts. In many radio loud AGNs, the large scale jets share some common features in their profiles. Firstly, there are compact and bright cores in the center of jets. Secondly, adjacent to the cores, the flux of the jets drops down very quickly, even form some gaps where the jets are undetectable. Thirdly, the jets will be re-brightened in the outer region.

Recently, the SWIFT found some interesting properties in GRBs. Five GRBs' X-ray light curves are characterized by a rapid fall-off for first few hundred seconds, followed

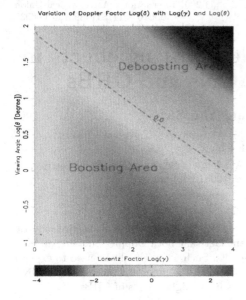

Figure 1. The function of Doppler factor in terms of viewing angle and Lorentz factor. Doppler factor, Lorenz factor and viewing angle are all plotted in the logarithmic scale. A contour line, corresponding to $\log \delta = 0$, is shown. This line cuts the figure into two areas, i.e. Doppler boosting area and de-boosting area.

by a less rapid decline lasting several hours. The light curves also show discontinuity (Tagliaferri *et al.* (2005)). Burrows *et al.* (2005) found that there were bright X-ray flares in GRB afterglows.

Explanation. To qualitatively explain the profiles of the jets in AGNs as well as the light curves in GRBs, we divide the evolution of the jets into four stages :

• Stage I : The jets are accelerating and boosting, which relates to the bright cores of AGNs and the bursts of GRBs. The initial acceleration of relativistic jets has been detected in 3C 273 by Krichbaum *et al.* (2001) and modeled by Zhou *et al.* (2004). In this stage, γ is usually less than a few tens, and δ increase very quickly.

• Stage II : The jets are accelerating but de-boosting. As the acceleration continues, γ will be very large (> 100). Therefore, the jets will enter into the de-boosting area, i.e. $\delta < 1$. In this stage, the observed flux of the jets decreases very quickly, and often forms the gaps between inner and outer jets in AGNs and the rapid decays and discontinuous light curves in GRBs.

• Stage III : The jets are decelerating and boosting. The acceleration, however, won't last forever because of radiation loss and the interaction between the jets and the surrounding medium. Thus, the jets will decelerate and their δ will increase again. Consequently, the Doppler boosted jets will appear again in the profiles of AGNs or in the light curves of GRBs.

• Stage IV : The jets are decelerating and de-boosting. Finally, due to the same reasons as in stage III, the jets will gradually disappear.

References

Blandford, R. D & Königl, A. 1979, ApJ, 232, 34

Burrows, D. N., Romana, P., Falcone, A. *et al.* 2005, Science, 309, 1833

Krichbaum, T. P., Graham, D. A., Witzel, A. *et al.* 2001, Particles and Fields in Radio Galaxies, ASP Conference Series

Tagliaferri, G., Goad, M., Ghincarini, G. *et al.* 2005, Nature, 436, 985

Zhou, J. F., Zheng, C., Li, T. P., Su, Y. & Venturi, T. 2004, ApJ, 616, L95

Author Index

Abolmasov, P. K. – 225, 229
Agapitov, A. – 395
Agüero, M. P. – 277
Angelini, L. – 23
Arévalo, P. – 319
Arnaboldi, M. – 467
Axon, D. – 365

Baes, M. – 321
Baganoff, F. K. – 181, 195
Bannikova, E. Yu. – 323
Barbosa, F. K. B. – 457
Barnes, A. D. – 219
Bartolini, C. – 447
Baum, S. – 365
Beifiori, A. – 349
Bertola, F. – 349
Beskin, G. – 159, 391
Bian, W. – 325
Bičák, J. – 139
Bicknell, G. V. – 247
Birkinshaw, M. – 443
Biryukov, A. – 159
Blanford, R. D. – 291
Bochkarev, N. G. – 327
Bon, E. – 329
Brusa, M. – 287
Burenkov, A. N. – 383
Bursa, M. – 333
Bursov, N. N. – 463

Callanan, P. J. – 361, 435
Campbell, R. M. – 437
Capak, P. – 287
Capetti, A. – 365
Cappellari, M. – 331
Cappi, M. – 429
Cardaci, M. V. – 335
Carrera, F. J. – 59
Casares, J. – 3, 219, 413
Casey, S. – 437
Charles, P. A. – 219
Chatterjee, T. K. – 337
Chavushyan, V. H. – 383
Chen, W.-C. – 339
Chen, X. – 341
Cheung, C. C. – 443
Chiaberge, M. – 273, 365
Ciroi, S. – 383
Clark, J. S. – 219
Coccato, L. – 349, 355
Coimbra-Araújo, C. H. – 343, 345, 427

Coker, R. F. – 347
Collin, S. – 111
Colpi, M. – 255
Copperwheat, C. – 251
Cornelisse, R. – 219
Corral, A. – 59
Corsini, E. M. – 349, 355
Côté, P. – 261, 351
Cottam, J. – 23
Cotton, W. D. – 93
Cropper, M. – 251
Cuadra, J. – 191
Czerny, B. – 99, 353, 411, 429

Dalla Bontà, E. – 349, 355
de Zeeuw P. T. – 331
Debur, V. – 159
Díaz, A. I. – 335
Díaz, R. D. – 277
Dodson, R. – 357
Done, C. – 23
Dotani, T. – 23
Dottori, H. – 277
Dovčiak, M. – 99, 181, 359, 407, 425
Dultzin-Hacyan, D. – 83
Duschl, W. J. – 153, 407

Eckart, A. – 181, 187, 407, 415
Edwards, P. – 357
Elebert, P. – 361
Elvis, M. – 287

Fabian, A. C. – 23, 129
Fabrika, S. N. – 225, 229, 461
Fender, R. P. – 437
Ferrarese, L. – 261, 355
Fiestas, J. – 361
Floyd, D. – 365
Fomalont, E. – 357
Freeland, M. C. – 247
Frey, S. – 357
Friaça, A. C. S. – 343
Fukazawa, Y. – 23
Fukumura, K. – 367

Gabor, J. – 287
Garrett, M. A. – 437
Gaskell, C. M. – 375
Gavrilović, N. – 369, 371
Gelbord, J. M. – 443
Genzel, R. – 173
Gerhard, O. – 467

Ghez, A. M. – 195
Gonçalves, A. C. – 453
González Hernández, J. I. – 43
González-Martín, O. – 373
Goosmann, R. W. – 99, 375, 429
Greene, J. E. – 87
Gu, Q. – 325
Guarnieri, A. – 447
Gurvits, L. – 357

Haas, J. – 201
Hameury, J.-M. – 297
Hanžl, D. – 447
Hara, T. – 377, 379
Harris, D. E. – 443
Hasinger, G. – 287
Heinzeller, D. – 153
Heinz, S. – 65
Hirabayashi, H. – 357
Ho, L. C. – 87, 385
Horák, J. – 381
Horiuchi, S. – 357
Hornstein, S. D. – 195
Huang, Y. F. – 403
Huchra, J. P. – 287
Hung, C. K. – 247
Hunstead, R. W. – 441

Ilić, D. – 329, 383
Imada, A. – 465
Impey, C. D. – 287
Inoue, H. – 385
Ishioka, R. – 465
Israelian, G. – 43
Ivanidze, R. Z. – 397

Jaffe, W. – 93
Janiuk, A. – 353
Jiménez-Bailón, E. – 373
Jorstad, S. G. – 443
Jovanović P. – 431

Kadlecová, H. – 387
Kajiura, D. – 377, 379
Kapanadze, B. Z. – 397, 399
Karas, V. – 99, 139, 173, 181, 201, 359,
    381, 407, 425, 429
Karitskaya, A. E. – 389
Karpov, S. – 159, 225, 391
Kato, T. – 465
Kawai, N. – 461
Kelleher, C. – 435
Kimeridze, G. N. – 397, 399
King, A. R. – 31
Knigge, C. – 219
Kohmura, T. – 23
Kollatschny, W. – 369

Komberg, B. V. – 393
Komossa, S. – 473
Kontorovich, V. M. – 323
Kotani, T. – 23, 461
Kryvdyk, V. – 395
Kubota, A. – 23
Kuncic, Z. – 247, 453
Kunitomo, S. – 377, 379
Kurtanidze, O. M. – 397, 399, 419

La Mura G. – 383
Landt, H. – 443
Lasota, J.-P. – 297
Lawrence, A. – 117
Ledvinka, T. – 139
Letelier, P. S. – 343, 401
Li, X.-D. – 339
Lilly, S. J. – 287
Lindfors, E. J. – 305
Liu, F. K. – 341
Lovell, J. – 357
Lu, J. R. – 195
Lu, Y. – 403

Macchetto, F. D. – 273, 365
Machida, M. – 37, 405, 421
Madau, P. – 73
Madrid, J. – 365
Maeda, K. – 241
Majorova, E. K. – 463
Makishima, K. – 23, 209
Malkan, M. A. – 291
Marcado, A. – 383
Markoff, S. – 145
Márquez, I. – 373
Marscher, A. P. – 443
Martínez-Pais, I. G. – 413
Marziani, P. – 83
Masegosa, J. – 373
Mast, D. – 277
Mateos, S. – 59
Matsumoto, R. – 37, 405, 421
Matt, G. – 359
Matthews, K. – 195
McCarthy, P. J. – 287
McHardy, I. – 319
Meisenheimer, K. – 93
Merloni, A. – 65
Meyer, L. – 181, 407, 415
Mickaelian, A. – 371
Miley, G. – 365
Miller-Jones, J. C. A. – 437
Mineshige, S. – 153
Mirabel, I. F. – 19, 309
Miralda-Escudé, J. – 355
Mitra, A. – 409
Moellenbrock, G. – 357

Molteni, D. – 423
Morris, M. R. – 181, 195
Moscibrodzka, M. – 411
Mouchet, M. – 99
Muñoz-Darias, T. – 413
Murata, Y. – 357
Murphy, K. D. – 123
Mužić, K. – 415

Nakamura, K. E. – 417, 421
Namiki, M. – 461
Nayakshin, S. – 191
Negrete, C. A. – 83
Niedźwiecki, A. – 13
Nikolashvili, M. G. – 397, 399, 419
Nizhelskij, N. A. – 463
Nogami, D. – 465
Nomoto, K. – 241

Oda, H. – 421
O'Dea, C. – 365
Ohkubo, T. – 241
Ohsuga, K. – 153, 301
Okuda, T. – 423

Page, K. L. – 105
Page, M. – 59
Paragi, Z. – 357, 437
Pecháček, T. – 425
Peck, A. B. – 269
Pedron, I. T. – 427
Perlman, E. – 365, 443
Petit, C. – 371
Piasecki, S. – 321
Piccioni, A. – 447
Pittard, J. M. – 347
Pizzella, A. – 349, 355
Plokhotnichenko, V. – 159
Podolský, J. – 387
Pollack, L. K. – 269
Ponti, G. – 99, 429
Pooley, G. G. – 437
Popović, L. Č. – 329, 369, 371, 383, 431
Pounds, K. A. – 105
Požár, N. – 433
Proga, D. – 165, 411
Prugniel, P. – 371

Quillen, A. – 365

Rafanelli, P. – 383
Rahvar, S. – 439
Rebolo, R. – 43
Rees, M. J. – 51, 241
Reeves, J. N. – 105
Reynolds, C. – 435, 437

Riffel, R. A. – 457
Rodrigues, I. – 277
Rodriguez, C. – 269
Romani, R. W. – 269
Rushton, A. – 437
Russell, L. – 361

Šácha, J. – 433
Safonova, M. – 439
Sakai, K. – 377, 379
Santos-Lleó, M. – 335
Saripalli, L. – 441
Sarzi, M. – 349
Schinnerer, E. – 287
Schödel, R. – 181, 187, 407, 415
Schwartz, D. A. – 443
Scott, W. – 357
Scoville, N. Z. – 287
Semerák, O. – 433
Shalyapin, V. N. – 445
Shapovalova, A. I. – 327, 383
Shaw, S. E. – 361
Shen, Z. – 357
Shoji, M. – 375
Sholukhova, O. N. – 229
Siemiginowska, A. – 443
Sigua, L. A. – 397, 399
Šimon, V. – 447
Slaný, P. – 449, 459
Sobolewska, M. A. – 13
Sobouti, Y. – 451
Soria, R. – 235, 247, 251, 453, 455
Sparks, B. – 365
Spencer, R. E. – 437
Spurzem, R. – 361
Stawarz, L. – 443
Steeghs, D. – 219
Storchi-Bergmann, T. – 283, 457
Straubmeier, C. – 181
Strong, M. – 437
Stuchlík, Z. – 449, 459
Šubr, L. – 201
Subrahmanyan, R. – 441
Sulentic, J. W. – 83
Su, Y. – 477
Suzuki, T. – 241
Szomoru, A. – 437

Takahashi, H. – 23
Takahashi, M. – 367
Taylor, G. B. – 269
Tedds, J. A. – 59
Terashima, Y. – 123, 385
Teresi, V. – 423
Tinarelli, S. – 365
Török, G. – 459

Treu, T. – 291
Trinchieri, G. – 255
Tristram, K. R. W. – 93
Trump, J. R. – 287
Trushkin, S. A. – 461, 463
Tsuruta, S. – 241, 367
Tudose, V. – 437
Türler, M. – 305

Ueda, Y. – 23
Uemura, M. – 465
Ulubay-Siddiki, A. – 467
Umeda, H. – 241
Uttley, P. – 319

van de Ven, G. – 331
van den Bosch, R. C. E. – 331
Viallet, M. – 297
Vilkoviskij, E. – 469
Volonteri, M. – 51

Watarai, K. – 471
Watson, M. G. – 59
Wiik, K. – 357
Winge, C. – 457

Wolter, A. – 255
Wong, D. S. – 455
Woo, J.-H. – 291
Worrall, D. M. – 443
Wu, K. – 251

Xu, D. – 473

Yamada, S. – 23
Yamaoka, K. – 23
Yaqoob, T. – 123
Yasuda, T. – 23
Yefimov, S. – 469

Žáček, M. – 433
Zakharov, A. F. – 475
Zavala, R. T. – 269
Zellerin, T. – 433
Zensus, A. – 415
Zhang, S. N. – 403
Zhao, Y. – 325
Zheng, Z. – 403
Zhou, J. – 477
Życki, P. T. – 13

Black Holes
From Stars to Galaxies
across the range of masses

IAU Symposium 238
Prague, 21–25 August 2006

astro.cas.cz/iaus238

Monday, August 21

Session I. Stellar-mass black holes (chair: P. Charles)
14:00 Observational evidence for stellar-mass black holes (Invited Review); J. Casares
14:35 X-ray energy spectra of low and high-frequency quasi-periodic oscillations in accreting black holes; P. T. Zycki
14:55 Microquasars: disk-jet coupling in stellar-mass black holes; F. Mirabel
15:10 Suzaku observation of the black hole transient 4U1630-472; A. Kubota
15:30 Poster viewing (South Hall), Coffee

Session II. Stellar-mass black holes (chair: R. Narayan)
16:00 Matter accretion and ejection in black-hole systems (Invited Review); A. R. King
16:35 Formation of rapidly rotating black holes in massive binary stellar systems; P. C. Joss
16:55 Sawtooth-like oscillations of black hole accretion disks; R. Matsumoto
17:10 On the origin of the black hole in the X-ray binary XTE J1118+480; J. I. Gonzalez Hernandez
17:30 Poster viewing (South Hall)

Invited Discourse
18:15 Similar phenomena at different scales: black holes, Sun, supernovae, galaxies and galaxy clusters; Shuang Nan Zhang (Congress Hall)

Tuesday, August 22

Session III. Formation and evolution of massive black holes (chair: K. Pounds)
09:00 Massive black holes (Invited Review); M. J. Rees
09:35 The cosmic evolution of black hole accretion; G. Hasinger
09:55 Statistical analysis of optical emission line properties of 2dF type-1 AGN; S. Mateos
10:10 Cosmological growth of supermassive black holes: the kinetic luminosity function of AGN; A. Merloni
10:30 Poster viewing (South Hall), Coffee

Session IV. Formation and evolution of massive black holes (chair: F. Macchetto)
11:00 Formation and evolution of supermassive black holes, black-hole binary merging (Invited Review); P. Madau
11:35 Quasar evolution: black hole mass and accretion rate determination; D. Dultzin-Hacyan
11:55 X-ray spectral evolution of quasars; G. Chartas
12:10 Black hole growth in the local universe; J. E. Greene
12:30 Lunch

Session V. Active galactic nuclei (chair: A. Celotti)
14:00 Supermassive black holes: accretion and outflows (Invited Review); M. C. Begelman
14:35 Hard X-ray spectra of AGN observed with Suzaku; H. Kunieda
14:55 Mapping the circumnuclear dust in nearby AGN with the mid-infrared interferometric instrument MIDI; K. R. W. Tristram
15:10 X-ray variability in AGN: implications of magnetic flares; R. W. Goosmann
15:30 Poster viewing (South Hall), Coffee

Session VI. Active galactic nuclei (chair: G. Matt)
16:00 Cumulative effects of outflows on the X-ray spectra of AGN; K. A. Pounds
16:20 Uncertainties on the black hole masses in AGN and consequences on the Eddington ratios; S. Collin
16:35 Warped disks and the Unified Scheme; A. Lawrence
16:55 An accretion disk laboratory in the Seyfert galaxy NGC 2992; T. Yaqoob
17:10 Cosmological evolution of active galactic nuclei X-ray luminosity function; Y. Ueda
17:30 Poster viewing (South Hall)

Invited Discourse
18:15 The power of new experimental techniques in astronomy: zooming in on the black hole in the center of the Milky Way; Reinhard Genzel (Congress Hall)

Wednesday, August 23

Session VII. Physical processes near black holes (chair: V. Karas)
09:00 Strong-gravity effects: X-ray spectra, timing, polarimetry (Invited Review); A. C. Fabian
09:35 GRS 1915+105: a near-extreme Kerr black hole; R. Narayan
09:55 Black holes and magnetic fields; J. Bicak
10:10 Constraining jet physics in weakly accreting black holes; S. B. Markoff
10:30 Poster viewing (South Hall), Coffee

Session VIII. Physical processes near black holes/The Galactic Center (chair: Z. Stuchlik)
11:00 Black hole accretion: theoretical limits and observational implications; D. Heinzeller
11:20 Search for the event horizon evidences by means of optical observations with high temporal resolution; G. Beskin
11:35 Dynamics of radiatively inefficient flows accreting onto radiatively efficient black hole objects; D. Proga
11:55 The Galactic Center (Invited Review); R. Genzel
12:30 Lunch

Session IX. The Galactic Center (chair: T. Yaqoob)
14:00 The simultaneous radio to X-ray observations and polarized NIR emission from Sagittarius A*; A. Eckart
14:20 The structure of the nuclear stellar cluster of the Milky Way; R. Schoedel
14:35 Variable accretion of stellar winds onto Sagittarius A*; J. Cuadra
14:55 The character of the short-term variability of Sagittarius A* from the radio to the near-infrared; M. R. Morris
15:10 Stellar dynamics with Kozai's resonance in Sagittarius A*; L. Subr
15:30 Poster viewing (South Hall), Coffee

Session X. Ultraluminous X-ray sources (chair: M. Abramowicz)
16:00 Observational evidence for intermediate-mass black holes (Invited Review); K. Makishima
16:35 SS433-type X-ray binaries and the nature of ULXs; P. A. Charles
16:55 The supercritical accretion disk in SS433 and ultraluminous X-ray sources; S. N. Fabrika
17:10 The optical counterpart of an ultraluminous X-ray source NGC 6946 X-1; P. Abolmasov
17:30 Poster viewing

Thursday, August 24

Session XI. Ultraluminous X-ray sources (chair: B. Czerny)
09:00 Recipes for ULX formation: necessary ingredients and garnishments; R. Soria
09:20 Explosion of very massive stars and the origin of intermediate mass black holes; S. Tsuruta
09:35 Ultraluminous X-ray sources: X-ray binaries in a high/hard state? Z. Kuncic
09:55 On the nature of ultraluminous X-ray sources from optical/IR measurements; M. Cropper
10:10 Variability of ultraluminous X-ray sources in the Cartwheel Ring; A. Wolter
10:30 Poster viewing (South Hall), Coffee

Session XII. Supermassive black holes and their galaxies (chair: G. Hasinger)
11:00 The inner workings of early-type galaxies: supermassive black holes and stellar nuclei (Invited Review); L. Ferrarese
11:35 Imaging compact binary black holes with VLBI; G. B. Taylor
11:55 Radiatively inefficient accretion disks in low-luminosity active galactic nuclei; F. D. Macchetto
12:10 The central 80x200 parsecs of M83, how many black holes and how massive are they? H. A. Dottori
12:30 Black holes across the mass spectrum – from stellar-mass black holes to ultra-luminous X-ray sources and active galactic nuclei (Invited Review); R. Mushotzky
13:00 Lunch

Friday, August 25

Session XIII. Supermassive black holes and their galaxies (chair: S. Collin)
09:00 Inward bound: following the gas flow from nuclear spirals to the accretion disc; T. Storchi-Bergmann
09:20 Probing the coevolution of supermassive black holes and galaxies out to z=4.5 using gravitational lensing; C. Y. Peng
09:35 The smallest black holes in nearby active galactic nuclei; A. J. Barth
09:55 The evolution of supermassive black holes and galaxies in the COSMOS survey; C. D. Impey
10:10 Cosmic evolution of black holes and galaxies to z=0.4; J. Woo
10:30 Poster viewing (South Hall), Coffee

Session XIV. Black holes across the mass spectrum (chair: F. Mirabel)
11:00 The disc instability model for dwarf novae in the AGN context; J. M. Hameury
11:20 Radiation hydrodynamic simulations of super-Eddington accretion flows; K. Ohsuga
11:35 Synchrotron outbursts in galactic and extra-galactic jets, any difference? M. Tuerler
11:55 Symposium closing: Present status and future developments of black holes across the range of masses; F. Mirabel

IAU XXVIth
General
Assembly

SYMPOSIA

No. 235
Galaxy Evolution across the Hubble Time

No. 236
Near Earth Objects, our Celestial Neighbors:
Opportunity and Risk

No. 237
Triggered Star Formation in a Turbulent ISM

No. 238
Black Holes: from Stars to Galaxies -across
the Range of Masses

No. 239
Convection in Astrophysics

No. 240
Binary Stars as Critical Tools and Tests in Modern
Astrophysics

17 JOINT DISCUSSIONS
7 SPECIAL SESSIONS

MEETINGS: YOUNG ASTRONOMERS
WOMEN IN ASTRONOMY

View of Prague

Printed in the United States
By Bookmasters